PEM Electrolysis for Hydrogen Production

Principles and Applications

PEM Electrolysis for Hydrogen Production

Principles and Applications

Edited by
Dmitri Bessarabov
Haijiang Wang
Hui Li
Nana Zhao

CRC Press
Taylor & Francis Group
Boca Raton London New York

CRC Press is an imprint of the
Taylor & Francis Group, an **informa** business

CRC Press
Taylor & Francis Group
6000 Broken Sound Parkway NW, Suite 300
Boca Raton, FL 33487-2742

First issued in paperback 2017

© 2016 by Taylor & Francis Group, LLC
CRC Press is an imprint of Taylor & Francis Group, an Informa business

No claim to original U.S. Government works

ISBN-13: 978-1-4822-5229-3 (hbk)
ISBN-13: 978-1-138-77549-7 (pbk)

Visit the Taylor & Francis Web site at
http://www.taylorandfrancis.com

and the CRC Press Web site at
http://www.crcpress.com

Contents

Preface

Currently, there are two distinct commercial water electrolysis technologies that operate at low temperatures—alkaline and PEM (proton-exchange membranes) electrolyzers. Alkaline electrolyzers, a commercially more mature technology among the two, contain two electrodes immersed in a liquid alkaline electrolyte consisting of a concentrated KOH solution. In contrast, PEM electrolyzers use a solid proton-conducting polymer as the electrolyte and deionized water. As a result, PEM electrolyzers have many advantages over alkaline electrolyzers, such as a relatively simple system design and being able to operate safely at higher current densities. A third technology, currently at the precommercial stage, alkaline exchange membrane (AEM) systems, has the potential to place water electrolysis on a new cost reduction trajectory.

PEM water electrolysis has been known for many years; however, due to expensive components, such as membranes and bipolar plate materials and limited lifetime, PEM electrolyzers became established only in relatively small-scale niche applications, such as laboratory hydrogen and oxygen generators, life support systems, fuel supply for small fuel cell systems, etc. In general, PEM water electrolysis systems can provide a relatively simple, scalable, and easily deployable source of high-purity hydrogen for smaller retail and commercial applications near the point of consumption.

In recent years, hydrogen PEM fuel cells made significant progress toward commercialization, resulting in growing interest in technologies for hydrogen on-site production, such as PEM water electrolysis. Thus, the use of PEM water electrolysis for hydrogen fuel production became a vector of interest for fuel cell deployment opportunities in such sectors as sustainable mobility, material handling, and back-up power.

The rapid development of relatively small-scale PEM fuel cell technology also contributed to a "leapfrog" effect in the fundamental understanding of the requirements and functionalities of certain components and attributes of the PEM electrolysis technology that are both common for PEM fuel cells and electrolyzers, such as manufacturing aspects, components (membranes, plates, catalyst), flow-field design, etc. However, new trends in PEM water electrolysis systems development opened up new technology gaps and requirements that have not been discussed before with respect to PEM water electrolysis. For example, hydrogen is considered as one of the best solutions for large-scale energy storage that comes from renewable and intermittent power sources such as wind and solar electricity. If zero-carbon power sources, such as renewable or nuclear power, are used in combination with large-scale PEM water electrolysis, the resulting system will become suitable for large-scale clean and economically attractive hydrogen production and energy storage applications. Water electrolysis provides a sustainable solution for hydrogen production and is very well suited to be coupled with renewable energy sources. Thus, yet another vector of hydrogen applications for energy storage, called power-to-gas, is emerging and large utility companies are becoming involved.

To address technology gaps for large-scale PEM water electrolysis systems, the following areas require additional development: improved stack performance, scale up to megawatt size, grid integration, high pressure operation, high current density operation, degradation of components associated with transient operation, and a variety of market issues. All of these gaps relate directly to increased participation of PEM water electrolysis systems in hydrogen markets for various applications, not limited to fuel cells only. Megawatt scale-up, needed for such applications as power-to-gas and on-site refueling stations, includes requirements to reduce capital costs by 50% on a per kilowatt basis and availability of low-cost testing facilities; for example, electricity costs for PEM electrolysis megawatt testing can alone be a great challenge. Another challenge that the PEM water electrolysis industry has hardly discussed before is the large-scale manufacturing of cathode catalyst-coated membranes and stack components, availability of iridium, etc.

It is expected that demand for hydrogen as a fuel for fuel cells in both transport and stationary applications will continue to grow, alongside hydrogen for energy storage (the power-to-gas vector), thus generating more and more demands for PEM water electrolysis systems of large capacities.

It is well recognized that PEM water electrolysis systems are robust and dynamic. These systems can offer a fast response to volatile renewable energy sources. Due to the use of a dense proton-exchange membrane, PEM water electrolysis systems are capable of producing hydrogen at relatively high and practical discharge pressure, suitable, for example, for the injection of hydrogen into the grid of natural gas pipes. PEM water electrolyzers can also be scaled up to address various demands for energy storage.

Addressing climate change and the associated need for increasing renewable energy supply makes energy storage a critical technological component of the future energy landscape. PEM water electrolysis when coupled with renewable energy sources and when electrolytic hydrogen is used to capture CO_2 to produce synthetic methane via the Sabatier reaction can also be attractive as an additional power-to-gas application reducing CO_2 emissions.

Due to the ever-increasing desire for green energy, the last decade has seen regained research interest in PEM electrolysis. However, significant challenges still remain for PEM electrolysis to be a commercially feasible large-scale hydrogen production solution. These challenges include the insufficient durability of the catalysts and membrane, high cost associated with the use of platinum group metal-based catalysts, corrosion of the current collectors and separator plates, and the development of a stack concept for the megawatt power range.

The intention of this book is to provide a comprehensive research source for PEM electrolysis, discuss fundamental aspects as well as examples of applications, provide a review of the state-of-the-art technologies and challenges related to each of the components of the PEM electrolysis, identify various failure modes and failure mechanisms, and discuss component degradation testing methods and protocols.

This book provides researchers and technology engineers with the most comprehensive and updated knowledge on PEM electrolysis technology, thus helping them identify technology gaps and develop new materials and novel designs that lead to commercially viable PEM electrolysis systems. We believe that students and professionals in disciplines such as electrochemical engineering, electrochemistry, material science in electrocatalyst development, material science in polymer development, and chemical and mechanical engineers working on energy storage and clean technologies will find this book useful.

Editors

Dr. Dmitri Bessarabov joined the DST HySA Infrastructure Center of Competence at North-West University (NWU) and Council for Scientific and Industrial Research (CSIR) in 2010. He is an internationally recognized scientist with academic and industrial decision-making experience in the area of hydrogen and electrocatalytic membrane systems for energy applications and fuel cells. Dr. Bessarabov has more than 15 years of progressively increasing responsibility in academic and industrial R&D environment and leadership roles in the hydrogen energy sector in Canada and South Africa. His current responsibilities include leadership in the National Hydrogen and Fuel Cell Program (HySA). He is currently also leading PEM electrolyzer development projects at the HySA Infrastructure, which includes the establishment of technology platforms for electrolyzer development, related characterization tools, electrochemical hydrogen compression, and hydrogen production using renewables.

Dr. Bessarabov received his fundamental training in chemistry at the renowned Lomonosov Moscow State University in Russia (MSc, 1991). He continued further education at the Russian Academy of Sciences in membrane gas separations at the Topchiev Institute of Petrochemical Synthesis of the Russian Academy of Sciences. In 1993, he joined the PhD program at the Institute for Polymer Science, University of Stellenbosch in South Africa. He earned his PhD in 1996, specializing in membrane technology for gas separation. His postdoctoral research at the University of Stellenbosch was in the area of electrocatalytic membrane systems and electrochemical ozone generation (1997–1998), for which NRF granted him a "Y" rating. In 1999, Dr. Bessarabov was appointed senior lecturer at the University of Stellenbosch's Chemistry Department. In 2001, he joined Aker Kvaerner Chemetics in Vancouver, Canada, to work in the area of membrane technology for the chloralkali industry. In 2006, he joined Ballard Power Systems in Canada (and afterwards AFCC, Automotive Fuel Cell Cooperation Corp.), where he was leading an R&D group on MEA integration and evaluation. His main areas of professional interest include fuel cells, PEM electrolysis, hydrogen energy, hydrogen storage, hydrogen infrastructure, membranes, separations, applied electrochemistry, applied polymer science, environmental technologies, and water treatment. To date, he has published more than 100 journal papers and 14 conference papers, and has been issued three patents.

Dr. Haijiang Wang is a senior research officer and project manager in the National Research Council of Canada (NRC). He has been with NRC for 10 years. His research covers PEM fuel cell, electrolyzer, metal-air battery, microbial fuel cell, and lithium-sulfur battery. Dr. Wang earned his PhD in electrochemistry from the University of Copenhagen, Denmark, in 1993. He then joined Dr. Vernon Parker's research group at Utah State University as a postdoctoral researcher to study electrochemically generated anion and cation radicals. In 1997, he began working with Natural Resources Canada as a research scientist to carry out research on fuel cell technology. In 1999, he joined Ballard Power Systems as a senior research scientist to continue his investigations. After spending five years with Ballard Power Systems, he joined NRC in 2004. He is currently adjunct professor at five universities, including the University of British Columbia and the University of Waterloo. Dr. Wang has 30 years' professional research experience in electrochemistry and fuel cell technology. To date, he has published more than 160 journal papers, three books, 10 book chapters, 40 industrial reports, and 30 conference papers or presentations and has been issued five patents.

Dr. Hui Li is a research officer and a master project lead under the energy storage program at the National Research Council of Canada—Energy, Mining and Environment Portfolio (NRC-EME, which used to be the Institute for Fuel Cell Innovation). Dr. Li earned her BS and MSc in chemical engineering from Tsinghu University in 1987 and 1990, respectively. After completing her MSc, she joined Kunming Metallurgical Institute as a research engineer for four years and then took a position as an associate professor at Sunwen University for eight years. In 2002, she started her PhD program in electrochemical engineering at the University of British Columbia (Canada). After earning her PhD in 2006, she carried out one term of postdoctoral research at the Clean Energy Research Centre (CERC) at the University of British Columbia with Professors Colin Oloman and David Wilkinson. Since joining NRC in 2007, Dr. Li has been working on PEM fuel cell contamination and durability, PEM electrolysis, and zinc-air batteries. Dr. Li has many years of research and development experience in theoretical and applied electrochemistry and in electrochemical engineering. Her research is based on PEM fuel cell contamination and durability testing, preparation and development of electrochemical catalysts with long-term stability, catalyst layer/cathode structure, and catalyst layer characterization and electrochemical evaluation, failure diagnosis and mitigation for PEM fuel cells and electrolyzers, and air-cathodes for zinc-air battery. Dr. Li has coauthored more than 30 research papers published in refereed journals and coedited three books related to PEM fuel cells. Dr. Li has two granted patents and one technology licensed to the Mantra Energy Group.

Dr. Nana Zhao is a research scientist at Vancouver International Clean-Tech Research Institute Inc. (VICTRII), Burnaby, British Columbia, Canada. Her research interests include synthesis, evaluation and characterization of PFSA and hydrocarbon ionomer and membranes; MEA design and fabrication; characterization and electrochemical evaluation of catalyst layer; membrane and catalyst layer durability testing and diagnosis; synthesis and characterization of CO_2 separation membranes; preparation and characterization of subnanometer porous membrane for proton transportation and gas transport; and inorganic nanocrystals synthesis and application. Dr. Zhao received her BS in polymer chemistry and physics from Beijing Normal University in 2000. After that, she joined Changchun University of Science and Technology as a teaching assistant for two years and then started her PhD program on polymer chemistry and physics at Changchun Institute of Applied Chemistry, Chinese Academic Sciences, in 2002. After earning her PhD in 2008, she joined Professor Ting Xu's group as a postdoctoral fellow in material science and engineering at the University of California, Berkeley. At the same time, she also worked at Professor Frantisek Svec's team in Molecular Foundry at the Lawrence Berkeley National Laboratory. After one term of postdoctoral research, she took a position as a research associate at the National Research Council of Canada—Energy, Mining and Environment Portfolio (NRC-EME, which used to be the Institute for Fuel Cell Innovation) for two years. In 2013, she began working at VICTRII as a research scientist. Currently, she is taking a lead role in several collaborative PEM fuel cell projects.

Contributors

Rami Abouatallah
Hydrogenics Corporation
Mississauga, Ontario, Canada

Everett Anderson
Proton OnSite
Wallingford, Connecticut

Vincenzo Antonucci
CNR-ITAE
Via Salita S. Lucia sopra Contesse
Messina, Italy

Kathy Ayers
Proton OnSite
Wallingford, Connecticut

Dmitri Bessarabov
Faculty of Engineering
HySA Infrastructure Center of Competence
North-West University
Potchefstroom, South Africa

Peter Bouwman
Hydrogen Efficiency Technologies
Leemansweg, the Netherlands

Nicola Briguglio
CNR-ITAE
Via Salita S. Lucia sopra Contesse
Messina, Italy

Joseph Cargnelli
Hydrogenics Corporation
Mississauga, Ontario, Canada

Marcelo Carmo
Institute of Energy and Climate Research
Forschungszentrum Jülich GmbH
Jülich, Germany

Pyoungho Choi
Department of Chemical Engineering
Fuel Cell Center
Worcester Polytechnic Institute
Worcester, Massachusetts

Ravindra Datta
Department of Chemical Engineering
Fuel Cell Center
Worcester Polytechnic Institute
Worcester, Massachusetts

Yan Dong
Department of Chemical Engineering
Fuel Cell Center
Worcester Polytechnic Institute
Worcester, Massachusetts

Sergey Grigoriev
Moscow Power Engineering Institute
National Research University
and
National Research Center
Kurchatov Institute
Moscow, Russia

Rob Harvey
Hydrogenics Corporation
Mississauga, Ontario, Canada

Hiroshi Ito
Energy Technology Research Institute
National Institute of Advanced Industrial Science and
    Technology
Tsukuba, Japan

Krzysztof Lewinski
M Corporation
3M Center
St. Paul, Minnesota

Thomas Lickert
Fraunhofer Institute for Solar Energy Systems ISE
Division Hydrogen Technologies
Freiburg, Germany

Wiebke Lüke
Institute of Energy and Climate Research
Forschungszentrum Jülich GmbH
Jülich, Germany

Drew J. Martino
Department of Chemical Engineering
Fuel Cell Center
Worcester Polytechnic Institute
Worcester, Massachusetts

Pierre Millet
Department of Chemistry
Institute of Molecular Chemistry and Material Science
University Paris-Sud
Orsay, France

Emile Tabu Ojong
Fraunhofer Institute for Solar Energy Systems ISE
Division Hydrogen Technologies
Freiburg, Germany

Artem Pushkarev
Moscow Power Engineering Institute
National Research University
and
Kurchatov Institute
National Research Center
Moscow, Russia

Irina Pushkareva
Moscow Power Engineering Institute
National Research University
and
National Research Center
Kurchatov Institute
and
Department of Technology of Isotopes and Hydrogen
    Energetics
Institute of Modern Energetic Materials and
    Nanotechnology
D. Mendeleyev University of Chemical Technology of
    Russia
Moscow, Russia

Julie Renner
Proton OnSite
Wallingford, Connecticut

Mikhail Rozenkevich
Department of Technology of Isotopes and Hydrogen
    Energetics
Institute of Modern Energetic Materials and
    Nanotechnology
D. Mendeleyev University of Chemical Technology of
    Russia
Moscow, Russia

Tom Smolinka
Fraunhofer Institute for Solar Energy Systems ISE
Division Hydrogen Technologies
Freiburg, Germany

Detlef Stolten
Institute of Energy and Climate Research
Forschungszentrum Jülich GmbH
Jülich, Germany

and

Chair for Fuel Cells
RWTH Aachen University
Aachen, Germany

Svein Sunde
Department of Materials Science and Engineering
Norwegian University of Science and Technology
Trondheim, Norway

Magnus Thomassen
Department of New Energy Solutions
SINTEF Materials and Chemistry
Trondheim, Norway

Conghua "CH" Wang
TreadStone Technologies, Inc.
Princeton, New Jersey

1

Overview of PEM Electrolysis for Hydrogen Production

Nicola Briguglio and Vincenzo Antonucci

CONTENTS

1.1 Introduction

The expression "hydrogen economy" is used to indicate the role of hydrogen in the future energy scenario. Interest in hydrogen, as an energy carrier, has been growing in the recent years due to heightening of air pollution in the world. Hydrogen is a clean and flexible energy carrier that can be used to provide both power and heat across all end-use sectors. Vehicles and stationary power generation fed by hydrogen are local zero emission technologies. Hydrogen can be produced from both traditional fossil fuel and carbon-free energy sources, which are used to store energy and to provide response management to electricity grid. Today, only 4% of hydrogen is produced from electrolysis; other lower-cost methods are preferred, such as steam reforming of natural gas or refinery gas. However, in the next future, the renewable energy sources (RES) will take up an important portion of electric energy produced. In this context, the energy storage is expected to play a key role in the future as "Smart Grid." The future energy storage technologies should be more flexible and able to balance the grid, ensuring stability and security. Large-scale deployment of variable renewable source (primary wind and solar energy) will be required to store energy to avoid the RES curtailment. Electrolysis is considered as the cleanest way to produce hydrogen using RES and has (along with other storage technologies) the potential

as "energy storage" in this sector. In particular, bulk energy storage technologies are expected to have a key role for the integration of large amount of electricity produced from RES. This sector is dominated by pumped hydro as energy storage (PHES) in the world due to its large unit sizes and history. Anyway, long construction times and high uncertainty of future electricity price developments make PHES systems risky investments. Furthermore, constructions of PHES systems are strongly dependent on certain geographic requirements and topographical conditions.

Other technologies are compressed air energy storage (CAES), thermal energy storage, batteries, and flywheels. Anyway, the selection of technology depends on key parameters such as energy capacity and discharge time.

An interesting emerging application of electrolyzers is in the sector "Power to Gas." The hydrogen produced by electrolyzers, connected to RES, is injected in the gas network. This approach permits to use gas pipelines as large "storage tanks" avoiding construction of new infrastructures. The amount of hydrogen injected depends on the countries' regulations. This problem can be overcome through methanation, in which hydrogen and carbon monoxide/dioxide are converted in sustainable methane. The hydrogen stored in the gas infrastructure could be used for heating, in transportation, or reconverted in electricity.

Refueling stations with on-site hydrogen production are another application for electrolyzers. However, other technologies could be more cost-effective (i.e., steam methane reforming) than electrolysis. The choice of using electrolyzers will depend on local strategies, electricity price, etc. Uses of electrolyzers are not easily predictable in this sector.

Literature, conferences, and industry stakeholders on the energy sectors show that some issues must be resolved before electrolysis can be a competitive alternative, such as capital costs, lifetime, and system integration with energy services. Electrolyzers employed to store energy produced by intermittent RES need to operate dynamically with a lot of starts/stops. Pilot demonstration projects around the world on electrolyzers connected to RES are in progress, but the effects of these conditions on the stack and system durability have yet to be explored in depth. In fact, electrolyzers are usually designed to operate at constant conditions for industrial applications (i.e., chemical and food industry).

A recent study commissioned by the FCHJU (FCH JU report 2014) showed that electricity represents 70%–90% of the cost of a kilogram of hydrogen produced from electrolysis. As a consequence, the competitiveness of electrolysis will strictly relate to electricity prices. However, electrolysis can compete in circumstances where regulations permit a payment for grid services by operating as a controllable load or where policy context creates conditions to significantly reduce the cost of electricity.

1.2 Hydrogen Production

Water electrolysis is the process whereby water is split into hydrogen and oxygen through the application of electrical energy, as in the following equation:

$$H_2O + electricity \rightarrow H_2 + \frac{1}{2}O_2 \qquad (1.1)$$

Among the electrolysis technologies, alkaline electrolysis is at a mature stage, but some improvements as current density and working pressure can be made.

PEM electrolysis is in a precommercial phase in small-scale application. The research is mainly addressed toward the cost reductions of stack components and catalysts, and integration with RES.

Solid oxide electrolysis (SOE) is a technology under development and is not yet commercialized, but its technical advantages include potentially higher efficiency than other electrolysis technologies. SOE has also been used to produce CO from CO_2 and syngas H_2/CO from H_2O/CO_2.

Another interesting technology is the anion exchange membrane (AEM) electrolysis. This technology is in an early stage of development. The main advantage is the lower cost of catalysts than traditional PEM electrolysis, whereas the drawback is the low ion conductivity of membrane.

1.2.1 Solid Oxide Electrolysis

In an SOE, electricity is provided to electrochemically convert steam or CO_2 into hydrogen and CO. In the 1980s, this technology received great interest due to a study carried out by Donitz and Erdle, where a tubular SOE was demonstrated within the HotElly project. Following this, Westinghouse developed a tubular electrolysis based on the design used for SOFC. Interest in energy storage through hydrogen has led to new research activities on SOE in the last decade (Figure 1.1).

Examples of this renewed interest are the European project "Relhy" (http://www.relhy.eu/), which focuses on the development of low-cost materials and new components for SOE, and the research activities performed by universities in France, Germany, Denmark, and Japan.

The scheme of solid oxide electrolysis is shown in Figure 1.2.

Water is fed to cathode and oxygen ions are transported to the anode side through the electrolyte, and hydrogen is produced at the cathode side. If the cathode is also fed with CO_2, the overall electrochemical reactions of electrolysis are as follows:

$$\begin{cases} H_2O + 2e^- \rightarrow H_2(g) + O^{2-} \\ CO_2 + 2e^- \rightarrow CO + O^{2-} \end{cases} \quad (cathode) \qquad (1.2)$$

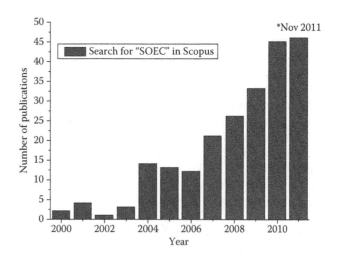

FIGURE 1.1
Number of publications per year in SOEC according to Scopus database. (Reprinted from *Journal of Power Sources*, 203, Laguna-Bercero, M.A., Recent advances in high temperature electrolysis using solid oxide fuel cells: A review, 4–16, Copyright [2012], with permission from Elsevier.)

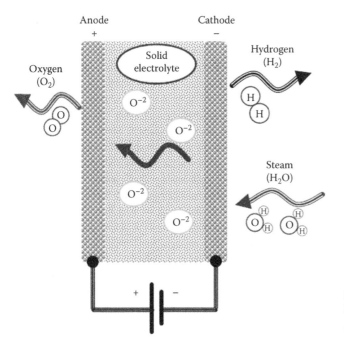

FIGURE 1.2
Sketch of an SOEC.

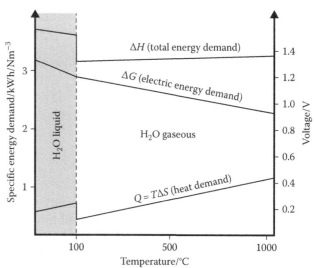

FIGURE 1.3
Electrolysis energy demand.

$$2O^{2-} \rightarrow O_2(g) + 4e^- \quad \text{(anode)} \quad (1.3)$$

Materials and technology used in SOE are based on that of solid oxide fuel cells (Laguna-Bercero 2012). The most common electrolyte material is YSZ (yttria-stabilized zirconia). This material shows high ionic conductivity, chemical, and thermal stability at working temperature (800–1000°C). Other potential candidates are ScSZ (scandia-stabilized zirconia), LSGM (lanthanum, strontium, gallium, manganite), and GDC (gadolinium-doped ceria) (Moyer et al. 2010). Materials for cathode include porous cermet of YSZ and metallic nickel (Ni–YSZ), samarium-doped ceria (SDC), titania/ceria composites, and various LSCM (lanthanum, strontium, chromium, manganite)-doped perovskites.

For anode, the most common material is the LSM (lanthanum, strontium, manganite), but other materials have also been proposed such as LSF (lanthanum, strontium, ferrite) or LSCo (lanthanum, strontium, cobaltite).

High temperature permits to reduce the electrical power demand as electrolysis process is increasingly endothermic with increasing temperature (Figure 1.3). The step of enthalpy curve at 100°C is due to the evaporation heat of water. After 100°C, the total energy demand remains quite constant, whereas the electric energy decreases and the heat demand increases. As a consequence, SOE could considerably reduce the cost of hydrogen if heat is supplied by waste heat.

The current research activities are focused on development of new materials stable at high temperature.

Today, the commercial products of SOE are not available, but the technology has been proven at laboratory scale.

1.2.2 Alkaline Water Electrolysis

Alkaline electrolyzers use an aqueous caustic solution of 20%–30% KOH as electrolyte. Today, it is the most mature technology up to megawatt range at commercial level. A description of process is shown in Figure 1.4. The cell is composed of two electrodes separated by a diaphragm used to avoid the recombination of hydrogen and oxygen produced. The diaphragm should have

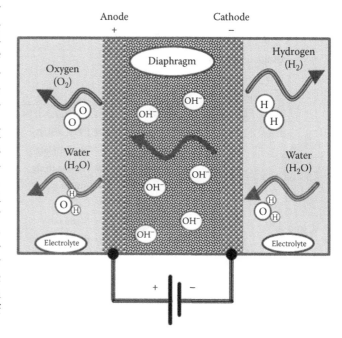

FIGURE 1.4
Alkaline electrolysis process description.

high chemical and physical stability with a high ionic conductivity. Typical operation temperature range is from 65°C to 100°C.

The basic reactions that take place in alkaline electrolysis are reported in Equations 1.4 and 1.5.

$$2OH^- \rightarrow \frac{1}{2}O_2(g) + H_2O + 2e^- \quad \text{(anode)} \qquad (1.4)$$

$$2H_2O + 2e^- \rightarrow H_2(g) + 2OH^- \quad \text{(cathode)} \qquad (1.5)$$

Hydrogen and oxygen gas bubbles, produced at cathode and anode, respectively, increase the cell resistance reducing the contact between the electrodes and electrolyte. As a consequence, efficiency is reduced. Particular attention should be given to cell design to maximize the contact between the electrode and the liquid electrolyte. Advanced alkaline cell uses a "zero gap configuration" to reduce the impact of gas bubbles and the ohmic losses by reducing the space between the electrodes (FCH JU report 2014; Ursú et al. 2012).

The main drawbacks related to alkaline technology are the corrosive environment, the low current density, and the limited production rate of 25%–100% due to the diffusion of gases through the diaphragm at partial load. Typical operative pressure of this technology is 25–30 bar, but electrolyzers up to 60 bar are available.

Concerning the stack cost breakdown, the alkaline technology shows that the components have the main contribution in terms of size and weight due to low current density, refer Figure 1.5 (FCH JU report 2014). However, the cost of stack strictly depends on manufacturers.

There are two cell structures of alkaline electrolysis on market, unipolar and bipolar as reported in Figure 1.6.

In the monopolar configuration, the cells are connected in parallel and the electrodes are connected to the corresponding DC power supply. The total voltage applied to the stack is the same of that applied to an individual cell, and the electrodes have a single polarity.

Monopolar configuration is simpler from a fabrication point of view and permits a more easy maintenance and reparation without shutting down of the whole stack. Anyway, the disadvantage is that it usually operates at lower current densities and lower temperatures.

In the bipolar configuration, each electrode has two polarities and the cells are connected in series. The voltage applied to the stack is the sum of single-cell voltage. The current that flows through the stack is the same for cells.

The advantages of the bipolar design are higher current densities, which have capacity to produce higher pressure gas, and a more compact stack than unipolar design. The disadvantage is that it cannot be repaired without servicing the entire stack.

1.2.3 PEM Water Electrolysis

PEM water electrolyzers use a polymer electrolyte membrane (or proton exchange membrane) as ionic conductor. The scheme of operation of a PEM water electrolyzer is shown in Figure 1.7. Water is oxidized at the anode according to Equation 1.6 to produce oxygen and hydrogen evolves at the cathode according to Equation 1.7.

$$H_2(l) \rightarrow \frac{1}{2}O_2(g) + 2H^+ + 2e^- \quad \text{(anode)} \qquad (1.6)$$

$$2H^+ + 2e^- \rightarrow H_2(g) \quad \text{(cathode)} \qquad (1.7)$$

The electrolyte consists of a thin, solid ion-conducting membrane instead of the aqueous solution used in the alkaline electrolyzers. The membrane transfers the H$^+$ ion (i.e., proton) from the anode to the cathode side and separates the hydrogen and oxygen gases. The most commonly used membrane material is Nafion® from DuPont.

The main advantages of PEM electrolysis over the alkaline are related to greater safety and reliability because no caustic electrolyte is used. Besides, the possibility to operating at high differential pressure across the membrane avoids the oxygen compression. PEM electrolysis has faster ion transportation than alkaline due to the solid and thin membrane. In fact, liquid electrolyte has more inertia in transportation of ions (Rajeshwar et al. 2008). This aspect is particularly important when an electrolyzer operates under fluctuating conditions. Alkaline electrolyzer suffers from delayed reaction and difficulty in starting the system after shutdown. Besides, this technology can operate at higher current density (2 A cm^{-2} at 2.1 V and 90°C [Millet et al. 2011]) than alkaline electrolyzers.

Alkaline stack cost breakdown

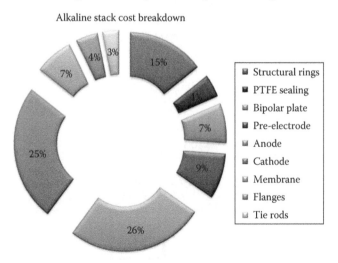

15%
4% 3%
7%
1%
7%
25%
9%
26%

- ■ Structural rings
- ■ PTFE sealing
- ■ Bipolar plate
- ■ Pre-electrode
- ■ Anode
- ■ Cathode
- ■ Membrane
- ■ Flanges
- ■ Tie rods

FIGURE 1.5
Stack cost breakdown for alkaline technology. (Modified from FCH JU report 2014, Study on development of water electrolysis in the EU 2014.)

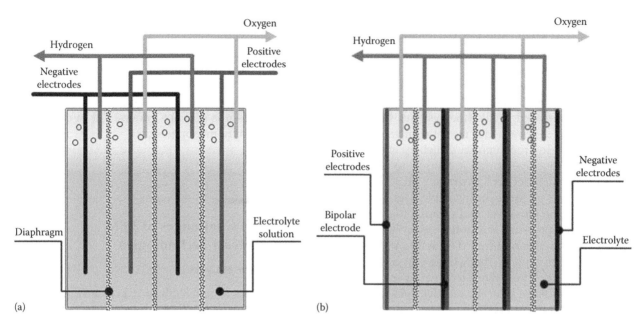

FIGURE 1.6
(a) Unipolar and (b) bipolar stack design.

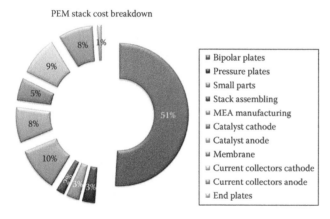

PEM stack cost breakdown

- Bipolar plates
- Pressure plates
- Small parts
- Stack assembling
- MEA manufacturing
- Catalyst cathode
- Catalyst anode
- Membrane
- Current collectors cathode
- Current collectors anode
- End plates

FIGURE 1.7
Scheme of a PEM water electrolyzer.

TABLE 1.1

Commercial Alkaline and PEM Electrolysis Technical Specifications

Specifications	PEM	Alkaline
Cell temperature [°C]	50–80	60–80
Rated production [N m³h⁻¹]	0.265–30	1–760
Rated power [kW]	1.8–174	2.8–3534
Specific energy consumption [kWh/N m⁻³]	5.8–7.3	4.5–7.5
Efficiency [%] HHV	48.5–65.5	50–70.8
Maximum pressure [bar]	7.9–85	Up to 30
Hydrogen purity [vol.%]	99.999	99.3–99.999
System cost [€/kg⁻¹]	1900–2300	1000–1200
System lifetime [year]	10–20	20–30

Source: Data elaborated by Ursú, A. et al., Proc. *IEEE*, 100, 410, 2012; FCH JU report, 2014; Smolinka, T. et al., NOW-Studies: "Stand und Entwicklungs Entwicklungspotenzialder Wasserelektrolyse zur Herstellung von Wasserstoff aus regenerativenEnergien," Technical report, Fraunhofer ISE, 2011.

The costs of some components (i.e., current collectors) and the use of noble-based catalyst (i.e., platinum and iridium) are the main drawbacks of PEM electrolyzers (Ayers et al. 2010).

1.2.4 Comparison of the Electrolysis Technologies at System Level

Currently, both PEM and alkaline electrolyzers are commercially available. Electrolysis based on AEM has limited products, whereas SOEC technology is at the early stage of development. In Table 1.1, the technical specifications of commercial alkaline and PEM electrolysis are summarized. The data are referred to maximum and minimum values reported by the manufacturers of electrolyzers.

Alkaline electrolyzers have the ability to operate high production rate but in range of 25%–100% of the nominal power. The products on the market can reach the maximum pressure of 30 bar even if some manufacturers, such as Avalence and Accagen, claim the possibility to work at 448 bar and 200 bar, respectively. The low current density (0.4 mA cm⁻²) and the purity of gas produced are the main drawbacks of this technology. High levels of purity can be obtained along with auxiliary purification unit.

Presently, few PEM electrolyzer industrial makers are available on market (ITM, Hydrogenics, GE, Giner, and Proton). Commercially available devices have at low flow rate and MW scale of PEM systems consists of multiple individual modules wired in series or parallel (about 150 kWe). On the contrary, MW alkaline systems are composed of large stacks. Anyway, stacks at MW scale are expected by 2015 for PEM technology (FCH JU report 2014). Compared to alkaline systems, PEM electrolyzer systems show a wider operational range (5%–100%) and higher hydrogen purity (Ulleberg et al. 2005).

Concerning the operating pressure, today PEM electrolysis usually delivers hydrogen at 30 bar, but products at higher pressure are available. The solid electrolyte reduces the crossover and permits to operate at high differential pressure across the electrolyte even if a thicker membrane is required. The ability of PEMWE systems to pressurize only the hydrogen compartment is a great advantage from system's point of view because oxygen under pressure is a very reactive substance that can be difficult to handle. Today, the manufacturing trend is to pressurize the hydrogen with stack up to 30–80, eliminating the first stage of external compression. Electrochemical compression of hydrogen is more energy efficient than mechanical compression, but the benefits are more difficult to estimate at the system level (FCH JU report 2014).

The costs breakdown for PEM and alkaline electrolyzer systems are reported in Figure 1.8. In both the cases, the costs are driven by stack components, but the percentages can change as a function of manufacture. For example, the contribution of stack can reach the 53% of system overall costs for PEM electrolysis (Anderson 2012).

Cost reductions are mainly expected in stack and system engineering rather than technology for alkaline electrolysis. The situation is different for PEM electrolysis where cost reduction is predicted both in the

materials and in stack components (i.e., bipolar plates design and current collectors). Another important aspect is the integration of PEM electrolysis system into real-world conditions. In fact, even if electrolyzer systems have been demonstrated in the lab (Briguglio et al. 2013) and in pilot demonstration projects, the degradation mechanism is still not understood in depth, especially under dynamic operation.

1.3 Overall Challenges of PEM Electrolysis Technologies

1.3.1 Electrocatalysts

Expensive noble materials are typically used today as electrocatalyst in PEM electrolysis. Palladium or platinum at the cathode for the hydrogen evolution reaction (HER) and iridium or ruthenium oxides at the anode for the oxygen evolution reaction (OER) are most commonly used (Marshall et al. 2007). IrO_2 exhibits higher corrosion resistance than RuO_2, but it shows an inferior activity as OER. RuO_2 performs more in the low over potential range, but the problems of stability have precluded practical application. The use of binary IrO_2–RuO_2 solid solutions can slightly improve the stability of RuO_2 (Aricò et al. 2013). Using IrO_2 with small particle size (2–3 nm) can reduce the noble metal loading, maintaining similar performance (Siracusano et al. 2010).

Solution with nonnoble metal formulation, as transition metal oxide, is interesting for OER but conductivity, electrocatalytic activity, and stability are challenging aspects (Cavaliere et al. 2011).

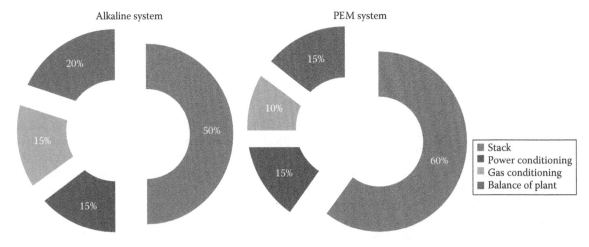

FIGURE 1.8
Cost breakdowns for alkaline and PEM electrolyzer systems. (Modified from FCH JU report 2014, Study on development of water electrolysis in the EU 2014.)

As reported earlier, platinum electrocatalysts are usually used in PEM water electrolyzers as HER because both have best catalytic activity and high corrosion resistance. The cathode loads range from 0.5 to 1 mg/cm^{-2} (Carmo et al. 2013). In particular, platinum nanoparticles supported on carbon black (Pt/C) are used as standard catalysts for the HER. These catalysts are very similar to that used for PEM fuel cell. Reduction of HER catalysts' costs is target for the researchers. Pd, for example, is less expensive than Pt and has appropriate electrolytic activity if dispersed as nanosized particles on carbon black. However, in order to obtain performance comparable to Pt, the Pd/C electrocatalysts need catalytic enhancers that often suffer from leaching phenomena during prolonged operation.

Core-shell structures with Pt on the surface and Pd or Cu as metallic core substrate are interesting alternative catalysts for HER (Aricò et al. 2013; Carmo et al. 2013). The aim of these studies is to increase the surface area and the reduction of noble metal loading.

1.3.2 Membrane

In PEM electrolysis, a perfluorosulfonic acid membrane (PFSA) is used as a solid electrolyte. Nafion (Dupont trademark) is the common commercial membrane used for this application. This membrane is characterized by excellent chemical stability, high proton conductivity (0.15 S cm^{-1} at 80°C), and an area-specific resistance of about 0.15 Ω cm^2 (Nafion 117). The thickness of membrane is a compromise among area-specific resistance, low crossover, and mechanical strength. A drawback of Nafion is the cost, and many studies are focused on cheaper membranes. Alternative membranes PFSA with higher ionic conductivity than Nafion have been developed by Dow, 3M, Gore, Asahi glass, and Solvay Specialty Polymers (Ghielmi et al. 2005).

Important properties of a PEM electrolysis membrane are low crossover, the ability to work at high temperature (>100°C), and high mechanical resistance. The crossover in PEMWE can destroy the membrane and cause the failure of a stack. Hydrogen and oxygen reaction is extremely exothermic and causes local heating that can destroy the membrane with time. This problem is particularly important when the electrolyzer works at high pressure (up to 350 bar). The possibility to operate at high pressure permits to reduce the mechanical energy for gas pressurization. Low levels of crossover are necessary in those applications, and a proper thickness of polymer membrane is required. Another important mechanical property of polymer membrane is the tear resistance. In fact, strong stress can occur during

the stack assembly, especially between the edge of electrode and gasket. Good tensile properties and low tear propagation resistance are key characteristics for a polymer membrane in PEM electrolyzer. Usually, composite or reinforced membranes are used for operating at high pressure and high temperature (Antonucci et al. 2008). Operation of PEM electrolyzers at high temperatures (>100°C) reduces the Gibb's free energy change and increases the reaction kinetics. Anyway, a drawback of Nafion is that the conductivity decreases at temperature above 100°C because of membrane dehydration. Composite membranes are already demonstrated in fuel cells application up to 150°C (Antonucci et al. 2008; Baglio et al. 2009).

Hydrocarbon membranes are an alternative to Nafion. These membranes are nonperfluorinated electrolytes and have appropriate conductivity and a better resistance than Nafion to crossover. Besides, their low cost makes them a real and attractive for PEM electrolyzers (Ng et al. 2011).

1.3.3 Stack Components

The components used in PEM electrolysis are vital to have a stack with good performance and durability. Key components include the bipolar plates, current collectors, and membrane electrode assembly (MEA). The bipolar plates have the task of distributing uniformly the water on the current collectors (or diffusion media) inside the cells.

Due to the presence of oxygen, acid environment, and high overvoltage, the bipolar plates' characteristic requirements are challenging for any class of materials. The design and the choose of materials for bipolar plates in PEMWE are crucial issues for scientists and engineers.

The main functions of the bipolar plate in PEMWE are as follows:

1. Conducting electrons to complete the circuit and connecting individual cells in a series to form a stack
2. Providing a flow path for water distribution over the current collector (or diffusion media) uniformly
3. Separating oxygen and hydrogen
4. Providing structure to the cell to support the membrane and electrodes
5. Providing thermal conduction to manage the electrolyzer temperature

To perform these important functions, the materials of the bipolar plate must have particular properties such as high electrical and thermal conductivity, low gas permeability, high mechanical strength, and corrosion resistance.

Several types of materials are currently used in bipolar plates, including titanium, graphite, and coated stainless steel. Anyway, graphite does not guaranty high performance in PEWE due to high corrosion rates and low mechanical strength. Titanium has excellent proprieties but forms a passive oxide layer that decreases its electric contact resistance and thermal conductivity. This phenomenon decreases the stack performance over the time. Gold-coated titanium bipolar plates have been also investigated with good results in terms of stability (Jung et al. 2009). However, this process increases the cost of the component. Coated stainless steel is a less expensive alternative to titanium, but low imperfections in the coating expose the metal to corrosion. Corrosion of base metal causes the destruction of coating increasing the ohmic resistance (Carmo et al. 2013). Besides, the dissolved metal ions diffuse into the PEM membrane and decrease ionic conductivity, leading to increased membrane degradation (Hung et al. 2005).

Bipolar plates (separator plates) and current collectors are the more expensive components in PEMWE for the reasons mentioned earlier, as shown in Figure 1.9.

The current collectors work in the same environment conditions (acid environment, oxygen, and high overvoltage) of bipolar plates. Thus, finding materials that fit the required characteristics exactly is not an easy task. The current collectors have the function to travel the electrons from the catalytic layer to bipolar plates and remove the gases (hydrogen and oxygen) from catalytic layer. Therefore, the porosity and electric conductivity are important qualities for a current collector. The pore size and structure are key factors because the gases must be expelled from the cell and the water is maintained close to the catalytic layer. An inadequate pore structure can increase the cell voltage up to 100 mV at 2 A cm^{-2} (Grigoriev et al. 2009). Besides, a good contact between

current collector and catalytic layer permits to avoid the formation of hot spot that can damage the membrane.

Currently, Ti grid, stainless steel grid, or sintered Ti particles are used as current collectors.

1.4 Summary

In this chapter, the main issues of PEMWE have been discussed. The renewed interest on this technology over the past decades is strictly connected to the exponential growth of RES due to the increase in greenhouse gas emissions. Unfortunately, the RES, such as wind and solar, delivers fluctuating and intermittent electricity. Therefore, large percentage of RES in the electricity networks needs energy storage. In this contest, hydrogen produced by electrolysis can be a valid possibility to store large amount of energy. Today, two electrolysis technologies are available on the market: alkaline and polymer exchange membrane electrolysis. The comparison between these technologies has pointed out that even if the alkaline electrolysis is a mature technology, some drawbacks exist under fluctuating operation. A limited operational range, delayed reaction, and difficulty in starting the system after shutdown are some problems of alkaline electrolyzers. Anyway, up to date PEMWE is the only technology for large hydrogen production.

The PEMWE has great potentials as hydrogen generator in conjunction with RES, but some critical points must be solved. The cost of manufacturing is a key point when a technology approaching to the market. In PEMWE, the use of noble metals, such as titanium, iridium, ruthenium, and platinum, is an important issue. The bipolar plates and current collectors represent 48% of the stack cost. Therefore, alternative less expensive materials are important research areas for the stack components as well as for the catalysts and membranes. Concerning the performances, currently the main challenges are the degradation mechanism under fluctuating condition that limits the lifetime of stacks and the low capacity of PEM electrolyzers.

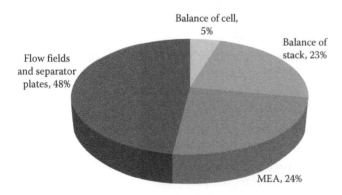

FIGURE 1.9
Stack cost breakdown for PEMWE. (Modified from FCH JU report 2014, Study on development of water electrolysis in the EU 2014.)

References

Anderson E. 2012. *Recent Advances in PEM Electrolysis and their Implications for Hydrogen Energy Market.* International Water Electrolysis Symposium 2012, Copenhagen, Denmark.

Antonucci V., Di Blasi A., Baglio V., Ornelas R., Matteucci F., Ledesma-Garcia J., Arriaga L. G., and Aricò A. S. 2008. High temperature operation of a composite membrane-based solid polymer electrolyte water electrolyser. *Electrochimica Acta* 53:7350–7356.

Aricò A. S., Siracusano S., Briguglio N. et al. 2013. Polymer electrolyte membrane electrolysis: Status of technologies and potential applications in combination with renewable power sources. *Journal of Applied Electrochemistry* 43:107–118.

Ayers K. E., Anderson E. B., Capuano C. et al. 2010. Research advances towards low cost, high efficiency PEM electrolysis. *ECS Transactions* 33:3–15.

Baglio V., Ornelas R., Matteucci F., Martina F., Ciccarella G., Zama I., Arriaga L G., Antonucci V., and Aricò A. S. 2009. Solid polymer electrolyte water electrolyser based on Nafion-TiO₂ composite membrane for high temperature operation. *Fuel Cells*, 9(3):247-252.

Briguglio N., Brunaccini G., Siracusano S. et al. 2013. Design and testing of a compact electrolyzer system. *International Journal of Hydrogen Energy* 38:11519–11529.

Carmo M., Fritz D. L., Mergel J., and Stolten D. 2013. A comprehensive review on PEM water electrolysis. *International Journal of Hydrogen Energy* 38:4901–4934.

Cavaliere S., Subianto S., Savych I., Jones D. J., and Rozière J. 2011. Electrospinning: designed architectures for energy conversion and storage devices. *Energy & Environmental Science* 4:4761–4785.

E4tech, Element Energy 2014. Study on development of water electrolysis in the EU. Fuel Cells and Hydrogen Joint Undertaking.

Ghielmi A., Vaccarone P., Troglia C., and Arcella V. 2005. *Journal of Power Source* 145:108–115.

Grigoriev S. A., Millet P., Volobuev S. A., and Fateev V. N. 2009. Optimization of porous current collectors for PEM water electrolysers. *International Journal of Hydrogen Energy* 34:4968–4973.

Hung Y., El-Khatib K. M., and Tawfik, H. 2005. Corrosion-resistant lightweight metallic bipolar plates for PEM fuel cells. *Journal of Applied Electrochemistry* 35:445–447.

Jung H. Y., Huang S. Y., Ganesan P., and Popov B. N. 2009. Performance of gold-coated titanium bipolar plates in unitized regenerative fuel cell operation. *Journal of Power Sources* 194:972–975.

Laguna-Bercero M. A. 2012. Recent advances in high temperature electrolysis using solid oxide fuel cells: A review. *Journal of Power Sources* 203:4–16.

Marshall A., Borresen B., Hagen G., Tsypkin M. et al. 2007. Hydrogen production by advanced proton exchange membrane (PEM) water electrolysers—Reduced energy consumption by improved electrocatalysis. *Energy* 32:431–436.

Millet P., Mbemba N., Grigoriev S. A., Fateev V. N., Aukulauoo A., and Etiévant C., 2011. Electrochemical performances of PEM water electrolysis cells and perspectives. *International Journal of Hydrogen Energy* 36:4134–4142.

Moyer C. J., Ambrosini A., Sullivan N. P. et al. 2010. *Solid Oxide Electrochemical Reactor Science.* Sandia Report, Sandia National Laboratories, New Mexico.

Ng F., Peron J., Jones D. J., and Roziere J. 2011. Synthesis of novel proton conducting highly sulfonated polybenzimidazoles for PEMFC and the effect of the type of bisphenyl bridge on properties. *Journal of Polymer Science. Part A: Polymer Chemistry* 49:2107–2117.

Patyk Andreas, 2008. Innovative Solid Oxide Electrolyser Stacks for Efficient and Reliable Hydrogen Production (RelHy). http://www.relhy.eu/.

Rajeshwar K., McConnell R., and Licht S. (Ed.) 2008. *Solar Hydrogen Generation. Toward A Renewable Energy Future.* Springer.

Siracusano S., Baglio V., Di Blasi A., Briguglio N. et al. 2010. Electrochemical characterization of single cell and short stack PEM electrolyzers based on a nanosized IrO₂ anode electrocatalyst. *International Journal of Hydrogen Energy* 35:5558–5568.

Smolinka T., Gunther M., and Garche J. 2011. NOW-Studies: "Stand und Entwicklungs Entwicklungspotenzialder Wasserelektrolyse zur Herstellung von Wasserstoff aus regenerativenEnergien". Technical report. Fraunhofer ISE.

Ulleberg O., Nakken T., and Ete A. 2010. The wind/hydrogen demonstration system at Utsira in Norway: Evaluation of system performance using operational data and updated hydrogen energy system modeling tools. *International Journal of Hydrogen Energy* 35:1841–1852.

Ursú A., Gandıa L. M., and Sanchis P. 2012. *Hydrogen Production from Water Electrolysis: Current Status and Future Trends.* Proceedings of the IEEE 100:410–426.

2

Fundamentals of PEM Water Electrolysis

Tom Smolinka, Emile Tabu Ojong, and Thomas Lickert

CONTENTS

Nomenclature

a	Tafel slope [mV dec^{-1}]	ΔH_V	Enthalpy of evaporation [kJ mol^{-1}]
$a\{i\}$	(Molar) Activity of substance i [mol l^{-1}]	HHV	Higher heating value [kJ mol^{-1}]
A	Surface or area [m^2]	i	Current density [A m^{-2}]
b	Tafel constant [A m^{-2}]	i_o	Exchange current density [A cm^{-2}]
c_i	Molar concentration of substance i [mol m^{-3}]	I	Electrical current [A]
E_a	Activation energy [kJ mol^{-1}]	k	Reaction rate constant [L mol^{-1} s^{-1}]
F	Faraday constant [96,485 C mol^{-1}]	l	Membrane thickness [m]
ΔG	Molar change in Gibbs free energy [kJ mol^{-1}]	LHV	Lower heating value [kJ mol^{-1}]
h_i	Molar enthalpy of substance i [kJ mol^{-1}]	m	Mass [g]
H	Molar enthalpy change [kJ mol^{-1}]	M	Molecular mass [g mol^{-1}]
$\Delta \dot{H}_i$	Enthalpy flow of substance i [kJ s^{-1}]	n	Amount of substance [mol]
		$\dot{n}_i$	Molar flow rate of substance i [mol s^{-1}]
		p	Pressure [MPa]

p_i	Partial pressure of substance i [MPa]
P	Electrical power [W]
Q	Electrical charge [C]
$\dot{Q}$	Heat flow [kJ s^{-1}]
r	Reaction rate per unit area [mol s^{-1} cm^{-2}]
R	Universal gas constant [8.314 kJ kmol^{-1} K^{-1}]
R_i	Ohmic resistance of cell component i [Ω]
S	Electrochemical active site [—]
ΔS	Molar change in entropy [kJ K^{-1} mol^{-1}]
t	Time interval [s]
T	Temperature [K]
ΔT	Temperature difference [K]
V	Voltage [V]
z	Charge number [—]
α	Heat transfer coefficient [W m^{-2} K^{-1}]
β	Symmetry factor [—]
ε	Efficiency [—]
η	Overpotential [V]
λ	Factor for the membrane hydration [–]
ν_i	Stoichiometric factor of substance i [–]
σ	Conductivity [S m^{-1}]

pump	Pump
R	Reaction
rct	Reactant
rect	Rectifier
rev	Reversible
rp	reference position
stack	Electrolysis stack
surr	Surrounding
th	Thermoneutral
theor	Theoretical
V	Voltage

Subscripts and Superscripts

0	Standard state for temperature and pressure (1 atm, 298.15 K)
AC	Alternating current
act	Activation
ads	Adsorbed
an	Anode
b	benchmark
BoP	Balance of plant
bub	Bubbles
cath	Cathode
cell	Cell
compr	Compressor
DC	Direct current
diff	Diffusion
en	Energy
f	Formation
(g)	Gaseous state
he	Heat exchanger
i	Arbitrary species
I	Current
(l)	Liquid state
loss	Thermal losses to the surrounding
mem	Membrane
mod	Module
ohm	Ohmic
op	Operating condition
pdt	Product
per	Periphery

Abbreviations

AC	Alternating current
ASR	Area-specific resistance
BoP	Balance of plant
BPP	Bipolar plate
CC	Current collector
DC	Direct current
EL	Electrolysis
GDE	Gas diffusion electrode
HER	Hydrogen evolution reaction
HHV	Higher heating value
HP	High pressure
HT	High temperature
KPI	Key performance indicator
LHV	Lower heating value
LP	Low pressure
LT	Low temperature
MEA	Membrane electrode assembly
OCV	Open cell voltage
OER	Oxygen evolution reaction
PEM	Proton exchange membrane or polymer electrolyte membrane
PFSA	Perfluorinated sulfonic acid
PGM	Platinum group metals
PSA	Pressure swing adsorption
SPE	Solid polymer electrolyte
STP	Standard conditions for temperature and pressure

2.1 Introduction

Water electrolysis is an electrochemical process in which electricity is applied to split water into hydrogen and oxygen. It represents one of the simplest approaches to produce hydrogen and oxygen in a zero-pollution process and has already been known for more than

200 years (Kreuter and Hofmann 1998). In particular, alkaline water electrolyzers have been in use for more than 100 years in industrial applications (LeRoy 1983), but due to its several advantages, the proton exchange membrane (PEM) electrolyzer has become an emerging technology with a growing market share; see Chapter 1.

This chapter provides a general overview about the fundamentals of PEM water electrolysis.

2.1.1 General Principle

The general principle of water electrolysis can be expressed by the following equation:

$$H_2O_{(l)} + \Delta H_R \rightarrow H_{2(g)} + \frac{1}{2}O_{2(g)} \tag{2.1}$$

In Equation 2.1, ΔH_R denotes the change in the reaction enthalpy for this endothermic reaction. Water is fed as liquid reactant as the PEM electrolysis cell is operated mostly below the boiling point of water. The overall electrolysis reaction is the sum of the two electrochemical half reactions, which takes place at the electrodes in an acidic environment according to the following equations:

$$H_2O_{(l)} \rightarrow \frac{1}{2}O_{2\ (g)} + 2H^+ + 2e^- \tag{2.2}$$

$$2H^+ + 2e^- \rightarrow H_{2(g)} \tag{2.3}$$

Equation 2.2 represents the anode half reaction and Equation 2.3 the cathode half reaction. Due to the limited self-ionization of pure water, an acid is used as electrolyte and electrocatalysts are applied to lower the activation energy of the process, see Section 2.3.1. An electrical DC power source is connected to the electrodes, which are, in the simplest case, two plates made from inert metal such as platinum or iridium immersed in the aqueous electrolyte (Figure 2.1). The decomposition of water starts once a DC voltage higher than the thermodynamic reversible potential, see Section 2.2, is applied to the electrodes. However, the entropy change for the reaction is negative and various activation barriers have to be overcome. Therefore, a DC voltage higher than the thermoneutral potential is required to operate a PEM electrolysis cell, for more details see Section 2.3.

At the anode (positively charged electrode) water is oxidized, the electrons pass through the external electrical circuit and oxygen evolves as gas (oxygen evolution reaction, OER). Protons migrate through the acidic electrolyte from the anode to the cathode (negatively charged electrode) where they are reduced by the electrons from the external electrical circuit to hydrogen

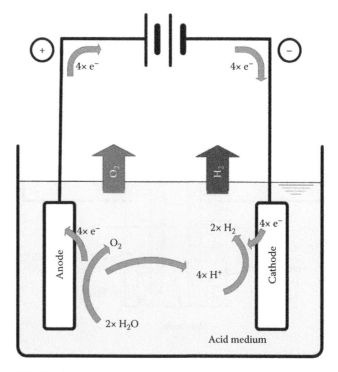

FIGURE 2.1
General principle of a water electrolysis cell with a watery acid as electrolyte. Oxyhydrogen is produced as the cell has no diaphragm to separate the two half-cells.

gas (hydrogen evolution reaction, HER). The amount of hydrogen generated is twice the amount of oxygen generated on the anode side if an ideal faradaic efficiency is assumed.

2.1.2 Main Cell Components

In PEM electrolysis cells, an acidic membrane is used as a solid electrolyte instead of a liquid electrolyte. The basic design of such a cell with its main components is depicted in Figure 2.2. The two half-cells are separated by the solid acidic membrane, which is commonly called proton exchange membrane or polymer electrolyte membrane (PEM). Sometimes, the expression solid electrolyte membrane (SPE) is used as well. In most cell designs, the electrodes are deposited directly onto the membrane, creating the key component of a PEM electrolysis cell, the membrane electrode assembly (MEA).

The MEA is sandwiched between porous current distributors/collectors (in this chapter only denoted as current collectors, CC). In some cell designs, the electrocatalysts are coated onto the current collectors. Following the nomenclature known from fuel cells, this combination is called gas diffusion electrode (GDE). The current collectors enable an electric current to flow from the bipolar plates to the electrodes and, simultaneously, the supply of reactant water to and the removal of the

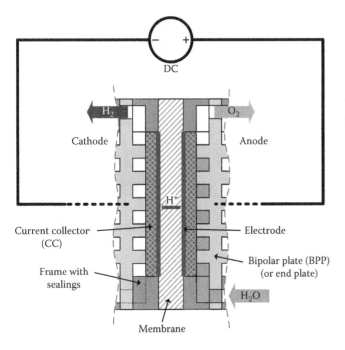

FIGURE 2.2
Simplified basic design of a single cell in a PEM water electrolysis stack.

generated gas bubbles from the electrodes. The bipolar plates encase the two half-cells and provide the electrical contact to the external power supply. Usually, they include flow field structures to enhance the transport of liquid water to the electrodes and oxygen and hydrogen out of the cell. Frames with sealing elements or gaskets tighten the half-cells to prevent gas and water leakage from the inside to the environment.

All in all, the main features of the design and components used for PEM electrolysis cells offer several advantages over other electrolysis technologies:

- The membrane as the solid electrolyte is very thin, allowing for a shorter proton transport pathway and thus lower ohmic loss.
- Electrocatalysts made from the elements of the platinum group metal (PGM) enable high efficiency and fast kinetics.
- Since the electrode is coated as thin layer directly on the membrane, the proton transport from the reaction sites to the solid electrolyte is facilitated, which minimizes the mass transport limitation.
- The electrolyte is immobilized in the membrane and cannot be leached out of the membrane or contaminate the produced gases.
- The membrane provides high gas tightness.
- The cell design is very compact resulting in low thermal masses and fast heat-up and cooling-off times and in combination with the fast kinetics

of the electrocatalysts, in a very fast response time even at ambient conditions.
- Only pure water is fed to the cell that entails a simple system design.

The main disadvantages of the PEM water electrolysis technology can be summarized as follows:

- The membrane as the solid electrolyte is more expensive than the liquid electrolyte of the alkaline technology.
- The membrane is a thin and sensitive foil that can be mechanically damaged by inappropriate cell design and operation.
- Chemical degradation may occur depending on the used materials.
- The corrosive nature of the membrane requires more expensive metal components for BPP, CC and PGM electrocatalysts.
- The electrodes are mostly catalyst systems with structures at the nanoscale that requires a comprehensive understanding for a high electrochemical durability.

Detailed information about the choice of materials, design of the components, and their advantages and disadvantages are provided in Chapters 3 through 8.

To yield higher hydrogen flow rates, several single cells are connected electrically in series and hydraulically in parallel as a stack. Here, a bipolar plate (BPP) separates two adjacent cells, so that it simultaneously acts as the anode of one cell and the cathode of the adjacent cell. The most common stack type is the filter press bipolar pressurized cell stack for PEM water electrolysis. Pressure plates (not shown in Figure 2.2) fix the components of the cells and provide the clamping force by thread bolts and nuts. Details on stack design can be found in Chapter 9.

2.1.3 General System Layout

The PEM stack is the core component of an electrolysis (EL) system to produce hydrogen and oxygen from water. However, operation of the stack is only possible with the help of several additional components of the EL module and further subsystems. The basic layout of a PEM electrolysis system is presented in Figure 2.3. There is no general standard for a layout of PEM electrolysis systems, but the figure comprises all relevant components and subsystems of such a system. It is similar to that of alkaline electrolysis system (Smolinka et al. 2014). But due to the absence of lye as the liquid electrolyte, components such as gas scrubbers are not needed and the PEM system is less complex.

Although every EL system differs from others in the layout, it is worthwhile to discuss the typical

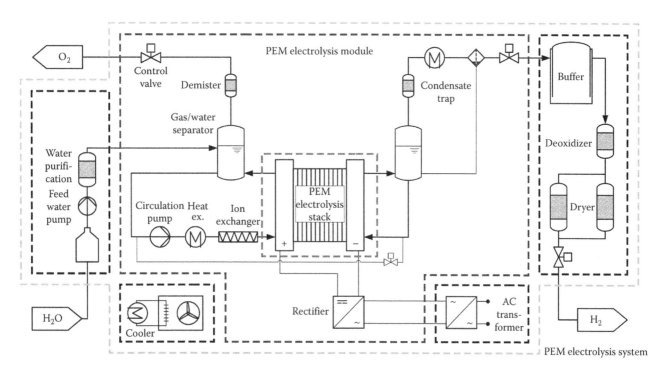

FIGURE 2.3
General layout of a PEM electrolysis system consisting of the PEM stack and module with power electronics and the EL subsystems for water purification, gas drying, and fine purification and cooling unit.

system boundaries of an EL system and its subsystems. Otherwise, it is not possible to provide a clear definition of the mole and energy balances (Section 2.2.3) and key performance indicators (KPIs) (Section 2.4).

According to Figure 2.3, an EL system is made up of three layers. The PEM stack itself is the chemical reactor in which water is decomposed with the help of a DC current into hydrogen and oxygen. The PEM module comprises all peripheral components to operate the stack properly at the desired operating conditions (temperature and pressure) and to provide the stack with the reactants and to remove the products. The rectifier converts the incoming AC power into a regulated DC current. The anode side consists at least of a circulation pump, a heat exchanger, an ion exchanger (mostly in a bypass), a gas/water separator, a demister, and a control valve:

- Feed water supply is connected to the anode since water is consumed on this side.
- In most cases, the natural convection of the two-phase flow forced by the ascending gases is sufficient to supply the stack with water. But a defined water flow by the circulation pump is essential to ensure stack cooling.
- The ion exchanger is needed to trap heavy metal cations such as iron, chromium, and

nickel, which may originate from the balance of plant (BoP) or as corrosion product from metal components within the stack.

- The gas/water separator is positioned above the stack and separates the two-phase flow of oxygen and water coming from the stack outlet. Depending on the system design inside the gas/water separator, heat exchangers and baffles are installed to adjust the temperature of the water and to reduce the aerosol content of oxygen.
- Subsequently, the gases flow through demisters (coalescent filter) in order to retain fine droplets of liquid water in the oxygen flow.
- The control valve after the demister regulates the pressure on the oxygen side.

In most cases, a circulation pump is not applied at the cathode side. But water that is transported by an electro-osmotic drag from the anode to the cathode needs to be separated from the hydrogen and be collected at the cathode side. For this purpose, a smaller gas/water separator with demister is installed at the cathode. Subsequently, a heat exchanger and a condensate trap to reduce the dew point are placed before the control valve. Via a drain valve, water is transported back to the anode according to the level control.

The system level encloses further auxiliary subsystems that are required according to the installation site and the application:

- Depending on the available water quality at the installation site, a water purification stage (e.g., reverse osmosis) is needed to purify the feed water and to prevent fouling in the system and degradation in the stack. A feed water pump controls the water level in the anode circulation loop.

- Most PEM electrolysis modules have additional heat exchangers for water and gas cooling, for example, in front of the condensate trap. A cooling unit provides the cooling energy for these different heat exchangers.

- Often, a buffer as gas reservoir for hydrogen is installed after the control valve to guarantee a constant hydrogen flow for the downstream application.

- A hydrogen purification unit purifies the gas to the purity grade demanded by the application. Mostly, a two-stage purification unit is realized. In the first stage, the remaining oxygen and hydrogen recombines catalytically to water. In the second stage, hydrogen is dried to the required dew point by removing the moisture, for example, in adsorption columns (pressure swing adsorption [PSA]). In particular, gas drying is an energy-consuming process. Partial flow of the dried hydrogen is required to generate the adsorption columns, which accounts to a loss of the produced hydrogen.

It should be kept in mind that this system layout is only a general description of a PEM electrolysis system. Further discussion and examples of developed PEM electrolysis system are provided in Chapter 9.

2.1.4 PEM Electrolysis Operation

Operating a PEM electrolysis cell under atmospheric conditions is a quite simple procedure. Once water is available at the anode and the cell voltage is higher than the thermoneutral cell voltage at ambient temperature, hydrogen and oxygen will be generated at the electrodes. However, in technical installations, PEM electrolysis cells are operated under pressure to provide compressed hydrogen and/or oxygen with higher energy density for the subsequent application. By pressurizing the hydrogen and/or oxygen, the need for an external gas compressor is eliminated or at least the energy demand is reduced. Only compression power

for pumping liquid water at system pressure is required, which is considerably less than compressing the product gas to the desired pressure (Onda et al. 2004; Bensmann et al. 2013). For this reason, the development of (high-) pressurized EL systems is in the focus of several manufacturers to gain further overall efficiency improvement. Operation (and thus stack design) of a pressurized PEM module can be classified according to

- The pressure regime: pressure balanced or differential pressure operation
- The water supply: anode or cathode feed
- The thermal management: water circulation at the anode and/or cathode side

In general, a PEM electrolysis cell/stack can be operated at balanced pressure and with differential pressure from anode to cathode (LaConti and Swette 2003). In the former case, both sides of the electrolysis cell are operated at the same pressure, which is regulated by an oxygen and hydrogen control valve. In the latter case, mostly, the hydrogen side works under high pressure (HP), whereas the oxygen side is not or only slightly pressurized. In some niche applications, it could be the reverse (Roy et al. 2011).

Differential operation demands a well-developed stack design, as the membrane has to withstand the differential pressure. Nevertheless, PEM electrolysis modules for differential pressure operation offer several advantages:

- Investment costs are reduced as components of the low pressure (LP) side do not need to be compression proof. In particular, the more complex circulation loop of the oxygen side and the feed water pump could be designed for LP operation.
- Purity of the product gas is higher as gas contamination from the LP side is lower (diffusion and/or gas leakage).
- Pressure control is less sophisticated and the stack should be more tolerant against pressure buildup and release.

The drawback of the differential pressure operation is associated with the higher volumetric gas flow at the LP side, which entails a larger steam mass fraction (higher effort for cooling, water recovery, and drying). Moreover, larger gas bubbles formed at the electrode increase the risk of mass transport limitation, see Section 2.3.3.

Balanced pressure operation is known from alkaline electrolyzers and it is a well-proven concept. But appropriate pressure control is challenging in particular for fast pressure release at HP operation. Accuracy of the two control valves can lead to small, fluctuating pressure differences between the two half-cells, which cause mechanical stress to the membrane.

Water supply of a PEM electrolysis cell/stack can be realized either as anode feed or as cathode feed. Most PEM electrolysis modules are operated with an anode feed as water is consumed on this side. For some special applications, a cathode feed can be realized as well, for example, for manned aerospace application (Sakurai et al. 2013) and medical isotope production (cathode water vapor feed) (Greenway et al. 2009). In the case of a cathode feed, water has to migrate through the membrane to be consumed at the anode. Water diffusion becomes the rate-determining step, which prevents operation at high current densities and results in larger cell areas for the same hydrogen production rate.

Thermal management of the PEM electrolysis stack can be realized at either the anode or cathode side. Apart from PEM electrolyzers in the very small power range for special applications (Wittwer et al. 2004), passive thermal management only by surface cooling is not reported in the literature. The most common is an active thermal management by a circulation loop with circulation pump, subsequent heat exchanger, and a water supply at the anode side as depicted in Figure 2.3.

Forced convection enhances water transport to and gas removal from the electrode and keeps the temperature gradient over the stack low. Without forced convection, gas bubbles formed at the electrodes and trapped within the stack can inhibit heat convection, leading to hot-spot formation and consequent MEA degradation or other destructive mechanism. Overpotentials and thus heat release are higher at the anode than at the cathode (refer to Section 2.3). For this reason, it is preferred to place the circulation pump on the anode side. Sometimes, circulation loops on both sides are implemented in a PEM electrolysis module to keep the stack at a well-defined temperature level (Yamaguchi et al. 1997).

2.2 Thermodynamics

2.2.1 Heat of Reactions and Nernst Equation

For the decomposition of water, external energy is needed to split up the molecules to its components. The amount of energy required or the energy that is released by building up or breaking up a chemical bond is called the reaction enthalpy ΔH_R. In general, this quantity is calculated from the sum of the enthalpies of formation of the reactants $\Delta H_{f,rct}$ minus the sum of enthalpies of formation of the products $\Delta H_{f,pdt}$, as seen in the following equation:

$$\Delta H_R = \sum_{pdt} v_{pdt}\Delta H_{f,pdt} - \sum_{rct} v_{rct}\Delta H_{f,rct}$$
$$= -\Delta H_{f,H_2O} \tag{2.4}$$

For a balanced reaction equation, the stoichiometric factors v_{pdt} for the product (H_2O) and v_{rct} for the reactants (H_2, O_2) are introduced. For the net water electrolysis reaction in Equation 2.1, the stoichiometric factors are $v_{H_2O} = 1$, $v_{H_2} = 1$, and $v_{O_2} = 1/2$. Since hydrogen and oxygen exist as molecules under ambient conditions, their enthalpy of formation is zero by definition. The enthalpy of reaction is, therefore, given solely by the enthalpy of formation of water. The splitting of water is driven by electrical and thermal energy input. ΔH_R can be rewritten as the sum of the contribution of these driving forces:

$$\Delta H_R = \Delta G_R + T \cdot \Delta S_R \tag{2.5}$$

The entropy is a measure of the disorder for a thermodynamic system. The change in entropy (ΔS_R) due to the chemical reaction is given by the difference in entropy of the product and the sum of the entropies of the reactants, with the associated stoichiometric factors:

$$\Delta S_R = \sum_{pdt} v_{pdt}\Delta S_{R,pdt} - \sum_{rct} v_{rct}\Delta S_{R,rct} \tag{2.6}$$

Multiplied by the system temperature T, the term $T \cdot \Delta S_R$ is called the entropy term and represents the thermal input needed for the considered chemical reaction, the splitting of water in this case. The second term ΔG_R is another thermodynamic quantity, called the change in Gibbs free energy or free enthalpy of reaction. It can be regarded as the maximum work that can be extracted from the thermodynamic system, without the volume work. Using Equations 2.4 through 2.6, the Gibbs free energy of reaction ΔG_R can be written as

$$\Delta G_R = \sum_{pdt} v_{pdt}\Delta H_{f,pdt} - \sum_{rct} v_{rct}\Delta H_{f,rct}$$
$$- T \cdot \sum_{pdt} v_{pdt}\Delta S_{R,pdt} - \sum_{rct} v_{rct}\Delta S_{R,rct} \tag{2.7}$$

For $\Delta G_R < 0$, the reaction occurs spontaneously under the considered conditions, without any external energy. For $\Delta G_R > 0$, the reaction is a nonspontaneous process as it is the case for the electrolysis of water. Energy from an external source is needed, which can be supplied to the system either by thermal energy or by electrical energy. At standard state ($p = 1$ atm, $T = 298.15$ K), the Gibbs free

TABLE 2.1

Enthalpy of Reaction for Liquid Water and Entropy of Reaction Values for Water, Hydrogen, and Oxygen under Standard Conditions

Substance (State of Matter)	$\Delta H_R^0 \left(\mathrm{kJ\ mol^{-1}} \right)$	$S_R^0 \left(\mathrm{J\ mol^{-1}\ K^{-1}} \right)$
H_2O (l)	−285.83	69.942
H_2 (g)	0	131.337
O_2 (g)	0	205.817

Source: McBride, B.J. et al., *NASA Glenn Coefficients for Calculating Thermodynamic Properties of Individual Species*, NASA Glenn Research Center, Cleveland, Ohio, 2002.

energy of reaction ΔG_R^0 is 236.483 kJ mol^{-1}. The standard values of change in enthalpy and entropy for H_2O, H_2, and O_2 are listed in Table 2.1.

State-of-the-art PEM electrolysis modules operate at low temperatures (<373.15 K). Therefore, the introduction of thermal energy is considered to be small, which means all the energy required must be applied by electrical energy. Figure 2.4 shows the temperature dependencies of the three thermodynamic quantities mentioned at standard pressure and for a temperature range from 273.15 to 1000 K. For PEM water electrolysis, the conditions in the region of liquid state (until 373.15 K) are of main interest. In this region, the demand of electric energy decreases with increasing temperature, while the thermal energy demand increases. The total energy demand to split water molecules decreases slightly with increasing temperature until the boiling point, where water gets in its gaseous state and the total energy demand increases again but stays below the values of liquid state. Figure 2.4 confirms the thermodynamic

benefit of operating an electrolysis cell at elevated temperatures. Beyond the boiling point, the slight increase in total energy demand can be compensated more by the thermal input, the higher the temperature is. This concept is made useful in the operation of high-temperature (HT) electrolysis, in which other materials are applied to split water in the form of steam at temperatures up to some 1300 K (Smolinka et al. 2010).

If the electrolysis process occurs under reversible conditions (without losses), the potential difference at the electrodes is called the reversible cell voltage V_{rev}. It is the minimal electrical work that is needed to split up water if the requisite contribution of thermal energy is present. Using the defined ΔG_R at standard state, the Faraday constant F and the amount of charges z (electrons) transferred during the reaction, V_{rev}^0 can be calculated by

$$V_{rev}^0 = \frac{\Delta G_R^0}{z \cdot F} = 1.229\ \mathrm{V} \tag{2.8}$$

Without having an external heat source, the entire energy for the reaction to take place $\left(\Delta H_R^0 \right)$ must be delivered by electrical energy. Hence, the voltage required is higher than V_{rev}^0 and is called the thermoneutral voltage V_{th}^0 at standard state

$$V_{th}^0 = \frac{\Delta H_R^0}{z \cdot F} = 1.481\ \mathrm{V} \tag{2.9}$$

As mentioned, for state-of-the-art materials for PEM electrolysis cells, the operating range is limited

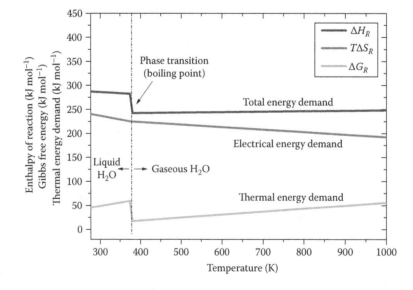

FIGURE 2.4
Temperature dependency of the total energy demand (enthalpy of reaction ΔH_R) for the electrolysis of water with its electrical (ΔG_R) and thermal fraction ($T \cdot \Delta S_R$). (Equilibrium data taken from McBride, B.J. et al., *NASA Glenn Coefficients for Calculating Thermodynamic Properties of Individual Species*, 2002.)

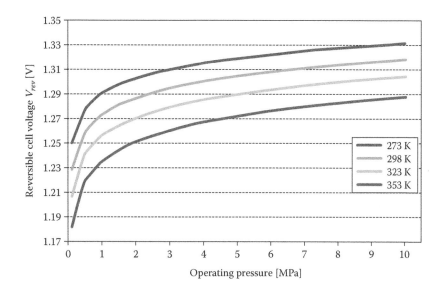

FIGURE 2.5
Reversible cell voltage V_{rev} as a function of pressure for four different temperatures (273–353 K).

to temperatures where water exists in its liquid state. Thus, the thermodynamic benefit of temperatures up to about 373.15 K is rather small and the minimum required cell voltage is near V_{th}^0.

As ΔG_R and ΔH_R are not only functions of temperature but of pressure as well, the Nernst equation links the concentration (or activity) of the reactants and products to the potential difference of the electrodes. The general expression of the Nernst equation for the reversible cell voltage of the water splitting process is given by the following equation:

$$V_{rev} = V_{rev}^0 - \frac{R \cdot T}{2 \cdot F} \cdot \ln\left(\frac{a\{H_2O\}}{a\{H_2\} \cdot a\{O_2\}^{1/2}}\right) \quad (2.10)$$

With the reversible cell voltage at standard state V_{rev}^0, the universal gas constant R, the Faraday constant F, and the activities of the reactant $a\{H_2O\}$ and products $a\{H_2\}$ and $a\{O_2\}$ with the respective stoichiometric factors. Replacing the activities by the partial pressures at standard temperature in reference to the operation pressure p_{op}, Equation 2.10 can be written as

$$V_{rev} = V_{rev}^0 - \frac{R \cdot T}{2 \cdot F} \cdot \ln\left(\frac{p_{H_2O}/p_{op}}{\left(p_{H_2}/p_{op}\right) \cdot \left(p_{O_2}/p_{op}\right)^{1/2}}\right)$$

$$= V_{rev}^0 - \frac{R \cdot T}{2 \cdot F} \cdot \ln\left(\frac{1}{\left(p_{H_2}/p_{op}\right) \cdot \left(p_{O_2}/p_{op}\right)^{1/2}}\right) \quad (2.11)$$

It can be seen from Equation 2.11 that the reversible cell voltage V_{rev} decreases with increasing temperature. With the reversible cell voltage at standard state $\left(V_{rev}^0\right)$, the ideal gas constant R, the Faraday constant F and the partial pressures of hydrogen p_{H_2} and oxygen p_{O_2} in the gas phase, the pressure dependencies of the reversible cell voltage are given in Figure 2.5.

2.2.2 Faraday's Law

The amount of gas produced by an electrochemical process can be related to the electrical charge consumed by the cell, which is described by Faraday's law.

$$Q = n \cdot z \cdot F \quad (2.12)$$

With a modification of this, the electrolytically converted mass per time interval t can be written as

$$m = \frac{I \cdot t \cdot M}{z \cdot F} \quad (2.13)$$

For an ideal water electrolysis process, the electric charge, passing through the cell, is a direct measure of the amount of hydrogen and oxygen produced. From this, it is possible to measure the effectiveness of the real process by comparing the charges fed to the system and the amount of hydrogen (or oxygen) that was produced. This parameter is called the Faradic efficiency and is introduced in Section 2.2.4.

2.2.3 Mole and Energy Balances

This section provides an overview of the mole and energy balances on stack and module level as an introduction to the definitions of efficiencies and subsequently for the discussion of the KPIs in Section 2.4. The concept of mole and energy balances is often used as basis for dynamic modeling of electrolysis cells, stack, or systems (Ni et al. 2008; Marangio et al. 2009; Awasthi et al. 2011).

To give an overview of the concept of mole and energy balances, the considerations done herein are made for systems in steady state. Dynamic behavior like load, temperature, or pressure change is not included. Furthermore, internal loss mechanisms inside the stack (permeation of the product gases through the membrane) are not considered.

A PEM electrolysis stack or module changes mass and energy with its surrounding. From a thermodynamic point of view, it is per definition an open system. The concept of open systems is used for both stack and module level. Two main flow rates are defined in the following. The flow of a substance "i" is given by the molar flow rate $\dot{n}_i$. The flow of energy is expressed by the enthalpy flow rate $\dot{H}_\iota$, which is defined as the product of the molar flow rate and the molar enthalpy of this substance "i".

$$\dot{H}_\iota = \dot{n}_i \cdot h_i \qquad (2.14)$$

2.2.3.1 Stack Level

For an electrolysis stack, considered as black box with the system boundary as given in Figure 2.6, the mole balance for the stationary case is expressed by Equation 2.15 with stoichiometric factors according to Section 2.2.1. It should be noted that an incoming flow is accounted as positive, while an outgoing flow as negative.

$$0 = \sum \dot{n} = \nu_{H_2O}\,\dot{n}_{H_2O} + \nu_{H_2}\,\dot{n}_{H_2} + \nu_{O_2}\,\dot{n}_{O_2}$$

$$0 = \dot{n}_{H_2O} - \dot{n}_{H_2} - \frac{1}{2}\dot{n}_{O_2}$$

$$0 = \dot{n}_{H_2O(l),in}^{an} - \dot{n}_{H_2(g),out}^{cath}$$

$$- \dot{n}_{H_2O(g),out}^{cath} - \dot{n}_{H_2O(l),out}^{cath} - \dot{n}_{O_2(g),out}^{an} - \dot{n}_{H_2O(g),out}^{an} - \dot{n}_{H_2O(l),out}^{an}$$

$$(2.15)$$

The energy balance with the enthalpy flows $\dot{H}_\iota$, the flow of heat or irreversible heat loss to the surrounding $\dot{Q}_{loss}$, and the electrical power P_{DC} fed to the stack is given by

$$0 = \sum P + \sum \dot{Q}_i + \sum \dot{H}_\iota$$
$$0 = P_{DC} - \dot{Q}_{loss} + \dot{H}_{H_2O} - \dot{H}_{H_2} - \dot{H}_{O_2} \qquad (2.16)$$

Here, the electrical power is defined as the product of the stack voltage and the electrical current passing through the stack ($P_{DC} = V_{stack} \cdot I_{stack}$). The thermal loss to the surrounding $\dot{Q}_{loss}$ depends on the temperature difference between the stack (operating temperature), the temperature of the surrounding, the heat transfer coefficient α, and the surface of the stack A_{stack}.

$$\dot{Q}_{loss} = \alpha \cdot A_{stack} \cdot \Delta T = \alpha \cdot A_{stack} \cdot \left(T_{stack} - T_{surr}\right) \qquad (2.17)$$

The considerations made for this chapter are valid for the cell and the stack level. But for simplicity, only the stack is mentioned herein. Figure 2.6 shows the considered molar flow rates $\dot{n}_i$, enthalpy flows $\dot{H}_i$, electrical power input P_{DC}, and irreversible heat loss $\dot{Q}_{loss}$.

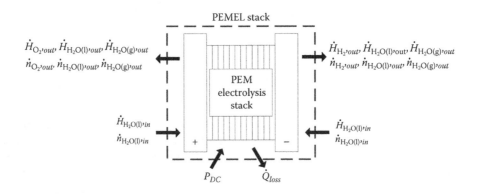

FIGURE 2.6

Schematic drawing of a PEM electrolysis stack as black box (dashed line) with the herein considered molar flow rates $\dot{n}_i$, the enthalpy flow rates $\dot{H}_i$, the irreversible heat loss $\dot{Q}_{loss}$, and the input of electrical power P_{DC}.

Liquid water can be fed either on the anode or on the cathode side, compared with Section 2.1.4. The electrical energy P_{DC} coming from the power electronics is needed for the electrochemical reaction.

As a result of the electrochemical process, hydrogen, oxygen, (excess) liquid water as well as water steam leave the system. Note that gas impurities such as hydrogen on the oxygen side and oxygen on the hydrogen side, caused by undesired permeation processes, are not considered herein (see Chapter 6 for more details).

For higher current densities, the irreversible losses, for example, Joule heating, become dominant, which makes a cooling system necessary. These processes determine the temperature of the stack.

2.2.3.2 Module Level

The mole and energy balances on the next level of complexity toward complete hydrogen production systems are done for the PEM electrolysis module, which is defined according to Section 2.1.3. As indicated by the dashed black line in Figure 2.7, the boundary of the PEM electrolysis module is set in this way that hydrogen (and

oxygen) can be produced at the intended pressure and temperature without liquid water in the product stream. Therefore, additional components, like gas/water separators, heat exchanger, ion exchanger, demister, valves, etc. are necessary.

Looking at the system boundaries in Figure 2.7, the same approach as for stack level can be applied with the difference that only oxygen, hydrogen, and water in gaseous state are considered to be in the outgoing streams (no liquid water). Hence, the mole balance for the EL module in the steady state is given by the following equation:

$$0 = \dot{n}^{an}_{H_2O(l)} - \dot{n}^{cath}_{H_2} - \dot{n}^{an}_{O_2} - \dot{n}^{cath}_{H_2O(g)} - \dot{n}^{an}_{H_2O(g)} \qquad (2.18)$$

The general stationary energy balance of an open system is already introduced by Equation 2.16. The sum of all power inputs is given by the electrical power supplied to the rectifier P_{AC}, to the circulation pump P_{pump}, and to the remaining BoP components P_{BoP}, which are negligible. The electrical power P_{AC} coming from the transformer is rectified and fed to the stack as DC

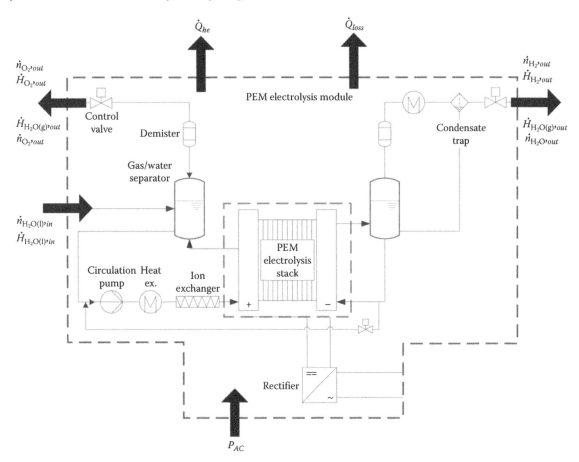

FIGURE 2.7
Schematic drawing of a PEM electrolysis module (dashed lines) with the herein considered molar flow rates $\dot{n}_i$, the enthalpy flows $\dot{H}_i$, the irreversible heat loss $\dot{Q}_{loss}$, and the input of electrical power P_{AC}.

power (P_{DC}). The efficiency of the rectification process is defined by the ratio of *DC* power and AC power to $\varepsilon_{rect} = P_{DC}/P_{AC}$. With this, the overall power consumption of the EL module can be derived:

$$P = P_{AC} + P_{pump} + P_{BoP} \approx P_{AC} + P_{pump}$$

$$P = \frac{P_{DC}}{\varepsilon_{rect}} + P_{pump} \tag{2.19}$$

The heat losses to the surrounding $\dot{Q}_{loss}$ are given by the sum of the heat losses of the stack, the rectifier, and all BoP components, which are heated above the ambient temperature by either the heated product streams or its own electrical losses. The heat exchanger(s) can be used to heat up or cool down the stack. Usually, the stack is operated at high current densities and cell voltages above V_{th}. In this case, the stack must be cooled down and a heat flow $\dot{Q}_{he}$ is dissipated in different heat exchangers. Therefore, $\dot{Q}_{loss}$ and $\dot{Q}_{he}$ account as negative flows and can be written as

$$\dot{Q}_{mod} = \dot{Q}_{he} + \dot{Q}_{loss} \tag{2.20}$$

$$\dot{Q}_{loss} = \dot{Q}_{stack} + \dot{Q}_{rect} + \dot{Q}_{BoP} \tag{2.21}$$

For the enthalpy flows, the mole balance (Equation 2.18) can be modified using Equation 2.14 to

$$0 = \dot{H}_{H_2O(l)}^{an} - \dot{H}_{H_2(g)}^{cath} - \dot{H}_{O_2(g)}^{an} - \dot{H}_{H_2O(g)}^{an} - \dot{H}_{H_2O(g)}^{cath} \tag{2.22}$$

With Equations 2.19 through 2.22, the total energy balance of a PEM electrolysis module in the steady state can be approximated by

$$0 = \frac{P_{DC}}{\varepsilon_{rect}} + P_{pump} - \dot{Q}_{he} - \dot{Q}_{stack} - \dot{Q}_{rect} - \dot{Q}_{BoP} + \dot{H}_{H_2O(l)}^{an}$$

$$- \dot{H}_{H_2}^{cath} - \dot{H}_{O_2}^{an} - \dot{H}_{H_2O(g)}^{an} - \dot{H}_{H_2O(g)}^{cath} \tag{2.23}$$

A detailed consideration of the mole and energy balances for the PEM EL system with its subsystems as water purification and gas drying is not provided here. The different possible technical solutions for the subsystems do not allow coherent mole and energy balances.

2.2.4 Efficiency of the PEM Water Electrolysis Process

In energy conversion systems, the ratio of useful energy output to the input of energy is defined as the energy (conversion) efficiency ε.* Based on this approach,

---

* Mostly, the Greek letter η is used for the term "energy efficiency," but in this chapter η already denotes the overpotential.

different definitions of an electrolyzer's efficiency can be found in the literature (Roy et al. 2006; Smolinka et al. 2010). For a low-temperature (LT) water electrolysis process, the useful output is the chemical energy of the produced hydrogen and the input is the electrical power. The efficiency is a KPI of electrolysis cells, stacks, and systems. The definition of efficiency for this section is based on the following assumptions:

- Only the production of hydrogen is regarded as the useful output of the system. The production of oxygen and the associated increase in the efficiency are not taken into account.
- The PEM module is operated at temperatures <373.15 K, with electricity as the only energy input. Thermal energy cannot be incorporated from the surroundings.
- The supplied water is always in the liquid state, which means the enthalpy of evaporation must be supplied by the electrical power input.

A general distinction in this section is made for the EL cell or stack as electrochemical conversion reactor and the PEM electrolysis module with peripheral components as BoP as described in Figure 2.7. Further subsystems as water purification and gas drying are not taken into account in this section.

2.2.4.1 Cell Level

On cell (or stack) level, the efficiency ε_{cell} is defined as the ratio of the molar flow rate $\dot{n}_{H_2}$ for the produced hydrogen multiplied by HHV or LHV and the electrical DC power $P_{DC} = I_{cell} \cdot V_{cell}$ supplied to the cell or stack. The question whether the HHV or LHV has to be used for calculating the efficiency of an EL process is discussed controversially in the literature. The use of HHV should be the preferred option to measure the efficiency of an electrolysis cell supplied with liquid water, since the enthalpy of evaporation ΔH_V has to be provided by the process. The efficiency ε_{cell}^{HHV} is a useful measure for the effectiveness of the applied electrochemical materials and chosen cell design and can be written as

$$\varepsilon_{cell}^{HHV} = \frac{\Delta H_R^0 \cdot \dot{n}_{H_2}}{P_{DC}} = \frac{\text{HHV} \cdot \dot{n}_{H_2}}{I_{cell} \cdot V_{cell}} \tag{2.24}$$

In principle, the cell efficiency can be calculated as well using the LHV instead of the HHV. However, with this definition, the maximum possible efficiency will always be <100% for an LT electrolysis process with liquid water supply, which would not be a practical evaluation criterion.

On the cell level, a simplified expression can be given, the so-called voltage efficiency ε_V. In general, ε_V is defined as the ratio of the thermoneutral voltage $V_{th}(T)$ at a given temperature and the real cell voltage $V_{op}(T)$ during operation at this temperature. But in most cases, the thermoneutral cell voltage V_{th}^0 at the standard state is taken for the calculation of the voltage efficiency ε_V^{HHV}:

$$\varepsilon_V^{HHV} = \frac{V_{th}^0}{V_{op}(T)} \quad (2.25)$$

Calculation of the cell efficiency according to Equation 2.25 is only valid under the assumption that the supplied current is converted completely into the electrochemical reaction of water splitting. This is practically not the case due to unmeant side reactions and stray current inside the cell. Moreover, permeation of the gases through the membrane and subsequent recombination to water and gas leakages lead to a hydrogen loss. To account for this mechanism, the Faradaic or current efficiency ε_I is defined as the ratio of the amount of actually produced hydrogen under real conditions and the amount of theoretically produced hydrogen according to Faraday's law (Section 2.2.2):

$$\varepsilon_I = \frac{\dot{n}_{H_2,op}}{\dot{n}_{H_2,theor}} = \frac{\dot{n}_{H_2,op}}{I \cdot (n \cdot F)^{-1}} \quad (2.26)$$

The product of the voltage efficiency ε_V^{HHV} and the Faradaic efficiency ε_I is the cell efficiency as introduced in the following equation:

$$\varepsilon_{cell}^{HHV} = \varepsilon_V^{HHV} \cdot \varepsilon_I \quad (2.27)$$

2.2.4.2 Module Level

Efficiency comparisons of different electrolysis modules require a clear setting of the system boundaries. In particular, efficiencies for EL modules and entire EL systems, as depicted in Figure 2.3, are mixed up in the literature or manufacturer's data. In this section, the definition of the module efficiency ε_{mod} will be provided.

The general definition of the energy efficiency can be applied as well to the module level including all peripheral components of the BoP, which have individual losses or need auxiliary power for operation and therefore reduce the overall module efficiency ε_{mod}. The sum of the electrical power demand for an EL module is given by Equation 2.19 and the useful output of an EL module is identical to that of the cell (stack).

Subsequently, the module efficiency ε_{mod} using the HHV is defined as

$$\varepsilon_{mod}^{HHV} = \frac{\Delta H_R^0 \cdot \dot{n}_{H_2}}{P_{AC} + P_{pump} + P_{BoP}} = \frac{HHV \cdot \dot{n}_{H_2}}{(P_{DC}/\varepsilon_{rect}) + P_{Pump}} \quad (2.28)$$

In comparison with the cell efficiency ε_{cell}^{HHV}, the term ε_{mod}^{HHV} is an adequate measure to evaluate the BoP performance of the module. Using the HHV in Equation 2.28 should be the preferred option for efficiency calculation and discussion of the EL module itself. However, on the module or system level it could be necessary to use the LHV instead of the HHV, albeit water is not fed as steam to the EL module. If hydrogen is converted back to electricity (fuel cell, gas turbine, etc.) or to heat in thermal processes, only the chemical energy according to the lower heating value is used in the back conversion steps. In these cases, ε_{mod} must refer to the LHV for the evaluation of the overall energy efficiency of the complete transformation chain.

$$\varepsilon_{mod}^{LHV} = \frac{(\Delta H_R^0 - \Delta H_V^0) \cdot \dot{n}_{H_2}}{P_{AC} + P_{pump} + P_{BoP}} = \frac{LHV \cdot \dot{n}_{H_2}}{(P_{DC}/\varepsilon_{rect}) + P_{Pump}} \quad (2.29)$$

In any case, it should be stated if the energy efficiency refers to the higher or lower heating value. To avoid any misunderstanding in this issue, energy efficiency is expressed commonly by the power consumption of the electrolysis module, which denotes how much electrical energy is consumed by the module to produce one normal cubic meter (kWh Nm^{-3} H$_2$) or kilogram of hydrogen (kWh kg^{-1} H$_2$).

2.3 Reaction Kinetics

2.3.1 Kinetic Losses inside a PEM Electrolysis Cell

As shown in Section 2.2.1, the theoretical open-cell voltage (OCV) for the electrolysis of water is $V_{rev}^0 = 1.229$ V and $V_{th}^0 = 1.481$ V for the LHV and the HHV, respectively. However, once the current is passed through the cell, the actual voltage for water splitting becomes considerably higher than the OCV values, due to irreversible losses within the cell. In this section, the different kinetic loss mechanisms in a PEM water electrolysis cell will be introduced. Practical methods for quantifying the contribution of each of these irreversibilities to the overall cell performance as well as measures to mitigate them will also be discussed.

There are three major mechanisms that lead to kinetics losses in a PEM electrolysis cell: activation losses due to slow electrode reaction kinetics, ohmic losses, and mass transfer losses. These can further be grouped under two categories—the faradaic and the non-faradaic processes. The activation losses are faradaic and result from the direct transfer of electrons between redox couples at the interface between the electrode and the electrolyte of the OER and the HER. This leads to irreversibilities on the anode and the cathode called anodic activation overpotential $\eta_{act,an}$ and cathodic activation overpotential $\eta_{act,cath}$, which when added together forms the total activation overpotential. The ohmic and mass transport losses on the other hand are due to non-faradaic loss mechanisms. Ohmic losses arise from resistance to electron flow through the electrodes and cell components as well as resistance to the flow of protons through the membrane. It is directly proportional to the amount of current passed through the cell according to Ohm's law. Activation losses are dominant at low current densities, while the ohmic overpotential becomes dominant at mid current densities. The evolution of mass transport losses takes two major forms: diffusion and bubbles overpotentials. Diffusion losses occur when gas bubbles partially blocks the pores network of current collectors and thereby limiting the supply of reactant water to the active sites, while bubbles overpotential arises when very large gas bubbles shield the electrochemical active area, reducing catalyst utilization. The actual cell voltage for water splitting is then the sum of the OCV plus, all irreversibilities within the cell. Figure 2.8 depicts a theoretical voltage–current density (V/i) curve of a PEM water electrolysis cell, showing the contribution of all three loss mechanisms: the activation, ohmic, and mass transport losses to the overall cell polarization.

2.3.2 Faradaic Losses

The electrochemical reaction that takes place at electrodes of the anode and cathode results in the transfer of one or more electrons between redox species. Once electric current is passed through the electrodes, there is a shift from thermodynamic equilibrium which slows down the reactions taking place on the electrodes' surface. Activation losses then set in, which depend on the rate of the electrochemical reaction that leads to water splitting. This is a faradaic mechanism because it is governed by Faraday's law, introduced in Section 2.2.2. The amount of substance involved in the electrochemical transformation during electrolysis is directly proportional to the amount of electric current transferred at the electrode and, from Equations 2.12 and 2.13, Faraday's law can thus be written as

$$I = \frac{dQ}{dt} = z \cdot F \cdot \frac{dn}{dt} = z \cdot F \cdot \dot{n}\,[A] \qquad (2.30)$$

where $dn/dt = \dot{n}\left[\text{mol s}^{-1}\right]$ is the rate of electrochemical reaction and it corresponds to the amount of reactant consumed, and it is equal in magnitude and opposite in sign to the amount of product formed.

Electrochemical reactions typically occur only at interfaces, so the current is usually directly proportional to the area at the interface. Therefore, the current density expressed as the current per unit area, Equation 2.31, is a more fundamental expression than the current.

$$i = \frac{I}{A}\left[\text{A cm}^{-2}\right] \qquad (2.31)$$

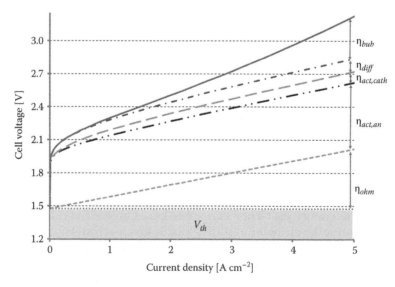

FIGURE 2.8
Simulated overpotentials in a conventional PEM water electrolysis cell.

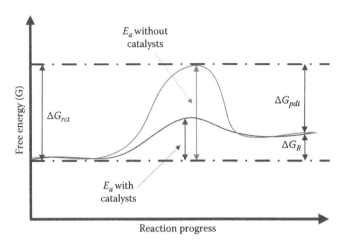

FIGURE 2.9
Activation energy against reaction progress.

Putting together Equations 2.30 and 2.31 gives Equation 2.32, which gives an expression for the net rate of the electrochemical reaction per unit area.

$$r = \frac{i}{z \cdot F \cdot A} = \frac{I}{z \cdot F} \left[\text{mol s}^{-1} \text{ cm}^{-2} \right] \qquad (2.32)$$

In order for the reaction to take place, an initial energy barrier (activation energy) must be overcome. The probability that this activation energy barrier can be surpassed determines the rate at which the water splitting reaction occurs. Energy must be supplied either in the form of heat or by the application of electrical potential on the electrodes to overcome the activation energy barrier. The applied electrical potential acts as the driving force for the electron transfer reaction and with the use of catalysts, the magnitude of the activation barrier can be reduced as shown in Figure 2.9.

In chemical kinetics, the reaction rate at thermodynamic equilibrium can be quantified with respect to the activation energy by the Arrhenius equation, and the reaction rate then becomes

$$r = e^{\frac{-E_a}{R \cdot T}} \qquad (2.33)$$

where E_a is the activation energy of reaction and depends on the rate-limiting step of the experimentally determined reaction mechanism.

As can be seen from Figure 2.8, the net contribution of the activation losses on the anode far outweighs that of the cathode. Therefore, the more critical electrode transfer reaction in a PEM water electrolysis cell is the net oxidation reaction (OER) that takes place at the anode. It is, therefore, necessary to look at some possible OER

mechanisms in order to understand the reaction rates and electrode kinetics for PEM water electrolysis.

The anode half reaction in Equation 2.2 is a complex four-electron multistep process involving adsorbed intermediates. Oxygen evolution on metal electrodes occurs at potentials higher than the OCV, a potential at which most metal surfaces in an aqueous electrolyte will be covered by a thin oxide layer. That makes oxide electrodes practical electrocatalysts for the OER. The general mechanism for the OER in PEM water electrolysis can follow several paths depending on the electrode material. The electrochemical oxide and the oxide pathways on ruthenium and iridium oxide electrodes (Bockris 1956) as well as the Krasilshchikov pathway (Dey et al. 1966) are some of the most considered reaction mechanism paths for the OER.

The electrochemical oxide pathway at an electrochemical active site S of the electrode involves the following reaction steps:

$$S + H_2O \rightleftarrows S - OH_{ads} + H^+ + e^- \qquad (2.34)$$

$$S - OH_{ads} \rightleftarrows S - O + H^+ + e^- \qquad (2.35)$$

$$2S - O \rightarrow 2S + O_{2(g)} \qquad (2.36)$$

The oxide path has the following steps:

$$S + H_2O \rightleftarrows S - OH_{ads} + H^+ + e^- \qquad (2.37)$$

$$2S - OH_{ads} \rightleftarrows S - O + S + H_2O \qquad (2.38)$$

$$2S - O \rightarrow 2S + O_{2(g)} \qquad (2.39)$$

The Krasilshchikov pathway involves the following steps:

$$S + H_2O \rightleftarrows S - OH_{ads} + H^+ + e^- \qquad (2.40)$$

$$S - OH_{ads} \rightleftarrows S - O^- + H^+ \qquad (2.41)$$

$$S - O^- \rightleftarrows S - O + e^- \qquad (2.42)$$

$$2S - O \rightarrow 2S + O_{2(g)} \qquad (2.43)$$

The slowest step in each reaction mechanism is the rate-controlling step and determines the reaction rate of the overall charge-transfer reaction. Once the rate-controlling step is known, the reaction rate of the forward reaction can then be described according to Equation 2.33 as

$$r_1 = e^{\frac{-\Delta G_{rct}}{R \cdot T}} \quad\quad (2.44)$$

And the backward reaction by

$$r_2 = e^{\frac{-\Delta G_{pdt}}{R \cdot T}} \quad\quad (2.45)$$

where $-\Delta G_{rct}$ and $-\Delta G_{pdt}$ are the molar changes in Gibbs free energies of the reactants and products, respectively, and $\Delta G_{pdt} = \Delta G_{rct} - \Delta G_R$ as can be deduced from Figure 2.9. The net reaction rate defined as $r = r_1 - r_2$ then becomes

$$r = \left\{ e^{\frac{-\Delta G_{rct}}{R \cdot T}} - e^{\frac{-(\Delta G_{rct} - \Delta G_R)}{R \cdot T}} \right\} \quad\quad (2.46)$$

The reaction rate is directly related to the current density according to Equation 2.32, so Equation 2.46 can be expressed in terms of the current density as

$$i = z \cdot F \cdot A \cdot k \cdot \left\{ e^{\frac{-\Delta G_{rct}}{R \cdot T}} - e^{-\left(\frac{\Delta G_{rct} - \Delta G_R}{R \cdot T} \right)} \right\} \quad\quad (2.47)$$

At thermodynamic equilibrium, the forward and reverse current densities balance out. Although there is no net current density, both the forward and backward reactions are occurring at a rate characterized by the so-called exchange current density, denoted i_0. The magnitude of the exchange current density determines the rate of the electrochemical reaction at equilibrium and depends on the electrode surface on which the reaction occurs. Since different materials can give different exchange current densities, the electrode material therefore has a strong effect on the kinetics of the OER. Table 2.2 summarizes the kinetic properties of some commonly used electrodes for the OER in PEM water electrolysis. Figure 2.10 simulates the evolution of activation overpotential of the different electrode materials based on the exchange current densities shown in Table 2.2. It can be seen that the most catalytically active electrode is ruthenium oxide with the smallest activation overpotential.

As mentioned earlier, the OER on the anode is the rate determinant of the overall water electrolysis process. The HER on the cathode side is some order of magnitudes faster than the OER on the anode as can be seen in Table 2.3, which summarizes the exchange current densities for typical electrode materials for the HER.

TABLE 2.2

Kinetic Properties of Some Commonly Used Electrocatalysts for PEM Water Electrolysis

Electrode Material	Exchange Current Density [A cm^{-2}]	Tafel Slope [mV dec^{-1}]
Ru-oxide	1.2×10^{-8}	65
Ir/Ru-oxide	3.2×10^{-9}	56
Ir-oxide	5×10^{-12}	50
Ir	9.4×10^{-10}	57
Rh-oxide	7.5×10^{-9}	120
Rh	5.1×10^{-9}	115
Pt	1.7×10^{-11}	130

Source: Ahn, J. and R. Holze, *J. Appl. Electrochem.*, 22, 1167, 1992.

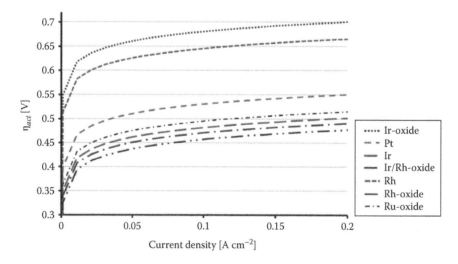

FIGURE 2.10
Simulated electrode kinetics performance of some known electrocatalysts for the OER, showing the evolution of activation overpotential with current density at 353 K.

TABLE 2.3

Exchange Current Densities of Typical Electrode Materials for the HER

Electrode Material	Exchange Current Density [A cm⁻²]
Pt on carbon support	2×10^{-1}
Ir on carbon support	5×10^{-2}
Pd on carbon support	5×10^{-3}

Source: Durst, J. et al., *Energy Environ. Sci.*, 7, 2255, 2014.

In terms of the exchange current density, the net reaction rate for infinitely fast mass transport then becomes

$$i = i_0 \cdot \left\{ e^{\frac{-\Delta G_{rct}}{R \cdot T}} - e^{\frac{-(\Delta G_{rct} - \Delta G_R)}{R \cdot T}} \right\} \tag{2.48}$$

The activation energy is sensitive to the applied voltage, so changing the voltage changes the free energy of the charged species involved in the reaction. Therefore, the size of the activation barrier E_a can be altered by varying the potential applied across the electrodes. Applying a finite potential difference across the electrode lowers the activation energy barrier by a fixed amount β, called the symmetry factor or the electron transfer coefficient $0 < \beta < 1$, which determines how the electrical energy input affects the redox process. Applying Equation 2.8 in the case of deviation from the thermodynamic equilibrium, the activation barrier of the forward reaction is therefore decreased by a factor of $z\beta F\eta$ and that of the reverse reaction is increased by $z(1-\beta)F\eta$. η is the activation overpotential and the difference between the applied voltage and the equilibrium Nernst potential

for the charge-transfer reaction. In terms of the overpotential and exchange current density, the net reaction rate becomes

$$i = i_0 \cdot \left\{ \exp^{\left(\frac{z\beta F\eta}{RT}\right)} - \exp^{-\left(\frac{z(1-\beta)F\eta}{RT}\right)} \right\} \tag{2.49}$$

Equation 2.49 is the so-called Butler–Volmer equation for charge-transfer kinetics. It indicates that the current produced (or consumed) by an electrochemical reaction increases exponentially with the applied overpotential. Increasing the exchange current density hugely increases the electrode kinetics of the OER and thus water splitting. The exchange current density can be increased by decreasing the activation barrier. This can be achieved by increasing the temperature or by increasing the number of active reaction sites, by increasing the surface roughness of the electrodes and by the use of an appropriate catalyst material, refer to Figure 2.9.

As can be seen in Figure 2.11, when the overpotential is very high, the reverse reaction (with current density i_2) becomes insignificant and the second exponential term of the Butler–Volmer equation vanishes. The forward reaction (with corresponding current density i_1) thus dominates the reaction rate, leading to a completely irreversible reaction process. In this special case, the Butler–Volmer charge-transfer reaction can, therefore, be simplified to

$$i = i_0 \cdot \exp^{\left(\frac{z\beta F\eta}{RT}\right)} \tag{2.50}$$

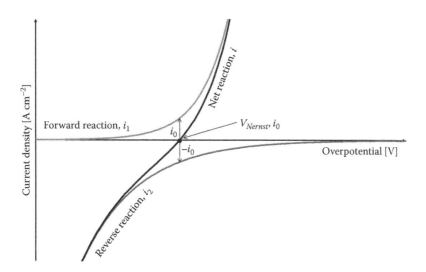

FIGURE 2.11
Butler–Volmer dependence of the electrode current on overpotential, with symmetry factor of $\beta = 0.5$.

TABLE 2.4

Tafel Slopes of Different Reaction Mechanism Paths of the OER on Platinum Catalyst

Reaction Mechanism	Reaction Step	Tafel Slope for Small η [mV dec^{-1}]	Tafel Slope Large η [mV dec^{-1}]
Electrochemical oxide path	Equation 2.34	120	120
	Equation 2.35	40	120
	Equation 2.36	15	∞
Oxide pathway	Equation 2.37	120	120
	Equation 2.38	30	∞
	Equation 2.39	15	∞
Krasilshchikov pathway	Equation 2.40	120	120
	Equation 2.41	60	60
	Equation 2.42	30	120
	Equation 2.43	15	∞

The relationship between the overpotential and the rate of the electrochemical reaction can be derived from Equation 2.50, and it is called the Tafel relationship with a and b as Tafel constants.

$$\eta = \frac{RT}{z\beta F} \cdot \ln(i) - \frac{RT}{z\beta F} \cdot \ln(i_0) = a \cdot \ln(i) - b \quad (2.51)$$

From the Tafel equation, the Tafel constant b can be extrapolated to determine the exchange current density i_0 and thus, the relative reaction rate at thermodynamic equilibrium. The Tafel slope a also gives insightful information on the reaction mechanism and an indication of the rate-determining step of the overall reaction pathway. The Tafel constants can only be determined experimentally and, in Table 2.4 the Tafel slopes of different reaction mechanism paths of the OER on a platinum catalyst are summarized (Bockris 1956; Dey et al. 1966).

As can be seen from the exchange current density values in Table 2.3, the fastest electrokinetics for the HER is achieved by the use of platinum as catalyst. For the OER, ruthenium oxide has been shown to have the best electrode kinetics performance compared to other known catalysts such as iridium and rhodium and their oxides and platinum. However, ruthenium is highly unstable (Marshall et al. 2007), so iridium and iridium oxide or their mixtures are widely used because of their better stability.

More details on the reaction mechanisms for the OER and the HER can be found in Chapters 3 and 4, respectively.

2.3.3 Non-Faradaic Losses

Non-faradaic losses are those that do not result from the direct transfer of electrons across the electrodes due to electrochemical reaction. These are mainly irreversibilities caused by mass transport and the resistance to the flow of electric current through the cell components and the flow of protons through the solid electrolyte membrane.

The electron transfer mechanism described in Section 2.3.2 assumes that the supply of reactant to and removal of product species from the reaction interface is not limited, and thus, the reaction kinetics is strictly faradaic in nature. The rate of transport of the reactant species to and the removal of products from the reaction sites can, however, affect or even dominate the kinetics when the reaction shifts from thermodynamic equilibrium. Different forms of transport phenomena that influence the reaction rate are diffusion, convection, and migration.

When the electrode kinetics is infinitely fast, there is zero accumulation of reactants on the electrode surface as it is being quickly used up, and the reaction is mass transport controlled. For the reaction to be sustained, reactants need to be supplied to the reaction interface at an appropriate rate. The rate of reaction will then be determined by the rate of supply of the reactants. Since the half-cell reaction occurs on porous electrode surfaces and since there are no more than two component mixtures for the anode and cathode reactions, Fick's diffusion is assumed to be the dominant mass transport mechanism.

The Nernst equation for determining the nonequilibrium electric potential of the electrochemical reaction is given by (Perez 2004)

$$V = V^0 + \frac{RT}{nF} \cdot \ln(k) \quad (2.52)$$

where
k is the reaction rate constant at equilibrium
$V - V^0 = \eta$ is the overpotential

$$\eta = \frac{RT}{nF} \cdot \ln(k) \quad (2.53)$$

With the rate constant given by $k = c_i/c_b$ (Tobias et al. 1952), the concentration overpotential can be expressed as

$$\eta_{con} = \frac{RT}{nF} \cdot \ln\left(\frac{c_i}{c_{rp}}\right) \quad (2.54)$$

where c_i and c_b are the species concentration at the electrode interface and a reference position, respectively. Equation 2.54 can be separately applied to the anode and the cathode sides. This indicates that the overpotential due to mass transport limitation increases with increase in product species at the reaction interface.

Diffusion dominates mass transport losses only at low current densities. PEM electrolysis cells can be operated at high current densities, sometimes well above $i = 2$ A cm^{-1}. At such high current densities, the production of the gas phase in the form of bubbles shields the active area, distorts the contact between the electrode and the electrolyte, and reduces the catalyst utilization. This results in an increase in the local current density and the so-called bubbles overpotential, which increases exponentially with increasing current densities (Fritz et al. 2014) as is shown in Figure 2.8.

Ohmic overpotentials, on the other hand, are a form of non-faradaic losses resulting from the resistance of the flow of electrical currents through the cell components and the flow of protons through the polymer electrolyte membrane. The total ohmic losses (ohmic overpotential) are determined by the application of Ohm's law:

$$\eta_{ohm} = i \cdot \text{ASR} = \frac{I}{A} \cdot \sum R_i \qquad (2.55)$$

where ASR is the overall area-specific resistance of the cell, which is the sum of all electrical and ionic (ohmic) resistances R_i of the cell components because they are connected electrically in series. The electron-conducting cell components such as the BPPs and CCs are usually made of highly conducting and electrochemically stable metals such as titanium. If all design tolerances are met, it is often considered that electron transport is infinitely fast so that the effect of ohmic losses due to electron transport becomes negligible compared to ohmic losses due to proton transport through the polymer electrolyte membrane.

The mechanisms of proton transport through a perfluorinated sulfonic acid (PFSA) membrane like Dupont's Nafion® have been described by Choi et al. (2005), and the main factors affecting the proton conductivity of such a membrane are temperature and the membrane hydration. Equation 2.56 is the semiempirical equation that is often used to determine the membrane conductivity and hence its resistance to proton transport (Springer et al. 1991; Choi et al. 2004; Santarelli and Torchio 2007):

$$\sigma_{mem} = \left(0.005139 \cdot \lambda - 0.00326\right) \cdot \exp\left[1268 \cdot \left(\frac{1}{303} - \frac{1}{T}\right)\right] \qquad (2.56)$$

where T is the operating cell temperature and λ is the degree of membrane hydration expressed as the mole of H$_2$O per mole of SO$_3^-$ in the Nafion membrane. The λ value ranges from 14 to 25, depending on the membrane hydration, with 14 for very poorly hydrated membrane and 25 when it is fully hydrated. For PEM water electrolysis operating at optimal design water flow rates, it is considered that the membrane is fully hydrated. The ohmic overpotential due to proton transport through the membrane is given in terms of membrane thickness l_{mem} and proton conductivity σ_{mem} as

$$\eta_{ohm,mem} = \frac{l_{mem} \cdot i}{\sigma_{mem}} \qquad (2.57)$$

2.3.4 Polarization Curves

Since the cell components are electrically arranged in a series connection, all voltage losses from the anode and the cathode sides can be summed up to give the total cell polarization. The cell voltage V_{cell}, which is a measure of the total amount of electrical energy demand for water decomposition, then results from the sum of the reversible cell voltage V_{rev} and all irreversible losses within the cell.

$$V_{cell} = V_{rev} + \left|\eta_{act,an}\right| + \left|\eta_{act,cath}\right| + \left|\eta_{diff}\right| + \left|\eta_{bub}\right| + \left|\eta_{ohm}\right| \qquad (2.58)$$

The contribution of the anode and cathode activation overpotentials, the reversible cell voltage, and the ohmic overpotential to the total electrical energy demand for water splitting by a PEM water electrolysis cell is depicted in Figure 2.8. It is based on an electrochemical model with data input on real PEM electrolysis cell operation.

2.3.5 Measures to Improve Electrolysis Cell Performance

At low current densities, the kinetics is controlled by the exchange current density. A high exchange current density is required in order to reduce the activation losses and improve the electrolysis performance. The exchange current density can be increased by increasing the operating temperature, by decreasing the activation barrier E_a and by increasing surface roughness or electrode area of the reaction interface and thereby increasing the number of possible reaction sites. Increasing the temperature exponentially increases the exchange current density and with it the EL performance. The activation barrier can be reduced by the use of highly catalytic electrodes. Some of the most active catalysts for PEM water electrolysis have been summarized in Tables 2.2 and 2.3 for the OER and the HER, respectively. For an extremely rough electrode surface, the electrochemically active surface area can be several orders of magnitude larger than the geometric electrode surface area and provides as a result many more sites for reactions to take place and thereby increasing the kinetics and the performance. Heavy metal cations such as nickel, iron, and chromium which are corrosion products from cell, stack, and system components may pollute the electrodes, reduce the effective number of active sites, and consequently lower the reaction rates. This effect can be mitigated by the use of ion exchangers to trap cations, preventing them from reaching the electrodes.

When ohmic overpotentials become the major source of losses at intermediate current densities, the performance of the electrolysis cell can be improved by improving the proton conductivity of the proton exchange membrane (PEM) and the electron conductivity of the BPP, porous current collectors, and the electrodes. The membrane must also be as thin as possible, see Equation 2.57, in order to reduce the diffusion pathway of protons and thus increase ionic conductivity. Metals with high electronic conductivity should be used for the BPPs, CCs, and electrodes to minimize electronic ohmic losses. Additionally, poor contact between adjacent cell components may lead to electrical contact resistances, leading to further ohmic losses. In order to minimize this effect, the design tolerances of the cell and its components must be very tightly met during components' manufacturing.

At high current density operations, the electrolysis performance can be improved by reducing the possibility of mass transport limitation. This can be achieved by optimizing the porous microstructure of the diffusion layer (current collectors), in order to facilitate the removal of product gases from, and the supply of reactant water to the electrochemical active sites. More information on current collectors is provided in Chapter 8. At elevated pressures, the size of bubbles become relatively smaller and so the effect of shielding and bubbles overpotential. Therefore, bubbles overpotential can also be mitigated by HP operation.

2.4 Key Performance Indicators

KPIs are used mostly in business administration for performance measurement. Continuous monitoring of KPI (e.g., turnover or sales forecasts) allows identifying gaps between current and desired performance and provides indication of progress toward closing the gaps. Performance measurement in engineering is commonly based on technical and quantitative indicators, for example, availability of the system or mean time between overhaul. In this section, KPIs are defined as a set of variables, which describe the main technical and economical features of electrolysis stacks and systems. Providing a clear definition for these KPI is the first step to comparing different EL systems and technologies on

a more even footing. Manufacturers often use different definition and system boundaries, which lead to misunderstanding. Discussion of KPIs will be oriented to system boundaries as presented in Figure 2.3. Again, it should be noted that a common standard for EL systems is not defined, and thus, other system boundaries and KPI definitions may exist in the literature.

2.4.1 Production Capacity, Power, and Gas Quality

The most important technical features of an EL system as a hydrogen generator are the hydrogen production rate and the electrical power demand. Gas quality and pressure of the produced hydrogen have an influence on these KPIs but are determined by the application.

- The **hydrogen production rate** is commonly expressed in normal cubic meters per hour [$Nm^3 \ h^{-1}$]. In the gas industry, a normal cubic meter is measured with reference to 273.15 K at 101.325 kPa—which are called standard conditions for temperature and pressure (STP)—and a relative humidity of 0%. STP should not be confused with the standard state commonly used in thermodynamic calculations (see Section 2.2). To avoid this confusion, the hydrogen production rate can be indicated as well in kilogram per hour [$kg \ h^{-1}$]. On the system level, the hydrogen production rate takes into account the hydrogen loss for O_2 fine purification and gas drying in order to obtain the required hydrogen purity. The electrical power demand of a PEMEL system includes all the electrical input to the EL module with stack and subsystems on the system level, see Figure 2.3.

- The **electrical power** is the only energy provided to the low-temperature PEM electrolysis system and thus essential for calculating the system efficiency, see Section 2.2.4. The electrical power demand refers to the nominal operation point where the EL system can be operated continuously at the desired hydrogen production rate, purity, and nominal pressure and temperature of the stack.

- The **hydrogen pressure** is defined at the outlet of the EL systems after the control valve, and after hydrogen has passed the gas purification and gas-drying stages (both provoke a pressure drop). The hydrogen outlet pressure is expressed in [bar] or in [MPa].

- The **hydrogen purity** coming out of the stack is not sufficient to certain standard as the SAE J2719 (Hydrogen Fuel Quality for Fuel Cell Vehicles). Main impurities are oxygen and water vapor in the hydrogen flow. Hydrogen purification requires additional electrical energy and/ or causes a hydrogen loss for gas drying, for example, in a pressure swing adsorption (PSA) unit. Mostly, the purity of the gas is given in decimal fraction, where the first digit represents the number of nines in the percentage value and the last digit represents the last digit of the percentage value. For example, a purity of 99.998% is abbreviated as purity 4.8.

2.4.2 Efficiency, Lifetime, and Degradation

The **efficiency** of an EL system belongs to the most important system parameters. However, it can be very sophisticated to exactly define this parameter due to the different systems designs and operating strategies. In any case, the efficiency should not be confused with the capacity factor, which is the ratio of the actual hydrogen output over a period of time to its potential output if operated at nominal power. The capacity factor takes into account part load and overloads operation, standby and start/stop processes during a period of time and therefore depends on the application and operating strategy. Mathematical description of the efficiency and discussion with respect to different system boundaries is already given in Section 2.2.4. The efficiency can be expressed as the ratio of the energy content of the produced hydrogen based on the LHV or HHV and the supplied electrical power. To avoid any confusion, the specific electrical power demand for the production of one normal cubic meter hydrogen [kWh Nm^{-3}] at nominal operation is commonly given by EL manufacturers.

Lifetime is another important KPI of an EL system, but this term is often rather imprecise. In general, the lifetime for technical systems refers to the time during which the devise or installation can be operated without replacement of core components or complete failure of the system. In this respect, different criteria exist, which can determine the life of a technical device (1) failure of a component which makes further operation impossible; (2) too high maintenance efforts which render further use uneconomical; and (3) system is designed for a limited useful life. According to the definition given in Section 2.1.3 and these criteria, it is necessary to distinguish between the lifetime of an electrolysis stack, module, and system.

- The lifetime of an EL stack is expressed in operating hours at nominal load until a stack replacement and subsequent overhaul by the manufacturer or recycling is required. In research and development, the lifetime of a stack is usually given as "mean lifetime," for

example, until failure of the membrane. But for commercial stacks, the lifetime should be specified as "service life" that expresses the acceptable period of use in service until replacement is needed. It is possible that the end of life for a stack is determined by the user himself, for example, by setting a threshold for the stack efficiency.

- The lifetime of an EL module is mostly given in (operating) years as it is common for technical installations, describing the time for which the module can be operated continuously around the clock. Before reaching this lifetime, however, maintenance of certain components is often necessary due to shorter service lifes.

- The lifetime of a complete EL system is expressed in years as well. But for practical reasons, it is not recommended to define one lifetime for all different subsystems.

Degradation in chemistry means the decomposition of compounds to intermediate products or a modification of the substance, which lowers the initial desired performance. In an electrolysis cell, various degradation mechanisms occur which minimizes the lifetime of the stack, refer to Chapter 11 for more details. The main indicator for degradation inside a cell is the *voltage decay*. This relates to an increase in the overpotential that has to be applied to the cell in order to maintain a constant hydrogen production rate. The voltage decay is expressed in μV per operating hour [μV h^{-1}] or in mV per thousand hours [mV (1000 h)$^{-1}$]. The decay rate considers mainly aging of the electrocatalysts and an increase in the cell impedance due to higher contact resistances (e.g., passivation processes) and lower conductivity of the membrane. For monitoring the internal resistance of the cell, the *high-frequency impedance* is another KPI but rather used in laboratory work. The *gas purity* of the produced hydrogen and oxygen is another KPI for degradation inside a cell and is affected primarily by the membrane. Thinning of the membrane due to chemical degradation or mechanical damaging of the sensitive foil reduces the gas tightness. Main impurities in the produced gases are hydrogen in oxygen at the anode side and oxygen in hydrogen at the cathode side caused by migration process through the membrane. Gas impurities are expressed either in parts per million [ppm] or in volume percent [vol-%] of the dry gas volume. Exact values at the outlets of the stacks are only measured in research and development work. In commercial products, mostly thresholds at the outlets of the EL module or after the fine purification on the system level are monitored.

2.4.3 Investment and Hydrogen Production Cost

The last two sections described technical KPI for the electrolysis process. In the same manner, economical KPIs are important for the end user, namely, investment and hydrogen production costs.

Investment costs or capital costs of an EL system are the one-time setup costs of the installation including design and planning costs. Development costs of the ground or installation site are not covered. If the EL system is delivered as container solution with integrated thermal management, safety concept, etc., housing is included in the investment costs as well. But enclosure and housing do not belong to the investment costs for larger electrolysis plant installations. Here, again it is important to consider the system boundaries for assessing different EL systems. Mostly, investment costs are expressed in Euro or U.S. dollars per installed nominal electrical power [€ kW^{-1} or USD kW^{-1}] of the EL module or system. When it comes to comparing different systems, this cost specification requires same electrolysis efficiencies that are mostly not the case. For this reason, it is recommended to refer investment costs to the nominal hydrogen production rate of the module or system [€ (Nm^3 H$_2$ h^{-1})$^{-1}$].

Investment costs do not include operational or running and maintenance costs.

Hydrogen production costs specify all costs to bring out one unit of hydrogen (volume or mass) at the installation site. Probably this is the most important KPI for an end user as it allows an economical evaluation against other hydrogen production technologies, for example, steam reforming or partial oxidation from fossil fuels. Main shares of the hydrogen production costs are as follows:

- Investment or capital costs through depreciation (CAPEX);
- Electricity costs to run the electrolysis process as main part of the variable costs;
- Remaining operating costs (OPEX) for water, operating resources (e.g., nitrogen for purging) and service, planned and unplanned maintenance, overhaul, rental charges, etc.

Storage, transportation, and distribution are not included in hydrogen production costs.

In industrial applications with a constant hydrogen demand, the electrolysis process runs continuously resulting in a high number of annual full load hours. In such an application, the hydrogen production costs are dominated by the costs for electricity and capital costs are only of minor importance. Thus, high efficiency of the electrolysis process is essential as it determines the electricity demand.

Applying the electrolysis process in the energy sector, for example, to enable a flexible load management, requires different features and operation strategies of an EL system. In such applications, the annual full load hours are considerably lower and capital costs drive the hydrogen production costs. Although a high efficiency is still important, reduction strategies for capital costs become more dominant.

A comprehensive overview about market trends and required features for PEM water electrolysis is provided in Chapter 14.

References

Ahn, J. and R. Holze. 1992. Bifunctional electrodes for an integrated water electrolysis bifunctional electrodes for an integrated water-electrolysis and hydrogen–oxygen fuel cell with a solid polymer electrolyte polymer electrolyte. *Journal of Applied Electrochemistry* 22: 1167–1174.

Awasthi, A., S. Keith, and S. Basu. 2011. Dynamic modeling and simulation of a proton exchange membrane electrolyzer for hydrogen production. *International Journal of Hydrogen Energy* 36: 14779–14786.

Bensmann, B., R. Hanke-Rauschenbach, Peña Arias, I. K., and K. Sundmacher. 2013. Energetic evaluation of high pressure PEM electrolyzer systems for intermediate storage of renewable energies. *Electrochimica Acta* 110: 570–580.

Bockris, J. O. 1956. Kinetics of activation controlled consecutive electrochemical reactions: Anodic evolution of oxygen. *The Journal of Chemical Physics* 24: 817–827.

Choi, P., D. G. Bessarabov, and R. Datta. 2004. A simple model for solid polymer electrolyte (SPE) water electrolysis. *Solid State Ionics* 175: 535–539.

Choi, P., N. H. Jalani, and R. Datta. 2005. Thermodynamics and proton transport in Nafion II. Proton diffusion mechanisms and conductivity. *Journal of the Electrochemical Society* 152: 123–130.

Dey, A., A. Damjanovic, and J. O. Bockris. 1966. Kinetics of oxygen evolution and dissolution on platinum electrodes. *Electrochimica Acta* 11: 791–814.

Durst, J., A. Siebel, C. Simon, F. Hasché, J. Herranz, and H. A. Gasteiger. 2014. New insights into the electrochemical hydrogen oxidation and evolution reaction mechanism. *Energy and Environmental Science* 7: 2255–2259.

Fritz, D. L., J. Mergel, and D. Stolten. 2014. PEM electrolysis simulation and validation. *ECS Transactions* 58: 1–9.

Greenway, S. D., E. B. Fox, and A. A. Ekechukwu. 2009. Proton exchange membrane (PEM) electrolyzer operation under anode liquid and cathode vapor feed configurations. *International Journal of Hydrogen Energy* 34: 6603–6608.

Kreuter, W. and H. Hofmann. 1998. Electrolysis: The important energy transformer in a world of sustainable energy. *International Journal of Hydrogen Energy* 23: 661–666.

LaConti, A. B. and L. Swette. 2003. Special application using PEM-technology. In Vielstich, W.; Gasteiger, H. A. et al. (Eds.). *Handbook of Fuel Cells*, Vol. 4, John Wiley & Sons, Ltd., Chichester, Chapter 55, pp. 745–761.

LeRoy, R. L. 1983. Industrial water electrolysis: Present and future. *International Journal of Hydrogen Energy* 8: 401–417.

Marangio, F., M. Santarelli, and M. Cali. 2009. Theoretical model and experimental analysis of a high pressure PEM water electrolyser for hydrogen production. *International Journal of Hydrogen Energy* 34: 1143–1158.

Marshall, A. T., S. Sunde, M. Tsypkin, and R. Tunold. 2007. Performance of a PEM water electrolysis cell using $IrxRuyTazO_2IrxRuyTazO_2$ electrocatalysts for the oxygen evolution electrode. *International Journal of Hydrogen Energy* 32: 2320–2324.

McBride, B. J., M. J. Zehe, and S. Gordon. 2002. *NASA Glenn Coefficients for Calculating Thermodynamic Properties of Individual Species*, NASA Glenn Research Center, Cleveland, Ohio.

Ni, M., M. K. Leung, and D. Y. Leung. 2008. Energy and exergy analysis of hydrogen production by a proton exchange membrane (PEM) electrolyzer plant. *Energy Conversion and Management* 49: 2748–2756.

Onda, K., T. Kyakuno, K. Hattori, and K. Ito. 2004. Prediction of production power for high pressure hydrogen by high-pressure water electrolysis. *Journal of Power Sources* 132: 64–70.

Perez, N. 2004. *Electrochemistry and Corrosion Science*. Boston, MA: Kluwer Academic Publishers.

Roy, A., S. Watson, and D. Infield. 2006. Comparison of electrical energy efficiency of atmospheric and high-pressure electrolysers. *International Journal of Hydrogen Energy* 31: 1964–1979.

Roy, R., J. Graf, T. Gallus, D. Rios, S. Smith, and S. Diderich. 2011. Development testing of a High Differential Pressure (HDP) water electrolysis cell stack for the High Pressure Oxygen Generating Assembly (HPOGA). *SAE International Journal of Aerospace* 4: 19–28.

Sakurai, M., Y. Sone, T. Nishida, H. Matsushima, and Y. Fukunaka. 2013. Fundamental study of water electrolysis for life support system in space. *Electrochimica Acta* 100: 350–357.

Santarelli, M. G. and M. F. Torchio. 2007. Experimental analysis of the effects of the operating variables on the performance of a single PEMFC. *Energy Conversion and Management* 48: 40–51.

Smolinka, T., E. T. Ojong, and J. Garche. 2014. Hydrogen Production from Renewable Energies – Electrolyzer Technologies. In Mosley, P.T.; Garche, J. (Eds.) 2015 – *Electrochemical Energy Storage for Renewable*, Elsevier B.V., Amsterdam, Chapter 8, pp. 103–128.

Smolinka, T., S. Rau, and C. Hebling. 2010. Polymer Electrolyte Membrane (PEM) Water Electrolysis. In Stolten, D. (Ed.) *2010 – Hydrogen and Fuel Cells*, Wiley-VCH Verlag GmbH & Co. KGaA, Weinheim, pp. 271–289.

Springer, T. E., T. A. Zawodzinski, and S. Gottesfeld. 1991. Polymer Electrolyte Fuel Cell Model. 138: 2334–2342.

Tobias, C. W., M. Eisenberg, and C. R. Wilke. 1952. Diffusion and convection in electrolysis – A theoretical review. *Electrochemistry of Ionic Crystals* 99: 359–365.

Wittwer, V., M. Datz, J. Ell, A. Georg, W. Graf, and G. Walze. 2004. Gasochromic windows. *Solar Energy Materials and Solar Cells* 84: 305–314.

Yamaguchi, M., K. Okisawa, and T. Nakanori. 1997. Development of high performance solid polymer electrolyte water electrolyzer in WE-NET. In *Energy Conversion Engineering Conference. IECEC-97. Proceedings of the 32nd Intersociety*, Vol. 3, Honolulu, HI: IEEE, pp. 1958–1965.

3

Electrocatalysts for Oxygen Evolution Reaction (OER)

Magnus Thomassen and Svein Sunde

CONTENTS

3.1 Introduction

The first electrode used for the evolution of oxygen from water was a thin gold wire, used by the two Dutchmen, Paets van Troostwijk and Deiman in 1789 (de Levie 1999). Two gold wires were placed close together in an inverted water filled glass cylinder and connected to a powerful electrostatic generator based on friction. The electric discharges caused gas evolution on both electrodes and when enough gas had been produced, the gap between the gold electrodes was above the water level, causing the next electrostatic discharge to form a spark between the wires. The spark ignited the oxygen–hydrogen mixture, recombining it to water and the experiment could then be repeated.

With the invention of the DSA (dimensionally stable anodes) in 1965, Henri Bernard Beer (1969) laid the foundation for the development of PEM (polymer–electrolyte membrane) electrolyzers as we see them today. First, Beer developed the thermochemical coating of titanium by a mixture of titanium oxide and ruthenium oxide, which resulted in a catalyst layer with extraordinary electrocatalytic activity, high mechanical and chemical stability in acidic environments, and high electronic conductivity. These DSA electrodes made it possible to introduce the new membrane cell design in the chloralkali industry (Trasatti 2000), opening a large-scale commercial market for the solid polymer membranes, on which all PEM fuel cells and electrolyzers are based. Thermally formed PGM (platinum group metal) oxides, primarily IrO_2, and Nafion membranes are now the state of the art for PEM water electrolyzers.

The performance, costs, and scarcity of iridium oxide are issues intrinsically linked with the acidic conditions under which current PEM technology operates. In principle, there are two approaches toward the shortcomings of iridium oxide: better utilization of the material or its replacement. The latter can be based on combinatorial approaches or a fundamental understanding of its mode of operation as a catalyst with the purpose of transferring these properties to other material combinations. The former is based on maximization of surface area under the constraints of achieving also a sufficient stability. Later, we review research directed toward the resolution of these issues, both theoretical and experimental investigations of the catalytic performance of iridium oxide and related materials for the oxygen evolution reaction (OER), and also synthesis and the application of catalysts supports are addressed.

Since the research areas of oxygen evolution electrocatalysis under alkaline and acidic conditions are necessarily connected, we will, when appropriate, refer to research results in the alkaline area when relevant also for PEM electrocatalysts, particularly in the area of theory of electrocatalysis for the OER and activity rationalizations.

3.2 Methods for Investigation of Electrocatalysts for the Oxygen Evolution Reaction

3.2.1 Electrochemical Methods

3.2.1.1 Steady-State Polarization

Steady-state polarization is an analysis method where the potential/current (E and I) relationship is measured. The electrode can be controlled either galvanostatically or potentiostatically and in order for the relationship to be considered steady state, the controlled variable (either E or I) must be kept constant for an extended period of time until the other variable reaches a constant value. Normally, a staircase type profile is used with a spacing of the control variable, which ensures sufficient detail in the resulting polarization curve for further analysis. The analysis of the polarization curve can be either to find kinetic parameters and any applicable reaction mechanisms or to find the overall performance of the catalyst/electrode by comparing results with other catalysts/electrodes at a specific overpotential or current density.

3.2.1.2 Cyclic Voltammetry

Cyclic voltammetry (CV) is one of the most useful electrochemical analysis tools used for characterizing anode electrocatalysts. CV is a very versatile technique that is used from initial electrochemical studies of new systems to obtaining information of fairly complex electrode reactions. Normally, it is performed by sweeping the potential of the electrode between two potential limits at a constant sweep rate and monitoring the current response of the electrode, often repeating the sweep several times. Often, cyclic voltammetry is used to study electrode reactions involving electroactive species present in the electrolyte, but it is also a powerful technique for studying the nature of an electrode material and its interaction with the electrolyte. In this section, we limit ourselves to the latter.

The voltammogram gives an electrochemical spectrum of the electrode surface, with information regarding any solid-state redox transitions, adsorption/desorption reactions, and charging/discharging of the double layer of the catalyst surface. From these results, properties such as the catalysts' active surface area and the electrode capacitance can be found.

Noble metal oxides such as IrO_2 and RuO_2 have quite characteristics voltammograms with several broad peaks or waves. As oxide surfaces in an aqueous environment will be covered by OH groups, they will act as Brønsted acids or bases (i.e., proton acceptors or donors) (Ardizzone and Trasatti 1996). This property, together with the metallic conductivity of the Ir and Ru oxides,

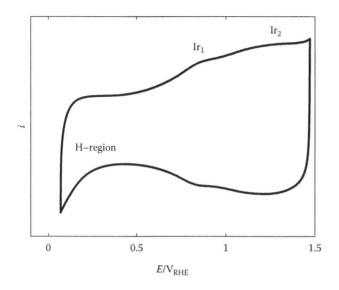

FIGURE 3.1
Cyclic voltammogram of IrO_2 prepared by standard DSA technique and measured in 0.5 mol dm^{-3} H_2SO_4 at room temperature. (Reprinted from Rasten, E., Electrocatalysis in water electrolysis with solid polymer electrolyte, Fakultet for naturvitenskap og teknologi, 2001. With permission.)

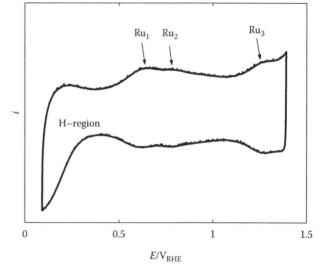

FIGURE 3.2
Cyclic voltammogram of RuO_2 prepared by standard DSA technique and measured in 0.5 mol dm^{-3} H_2SO_4 at room temperature. (Reprinted from Rasten, E., Electrocatalysis in water electrolysis with solid polymer electrolyte, Fakultet for naturvitenskap og teknologi, 2001. With permission.)

make RuO_2 and IrO_2 behave as "protonic condensers" (Ardizzone and Trasatti 1996) in the potential between oxygen and hydrogen evolution. Thus, the broad peaks originate from redox reactions of adsorbed oxygen species from the solution and solid-state redox reactions of the Ir or Ru oxides.

A typical voltammogram of iridium oxide is shown in Figure 3.1. The redox peaks are found around 0.8–0.9 and 1.25–1.35 V vs RHE (reversible hydrogen electrode) and correspond to the redox transitions of Ir^{3+}/Ir^{4+} and Ir^{4+}/Ir^{6+}, respectively, as specified in following equation (Michell et al. 1978):

$$IrO_x(OH)_y \leftrightarrow IrO_{x-\delta}(OH)_{y+\delta} + \delta H^+ + \delta e^- \quad (3.1)$$

The voltammogram of ruthenium oxide is shown in Figure 3.2. For this oxide, three redox peaks are present and are found at and correspond to the three redox transitions, Ru^{2+}/Ru^{3+}, Ru^{3+}/Ru^{4+}, and Ru^{4+}/Ru^{6+} as specified in the following equation (Ardizzone et al. 1990):

$$RuO_x(OH)_y \leftrightarrow RuO_{x-\delta}(OH)_{y+\delta} + \delta H^+ + \delta e^- \quad (3.2)$$

These processes can also be described as a redox pseudocapacitance (Conway et al. 1997), as the proton injection effectively stores charge in the electrode material. This differs from the double-layer capacitance as this pseudocapacitance is a faradaic process in which charge crosses the electrode double layer. This pseudocapacitance is frequently used in electrochemical capacitors or "supercapacitors" (Kötz and Carlen 2000).

The voltammogram of noble metal oxides gives vital information regarding the active area of these materials. It has been shown that the charge associated with the oxide region (normally 0.4–1.4 V) is directly proportional to the electrode roughness and loading (Savinell et al. 1990). It is also well known that RuO_2 and IrO_2 surfaces show a large dependence of the voltammetric charge (q^*) with sweep rate (ν), which yield important information regarding the morphology of the electrode surface. Trasatti and coworkers (Ardizzone et al. 1990; De Pauli and Trasatti 1995) have shown that the sweep-rate dependence of q^* can be related to less accessible surface regions that progressively become excluded from the redox reactions of Equations 3.1 and 3.2 as the sweep rate is increased. If q^* is divided into an "outer" and "inner" charge and it is assumed that at high sweep rates only the outer surface of the oxide material takes part in the charging process, whereas at very low sweep rates the "inner" surface is also accessible, it is possible to separate these two quantities. By plotting $1/q^*$ vs $\nu^{0.5}$ and q^* vs $\nu^{-0.5}$ and extrapolating the linear sections to the ordinate, the value of q^* at $\nu = 0$ and $\nu = \infty$ can be found, which corresponds to the total voltammetric charge q_t^* and the outer charge q_o^*. The inner charge, q_i^*, is found by subtracting the outer charge from the total charge.

3.2.1.3 Electrochemical Impedance Spectroscopy (EIS)

Electrochemical impedance spectroscopy (EIS) (Macdonald 1987; Orazem and Tribollet 2008) has been employed in various contexts at electrodes with

PEM-relevant catalysts. Silva et al. (1998) used impedance spectroscopy to determine the flatband potential (Memming 2000) of AIROF (anodically formed iridium oxide films). Similar studies have been performed for catalysts for alkaline water electrolysis (Bockris and Otagawa 1983, 1984; Bockris et al. 1983). These studies demonstrate the semiconducting nature of the films, and for AIROF in the reduced (trivalent) state band edge positions on the hydrogen electrode scale may also be determined.

Since films of catalysts such as iridium oxide and ruthenium oxide are expected to intercalate protons and other species when subjected to low potentials (Birss et al. 1997) and since iridium oxide displays electrochromism during the process (Gottesfeld and McIntyre 1979; Gottesfeld 1980; Granqvist 1994, 2014), impedance spectroscopy has been employed to quite some extent to characterize transport and other processes in AIROF and related systems. Glarum and Marshall (1980) modeled the intercalation process in AIROF and analyzed experimental data collected for such films in terms of the model, which basically was a model for ambipolar diffusion of counterions. At a more empirical level, however, they resolved their impedance data in a series capacitance and a series conductance. The latter displayed two orders of magnitude variation with potential, which is sometimes taken as an indication of the conductance changes one would expect in the film from a rigid band model for the intercalation of ions into AIROF (Kötz and Neff 1985). An attempt to find similar conductivity changes in nanostructured iridium oxide based on a model for particulate oxides with mixed conductivity (Sunde et al. 2009) concluded that these oxides do not display similar conductivity changes as AIROF (Sunde et al. 2010).

Data collected during oxygen evolution have also been collected and analyzed with models assuming specific reaction mechanisms, frequently taken to be the electrochemical oxide path (Bockris 1956; Ferrer and Victori 1994a,b; Hu et al. 2004). Based on a Tafel analysis displaying different Tafel slopes at low and high overpotential, Hu et al. (2004) concluded that the first reaction step of the electrochemical oxide path had to be substituted by a two-step reaction and established a simplified impedance model assuming that these two steps were the slow ones. The model was able to mimic the experimental data quite well.

3.2.2 Structure–Sensitive Methods

3.2.2.1 X-Ray-Based Methods

X-ray diffraction (XRD) is used routinely for the characterization of catalytic powders for the verification of the structure, assessment of the degree of crystallinity, assessment of whether solid solutions have formed in

the case of oxides with more than one metal atom in the lattice, and assessment of parameters such as axis lengths of the unit cell and crystallite size (Marshall et al. 2004, 2005, 2006b; Petrykin et al. 2009, 2011, 2013). Linear trends in unit-cell parameters with changes in composition are frequently taken as an indication of a solid solution (Owe et al. 2012). In some cases, more sophisticated XRD analysis such as pair-distribution function (PDF) analysis (Owe et al. 2012) is employed, but as this normally requires synchrotron radiation few examples are available in the literature.

X-ray photoelectron spectroscopy (XPS) is employed in the area of oxygen evolution catalysts for PEM water electrolysis in typically two ways. The first is related to the structural characterization of the electrocatalystic materials, and the other to the oxygen evolution reaction per se. Previously, the technique was strictly speaking employed as an ex situ technique (Hutchings et al. 1984; Marshall et al. 2005), but was in some cases extended to a quasi in situ technique by immersing the electrode containing the electrocatalytic material from the electrochemical cell at an applied potential. This was achieved by the use of specially designed sluices for rapid introduction of the sample into the vacuum chamber, allowing transfer to the spectrometer within minutes (Kötz et al. 1983a,b, 1984; Kötz and Neff 1985), see also Kötz (1991). Recently, XPS has been employed in water electrolysis research as an in situ technique (Sanchez Casalongue et al. 2014).

A typical application of XPS in catalyst characterization is the assessment of oxidation states of the catalyst's metallic component and also the changes in the oxidation state as a function of potential (Kötz et al. 1984; Sanchez Casalongue et al. 2014) (below the oxygen evolution potential). The oxidation states under potential evolution were probed by Kötz and coworkers for anodically formed IrO_2 films (AIROF) (Kötz et al. 1984), RuO_2 produced by thermal decomposition of $RuCl_3$ (Kötz et al. 1983b), and for ruthenium–iridium alloys (Kötz and Neff 1985), and more recently for IrO_2 by Sanchez Casalongue et al. (2014). For IrO_2, deconvolution of the oxygen O1s spectra into parts corresponding to oxide, water, and hydroxyl groups shows that the surface is predominantly a hydroxide (Kötz et al. 1984) at low potential (O1s binding energy of 531 eV), in agreement with calculations by Rossmeisl et al. (2007b). Some water is also apparent at the surface (O1s binding energy of 533.1 eV). Based on this, Kötz et al. (1984) proposed a mechanism for oxygen evolution corresponding to the electrochemical oxide path (Bockris 1956), in which oxygen is evolved on tetravalent IrO_2 ($IrO(OH)_2$) through pentavalent $IrO_2(OH)$) and hexavalent (IrO_3) surface states. Sanchez Casalongue et al. (2014) performed similar experiments for iridium oxide nanoparticles and with similar results. In this latter work, the authors utilized the dependence of the penetration depth on the photon energy to conclude that the hydroxide layer is limited to the surface of the oxide. These authors suggest that oxygen evolution takes place through the presence of a pentavalent (Ir(V)) hydroperoxy OOH* intermediate at the surface (corresponding to the pentavalent state in Kötz et al.'s work (1984)), but do not address the hexavalent state proposed by Kötz et al. (1984). Kötz et al. also proposed at reaction scheme for oxygen evolution a–RuO$_2$ (Kötz et al. 1983b). These authors also included the possibility of corrosion by-products in their analysis of both Ru and Ir oxides.

Another use of XPS is the assessment of surface segregation in mixed oxides. The key idea is that since the x-rays will have limited penetration depth in the oxides, in the order of one nanometer (Sanchez Casalongue et al. 2014), the relative (integrated) signals can be taken as a measure of the concentration of the surface of the oxide. This has been utilized for a number of oxides and anodic films to assess the presence of enriched layers at the surface (Hutchings et al. 1984; Kötz and Neff 1985; Trasatti 1999; Marshall et al. 2005; Petrykin et al. 2009). For iridium-containing catalysts, a common finding is that the surface is enriched in iridium (Hutchings et al. 1984; Kötz and Neff 1985; Pauli and Trasatti 1995; Owe et al. 2012). Petrykin et al. (2009) have pointed out that for small particles the penetration depth of the photon beam may actually amount to a substantial fraction of the particle size. By utilizing the emission angle dependence of the penetration depth of XPS, Kötz and Neff (1985) demonstrated that the surface segregation layer may be very thin and that XPS performed with an emission angle of 20° (and therefore relatively short penetration depth) indicates a stronger segregation than the measurements with 90° emission angle (and correspondingly larger penetration depth). Later, results of Kötz (1991), utilizing the still higher surface sensitivity of ultraviolet photoelectron spectroscopy (UPS), concluded that oxides formed at alloys of Ru$_{0.5}$Ir$_{0.5}$ are still more segregated than inferred from the low emission angle XPS measurements and correspond to an almost pure iridium oxide surface. This may be compared to the very minor segregation predicted even by the low emission angle XPS measurements in Kötz and Neff (1985).

X-ray absorption falls into two broad classes of analysis, those based on analysis of the x-ray absorption near edge structure (XANES) and those based on an analysis of the extended x-ray absorption fine structure (EXAFS) (Garrett and Foran 2003). XANES is typically associated with the determination of oxidation states, but would be probing the bulk rather than the surface as XPS would. The most important information gained from EXAFS is related to the local structure associated with the atom whose absorption edge is being probed, typically coordination numbers and bond lengths, and the analysis

may be performed including a number of coordination shells depending on the quality of the data and the needs.

XANES and EXAFS have been applied to materials relevant for electrocatalysis of the OER in PEM water electrolysis such as AIROF (Hüppauff and Lengeler 1993; Lengeler 2006), electrochemically deposited iridium oxide films (EIROF) (Mo et al. 2002), sputtered iridium oxide films (SIROF) (Pauporté et al. 1999), and doped iridium oxides (Petrykin et al. 2009, 2011, 2013). Both oxidation processes for below the oxidation potential of water and the oxygen evolution process itself have been probed. Whereas Hüppauff and Lengeler (1993) found that the oxidation state of AIROF varied from 3 to 4.8, Pauporté et al. (1999) reported the range 3 through 3.85 for SIROF. Hillman et al. (2011) found a voltammetric charge corresponding to only one electron per Ir atom in EIROF samples. In their analysis, they also utilized changes in the Debye–Waller factor, expressing disorder in the sample, with potential to conclude that both the oxidation voltammetry peaks of EIROF are associated with the same change in the oxidation state of iridium, but corresponding to different sites in the film. A somewhat related work is that of Yuen et al. (1983), who analyzed voltammetric peaks not in terms of sites but in terms of the density of states in AIROF films.

In a series of papers, Petrykin et al. (2009, 2011, 2013) analyzed the local structure in Ni- (Petrykin et al. 2009), Zn- (Petrykin et al. 2011), and cocontaining (Petrykin et al. 2013) iridium oxide catalysts. In general, they found the dopants to cluster within the host lattice. Ni and Co tended to distribute along the [111] direction of the unit cell, leading to clustering in shear planes in the case of Ni.

3.2.3 Microscopic Methods

3.2.3.1 Scanning Electron Microscopy

Scanning electron microscopy (SEM) is a convenient method to assess the morphology and element distribution of electrodes. Electron microscopy is based on using an electron beam to probe the surface (or bulk as in transmission electron microscopy [TEM]). In SEM, the primary electron beam is scanned across the surface of a sample and either the secondary or backscattered electrons emitted from the sample are measured. Secondary electrons arise due to nonelastic collisions between the primary electron beam and the outer electrons of the sample, with these outer electrons ejected from the sample as secondary electrons. These electrons have low energy < 50 eV and therefore must be close to the surface of the sample in order to overcome the surface energy barrier and escape to the detector.

3.2.3.2 Transmission Electron Microscopy

TEM utilizes the electrons transmitted through an ultrathin sample. As the electron beam passes completely through the very thin sample, both an image and diffraction pattern formed from the interaction of the electrons with the sample can be observed. TEM is a major analysis method in several scientific fields where detailed information of the nanostructure of the material is important. For detailed information of TEM imaging, please refer to, for example, Williams and Carter (2009).

TEM imaging has been extensively used to characterize the structure of catalysts for PEM fuel cells and the degradation mechanisms of these catalysts (Wang et al. 2006; Chatenet et al. 2010; Dubau et al. 2013; Ahluwalia et al. 2014; Lopez-Haro et al. 2014; Nikkuni et al. 2014), and much of the techniques developed in these studies can be applied to the detailed study of anode catalysts for PEM electrolyzers.

3.2.3.3 Spectroscopic Methods

Spectroscopic methods, quite widespread in other areas of electrocatalysis (Rodes and P'erez 2003), have not been extensively used in the study of the OER. However, Sivasankar et al. (2011) employed FTIR measurements to assess the nature of intermediates in the oxygen evolution at iridium oxide nanoparticles in solution (Sivasankar et al. 2011). Both the results of this work and those of a recent work employing potential-modulated reflectance spectroscopy (PMRS) (Kuznetsova et al. 2014) at iridium oxide electrodes may be interpreted to indicate that hydroperoxyl is an intermediate during oxygen evolution.

The electrochromic process in iridium oxide has been extensively studied by various spectroscopic methods, but typically emphasizing processes at potentials lower than those at which oxygen evolves (Beni and Shay 1978; Bock and Birss 1999).

3.3 Reaction Mechanisms of the Oxygen Evolution Reaction

3.3.1 Reaction Mechanisms

The OER is a catalyzed electrochemical reaction, its rates vary widely, and the reaction involves adsorption of intermediates at the catalyst surface. A large number of reaction mechanisms have been proposed, and Tafel slopes and reaction orders (which depend on the rate-determining step assumed) have been worked out for a number of these reaction mechanisms, see, for example,

Bockris (1956). Examples of suggested reaction mechanisms are the electrochemical oxide path (I) (Bockris 1956)

$$S + H_2O(l) \rightleftharpoons S - OH_{ads} + H^+(aq) + e^-$$
$$S - OH_{ads} \rightleftharpoons S - O + H^+(aq) + e^- \quad (3.3)$$
$$2S - O \rightleftharpoons 2S + O_2(g)$$

the oxide path (II) (Bockris 1956)

$$S + H_2O(l) \rightleftharpoons S - OH_{ads} + H^+(aq) + e^-$$
$$2S - OH_{ads} \rightleftharpoons S - O + S + H_2O \ (l) \quad (3.4)$$
$$2S - O \rightleftharpoons 2S + O_2(g),$$

and the Krasil'shchikov path (III) (Damjanovic et al. 1966)

$$S + H_2O(l) \rightleftharpoons S - OH_{ads} + H^+(aq) + e^-$$
$$S - OH_{ads} \rightleftharpoons S - O^- + H^+(aq)$$
$$S - O^- \rightleftharpoons S - O + e^- \quad (3.5)$$
$$S - O \rightleftharpoons 2S + O_2(g),$$

in which S denotes a (usually unspecified) surface (active) site on the catalyst such as a metal ion in valence state z^+. The Tafel slopes b_a and reaction orders with respect to protons (ν_{H^+}) corresponding to these proposed reaction mechanisms are reproduced in Table 3.1.

Tafel slopes between 40 and 60 mV are frequently observed at low overpotentials η. An example is given in Figure 3.3, showing a Tafel slope of 40 mV. Values as low as 15 mV have not been observed. The experimental values therefore appear to imply that a step related to the formation of oxygen from adsorbed hydroxyl at the surface, such as reactions (3.4) and (3.5), is the rate-determining step (rds) (Hoare 1968; Trasatti 1990a). This is also consistent with a reaction order ν_{H^+} equal to 1.5 (Trasatti 1990b), the fractional reaction order being ascribed to a double-layer effect. At high η, Tafel slopes of 120 mV are sometimes observed (Hu et al. 2004),

TABLE 3.1

Summary of Tafel Slopes for the Reaction Mechanisms in Equations 3.5 and 3.6

Path	Rate-Determining Step	b_a, low η (mV)	b_a, high η (mV)	ν_{H^+}
I	1	120	120	0
	2	40	120	−1
	3	15	∞	−4
II	4	120	120	0
	5	30	∞	−2
	6	15	∞	−4
III	7	120	120	0
	8	60	∞	−1
	9	40	120	−2
	10	15	∞	−4

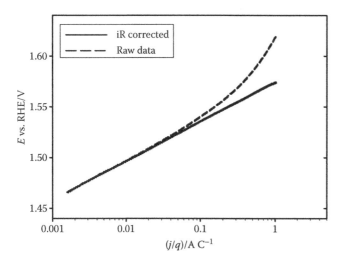

FIGURE 3.3
Polarization curves of a sample of pure iridium oxide in 0.5 mol dm^{-3} H$_2$SO$_4$ before iR correction (– – –) and after (——). The data were collected at 5 mV min^{-1}. The sample was produced by the hydrolysis method and annealed at 600°C. (The data in the figure were normalized with respect to the voltammetric charge obtained at a sweep rate of 200 mV s^{-1}.) (Reprinted from *Electrochimica Acta*, 70, Owe, L.-E. et al., Iridium–ruthenium single phase mixed oxides for oxygen evolution: Composition dependence of electrocatalytic activity,158–164, Copyright (2012), with permission from Elsevier.)

which would be consistent with all three pathways I through III. (Changes in Tafel slopes may also result from a change in the reaction pathway, a change in the electrode substrate, a change in the rds within the same mechanism, and an influence of the reaction conditions with potential [e.g., high coverage of intermediates [Kinoshita 1992, p. 87]].) However, if all possible number of intermediates is considered, a large number of reaction paths are possible (Milner 1964), which will not be unique in the value of kinetic parameters (Kinoshita 1992), p. 89. Additional evidence than that collected from polarization curves will be necessary (Hoare 1968). However, electrochemistry may sometimes be utilized in unconventional ways, such as in Hill and Hickling's consideration that a mechanism in which the combination of two oxygen atoms (Hickling and Hill 1950) is rate determining would correspond to unrealistically high oxygen pressures at high overpotentials. More trivially, for gas-evolving electrodes an apparent change in the Tafel slope may also be the result of poor gas removal.

More recently, Rossmeisl et al. (2007b) considered a series of one-electron step as a reaction mechanism for the OER

$$S + H_2O(l) \rightleftharpoons S - OH_{ads} + H^+(aq) + e^-$$
$$S - OH_{ads} \rightleftharpoons S - O + H^+(aq)$$
$$S - O + H_2O \rightleftharpoons S - OOH + H + (aq) + e^-$$
$$S - OOH \rightleftharpoons S + O_2(g) + H + (aq) + e^-$$

$$(3.6)$$

formulated in the same notation as the other mechanisms mentioned earlier. In this case, however, S is specified to be either a bridge site between two fourfold coordinated metal ions or a coordinatively unsaturated site (CUS) on top of a fivefold coordinated metal ion in rutile (110) surfaces. The mechanism in reaction (Equation 3.6) is a mononuclear mechanism since it assumes that only one site (one S) at the catalyst surface is involved in any of the elementary reactions comprising the mechanism. In a binuclear reaction mechanism the onset of oxygen evolution occurs when adsorbed species at two neighboring catalytic sites are oxidized to adsorbed oxygen (Steegstra et al. 2013), for example, when sites covered with oxo adsorbates (S–O) combine to adsorbed molecular oxygen (di-oxo-μ-peroxo mechanism [Busch et al. 2011b]), which we will consider as implied in the final step of the electrochemical oxide path here.

3.3.2 Spectroscopic Assessments of the Reaction Mechanism

Recent spectroscopic work identifies the hydroperoxyl intermediate at the surface of iridium oxide during oxygen evolution (Sivasankar et al. 2011; Kuznetsova et al. 2014; Sanchez Casalongue et al. 2014), and therefore appears to support the mechanism employed by Rossmeisl et al. (2007b).

3.3.3 By-Products of the OER

Since in PEM water electrolysis the electrolyte is solid and water is the only liquid phase, the number of by-products from PEM water electrolysis will be limited. Kuznetsova et al. (2014), however, found that during oxygen evolution at Nafion (Hu et al. 2004)-free IrO$_2$ and Ir$_{0.6}$Ru$_{0.4}$O$_2$ electrodes substantial amounts of by-products are formed and could be registered at the ring of a ring-disk electrode poised at a potential (1.2 V vs. RHE) above those that would correspond to reduction of oxygen formed at the disk. Both anodic and cathodic ring currents were observed. The ring currents were interpreted as being due to the formation of hydrogen peroxide and ozone during the OER at the disk electrode (Kuznetsova et al. 2014). The rates of formation were dependent on the dynamic conditions (sweep rate) under which the disk electrode operated and were also dependent on oxide composition. The more active (for OER) oxide (Ir$_{0.6}$Ru$_{0.4}$O$_2$) thus produced the lesser amount of by-products in general.

3.3.4 Rationalizing Trends in Electrocatalytic Activity

Various chemical aspects of the steps of the reaction mechanisms discussed have been used for rationalizing and explaining trends in catalytic activity of OER catalysts (alkaline and PEM), such as acid–base properties (Trasatti 1990b), redox potentials of the catalyst (Rasiyah and Tseung 1984), binding energies of the metal component to hydroxyl ions (Rüetschi and Delahay 1955), binding energies of the metal component of the catalyst to oxygen (Trasatti 1980), to the number of d-electrons (Bockris and Otagawa 1983), and geometrical factors (Kinoshita 1992).

The rationalization in terms of binding energies follows directly from the involvement of adsorbed species in all proposed reaction schemes, such as those discussed earlier. Thus, the rate of the reaction at a given potential depends on the balance between adsorption and desorption reactions, and the surface should not bind the adsorbates too weakly or too strongly (the Sabatier principle).

Rüetschi and Delahay (1955) thus correlated overpotential for the OER on metal electrodes in alkaline solutions with the binding energy for the bond strength for OH. The bond strength was approximated from three different reaction cycles. For the group of metals included, an approximately linear correlation of negative slope was found for the overpotential as a function of the binding energy. Trasatti (1980) pointed out that only if the third of the three methods of calculating the bond energies considered is employed the data will fit a straight line. Also, the assumed oxidation state of the metal will influence the results. Instead of the Rüetschi and Delahay (1955) correlation, Trasatti (1980) suggested in its place correlating the overpotential with the enthalpy associated with the transition $MO_x \rightarrow MO_{1+x}$, generally referred to as the "lower → higher oxide transition" and in which MO_x and MO_{1+x} are two forms of the oxide differing in the ratio from the metal (M) to oxygen content ($1:x \rightarrow 1:1 + x$). Trasatti (1980) was thus able to obtain a volcano-shaped curve (Parsons 2011) for the catalytic activity of a number of oxides vs the enthalpy of the lower → higher oxide transition with RuO_2 at its apex, refer Figure 3.4. (In our plot of these data shown in Figure 3.4 only the data for acid conditions are included.)

However, Bockris and coworkers (Bockris and Otagawa 1983, 1984; Bockris et al. 1983) correlated the activity of perovskite oxygen-evolving electrodes in alkaline solutions with the enthalpy of formation for the corresponding metal hydroxides and the bond strength of M–OH estimated from a modified version of one of Rüetschi and Delahay's (1955) methods. Noting that the type of A-ion in the perovskite ABO_3 compounds had little effect on the oxygen evolution activity of their catalysts (Bockris and Otagawa 1983), the correlation was based on the B-ion of the catalyst. As in Rüetschi and Delahay's (1955) treatment, not only a descending straight line was obtained for the activity vs the bond

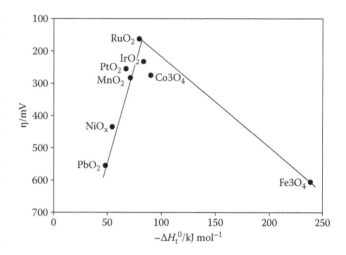

FIGURE 3.4
Volcano curve for oxygen-evolving catalysts showing overpotential as a function of enthalpy associated with the lower/higher oxide transition $-\Delta H_t^0$. Only data for acidic conditions are included. (Data taken from Trasatti, S., *J. Electroanal. Chem.*, 111, 125, 1980.)

energy, but also suggesting a hypothetical ascending part. Bockris and coworkers finally showed that the activity also correlates with the number of d-electrons on the trivalent B-ion in the perovskite (two in V, three in Cr, etc.), the higher the number of d-electrons the higher the activity. The latter correlation was rationalized by emphasizing that the d-electrons would enter the antibonding orbitals in the hydroxyl bond to the surface. Obviously, the larger the number of electrons in the antibonding orbital the weaker the bond and the higher the activity, in line with the descending activity–bond strength correlation presented.

Based on results for a series of perovskites in alkaline solution, Suntivich et al. (2011) located a maximum in the catalytic activity for oxygen evolution for a number of d-electrons corresponding to approximately one electron in the catalyst's e_g-band and argued that the number of electrons in the e_g-band rather than the overall number of d-electrons is a better descriptor for catalytic activity. Based on this, they successfully predicted an optimum composition ($Ba_{0.5}Sr_{0.5}Co_{0.8}Fe_{0.2}O_3$) and demonstrated that this compound displayed a catalytic activity at the apex of the volcano curve. Vojvodic and Nørskov (2011) generalized this descriptor to also include compounds containing only t_{2g} electrons, which is more relevant for PEM water electrolysis since the catalyst employed in PEM water electrolysis usually does not contain e_g electrons.

In a series of papers, Tseung (Tseung and Jasem 1977; Jasem and Tseung 1979; Rasiyah et al. 1982a,b; Rasiyah and Tseung 1984) and coworkers focused on an entirely different aspect of the OER, viz. that of the oxidation state of the cation in oxide catalysts. For catalysts at which the OER proceeds through any reaction mechanism similar to those presented earlier, the valence of the cation (assumed to be represented by the site "S") will change during the catalytic cycle. Rasiyah (1984) thus classified oxide catalysts into two groups, those for which the catalyst undergoes an increase in the cationic oxidation state before evolving oxygen and those that do not. The latter group tends to evolve oxygen at rather high potentials (potential around 1.77 V at 1 A cm^{-2}). For this group, the governing potential is the potential of the OH$^-$/HO$_2^-$ couple. At the former group, however, the oxygen evolution potential is lower and governed by the lower oxide/higher oxide couple. For example, for iridium oxide the governing potential is that of the IrO$_2$/IrO$_3$ couple (1.350 V, Rasiyah and Tseung 1984).

Yet, another aspect of the OER on oxides upon which correlation with catalytic activity has been sought is the acid–base properties of the catalysts. This is natural, since besides being descriptive of the surface chemistry of the oxide in general, the acid–base properties are directly related to the exchange of protons with the solution, obviously relevant for the oxygen evolution since proton (or hydroxyl) exchange is part of most of the proposed reaction mechanisms such as those explained earlier, for example, Equation 3.4. Trasatti has thus advocated the use of the point of zero charge (pzc) as a descriptor for the catalytic activity of oxides (Trasatti 1990b). This intensive (area-independent) parameter is defined as the solution pH at which the concentrations of H$^+$ and OH$^-$ at the oxide surface are the same, and may be determined by potentiometric titration (Ardizzone and Trasatti 1996). Thus, below the pzc adsorption at the oxide surface would be dominated by protons and above the pzc by hydroxyl. Since the pzc will be determined by the strength of the interaction between the catalyst and OH (the M–OH bond) (Ardizzone and Trasatti 1996), which is exactly the same as those expected to govern the electrocatalytic activity, the relation to oxygen evolution electrocatalysis seems obvious. It has thus been shown whereas changes in calcination temperature for thermal RuO$_2$ catalysts bring about a change in the Tafel slope (also an intensive parameter) *and* the pzc, neither of these quantities vary with calcination temperature for cobalt oxide (Trasatti 1990b). Also, both the pzc and the Tafel slope were found to depend in a nonlinear fashion on the amount of iridium oxide in iridium–ruthenium mixed oxides (Trasatti 1990b). While prediction of electrocatalytic behavior for a proposed oxide composition may appear difficult, the big advantage is that the pzc is an intensive property and may thus be employed as a tool to separate surface area effects from electronic factors. For example, for mixed oxides the pzc has been used to distinguish atomic mixtures from physical ones (Guerrini and Trasatti 2006), which may be a useful practical application.

The pzc for a number of relevant materials is listed in Ardizzone and Trasatti (1996).

Incidentally, the acid–base property of an oxide has been correlated with the electronegativity of the oxide (Butler and Ginley 1978), defined as the geometric mean of the atomic (Mulliken) electronegativities (Butler and Ginley 1978; Trasatti 1999). Since the electron affinity of a (semiconducting) oxide would be related to the oxidation potential of the oxide, the pzc picture appears to be related to the rationalization by oxidation potential advocated by Tseung et al. (Rasiyah and Tseung 1984).

More details on the early literature on oxygen evolution and properties of catalysts may be gathered from the books by Kinoshita (1992) and Hoare (1968) and reviews by Trasatti (Daghetti et al. 1983; Trasatti 1984, 1987, 1991, 1999; Guerrini and Trasatti 2006).

3.3.5 Atomistic Calculations for Electrocatalysts

It is fruitful to consider the band structure of oxides relevant for PEM electrolysis in some more detail, not only because of its relevance to electrocatalysis but also because it determines the conductivity of the oxide and also other properties of interest for correlating electrocatalysis with other research areas for a more complete understanding of the materials. A connecting feature of most of the relevant oxides is that the metal ions are octahedrally coordinated to oxygen, or at least approximately octahedrally coordinated. A schematic exposition of the band structure of rutile (and other structures with octahedrally bound metal atoms) oxides is shown in Figure 3.5. The electronic structure may be understood by considering the formation of the oxide as a result of a process whereby the d-levels of the transition metal are split by the crystal field in the octahedral environment. Starting with the individual d-states (fivefold degenerate), the electronic structure might be thought of as emerging from first placing them in an octahedral environment, leading to crystal field splitting into t$_{2g}$ and e$_g$ levels, lifting the initial degeneracy. These crystal-field split levels then form bands, and these are usually referred to as the t$_{2g}$ for the lower-lying band and e$_g$ for the higher-lying band, thus referring specifically to this picture of their genesis.

This electronic structure has served as a basis for rationalizing the electrochromic properties of RuO$_2$, IrO$_2$, and others (Granqvist 1994, 2014). Accordingly, within the rigid band approximation proposed, the anodic coloration of IrO$_2$ is explained by the oxide having a partially filled (conduction) band in the oxidized state (corresponding to stoichiometric IrO$_2$). When the oxide is reduced, the electrons enter, for example, the t$_{2g}$ band in oxides with a partially filled t$_{2g}$ band, such as RuO$_2$ or IrO$_2$, protons entering the oxide to compensate for the extra charge. For IrO$_2$, a transition from

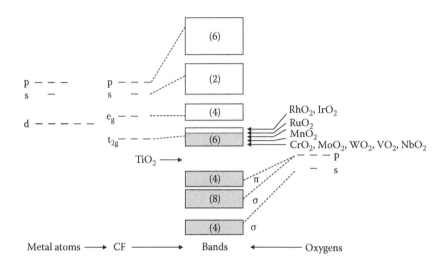

FIGURE 3.5
Schematic electronic structure of rutiles. (Adapted from Honig, J. M., Electronic Band Structure of Oxides with Metallic or Semiconduction Characteristics, in S. Trasatti (Ed.): Electrodes of conductive metallic oxides, Elsevier, Amsterdam (1980) pp. 1–95; J. B. Goodenough, Progress in Solid State Chemistry, Vol. 5, Pergamon, Oxford (1971) 145.)

conductive (oxidized state) to semiconducting (reduced state) behavior should be expected to emerge if the oxide is reduced by one elementary charge per formula unit, since the t_{2g} band of stoichiometric IrO_2 has available exactly one electron per formula unit before being filled up. Thus, one would expect reduced iridium oxide to behave as a semiconductor since it has got a filled band. Also, the optical properties change, introducing a band gap in the optical spectrum and transforming the oxide into a transparent one from its original blue color, the optical properties in the latter case being determined by intraband transitions. Also, one would expect a large leap in conductivity, as has been demonstrated directly for anodically formed IrO_2 (AIROF) by impedance measurements (Glarum and Marshall 1980). Measurements of changes in oxidation states (Michell et al. 1978; Kötz et al. 1983b, 1984; Hüppauff and Lengeler 1993; Petit and Plichon 1998; Pauporté et al. 1999; Lengeler 2006; Hillman et al. 2011) and the low diffusion coefficients measured for electrochromic IrO_2 (Granqvist 1995), which are orders of magnitude lower than those in electrolytes, support this picture. Ruthenium oxide, on the other hand, with one electron less than iridium oxide in the conduction band, is not electrochromic since transfer of an equivalent electronic charge to ruthenium oxide does not fill its conduction band, and the oxide would still be a conductor.

A very compelling evidence of this model for IrO_2 is provided by the ultraviolet photoelectron spectra (UPS) provided by Kötz and Neff (1985), who directly demonstrated how the Fermi level is shifted from inside the conduction band to the band edge.

The metal-semiconductor transition is interesting because it is in certain aspects easier to characterize semiconductors in situ than conductors, simply because much of the response to an electrochemical stimulus appears in the electrode itself rather than in interfacial layers of a few angstroms (c.f., e.g., Morrison 1980). Thus, Mott–Schottky analysis of capacitance data indicates that charging in these films is associated with space-charge layers within the oxides similar to those found for semiconductor electrodes (Bockris and Otagawa 1983, 1984; Bockris et al. 1983; Silva et al. 1998; Rasten et al. 2003). Another example can be found in Gottesfeld (1980), where the magnitude of faradaic currents appears to be solely determined by the overlap between bands in AIROFs and fluctuating redox levels in solutions of $Fe(CN)_6^{3-}/Fe(CN)_6^{4-}$. These measurements were later reproduced by Lervik et al. (2010) and Owe et al. (2012). Lervik et al. (2010) did not, however, find a similar behavior for nanostructured IrO_2 manufactured by hydrolysis. Yet, the charge-normalized activity for the OER was the same as for AIROF for these hydrolysis-produced catalysts, indicating that the electrocatalytic activity is determined by the local properties of the surface. A similar result for the OER activity was found in a comparison between AIROF and thermally produced iridium oxide by Comninellis and coworkers (Ouattara et al. 2009); the two different forms of the oxide display the same catalytic activity.

In addition to being of interest concerning the availability of experimental techniques, the metal-semiconductor transition also has bearings on the correlations listed at the beginning of this paragraph. Within the rigid band approximation referred as earlier (Granqvist 1994, 2014), the oxidation potential of the reduced semiconducting oxide (i.e., the potential of the lower metal oxide → higher metal oxide transition) is related to the position of the top of the valence band. (Valence band refers here to the reduced oxide, and is thus the same as the conduction band of the oxidized oxide.) This is because during a potential scan in the anodic direction, the electrons removed first reside at the top of this band, and this would define the oxidation potential. For a practical application, this picture would be complicated by the uncertainty in the share of the applied potential appearing inside the oxide.

In principle, the electron affinity EA, and thus the oxidation potential for the high and low oxide transition, would also depend on the band gap and density of states (DOS). The means for estimating these quantities are also available, such as DOS calculations for oxides and the means for combining them (Cyrot and Cyrot-Lackmann 1976; Mattheiss 1976; Griessen and Driessen 1984).

In summary, however, while the electronic structure referred to as earlier will continue to serve as a useful backdrop to the understanding of these oxides and their electrocatalytic activity, the results show that at least a qualitative understanding of catalytic properties would not necessarily have to include fundamental parameters beyond the nearest-neighbor interaction at the surface. In terms of the shape of the density of states, this means that moments higher than the third are presumably not very important.

The rate of reactions is usually associated with an activation energy, and by relating the enthalpy of adsorption of intermediate species of the reaction to the activation energy through the Brønsted relation it is possible to obtain volcano relationships for electrochemical reactions such as the hydrogen evolution reaction (Parsons 2011). By calculating the enthalpy of adsorption of intermediates in a reaction sequence and by identifying the largest one (neglecting here the effect of differences in frequency factors), one might hope to determine the rate-determining step and the catalytic activity (at least in relative terms). Implicit in this approach is the assumption that all reaction steps have negative-free energies and thus proceed spontaneously at rates determined by the kinetic factors. According to this, an electrochemical reaction would proceed in principle at any potential above the reversible potential for the reaction, although at minute rates for sluggish reactions.

However, Rossmeisl, Nørskov, and others (Rossmeisl et al. 2005, 2007a,b; Hansen et al. 2010), based on density-functional theory (DFT) calculations, demonstrated *inter alia* for the OER that one or more steps in the reaction may be thermodynamically infeasible at low overpotentials. Figure 3.6 shows the free energies associated with the reaction steps described (Equation 3.6) for three different applied potentials at the 110 surface of a RuO_2 catalyst. At 0 V, all reaction steps are uphill, as well as of course the total free energy of the reaction. At 1.23 V, the free energy of all intermediates and products is lowered by the applied potential. At this potential, the overall

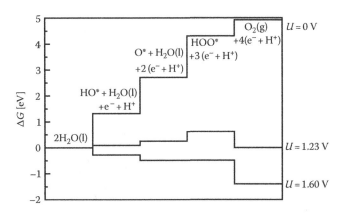

FIGURE 3.6
The free energies of the intermediates on an O*-covered RuO_2 at three different potentials ($U = 0$, $U = 1.23$, and $U = 1.60$ V) are depicted. At the equilibrium potential ($U = 1.23$ V), the reaction steps are uphill in free energy. At 1.60 V, all the reaction steps are downhill in free energy. (Reprinted from *Journal of Electroanalytical Chemistry*, 607, Rossmeisl, J. et al., Electrolysis of water on oxide surfaces, 607, 83–89, Copyright (2007), with permission from Elsevier.)

free energy of the total reaction is zero, but all reaction steps except the last (the formation of oxygen gas from the hydroperoxyl intermediate OOH*) are associated with a positive change in free energy. In consequence reaction, steps are thermodynamically blocked, and the overall reaction will not proceed. At 1.60 V, however, all free energies become negative and the reaction will commence.

The picture emerging from these calculations does not assume in principle that the overpotential necessary for all steps to become spontaneous is larger than zero. For some reactions, this may happen already at the reversible potentials, such as chlorine evolution (Hansen et al. 2010), given the right catalyst. However, a general finding is that the adsorption energies vary in proportion to one another for a given family of catalysts such as the rutile structure. Thus, the adsorption energy of hydroxyl OH* and hydroperoxyl (OOH*) is proportional to the binding energy of oxygen (O*) at the surface and therefore to one another (Rossmeisl et al. 2007b; Man et al. 2011). The practical implication of these so-called scaling relations is that the magnitude of the steps in the reaction cannot be varied independently, but will rather shift in concert when one catalyst is replaced with another one.

Obviously, the lower the potential at which the last step attains a negative free energy, the better the catalyst. Rossmeisl et al. (2007b) thus suggested to rank catalytic activity based on the potential for this potential-limiting step. Since the scaling relations imply that the free energies of all adsorbates can be plotted as a function of the same variable (such as the binding energy of oxygen, O*). Rossmeisl et al. (2007b) thus constructed a volcano curve based on the DFT

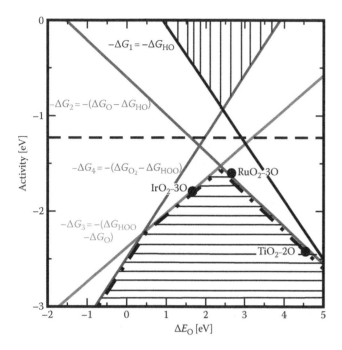

FIGURE 3.7
The theoretical activity of the four charge transferring steps of oxygen evolution is depicted as a function of the oxygen-binding energy ΔE_O. The vertical axis shows the negative of the free energies ΔG_1 through ΔG_4, which represent the free energies of reactions (Equation 3.6). (Reprinted from *Journal of Electroanalytical Chemistry*, 607, Rossmeisl, C. et al., Electrolysis of water on oxide surfaces, 83–89, Copyright (2007), with permission from Elsevier.)

calculations with the binding energy of oxygen, O* at the ordinate, which is reproduced in Figure 3.7. For strongly bound (small positive or negative ΔE_O) O*, the oxygen-evolving activity is limited by the formation of hydroperoxyl (HOO*, line labelled ΔG_3). For weakly bound (large positive ΔE_O) O* on the oxide surfaces, the oxygen-evolving activity is limited by the formation of oxygen (O*, line labelled ΔG_2). Thus, iridium oxide binds oxygen a little too strongly and ruthenium a little too weakly for the catalyst to perform optimally.

An ideal catalyst would have identical steps in free energy for each elementary reaction (Dau et al. 2010), as can be easily inferred from Figure 3.6. When the potential is increased, the magnitude of all steps decreases by the same amount (Figure 3.6), and this would clearly minimize the size of the largest step. The results of Rossmeisl et al. (2007b) actually indicate a suboptimal difference between ΔE_{OOH*} and ΔG_{OH*}, which should be 2.46 eV (Dau et al. 2010).

In a recent paper, Halck et al. (2014) explore theoretically and experimentally the design of catalysts for which the universality of the scaling relationship does not apply by the incorporation of dopant elements (exemplified by Ni and Co in RuO₂ surfaces) in the surface. These dopant elements activate bridging O atoms in the surface, which are deemed inactive in

the pure RuO₂ catalyst surface (Rossmeisl et al. 2007b). In the doped surface, however, the bridge O attracts protons in OH* and OOH* moieties adsorbed at the surface and lowers the energy of these states. The doping element therefore introduces a second tunable parameter of the surface. This reduces the free energy of the largest step (i.e., the "potential-determining step") to 1.49 and 1.33 eV for Ni- and Co-modified ruthenia, respectively. The previously theoretically predicted minimum overpotential of 0.4 V for ruthenia is correspondingly reduced to 0.3 V for Ni-doped and 0.1 V for Co-doped ruthenia.

Calculations similar to those explained have also been performed for the electrochemical oxide path (Bockris 1956), emphasizing the last elementary reaction of the mechanism, Equation 3.4 (Busch et al. 2011a, 2011b, 2013a, 2013b; Steegstra et al. 2013) which constitutes the main difference between the path employed by Rossmeisl et al. (2007b), Equation 3.6, and Bockris' electrochemical oxide path, Equation 3.3. Energy profiles corresponding to those of Rossmeisl's (2007b) are provided in Busch et al. (2011a,b, 2013b).

Steegstra et al. (2013) thus find experimental support for the binuclear mechanism (i.e., the electrochemical oxide path, the binuclear step but with a μ-peroxo intermediate implied) in the ratio of charges in voltammograms of electrochemically deposited iridium oxide (EIROF). The III/IV peak contains approximately twice as much charge as the IV/V peak in alkaline solutions. Attributing the redox peaks of iridium oxide to surface processes, this is considered as evidence of the precedence of a mixed oxidation state, Ir(IV)Ir(V), in the EIROF surface, stabilized through hydrogen bonding. Steegstra et al. (2013) also found indications of a larger amount of OER-suppressing Ir(V) electronic structures with lower pH, appearing due to the presence of protons bridging pentavalent surface iridium (Ir(V)–OH⁻–O–Ir(V)).

For the binuclear mechanism, Busch et al. (2013a) discovered a correlation between the energetics of the oxidation of hydroxyl species to di-oxo adsorbates (Equation 3.3) and that of the formation of the formation of μ-peroxo adsorbates from the di-oxo intermediates. Thus, oxides were thus classified into those for which the di-oxo-forming step was exothermic and the μ-peroxo-forming step (O–O bond forming step) was endothermic (±), and a class for which the di-oxo-forming step was endothermic and the μ-peroxo-forming step was exothermic (±). These authors suggested mixing oxides of the two classes to obtain a better catalyst and provided a simple relation for estimating the energetics of the two reactions for the mixed oxide.

Steegstra et al.'s (2013) interpretation of the redox peaks of the voltammograms as III/IV and IV/V and on which the analysis hinges appears to find support in available data for oxidation states in various forms of iridium oxide, including AIROF (Michell et al. 1978; Kötz et al. 1984; Hüppauff and Lengeler 1993; Lengeler 2006) and EIROF (Petit and Plichon 1998; Mo et al. 2002). Since Steegstra et al. (2013) relate oxygen evolution, which is a surface process, to voltammetry, another basic assumption behind this interpretation of voltammetric features is that whole charge is related to surface processes as well. By dissolving EIROF in boiling hydrochloric solutions, Petit and Plichon (1998) showed unequivocally that a total of two electrons per iridium are exchanged in the two redox peaks of voltammograms in alkaline media. The particle size of EIROF the samples employed by Steegstra et al. (2013) is in the order of 2 nm across (Steegstra and Ahlberg 2012) and the IrO_2 unit cell approximately $6 \cdot 10^{-2}$ nm^3 (Mattheiss 1976). A considerable number of the iridium atoms will therefore reside not in the surface but in the bulk, and consequently, one would from Petit and Plichon's measurements (1998) expect that a considerable fraction of the voltammetric charge reflects redox processes in which these bulk atoms take part.

While the interpretation of Vojvodic and Nørskov (2011) appears to rationalize electrocatalytic activity for oxides in terms of the number of d-electrons in the oxide, a very different model has been emerged from similar calculations and considerations for metallic electrocatalysts. Thus, the d-band energy center appears to play the center role (Ruban et al. 1997). A low-lying d-band will cause a large and occupied bonding adsorbate-projected density of states, whereas a higher-lying d-band will lead to a more weakly bonded adsorbate (Hammer 2006).

3.3.6 The Effect of Electrolyte on Reaction Mechanisms and Kinetics

The electrolyte may have a substantial effect on the oxygen evolution characteristics of a given electrode material (Hoare 1968). While PEM water electrolyzers are fed with pure water, the polymer electrolyte may contain soluble species. For example, the Nafion (Hu et al. 2004) membranes typically employed lose their mechanical properties (Jalani et al. 2005), and there is a risk of dehydration causing loss of conductivity at temperatures above 100°C (Jalani et al. 2005; Antonucci et al. 2008). A polybenzimidazole (PBI) membrane (Wainright et al. 1995; Wang et al. 1996) is one alternative that can be used at temperatures above 100°C. However, the PBI membranes must be doped with phosphoric acid in order to achieve a sufficient conductivity (Wainright et al. 1995).

Phosphates are known to adsorb strongly on ruthenium oxide (Daghetti et al. 1983), however. Owe et al. (2011) therefore conducted a study of the influence of the electrolyte on the oxygen evolution characteristics of iridium oxide (powder catalysts and AIROF) and concluded that the presence of phosphate significantly

reduced the activity of the oxide. The Tafel slope appeared to be quite similar to that in perchloric or sulfuric acid though. If the catalyst is exposed to phosphoric acid, for example, in a PBI cell, one should expect to see a significantly degraded performance. However, Huynh et al. (2014) recently found that the reaction order of the OER with respect to phosphate is zero, indicating that the phosphate dependence will have to be investigated for each and every catalyst.

Nafion, on the other hand, appears to be a very well-suited electrolyte for PEM water electrolysis from an electrocatalytic point of view. Simulating the interaction between iridium oxide and Nafion by employing aqueous solutions of trifluoromethanesulfonic acid (TFMSA) as the electrolyte rather than the sulfuric acid commonly employed, a significant increase in catalytic activity for the OER was observed on iridium oxide electrodes (Owe et al. 2010). This is slightly surprising given the ability of fluoride to adsorb at DSA electrodes in NH_4F-containing electrolytes documented by Fukuda et al. (1979).

3.3.7 Oxygen Evolution on Single-Crystal Electrodes

Measurements on well-defined single-crystal surface play an important role in metal electrocatalysis (Feliu and Herrero 2003; Iwasita 2003). However, little work appears to have been done in terms of the oxygen evolutions, and in particular on the OER at oxide surfaces. However, in a recent work, Stoerzinger et al (2014) investigated the electrocatalytic properties of the (100) and (110) surfaces of RuO_2 and IrO_2 films made by pulsed laser deposition. The (100) surface appears to be significantly more active than the (110) surface for both oxides. Stoerzinger et al. (2014) suggested the difference to be due to the difference in the density of catalytically active sites in the surface. These measurements were performed in alkaline solution, but are expected to be of quite some significance also for PEM water electrolysis.

3.4 Oxygen Evolution on Noble Metal Oxide Nanoparticles

The catalyst of choice for the anode in PEM electrolyzers has since the first journal publication by Russell et al. in 1973 been Ir black or IrO_2 due to its higher stability and corrosion resistance in PEMWE conditions compared to RuO that has an intrinsically higher electrochemical activity (Miles et al. 1978; Matsumoto and Sato 1986). Similarly, the anode catalyst loading in PEM electrolyzers has not changed significantly during the last decades due to the relatively low surface area and hence the low utilization of the iridium catalysts (Carmo et al. 2013).

Significant efforts have been made to increase the electrocatalytic activity of the anode catalysts by alloying with more active materials (Ru) or ternary (inactive) materials for improved stability and activity or by increasing the surface area by nanostructuring and use of support materials. In this section, we review the main results.

3.4.1 Ir and Ru

The use of pure IrO_2 as a catalyst in PEM electrolyzers is limited by high costs and relatively low activity toward oxygen evolution. By mixing RuO_2 and IrO_2, it could be possible to retain the high activity of RuO_2 and the high stability of IrO_2. Over the years, mixed Ir–Ru oxides have been prepared by a range of methods; powders by thermal decomposition (Angelinetta et al. 1986; Marshall and Haverkamp 2010), hydrolysis (Owe et al. 2012), the Adams fusion method (Tunold et al. 2010), and the sol–gel method (Murakami et al. 1994a); films have been obtained by reactive sputtering (Kötz and Stucki 1986) or spray deposition (Roller et al. 2013). It is evident that the structural properties of the Ir–Ru oxide mixtures, and thus also the electrocatalytic properties, depend on the method used to prepare these materials. For powders synthesized with chemical methods, the mixed oxides have been shown to be both homogeneous solid solutions (Cheng et al. 2009b; Owe et al. 2012; Roller et al. 2013) or as separate IrO_2 and RuO_2 phases (Murakami et al. 1994a). Furthermore, significant surface enrichment of Ir has been observed for oxides prepared both with chemical and hydrothermal methods (Angelinetta et al. 1986, 1989; Owe et al. 2012). Oxides prepared by physical methods, such as reactive sputtering, have however been shown not to present surface enrichment of Ir (Kötz and Stucki 1986). The surface composition of the catalyst is a very important parameter as it is only at the surface of a heterogeneous catalyst that reaction occurs.

In general, the studies performed on the stability and activity of Ir–Ru mixed oxides show that these materials have an activity between the activities of the pure oxides and stability higher than RuO_2 (Carmo et al. 2013). It is, however, several conflicting results on the nature of the interaction between Ru and Ir in these mixed oxides and how this interaction affects the catalytic activity and stability of these oxides. If there are no atomic interactions between iridium and ruthenium in mixed oxides, it is expected that the total current should be a linear superposition of the activity of the two pure oxides. However, if an interaction between the noble metals in the oxides exists, the resulting current should deviate from this (Owe et al. 2012). The same effect is also expected for the corrosion stability of the mixed oxide; an interaction between the constituents should be manifested by

a change in the overall corrosion stability of the oxide, predominantly by increasing the stability of RuO_2, the least stable element in the mixed oxide.

Kötz and Stucki (1986) observed that a modest addition of 20% IrO_2 to RuO_2 caused a drastic increase in the stability of the oxide, but with the cost of a lower electrocatalytic activity. The study found that the potential of the oxygen evolution reaction at a current density of 0.1 mA cm^{-2} increased linearly with increasing iridium content, indicating a nonlinear dependence of the catalytic activity on composition (the dashed line in Figure 3.8). The results of Owe et al. (2012) from mixtures of pure IrO_2 and RuO_2 and hydrothermally prepared iridium–ruthenium oxide catalysts, however, indicate a linear superposition of the activity of the two components (solid line in Figure 3.8), even though a solid solution was confirmed in the latter case. These conflicting results show that any synergistic interaction between iridium and ruthenium in mixed oxides is dependent on the preparation method and may be weaker than what has previously been suggested (Kötz and Stucki 1986).

However, as commented earlier, the surface layers in which the composition of the oxide differs from that of the bulk may not penetrate the oxide surface vary deeply and less than the typical penetration depth of techniques otherwise considered to be surface sensitive. Since a proper assessment of surface segregation is critical in this context, the differences between the various oxides may in fact be due, at least in part, to differences in the surface composition. If iridium is assumed to be the element typically segregated to the surface in $Ir_{1-x}Ru_xO_2$

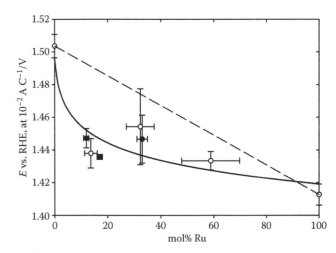

FIGURE 3.8

Potential for the OER at 10 mA C^{-1}. (•) Atomic mixtures and (○) physical mixtures. The potential calculated by assuming a linear combination of the properties of IrO_2 and RuO_2 is shown as the solid line. The straight dashed line connects the potentials for the pure IrO_2 and RuO_2 powders. (Reprinted from *Electrochimica Acta*, 70, Owe, L.-E. et al., Iridium–ruthenium single phase mixed oxides for oxygen evolution: Composition dependence of electrocatalytic activity, 158–164, Copyright (2012), with permission from Elsevier.)

and since the two endpoints of the curve in Figure 3.8 are fixed, any underestimate of the iridium segregation to the surface would tend to bring the data points closer to the dashed straight line. The true line would under these assumptions have all data points lying more to the left than the experimental values. The extent to which this is the case in the various data sets is not clear.

3.4.2 Effect of Ternary Doping

In order to improve the stability of the IrO_2–RuO_2-based electrocatalysts or to increase the utilization of the noble metals, several studies have been performed where inert, nonprecious oxides have been added to the Ir–Ru mixed oxides. Materials such as SnO_2 (Murakami et al. 1994b; Pauli and Trasatti 1995; Marshall et al. 2004, 2005), TiO_2 (Kameyama et al. 1993), Nb_2O_5 (Terezo et al. 2001), Ta_2O_5 (Murakami et al. 1994c), and their mixtures (Takasu et al. 1994; Ardizzone et al. 2006; Marshall et al. 2007; Cheng et al. 2009b; Kadakia et al. 2012) have been used as nonprecious metal oxide additions. Most of these studies show that the addition of a nonprecious metal oxide is primarily diluting the active catalyst materials (Ir–Ru) so that the mixed oxides have a similar catalytic activity to pure IrO_2 or RuO_2, resulting in a reduction in noble metal content without sacrificing activity. An example of the effect of ternary doping on electrolyzer performance is shown in Figure 3.9.

A limit of addition of the nonnoble oxides seems to be in the range of 50%–60%, after which additional dilution will reduce the activity of the catalyst (Marshall et al. 2005; Kadakia et al. 2012). Some studies report that addition of oxides such as Ta_2O_5 (Ardizzone et al. 2006) and Mo_2O_5 (Cheng et al. 2009b) increases the surface segregation of Ir and reduces the oxide particle size. Furthermore, most studies also show that the addition of inert oxides has a slight positive effect on the stability of the active IrO_2 and RuO_2 materials against corrosion. As a summary, the dilution of IrO_2 and/or RuO_2 with inert oxides contributes to an increase in the overall noble metal mass activity by essentially acting as a support or core material and increasing the dispersion of the active material. The inert metal oxides also contribute to the stability of the noble metal particles against corrosion. However, to date, the reasons for these phenomena are not completely clear (Carmo et al. 2013).

3.4.3 Effect of Size and the Use of Support Materials

In the two previous sections, we have shown that PGM oxides, and especially Ir- and Ru-based oxides, are the prevailing catalysts for the OER in acidic media. Nonnoble metal–based electrocatalysts are essentially unknown for this system because of the highly corrosive acidic environment and high anodic potentials. As the likelihood of replacement of the noble metals as catalysts for the OER is extremely low, we are left with the option of maximizing the utilization of these precious materials. As described in the previous section, significant efforts have already been applied toward increasing the utilization of Ir and Ru mixed metal oxides by the addition of inert, nonprecious oxides with increased catalytic surface areas and noble metal oxide surface segregation as results. However, in order to significantly reduce the noble metal content in PEM electrolyzers, the use of nanoscale catalyst structures such as supported metal oxide nanoparticles or extended thin film surfaces is necessary.

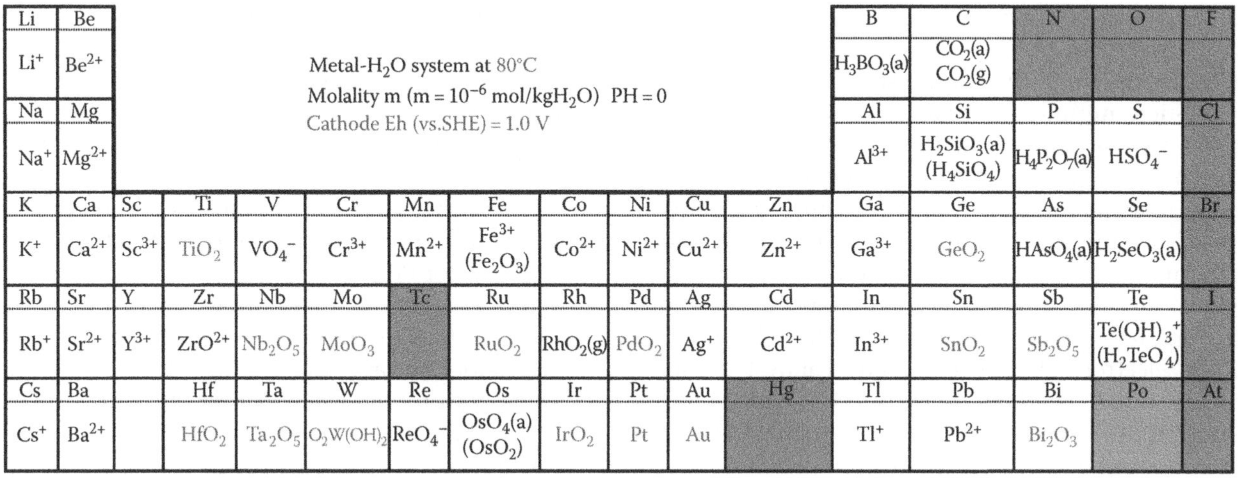

Li	Be										B	C	N	O	F
Li^+	Be^{2+}										$H_3BO_3(a)$	$CO_2(a)$ $CO_2(g)$			
Na	Mg										Al	Si	P	S	Cl
Na^+	Mg^{2+}										Al^{3+}	$H_2SiO_3(a)$ (H_4SiO_4)	$H_4P_2O_7(a)$	HSO_4^-	
K	Ca	Sc	Ti	V	Cr	Mn	Fe	Co	Ni	Cu	Zn	Ga	Ge	As	Se	Br
K^+	Ca^{2+}	Sc^{3+}	TiO_2	VO_4^-	Cr^{3+}	Mn^{2+}	Fe^{3+} (Fe_2O_3)	Co^{2+}	Ni^{2+}	Cu^{2+}	Zn^{2+}	Ga^{3+}	GeO_2	$HAsO_4(a)$	$H_2SeO_3(a)$	
Rb	Sr	Y	Zr	Nb	Mo	Tc	Ru	Rh	Pd	Ag	Cd	In	Sn	Sb	Te	I
Rb^+	Sr^{2+}	Y^{3+}	ZrO^{2+}	Nb_2O_5	MoO_3		RuO_2	$RhO_2(g)$	PdO_2	Ag^+	Cd^{2+}	In^{3+}	SnO_2	Sb_2O_5	$Te(OH)_3^+$ (H_2TeO_4)	
Cs	Ba		Hf	Ta	W	Re	Os	Ir	Pt	Au	Hg	Tl	Pb	Bi	Po	At
Cs^+	Ba^{2+}		HfO_2	Ta_2O_5	$O_2W(OH)_2$	ReO_4^-	$OsO_4(a)$ (OsO_2)	IrO_2	Pt	Au		Tl^+	Pb^{2+}	Bi_2O_3		

Metal-H_2O system at 80°C
Molality m (m = 10^{-6} mol/kgH$_2$O) PH = 0
Cathode Eh (vs.SHE) = 1.0 V

FIGURE 3.9
Periodic table showing thermochemically stable metal oxides in red, derived by theoretical calculations under the molarity = 10^{-6} mol kg^{-1}–H_2O, pH = 0, cathode Eh = 1.0 V vs. SHE. (Reprinted from Sasaki, K. et al., *ECS Trans.*, 33, 473, 2010. With permission from the Electrochemical Society.)

In this section, we review the recent research on the supported noble metal (oxide) nanoparticles and studies performed on the effect of the catalyst particle size on catalyst activity and stability. Compared to the research performed on catalysts for PEM fuel cells, supporting and nanostructuring of OER electrocatalysts are still in the early research stages. Evidently, nanoscale catalysts are of great advantage over larger catalyst particles due to a much higher surface area-to-volume ratio (Campelo et al. 2009), but other effects of nanostructuring such as catalyst-support interactions and particle size effects can have great impact on both activity and stability of these materials.

3.4.3.1 Size Effects in OER

In contrast to the research performed on catalyst size effects for the oxygen reduction reaction (ORR), which have been extensively studied for several decades (Kinoshita 1990; Mayrhofer et al. 2005), systematic studies focusing on the size effects of electrocatalysts for the OER are almost nonexisting, although some studies have been performed. Jirkovský et al. studied the effect of particle size and shape of nanocrystalline RuO_2 (Jirkovský et al. 2006) and $Ru_{0.8}Co_{0.2}O_{2-x}$ (Jirkovský et al. 2006a,b) prepared by a sol–gel method on the electrochemical activity and selectivity toward the OER and chlorine evolution reaction (CER) by combining diffraction, microscopic, and spectroscopic techniques. The authors found that the OER activity of pure nanocrystalline RuO_2 decreases with increasing particle size, while the CER is essentially unaffected by crystallite size and conclude that the OER is mainly occurring on crystal edges while the CER mainly proceeds on crystal faces. For the $Ru_{0.8}Co_{0.2}O_{2-x}$ nanoparticles however, the OER activity increases with increasing crystallite size, the complete opposite of pure RuO_2 and is by the authors explained by a specific effect of cobalt, which substitutes the Ru in the surface structure and changes the rate-determining step from a charge-transfer reaction to a slow surface recombination of oxygen.

Recently, Reier et al. (2012) performed a detailed comparative study of the activity and durability of nanoparticles ($d \sim$ 2–5 nm) and bulk materials of the PGM metals Ru, Ir, and Pt for the electrocatalytic OER. The electrochemical surface characteristics of the nanoparticles and bulk materials were investigated by cyclic voltammetry and the electrocatalytic activity and catalyst durability were investigated by linear sweep voltammetry. Significant differences in the surface oxidation chemistry were observed between the nanoparticles and bulk materials; Ru nanoparticles show lower passivation potentials, Ir nanoparticles show irreversible surface oxidation and completely lost their voltammetric metallic features during voltage cycling and Pt nanoparticles show an increased oxophilic nature compared to bulk Pt. For the OER activity, significant effects of nanoscaling were observed for Pt and Ru; the Ru nanoparticles suffered from severe corrosion at OER potentials as low as 1.4 V vs RHE and were completely unable to sustain the OER while Pt nanoparticles exhibited significantly reduced initial catalytic activity compared to bulk Pt with a complete deactivation due to surface oxidation during the OER experiment. In contrast to the two other metals, Ir nanoparticles showed comparable OER activity and durability to bulk Ir, strongly suggesting that supported Ir nanoparticles in the range of 2–5 nm are a suitable catalyst concept for anodes in PEM electrolyzers.

3.4.3.2 Supported Electrocatalysts for the OER

The simplest way to increase the mass activity of a catalyst is to divide the active material into smaller particles thus increasing the surface area per unit mass and the surface area-to-volume ratio. This approach is not always suitable; however, smaller particles may not be sufficiently stable, may lack the same specific activity as larger particles and, especially for gas evolution electrodes, may have too fine a structure to accommodate the necessary transport of liquid reactants to the catalytic sites and removal of the gaseous products. An alternative is to use a second material as a support, whereby the support determines the layer structure and usually allows electronic continuity with the supported electrocatalyst.

Conductive carbon, such as carbon blacks, nanofibers, or nanotubes, which is commonly used as catalyst supports in PEM fuel cells, will rapidly undergo electrochemical oxidation at the high anodic potentials (> 1.5 V vs SHE) in the PEM electrolyzer (Linse et al. 2011), thus long-term use of a carbon material as a catalyst support is not possible and in order to utilize nanostructured catalyst ($d <$ 10 nm) in an efficient manner, other more oxidation-tolerant support materials must be found. Any development of new candidate support materials for the anode of a PEM electrolyzer should be guided by the following requirements (a modified version of the requirement list presented by Rabis et al. (2012) for catalyst supports for PEM fuel cells:

1. The electronic conductivity of the support material should be above 0.01 S cm^{-1} at the operation temperature of the electrolyzer (40°C–80°C). This value is given by the assumption that the addition of a nanostructured electrocatalyst to the surface of the support will increase the overall conductivity of the electrocatalyst/support structure to at least 0.1 S cm^{-1}.

2. The support material must be stable under highly oxidative environments (potentials of ~2 V vs SHE and oxygen partial pressures of 10–50 bar) for periods of more than 10,000 h without significant reduction of conductivity or dissolution of metal cations.

3. The support should provide a BET surface area comparable to graphitized carbon blacks. Since most candidate materials will have a significantly higher density than carbon, it is important to normalize the surface area with the density of the material as it is the surface area per unit of volume, which is the important value to consider. As a reference number, we can take a carbon material of 100 cm^2 g^{-1} with a density of 2.2 g cm^{-3}, which gives a volume-specific surface area of 2200 cm^{-1}.

4. The support material should be able to form and maintain porous electrode structures with sufficient porosity and pore sizes to allow access of water to the reaction sites and transport of oxygen out of the electrode. Studies on the optimum pore size distribution in a PEM electrolyzer anode do not exist to our knowledge; however, some guidelines can be transferred from studies of gas-evolving electrodes in alkaline electrolysis (Heidrich et al. 1990; Rausch and Wendt 1992; Wendt 1994) (i) the bubble formation pressure in pores below 15 nm is extremely high, rising to values in the range of 80 MPa in a pore with a diameter of 4 nm; (ii) the effective utilization of any active catalytic sites in such small pores are strongly dependent on the exchange current density of the gas evolution reaction and will decrease with increasing exchange current density; and (iii) the effective utilization of small pores increases with increasing overvoltage. As the OER is highly irreversible with significant overvoltages in a PEM electrolyzer, it seems that an optimum pore size distribution should include significant amounts of pores in the range of 5–20 nm.

5. The support material should have a reasonable cost, but this can be significantly higher than the cost of support materials for PEM fuel cells as the capital cost target of a PEM electrolyzer is significantly higher than that of PEM fuel cells.

During the past decade, there has been an increased activity at several research groups in the search for and development of new catalyst support materials as an alternative to carbon in PEM fuel cells. A main driver for this research has been to increase the corrosion resistance of the support and thus increase the durability of PEM fuel cells. However, also other properties such as the multivalence of transition metal oxides, which potentially can enhance the catalytic activity of the catalyst particles, are of interest. The two main classes of materials that have been investigated are inorganic metal oxides and metal carbides and nitrides. Several excellent and relatively recent reviews of the status of the materials development are available (Antolini and Gonzalez 2009; Shao et al. 2009; Rabis et al. 2012), and several of the candidate materials presented in these reviews are also good candidates for anode catalyst supports in PEM electrolyzers.

In a PEM electrolyzer under operation, the environment is significantly more oxidative than in PEM fuel cells and any carbides or nitrides used as supports will, due to their thermodynamic instabilities in this environment, eventually be transformed to their respective metal oxide form which is the only thermodynamically stable state (Kimmel et al. 2014). Although several studies have shown good catalytic performance of TaC and TiC-supported Ir and IrO (Ma et al. 2008, 2009; Sui et al. 2009, 2011; Nikiforov et al. 2011; Polonsky et al. 2012, 2014; Fuentes et al. 2014), taking into account the argument as given earlier as, transition metal oxides seem to be a more suitable class of materials for electrocatalysts for long-term operation in PEM electrolyzers. Consequently, in this chapter, we present the studies performed on metal oxide-supported electrocatalysts.

Sasaki et al. (2010) evaluated the thermodynamic stability of a series of metal oxides under relevant conditions for PEM fuel cells. From these calculations (see Figure 3.9), the authors conclude that Sn, Ti, Nb, Ta, W, and Sb are stable as oxides, hydroxides, or metals. Most of these materials should also be stable under PEM electrolyzer conditions.

Most of the work in the recent years on oxide-supported electrocatalysts for the OER evolution has been carried out on either titanium oxides (Chen et al. 2002; Siracusano et al. 2009; Fuentes et al. 2011; Hu et al. 2014) or tin oxides (Marshall and Haverkamp 2010; Thomassen et al. 2011; Wu and Scott 2011; Xu et al. 2012; Avila-Vazquez et al. 2013; Puthiyapura et al. 2014). The studies performed on titanium oxides comprise supports based on pure anatase TiO_2 (Fuentes et al. 2011), substoichiometric titanium oxides, such as Ebonex (Chen et al. 2002; Siracusano et al. 2009), and Nb-doped TiO_2 (Chen et al. 2002; Hu et al. 2014) and all report an increased mass activity for the noble metal catalysts, which is explained by a better distribution of the catalyst particles and thus a higher utilization. Siracusano et al. (2009) also report a limited stability of the titanium suboxides under the conditions of oxygen evolution due to a gradual oxidation of the material and loss of electronic conductivity. The Nb-doped TiO_2, however, was resistant to electrochemical and thermal oxidation and the supported electrocatalyst retained its activity under polarization. Out of the six published studies on SnO_2 base supports, only four can be classified as real supported electrocatalysts, the catalysis in the two studies by Wu et al. (2011) and Avila-Vazquez et al. (2013) are more of a composite mixture of electrocatalyst and support particles of essentially equal size. Of the remaining four published studies, three groups have evaluated a commercially available Sb-doped SnO_2 nanopowder as a support. Marshall and Haverkamp (2010) prepared a series of mixed $Ir_xRu_{1-x}O_2$ clusters on the support material by thermal decomposition of chloride precursors, in which no specific size of the clusters was reported, but the authors state that the clusters should be amorphous with a thickness of a few nm. The electrocatalytic activity toward the OER was evaluated in 0.5 mol L^{-1} H_2SO_4 and compared to earlier works on IrO_2–RuO_2-based DSA layers and revealed a significant increase in the mass activity of the supported electrocatalysts. Wu et al. (2011) deposited nanoparticles of RuO_2 with a size of 10–15 nm on Sb-doped SnO_2 by a colloid method. The 20 wt% RuO_2/ATO (antimony-doped tin oxide) exhibited higher voltammetric charge than unsupported RuO_2, which indicates that the supported electrocatalysts increase the dispersion of the RuO_2 particles. The catalyst activity was evaluated in a PEM electrolyzer single cell with a noble metal loading of 2 mg RuO_2 on the anode. At 80°C, using a Nafion 212 membrane, the cell showed a voltage of 1.56 V at 1 A cm^{-2}. No evaluation of the long-term stability of the electrocatalyst was reported. In our group, we have developed a method of depositing Ir and Ir–Ru nanoparticles with a mean particle size of about 2 nm on Sb-doped SnO_2 by a modified polyol method giving a total noble metal loading of 20–30 wt%.

The deposited nanoparticles are metallic in nature, but are irreversibly oxidized within a short period of time at voltages above 1 V due to their small size, and exhibits the same voltammetric features (Figure 3.10) as thermally formed $Ir_xRu_{1-x}O_2$ particles, in agreement with the results from Reier et al. (2012).

The catalyst performance was evaluated in a PEM electrolyzer single cell using 4 mg cm^{-2} 20 wt% Ir/ATO (0.8 mg cm^{-2} Ir) on the anode, 0.5 mg cm^{-2} Pt/C on the cathode, and a Nafion 115 membrane. This cell achieved 1.65 V at 1 A cm^{-2} by exhibiting a mass activity of more than 2 times higher than previously reported results (Marshall et al. 2007) using unsupported electrocatalysts.

Although the number of studies of supported electrocatalysts for anodes in PEM electrolyzers is still low, the results published in the last 4–5 years are very

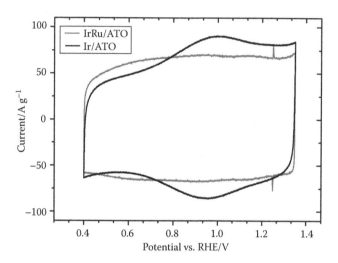

FIGURE 3.10
Cyclic voltammograms of 20 wt% Ir/ATO and $Ir_{0.5}Ru_{0.5}$/ATO in 0.5 mol dm^{-3} H_2SO_4 at 25°C, 300 mV s^{-1}.

promising and with the increasing interest in development of oxide supports for PEM fuel cells, there is a significant potential in reducing the use of noble metals in PEM electrolyzers to well below 1 mg cm^{-2} in the next few years.

3.5 Novel Catalyst Concepts

3.5.1 Core-Shell Catalysts

Particle architectures in which the core of the particle has been replaced by a different metal so that the metal in direct contact with the reactants of the reaction to be catalyzed constitute a thin shell over this core, so-called core-shell catalysts, have been under development for some years, and relatively simple synthesis methods exist for Ru–Pt catalystic particles of a few nm across (Alayoglu et al. 2008, 2009). Such catalysts have much changed properties from their monometallic counterparts, both for heterogeneous catalysis (Nilekar et al. 2010) and for electrocatalysis (Bernechea et al. 2011; Ochal et al. 2011).

The different behavior of core-shell catalysts has a simple explanation in the d-band theory (Ruban et al. 1997). If, for example, a metal atom of one type is placed on top of another and smaller atom and pseudomorphic growth is assumed, the compression of the top layer of atoms will lead to a broadening of the bands, a lower DOS by conservation of the number of states. Therefore, the topmost energy level in the surface atoms will experience a rise (for a more than half-filled band). This would cause a spillover of electrons into the core, and as a reaction to this a pile up of positive charge in the surface layer. To preserve electroneutrality (i.e., assuming a significant Hubbard U), the d-band level is thus shifted downward, and adsorbates would be expected to bond more weakly at a core-shell surface than at the corresponding shell-monometallic catalyst. Theoretical studies on overlayer systems, based on DFT, provide data of the changes in energy of d-band for a number of transition elements (Hammer and Nørskov 2000; Liao et al. 2000; Christoffersen et al. 2001; Ge et al. 2001; Koper et al. 2002; Davies et al. 2005; Hammer 2006; Tsuda and Kasai 2006). For example, according to Nørskov et al. for a Pt overlayer on a Ru substrate, the shift in the d-band center is estimated to the significant value of −0.61 eV excluding overlayer relaxation (Ruban et al. 1997) compared to that of a Pt bulk surface. Core-induced downward shifts in the d-band energy of Pt in Ru–Pt core-shell particles have been observed directly with synchrotron-based photoelectron spectroscopy (Goto et al. 2014), and catalytic activity has also been correlated with core-induced changes in bond length (Tsypkin et al. 2013).

Some works claiming core-shell structures have been performed also for oxygen evolution catalysts, synthesized by a polyol synthesis (Ir–Sn–Ru oxides) (Marshall et al. 2006a), sequential reduction of H_2IrCl_6 with $NaBH_4$ and reduction of Pt precursors for the shell by ascorbic acid (Ir–Pt) (Zhang et al. 2012), and dealloying (IrNi–IrO$_x$) (Nong et al. 2014). Recently, Akhade and Kitchin (2012) found an effect of strain on the second moment of the DoS (an expression of the width of the band) of 3d perovskites, similar to that observed in metals. They concluded that although d-band filling and oxidation state have the largest effect on the reactivity of the OER at 3d perovskite surfaces, strain can be used to modulate the reactivity in a fashion similar to that in metal electrocatalysis. While more work is clearly needed in this area, the results of Akhade and Kitchin (2012) appear to indicate that the electronic effects of core-shell architectures through strain may be more modest in the case of oxides.

3.5.2 Nanostructured Thin Films

Thin and well-controlled films of iridium or ruthenium oxide are simplest made by the oxidation of the base metal. However, this would be of little technological interest due to the rather large amount of iridium that would be needed. Thin films deposited on substrates other than the base metal may be achieved by electrode-position (Petit and Plichon 1998; Steegstra and Ahlberg 2012) or sputtering (Kötz and Slucki 1986; Pauporté et al. 1999). However, electrodeposition appears to result in separate particles, and the morphology may not be well controlled due to the indirect nature of the process,

whereas sputtering seems to produce rather dense films (Pauporté et al. 1999).

Recently, Smith et al. (2014) produced amorphous iridium oxide films from iridium acetylacetonate (Ir(acac)$_3$) precursors by spin casting followed by the application of UV (ultraviolet) light and annealing. Information on film thickness is scarce in the reference, and some problems associated with coverage appear to be associated with the method. However, a high activity for the OER was claimed as compared to a number of other synthesis methods when data were normalized with respect to geometric area. Amorphous films (annealed at 100°C) were more active than crystalline films achieved by annealing at 500°C (Smith et al. 2014).

3.5.3 Nonnoble Catalysts

Huynh et al. (2014) recently investigated manganese oxide MnO$_x$ as a catalyst under acidic conditions, which is obviously relevant for PEM water electrolysis. Manganese oxide is not as active as, for example, iridium oxide (Trasatti 1980), but the large quantity of manganese oxide being produced annually for battery applications (Andersen 1996) and the low cost of the material might easily compensate for this disadvantage. However, manganese oxide is not stable under acidic conditions. Huynh et al. (2014) showed that manganese oxide is self-repairing due to the potential for deposition of manganese oxide is lower than the potential at which the OER occurs. They predicted the self-healing process of manganese oxide to disappear only at pH values lower than 0.

3.6 Preparation of Catalysts for the Oxygen Evolution Reaction

3.6.1 Wet Chemical Methods

Wet chemical methods usually refer to preparation methods taking place in a liquid phase. Wet chemical methods have been widely used to prepare nanostructured materials for several applications and have a number of advantages. The precursors are mixed at a molecular level in a short period of time, and it is relatively easy to achieve uniform composition of the resulting product and reaction temperatures are comparably low.

3.6.1.1 Sol–Gel

The term "sol–gel" is used to describe the preparation of ceramic materials by a process that involves the preparation of a sol, the gelation of the sol, and removal of the liquid The sol is a stable dispersion of colloidal particles or molecular precursors in a solvent while the gel, gradually evolved from the sol, refers to a diphasic system containing a solid three-dimensional network that encloses a liquid phase. Precursors for preparation of the sol are usually inorganic salts such as metal alkoxides, which are subjected to hydrolysis and polymerization reactions. Controlling the pH of the solution is the typical way of selecting either the polymerization or hydrolysis processes, and thus synthesizing either small compact oxides or large inorganic polymer matrices. Upon drying the sols or gels, amorphous metal oxides form, which then can be further heat treated to crystalline oxides (Brinker and Scherer 1990; Wright and Sommerdijk 2003).

Noble metal oxides have been synthesized both as powders and as DSA-type layers by sol–gel techniques. Catalyst materials consisting of mixtures of RuO_2, IrO_2, SnO_2, TiO_2, and Ta_2O_5 have been produced by this method by dissolving the noble metal salt in anhydrous ethanol. Sodium ethylate was added, and the mixture was refluxed for 3–4 h at 70°C–80°C in a nitrogen atmosphere, forming the metal alkoxide. To this mixture, the second component was added as the metal tetraethoxide, and the mixture was hydrolyzed by the addition of an ammonia–ethanol solution and then H_2O_2. The product was separated from the liquid, dried, and calcined at high temperature to form the crystalline oxide particles (Kameyama et al. 1993; Ito et al. 1994; Murakami et al. 1994a–d; Takasu et al. 1994).

3.6.1.2 Polyol Method

The polyol method has been widely used to prepare metallic particles by reducing the precursor salts in polyols (Kurihara et al. 1995; Bonet et al. 1999; Marshall et al. 2006b). A polyol is an alcohol with two or more –OH groups, and in the polyol synthesis method this acts as both solvent and reducing agent to a metal precursor. In many cases, addition of a surfactant such as polyvinylpyrrolidone (PVP) is necessary to reduce the agglomeration of the resulting metallic particles. A commonly used polyol is ethylene glycol. The oxidation of ethylene glycol during the synthesis has been shown to mainly result in glycolic acid or, depending on the synthesis solution pH, the glycolate anion. The glycolate anion acts as a stabilizer for the noble metal colloids and the resulting particle size has been shown to be controlled via the synthesis solution pH for Pt and Ru (Bock et al. 2004). In order to produce metal oxide powders by this method, it is necessary to thermally oxidize the metallic particles after synthesis.

Both supported and unsupported metal catalyst particles can be produced by this method by selective introduction of a suitable support material to the noble metal colloid solution. It is of high importance that the noble metal particles are well dispersed on the support surface and that the degree of agglomeration is kept at a minimum. Good deposition of metal particles is obtained by tailoring the environment so that the two components (metal and support) have opposite surface charges and thus are attracted by each other. This can be done by adjusting the pH of the mixture (Brunelle 1979).

Although the polyol method has been used very frequently for the synthesis of supported electrocatalysts for PEM fuel cell, the use of this method for anode catalysts for PEM electrolyzers is much less common. Marshall et al. (2006b) used the method to investigate the electrochemical properties of nanocrystalline oxide powders of the type $Ir_xSn_{1-x}O_2$ by reducing precursors in ethylene glycol followed by thermal oxidation. The resulting crystallite size of the oxides was in the range of 3–15 nm. The method has also been used to produce supported electrocatalysts for the OER; Fuentes et al. (2011) produced nanoparticles of Pt, Ru, and Ir supported on anatase and rutile TiO_2 and in our own laboratories, we have synthesized a range of Ir and Ru nanoparticles supported on Sb-doped SnO_2 (Thomassen et al. 2011).

3.6.1.3 Aqueous Hydrolysis

This method is based on the hydrolysis of a noble metal salt in an aqueous solution. The resulting (hydr)oxide can be further oxidized by calcination or chemical oxidation. IrO_2 has been prepared from an iridium hydroxide hydrate precursor by the addition of $HClO_4$ to the solution and subsequent calcination of the precipitate at 400°C. The IrO_2 particle size was in the range of 30–50 nm (Ioroi et al. 2000). A similar method was used to produce IrO_2 at large scale. This involved dissolving Na_2IrCl_6 in a NaOH solution to produce a $Ir(OH)_4$ solution. After purification and the removal of NaCl, the solution was dried and pyrolysed at 200°C (Yamaguchi et al. 2000).

A series of $Ir_xRu_yTa_zO_2$, electrocatalyst materials were produced by Marshall et al. (2007) using the aqueous hydrolysis method followed by thermal oxidation at 400°C. By the addition of Ta, the oxide particle size increases and the redox peak of iridium oxide shifts slightly to lower potentials. At Ta contents above 20%, the conductivity of the oxide particles decreased significantly.

3.6.2 Thermal Decomposition

3.6.2.1 Adams Fusion

OER catalysts can be prepared by a modified Adams fusion method based on the oxidation of metal precursors by molten $NaNO_3$ (Adams and Shriner 1923). During the process, nitrogen dioxide is evolved and the noble metal oxide is formed. Noble metal oxide powder such as PtO_2 (Bruce 1936), IrO_2, IrO_2–Ta_2O_5, and RuO_2–IrO_2 (Rasten et al. 2003; Cheng et al. 2009a; Tunold et al. 2010), $Ir_xSn_{1-x}O_2$ (Marshall et al. 2005), and $Ir_{0.4}Ru_{0.6}Mo_xO_y$ (Cheng et al. 2009b) have been prepared by this method. More recently, a TaC-supported IrO_2 electrocatalyst was also produced using this method by the addition of TaC to the salt mixture before heat treatment (Polonsky et al. 2012).

3.6.2.2 Pechini Method

The Pechini method, similar to the sol–gel method, is based on thermal decomposition of polymeric precursors dissolved in a mixture of ethylene glycol and citric acid. It involves the polymerization of organic monomers in the presence of metal ions that homogeneously distributes the metal ions in a polymer gel. The gel is then calcined to form the metal oxide (Pechini 1967). The method has mainly been used not only to prepare Ti electrodes coated with RuO_2 (Terezo and Pereira 2002) and RuO_2–Ta_2O_5 (Terezo and Pereira 1999) as DSA electrodes, but also powder samples have been prepared (Ribeiro et al. 2008).

3.7 Degradation of Catalysts for the Oxygen Evolution Reaction

3.7.1 Mechanisms, Thermodynamics, and Kinetics of Catalyst Degradation

For any metal to be suitable as an OER electrocatalyst in a PEM electrolyzer, it must not only have high electrocatalytic activity and selectivity, but it must also be able to withstand the extremely harsh environment within the anode of the PEM electrolyzer for several tens of thousands of hours. The presence of strong oxidants, high temperature, low pH, and high anodic potentials on the oxygen electrode in a PEM electrolyzer leaves very few metals that are sufficiently noble to avoid dissolution and out of these, several are either catalytically inactive (Au), or forms insulating oxide films that inhibit the OER (Pt). Fortunately, both Ir and Ru form conductive oxides, which in turn also are surfaces that have high catalytic activity toward the evolution of oxygen, but the materials must also show long-term stability.

Pourbaix diagrams show the thermodynamic stability of metals in the voltage and pH range of the stability of water. When conditions move into areas of the diagram that represent a change of the most stable metallic form to an oxide or different oxidation states, then corrosion or passivation, the formation of a protective oxide layer on the metal surface, can occur.

Pourbaix et al. published the diagrams for Ir and Ru in 1959 with a consideration of their corrosion resistance based on these thermodynamic data (Figure 3.11). It can be seen that Ir is an extremely noble metal since its region of immunity extends over the greater part of the domain of stability of water, which is also reflected in that IrO_2 is the preferred catalyst for OER in PEM electrolyzers. The Pourbaix diagram for Ru is more complicated than for the other platinum metals due to the greater number of valence states in which it is able to exist. Although Ru is also classified as a noble metal, it is significantly less noble than the other four platinum metals and will be susceptible for corrosion at the elevated potentials in a PEM electrolyzer by the formation of the volatile RuO_4 species.

The electrocatalytic activity for the OER and durability of IrO_2, RuO_2, and their mixtures have been extensively studied during the last four decades and a review of the main results is given in Section 3.5. Regarding the durability of these catalysts in real systems, the high stability of IrO_2 and also the instability of RuO_2 predicted by the thermodynamic data have been confirmed in several studies (Kötz et al. 1984; Kötz and Stucki 1986; Angelinetta et al. 1989; Cheng et al. 2009a; Mamaca et al. 2012; Owe et al. 2012), which also have shown a significant increase in the stability of RuO_2 with only small (20%) addition of IrO_2. The reason for this stabilization effect has been proposed by Kötz and Stucki to be caused by the formation of a common band in the mixed oxides with the suppression of the formation of RuO_4 as a result (Kötz and Stucki 1986); however, more recent data from Owe et al. might suggest that any synergistic interaction between Ir and Ru in mixed oxides is weaker than previously assumed (Owe et al. 2012). Even though there has been a significant effort in studying and developing electrocatalysts for the OER, investigations of the long-term stability of these catalysts do not exist. Initial screening of a series of metallic Ru and Ir as well as their oxides was performed by Song et al (2008), with the stability increasing in the following series: $Ru < RuO_2 < Ir < IrO_2$. There also exist several studies on the stability of DSA-type catalysts and catalyst coatings used in the chloralkali industry, which have some relevance for OER catalysts in PEM electrolyzers. Krysa et al. (1996) evaluated the effect of coating thickness on the surface morphology and electrochemical properties of IrO_2–TaO_5 anodes. It was observed that for an iridium content of 0.45–1.2 mg cm^{-2} the service life is proportional to the iridium content in the coating, indicating a slow dissolution of the IrO_2. The same degradation mechanism was also seen by Hu et al. (2002).

There is clearly a need for further investigation into the mechanism of the interaction between iridium and ruthenium oxides and also any effects different preparation methods and the size of the catalyst particles

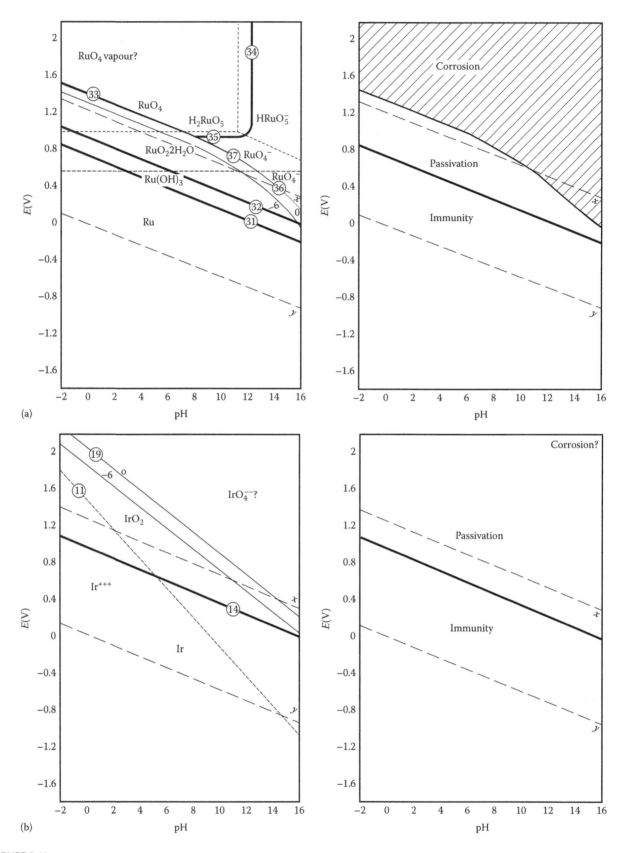

FIGURE 3.11
Pourbaix diagrams of Ir (a) and Ru (b). (Reprinted from Pourbaix, M.J.N. et al., *Platinum Met. Rev.*, 3, 100–106, 1959. With permission from Johnson Matthey.)

might have on the long-term durability in PEM electrolyzer environments.

Long-term stability studies of PEM electrolyzers are also almost nonexistant. Millet et al. (2010) constructed a GenHy1000 stack and measured its durability over a period of 800h and concluded that the main causes of performance degradation were due to the accumulation of metal ions in the membrane and a reduction in the conductivity as well as underpotential deposition of the metallic cations (iron, chromium, and nickel) on the Pt cathode catalyst. In addition, an irreversible degradation of the MEA was observed due to coalescence and loss of Pt nanoparticles in the cathode catalyst layer. No degradation of the anode catalyst was observed. Sun et al. (2014) tested a 9-cell electrolyzer stack for a period of 7800 h and observed an average degradation rate of 35.5 μV h^{-1}. The authors found that the degradation of the stack was mainly caused by a recoverable contamination of metal cations in the ion exchange sites in the polymer electrolyte in the membrane and catalyst layers, no degradation of the cathode or anode catalysts was observed. These electrolyzers, and also most other PEM electrolyzers operated today, are however using very high catalyst loadings on the anode (>1.5 mg cm^{-2} Ir) and are operated in galvanostatic or potentiostatic mode. Under future realistic operating conditions, that is, when the electrolyzer is supplied by an intermittent power source, it will experience extremely fluctuating loads and frequent startup/shutdowns and together with potentially lower loadings of anode catalysts, degradation of these catalysts are much more likely to occur. As an example, Kokoh et al. (2014) recently evaluated the durability of a PEM electrolyzer running on a 250 h solar cycle with highly fluctuating loads and frequent shutdowns. After the 250 h cycle, an increase in the cell voltage of 70 mV at 1 A cm^{-2} was recorded giving an average degradation rate of 280 μV h^{-1}.

It is, therefore, of high importance to investigate the degradation of electrocatalysts in situ in PEM electrolyzers as well as evaluating the conditions the catalysts are exposed to during fluctuating loads and intermittent operation and develop relevant accelerated degradation protocols based on this knowledge as has been done for PEM fuel cells (Yuan et al. 2011).

3.7.2 Development of Accelerated Degradation Protocols

In order to develop a relevant accelerated degradation protocol, it is important to investigate the conditions that the anode catalyst in a PEM electrolyzer is exposed to. A good starting point is the electrode potential. During normal operation, the potential of the anode of a PEM electrolyzer will be varying between 1.4 and 1.6 V vs SHE, which can either be calculated from a complete

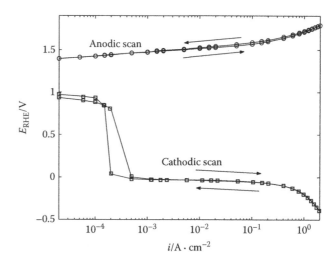

FIGURE 3.12
Forward and backward polarization of a PEM water electrolyzer MEA. (Reprinted from Rasten, E., Electrocatalysis in water electrolysis with solid polymer electrolyte, Fakultet for naturvitenskap og teknologi, 2001. With permission.)

PEM electrolyzer polarization curve or directly measured using a reference electrode, as demonstrated by Sun et al. (2014). What potential values will the anode of a PEM electrolyzer be exposed to during a shutdown from normal operation? This will of course significantly depend on the shutdown procedure used by the operator, but some measurements are available in the open literature, which can shed some light on this issue. Rasten (2001) performed a slow polarization curve of a PEM electrolyzer with an IrO$_2$ anode and Pt/C cathode from low current density 10^{-5} A cm^{-2} to 1 A cm^{-2} and back, measuring the anode and cathode simultaneously vs the same reference electrode. The cell was operated at 0.2 Acm^{-2} for 30 min before the scan was taken.

As can be seen from Figure 3.12, the potential variations in the cell at very low current densities are caused by the Pt cathode, which is caused by the presence of oxygen diffusing from the anode to the cathode. Since the IrO$_2$ electrocatalyst is not active toward oxygen reduction of hydrogen oxidation, the potential of the anode does not deviate from the potential window under normal operating conditions (1.4–1.6 V).

3.8 Summary

Iridium oxide appears to be the dominating electrocatalyst for PEM water electrolysis. Recent advances in the theory of electrocatalysis point to the binding energy of atomic oxygen to the surface as an appropriate descriptor of electrocatalytic activity. Experiment and theory indicate that iridium oxide does not quite meet the optimum criterion in terms of this descriptor. However, due to the universality of scaling relations, the OER is associated with a nonzero thermodynamic overpotential below which no single-phase oxide will be able to evolve oxygen. Some initial attempts of breaking the scaling relations by doping are promising and may result in the development of catalysts with lower thermodynamic overpotentials than those oxides for which the scaling relations apply. For iridium–ruthenium mixed oxides, however, previously reported synergistic effects are being challenged, and in some cases, the performance of the mixed oxide is similar to that of physical mixtures of the iridium and ruthenium oxides. Explanations for these inconsistencies may be sought in differences in surface segregation resulting from the difference in synthesis methods used to manufacture the oxides and also in limitations of the experimental assessment of segregation in themselves.

Within the realm of iridium- or ruthenium-based catalysts, a number of different synthesis techniques demonstrate different utilizations of the noble metals. Increasing the mass activity may be attempted through improved synthesis methods, application of catalyst supports, introduction of ternary dopants, and novel architectures such as core-shell catalysts.

References

Adams R and Shriner RL. 1923. Platinum oxide as a catalyst in the reduction of organic compounds. III. Preparation and properties of the oxide of platinum obtained by the fusion of chloroplatinic acid with sodium nitrate. *Journal of the American Chemical Society*. 45:2171–2179.

Ahluwalia RK, Arisetty S et al. 2014. Dynamics of particle growth and electrochemical surface area loss due to platinum dissolution. *Journal of the Electrochemical Society*. 161:F291–F304.

Akhade SA and Kitchin JR. 2012. Effects of strain, d-band filling, and oxidation state on the surface electronic structure and reactivity of 3d perovskite surfaces. *Journal of Chemical Physics*. 137:084703-084701–084703-084709.

Alayoglu S, Nilekar AU et al. 2008. Ru–Pt core–shell nanoparticles for preferential oxidation of carbon monoxide in hydrogen. *Nature Materials*. 7:333–338.

Alayoglu S, Zavalij P et al. 2009. Structural and Architectural Evaluation of Bimetallic Nanoparticles: A Case Study of PtRu CoreShell and Alloy Nanoparticles. *ACS Nano*. 3 10:3127–3137.

Andersen TN. 1996. The manganese dioxide electrode in aqueous solution. In: *Modern Aspects of Electrochemistry*. Springer, New York, pp. 313–413.

Angelinetta C, Trasatti S et al. 1986. Surface properties of RuO_2 + IrO_2 mixed oxide electrodes. *Journal of Electroanalytical Chemistry and Interfacial Electrochemistry*. 214:535–546.

Angelinetta C, Trasatti S et al. 1989. Effect of preparation on the surface and electrocatalytic properties of RuO_2 + IrO_2 mixed oxide electrodes. *Materials Chemistry and Physics*. 22:231–247.

Antolini E and Gonzalez ER. 2009. Ceramic materials as supports for low-temperature fuel cell catalysts. *Solid State Ionics*. 180:746–763.

Antonucci V, Blasi AD et al. 2008. High temperature operation of a composite membrane-based solid polymer electrolyte water electrolyser. *Electrochimica Acta*. 53:7350–7356.

Ardizzone S, Bianchi CL et al. 2006. Composite ternary SnO_2–IrO_2–Ta_2O_5 oxide electrocatalysts. *Journal of Electroanalytical Chemistry*. 589:160–166.

Ardizzone S, Fregonara G et al. 1990. Inner and outer active surface of RuO_2 electrodes. *Electrochimica Acta*. 35:263–267.

Ardizzone S and Trasatti S. 1996. Interfacial properties of oxides with technological impact in electrochemistry. *Advances in Colloid and Interface Science*. 64:173–251.

Avila-Vazquez V, Cruz JC et al. 2013. Electrochemical study of Sb-doped SnO_2 supports on the oxygen evolution reaction: Effect of synthesis annealing time. *International Journal of Electrochemical Science*. 8:10586–10600.

Beer HB, inventor. 1969. Improvements in or relating to electrodes for electrolysis. GB Patent 1147442(A).

Beni G and Shay JL 1978. Electrochromism in anodic iridium oxide films. PH effects on corrosion stability and the mechanism of coloration and bleaching. *Applied Physics Letters*. 33:208–210.

Bernechea M, Garcia-Rodriguez S et al. 2011. Synthesis of core-shell PtRu dendrimer-encapsulated nanoparticles. Relevance as electrocatalysts for CO oxidation. *Journal of Physical Chemistry*. 115:1287–1294.

Birss VI, Bock C et al. 1997. Hydrous Ir oxide films: The mechanism of the anodic prepeak reaction. *Canadian Journal of Chemistry*. 75:1687–1693.

Bock C and Birss VI. 1999. Irreversible decrease of Ir oxide film redox kinetics. *Journal of the Electrochemical Society*. 146:1766–1772.

Bock C, Paquet C et al. 2004. Size-selected synthesis of PtRu nano-catalysts: Reaction and size control mechanism. *Journal of the American Chemical Society*. 126:8028–8037.

Bockris J and Otagawa T. 1984. The electrocatalysis of oxygen evolution on perovskites. *Journal of the Electrochemical Society*. 131:290–302.

Bockris JO. 1956. Kinetics of activation controlled consecutive electrochemical reactions: Anodic evolution of oxygen. *Journal of Chemical Physics*. 24:817–827.

Bockris JOM and Otagawa T. 1983. Mechanism of oxygen evolution on perovskites. *Journal of Physical Chemistry*. 87:2960–2971.

Bockris JOM, Otagawa T et al. 1983. Solid state surface studies of the electrocatalysis of oxygen evolution on perovskites. *Journal of Electroanalytical Chemistry*. 150:633–643.

Bonet F, Delmas V et al. 1999. Synthesis of monodisperse Au, Pt, Pd, Ru and Ir nanoparticles in ethylene glycol. *Nanostructured Materials*. 11:1277–1284.

Brinker CJ and Scherer GW. 1990. *Sol-Gel Science: The Physics and Chemistry of Sol-Gel Processing*. Boston, MA: Academic Press.

Bruce WF. 1936. The preparation of platinum oxide for catalytic hydrogenations. *Journal of the American Chemical Society.* 58:687–688.

Brunelle JP. 1979. Preparation of catalysts by adsorption of metal complexes on mineral oxides. In: *Studies in Surface Science and Catalysis.* Elsevier, Amsterdam, pp. 211–232.

Busch M, Ahlberg E et al. 2011a. Electrocatalytic oxygen evolution from water on a Mn(III–V) dimer model catalyst – A DFT perspective. *Physical Chemistry Chemical Physics.* 13:15069–15076.

Busch M, Ahlberg E et al. 2011b. Hydroxide oxidation and peroxide formation at embedded binuclear transition metal sites; TM = Cr, Mn, Fe, Co. *Physical Chemistry Chemical Physics.* 13:15062–15068.

Busch M, Ahlberg E et al. 2013a. Validation of binuclear descriptor for mixed transition metal oxide supported electrocatalytic water oxidation. *Catalysis Today.* 202:114–119.

Busch M, Ahlberg E et al. 2013b. Water oxidation on MnOx and IrOx: Why similar performance? *Journal of Physical Chemistry C.* 117:288–292.

Butler MA and Ginley DS. 1978. Prediction of flatband potentials at semiconductor-electrolyte interfaces from atomic electronegativities. *Journal of the Electrochemical Society.* 125:228–232.

Campelo JM, Luna D et al. 2009. Sustainable preparation of supported metal nanoparticles and their applications in catalysis. *ChemSusChem.* 2:18–45.

Carmo M, Fritz DL et al. 2013. A comprehensive review on PEM water electrolysis. *International Journal of Hydrogen Energy.* 38:4901–4934.

Chatenet M, Guetaz L et al. 2010. Electron microscopy to study membrane electrode assembly (MEA) materials and structure degradation. In: *Handbook of Fuel Cells.* John Wiley & Sons, Ltd., Hoboken, NJ.

Chen GY, Bare SR et al. 2002. Development of supported bifunctional electrocatalysts for unitized regenerative fuel cells. *Journal of the Electrochemical Society.* 149:A1092–A1099.

Cheng J, Zhang H et al. 2009a. Study of Ir$_x$Ru$_{1-x}$O$_2$ oxides as anodic electrocatalysts for solid polymer electrolyte water electrolysis. *Electrochimica Acta.* 54:6250–6256.

Cheng J, Zhang H et al. 2009b. Preparation of Ir$_{0.4}$Ru$_{0.6}$Mo$_x$O$_y$ for oxygen evolution by modified Adams' fusion method. *International Journal of Hydrogen Energy.* 34:6609–6624.

Christoffersen E, Liu P et al. 2001. Anode materials for low-temperature fuel cells: A density functional theory study. *Journal of Catalysis.* 199:123–131.

Conway BE, Birss V et al. 1997. The role and utilization of pseudocapacitance for energy storage by supercapacitors. *Journal of Power Sources.* 66:1–14.

Cyrot M and Cyrot-Lackmann F. 1976. Energy of formation of binary transition alloys. *Journal of Physics F: Metal Physics.* 6:2257–2265.

Daghetti A, Lodi G et al. 1983. Interfacial properties of oxides used as anodes in the electrochemical technology. *Materials Chemistry and Physics.* 8:1–90.

Damjanovic A, Dey A et al. 1966. Electrode kinetics of oxygen evolution and dissolution on Rh, Ir, and Pt-Rh alloy electrodes. *Journal of the Electrochemical Society.* 113:739–746.

Dau H, Limberg C et al. 2010. The mechanism of water oxidation: From electrocatalysis via homogeneous to biological catalysis. *ChemCatChem.* 2:724–761.

Davies JC, Bonde J et al. 2005. The ligand effect: CO desorption from Pt/Ru catalysts. *Fuel Cells.* 5:429–435.

De Levie R. 1999. The electrolysis of water. *Journal of Electroanalytical Chemistry.* 476:92–93.

De Pauli CP and Trasatti S. 1995. Electrochemical surface characterization of IrO$_2$ + SnO$_2$ mixed-oxide electrocatalysts. *Journal of Electroanalytical Chemistry.* 396:161–168.

Dubau L, Castanheira L et al. 2013. An identical-location transmission electron microscopy study on the degradation of Pt/C nanoparticles under oxidizing, reducing and neutral atmosphere. *Electrochimica Acta.* 110:273–281.

Feliu JM and Herrero E. 2003. Formic acid oxidation. In: *Handbook of Fuel Cells – Fundamentals, Technology and Applications.* John Wiley & Sons, Chichester, pp. 625–634.

Ferrer JE and Victori L. 1994a. Oxygen evolution reaction on the iridium electrode in basic medium studied by electrochemical impedance spectroscopy. *Electrochimica Acta.* 39:581–588.

Ferrer JE and Victori L. 1994b. Study of the oxygen evolution reaction on the iridium electrode in acid medium by EIS. *Electrochimica Acta.* 39:667–672.

Fuentes RE, Colon-Mercado HR et al. 2014. Pt-Ir/TiC electrocatalysts for PEM fuel cell/electrolyzer process. *Journal of the Electrochemical Society.* 161:F77–F82.

Fuentes RE, Farell J et al. 2011. Multimetallic electrocatalysts of Pt, Ru, and Ir supported on anatase and rutile TiO$_2$ for oxygen evolution in an acid environment. *Electrochemical and Solid State Letters.* 14:E5–E7.

Fukuda K, Iwakura C et al. 1979. The mechanism of the anodic formation of S$_2$O$_2$ ions on a Ti-supported IrO$_2$ electrode in mixed aqueous solutions of H$_2$SO$_4$, (NH$_4$)$_4$SO$_4$ and NH$_4$F. *Electrochimica Acta.* 24:363–365.

Garrett RF and Foran GJ. 2003. EXAFS. In: *Surface Analysis Methods in Materials Science.* O'Connor, John, Sexton, Brett, Smart, Roger SC. (Eds.), Springer-Verlag, Berlin, pp. 347–373.

Ge Q, Desai S et al. 2001. Periodic Density Functional Study of CO and OH Adsorption on Pt–Ru Alloy Surfaces: Implications. *Journal of Physical Chemistry B.* 105:9533–9536.

Glarum SH and Marshall JH. 1980. The A–C response of iridium oxide films. *Journal of the Electrochemical Society.* 127:1467–1474.

Goto S, Hosoi S et al. 2014. Particle-size and Ru-core-induced surface electronic states of Ru-core/Pt-shell electrocatalyst particles. *Journal of Physical Chemistry C.* 118:2634–2640.

Gottesfeld S. 1980. Faradaic processes at the Ir/Ir oxide electrode. *Journal of the Electrochemical Society.* 127:1922–1925.

Gottesfeld S and McIntyre JDE. 1979. Electrochromism in anodic iridium oxide films. 2. *Journal of the Electrochemical Society.* 126:742–750.

Granqvist CG. 1994. Electrochromic oxides: A bandstructure approach. *Solar Energy Materials and Solar Cells.* 32:369–382.

Granqvist CG. 1995. *Handbook of Inorganic Electrochromic Materials.* Elsevier, Amsterdam.

Granqvist CG. 2014. Oxide electrochromic: An introduction to devices and materials. *Solar Energy Materials and Solar Cells.* 99:1–13.

Griessen R and Driessen A. 1984. Heat of formation and band-structure of binary and ternary metal-hydrides. *Physical Reviews B*. 30:4372–4381.

Guerrini E and Trasatti S. 2006. Recent developments in understanding factors of electrocatalysis. *Russian Journal of Electrochemistry*. 42:1017–1025.

Halck NB, Petrykin V et al. 2014. Beyond the volcano limitations in electrocatalysis – Oxygen evolution reaction. *Physical Chemistry Chemical Physics*. 16:13682–13688.

Hammer B. 2006. Special sites at noble and late transition metal catalysts. *Topics in Catalysis*. 37:3–16.

Hammer B and Nørskov J. 2000. Theoretical surface science and catalysis – Calculations and concepts. *Advances in Catalysis*. 45:71–129.

Hansen HA, Man IC et al. 2010. Electrochemical chlorine evolution at rutile oxide (110) surfaces. *Physical Chemistry Chemical Physics*. 12:283–290.

Heidrich HJ, Muller L et al. 1990. The influence of electrode porosity and temperature on electrochemical gas evolution at platinum and rhodium. *Journal of Applied Electrochemistry*. 20:686–691.

Hickling A and Hill S. 1950. Oxygen overvoltage. *Discussions of the Faraday Society*. 1:557–559.

Hillman AR, Skopeka MA et al. 2011. X-ray spectroscopy of electrochemically deposited iridium oxide films: Detection of multiple sites through structural disorder. *Physical Chemistry Chemical Physics*. 13:5252–5263.

Hoare JP. 1968. *The Electrochemistry of Oxygen*. Interscience Publishers (John Wiley & Sons, New York).

Honing C, Raistrick ID et al. 1980. Application of A–C techniques to the study of lithium diffusion in tungsten trioxide thin films. *Journal of the Electrochemical Society*. 127:343–350.

Hu JM, Meng HM et al. 2002. Degradation mechanism of long service life Ti/IrO$_2$–Ta$_2$O$_5$ oxide anodes in sulphuric acid. *Corrosion Science*. 44:1655–1668.

Hu J-M, Zhang J-Q et al. 2004. Oxygen evolution reaction on IrO$_2$-based DSA® type electrodes: Kinetics analysis of Tafel lines and EIS. *International Journal of Hydrogen Energy*. 29:791–797.

Hu W, Chen S et al. 2014. IrO$_2$/Nb–TiO$_2$ electrocatalyst for oxygen evolution reaction in acidic medium. *International Journal of Hydrogen Energy*. 39:6967–6976.

Hüppauff M and Lengeler B. 1993. Valency and structure of iridium in anodic iridium oxide films. *Journal of the Electrochemical Society*. 140:598–602.

Hutchings R, Muller K et al. 1984. A structural investigation of stabilized oxygen evolution catalysts. *Journal of Material Science*. 19:3987–3994.

Huynh M, Bediako DK et al. 2014. A functionally stable manganese oxide oxygen evolution catalyst in acid. *Journal of the American Chemical Society*. 136:6002–6010.

Ioroi T, Kitazawa N et al. 2000. Iridium oxide/platinum electrocatalysts for unitized regenerative polymer electrolyte fuel cells. *Journal of the Electrochemical Society*. 147:2018–2022.

Ito M, Murakami Y et al. 1994. Preparation of ultrafine RuO$_2$–SnO$_2$ binary oxide particles by a sol-gel process. *Journal of the Electrochemical Society*. 141:1243–1245.

Iwasita T. 2003. Methanol and CO electrooxidation. In: *Handbook of Fuel Cells – Fundamentals, Technology and Applications*. John Wiley & Sons, Chichester, pp. 603–624.

Jalani NH, Dunn K et al. 2005. Synthesis and characterization of Nafion (R)-MO$_2$ (M = Zr, Si, Ti) nanocomposite membranes for higher temperature PEM fuel cells. *Electrochimica Acta*. 51:553–560.

Jasem S and Tseung ACC. 1979. A potentiostatic pulse study of oxygen evolution on Teflon-bonded nickel–cobalt oxide electrodes. *Journal of the Electrochemical Society*. 126:1353–1360.

Jirkovský J, Hoffmannova H et al. 2006a. Particle size dependence of the electrocatalytic activity of nanocrystalline RuO$_2$ electrodes. *Journal of the Electrochemical Society*. 153:E111–E118.

Jirkovský J, Makarova M et al. 2006b. Particle size dependence of oxygen evolution reaction on nanocrystalline RuO$_2$ and Ru$_{0.8}$Co$_{0.2}$O$_{2-x}$. *Electrochemistry Communications*. 8:1417–1422.

Kadakia K, Datta MK et al. 2012. Novel (Ir,Sn,Nb)O$_2$ anode electrocatalysts with reduced noble metal content for PEM based water electrolysis. *International Journal of Hydrogen Energy*. 37:3001–3013.

Kameyama K, Shohji S et al. 1993. Preparation of ultrafine RuO$_2$–TiO$_2$ binary oxide particles by a sol-gel process. *Journal of the Electrochemical Society*. 140:1034–1037.

Kimmel YC, Xu XG et al. 2014. Trends in electrochemical stability of transition metal carbides and their potential use as supports for low-cost electrocatalysts. *ACS Catalysis*. 4:1558–1562.

Kinoshita K. 1990. Particle-size effects for oxygen reduction on highly dispersed platinum in acid electrolytes. *Journal of the Electrochemical Society*. 137:845–848.

Kinoshita K. 1992. *Electrochemical Oxygen Technology*. John Wiley & Sons, New York.

Kokoh KB, Mayousse E et al. 2014. Efficient multi-metallic anode catalysts in a PEM water electrolyzer. *International Journal of Hydrogen Energy*. 39:1924–1931.

Koper MTM, Shubina TE et al. 2002. Periodic density functional study of CO and OH adsorption on Pt-Ru alloy surfaces: Implications for CO tolerant fuel cell catalysts. *Journal of Physical Chemistry B*. 106:686–692.

Kötz ER and Neff H. 1985. Anodic iridium oxide films: An UPS study of emersed electrodes. *Surface Science*. 160:517–530.

Kötz R. 1991. Ultraviolet photoelectron spectroscopy (UPS) of anodic oxide films on Au, Pt, Ru and Ru_0.5Ir_0.5 alloy. *Surface Science*. 47:109–114.

Kötz R and Carlen M. 2000. Principles and applications of electrochemical capacitors. *Electrochimica Acta*. 45:2483–2498.

Kötz R, Lewerenz HJ et al. 1983a. Oxygen evolution on Ru and Ir electrodes. XPS-studies. *Journal of Electroanalytical Chemistry*. 150:209–216.

Kötz R, Lewerenz JH et al. 1983b. XPS studies of oxygen evolution on Ru and RuO$_2$ anodes. *Journal of the Electrochemical Society*. 130:825–829.

Kötz R, Stucki S et al. 1984. In-situ identification of RuO$_4$ as the corrosion product during oxygen evolution on ruthenium in acid media. *Journal of Electroanalytical Chemistry and Interfacial Electrochemistry*. 172:211–219.

Kötz R and Stucki S. 1986. Stabilization of RuO_2 by IrO_2 for anodic oxygen evolution in acid media. *Electrochimica Acta*. 31:1311–1316.

Krysa J, Kule L et al. 1996. Effect of coating thickness and surface treatment of titanium on the properties of IrO_2–Ta_2O_5 anodes. *Journal of Applied Electrochemistry*. 26:999–1005.

Kurihara LK, Chow GM et al. 1995. Nanocrystalline metallic powders and films produced by the polyol method. *Nanostructured Materials*. 5:607–613.

Kuznetsova E, Cuesta A et al. 2014. Identification of the byproducts of the oxygen evolution reaction on rutile-type oxides under dynamic conditions. *Journal of Electroanalytical Chemistry*. 728:102–111.

Lengeler B. 2006. Extended X-ray absorption fine structure. In: *Neutron and X-ray Spectroscopy*. Hippert, F., Geissler, E., Hodeau, J.L., Lelièvre-Berna, E., Regnard, J.-R. (Eds.) Springer, Dordrecht, pp. 131–168.

Lervik IA, Tsypkin M et al. 2010. Electronic structure vs. electrocatalytic activity of iridium oxide. *Journal of Electroanalytical Chemistry*. 645:135–142.

Liao M-S, Cabrera CR et al. 2000. A theoretical study of CO adsorption on Pt, Ru and Pt-M (M = Ru, Sn, Ge) clusters. *Surface Science*. 445:267–282.

Linse N, Gubler L et al. 2011. The effect of platinum on carbon corrosion behavior in polymer electrolyte fuel cells. *Electrochimica Acta*. 56:7541–7549.

Lopez-Haro M, Dubau L et al. 2014. Atomic-scale structure and composition of Pt_3Co/C nanocrystallites during real PEMFC operation: A STEM-EELS study. *Applied Catalysis B: Environmental*. 152:300–308.

Ma L, Sui S et al. 2008. Preparation and characterization of Ir/TiC catalyst for oxygen evolution. *Journal of Power Sources*. 177:470–477.

Ma L, Sui S et al. 2009. Investigations on high performance proton exchange membrane water electrolyzer. *International Journal of Hydrogen Energy*. 34:678–684.

Macdonald JR. 1987. *Impedance Spectroscopy, Emphasizing Solid Materials and Systems*. John Wiley & Sons, New York.

Mamaca N, Mayousse E et al. 2012. Electrochemical activity of ruthenium and iridium based catalysts for oxygen evolution reaction. *Applied Catalysis B: Environmental*. 111–112:376–380.

Man IC, Su H-Y et al. 2011. Universality in oxygen evolution electrocatalysis on oxide surfaces. *ChemCatChem*. 3:1159–1165.

Marshall A, Borresen B et al. 2005. Preparation and characterisation of nanocrystalline $Ir_xSn_{1-x}O_2$ electrocatalytic powders. *Materials Chemistry and Physics*. 94:226–232.

Marshall A, Børresen B et al. 2006a. Iridium oxide-based nanocrystalline particles as oxygen evolution electrocatalysts. *Russian Journal of Electrochemistry*. 42:1134–1140.

Marshall A, Børresen B et al. 2006b. Electrochemical characterisation of $Ir_xSn_{1-x}O_2$ powders as oxygen evolution electrocatalysts. *Electrochimica Acta*. 51:3161–3167.

Marshall A, Tsypkin M et al. 2004. Nanocrystalline $Ir_xSn_{(1-x)}O_2$ electrocatalysts for oxygen evolution in water electrolysis with polymer electrolyte – Effect of heat treatment. *Journal of New Materials for Electrochemical Systems*. 7:197–204.

Marshall AT and Haverkamp RG. 2010. Electrocatalytic activity of IrO_2–RuO_2 supported on Sb-doped SnO_2 nanoparticles. *Electrochimica Acta*. 55:1978–1984.

Marshall AT, Sunde S et al. 2007. Performance of a PEM water electrolysis cell using $Ir_xRu_yTa_zO_2$ electrocatalysts for the oxygen evolution electrode. *International Journal of Hydrogen Energy*. 32:2320–2324.

Matsumoto Y and Sato E. 1986. Electrocatalytic properties of transition metal oxides for oxygen evolution reaction. *Materials Chemistry and Physics*. 14:397–426.

Mattheiss LF. 1976. Electronic structure of RuO_2, OsO_2, and $CeIrO_2$. *Physical Reviews B: Condensation Matter*. 13:2433–2451.

Mayrhofer KJ, Blizanac BB et al. 2005. The impact of geometric and surface electronic properties of Pt-catalysts on the particle size effect in electocatalysis. *Journal of Physical Chemistry B*. 109:14433–14440.

Memming R. 2000. *Semiconductor Electrochemistry*, Wiley-VCH, Weinheim, Germany.

Michell D, Rand DAJ et al. 1978. A study of ruthenium electrodes by cyclic voltammetry and X-ray emission spectroscopy. *Journal of Electroanalytical Chemistry and Interfacial Electrochemistry*. 89:11–27.

Miles M, Klaus E et al. 1978. The oxygen evolution reaction on platinum, iridium, ruthenium and their alloys at 80°C in acid solutions. *Electrochimica Acta*. 23:521–526.

Millet P, Ngameni R et al. 2010. PEM water electrolyzers: From electrocatalysis to stack development. *International Journal of Hydrogen Energy*. 35:5043–5052.

Milner PC. 1964. The possible mechanisms of complex reactions involving consecutive steps. *Journal of the Electrochemical Society*. 111:228–232.

Mo Y, Stefan IC et al. 2002. In situ iridium LIII-edge X-ray absorption and surface enhanced Raman spectroscopy of electrodeposited iridium oxide films in aqueous electrolytes. *Journal of Physical Chemistry B*. 106:3681–3686.

Morrison SR. 1980. *Electrochemistry at semiconductor and oxidized metal electrodes*, Plenum Press, New York.

Murakami Y, Miwa K et al. 1994a. Morphology of ultrafine RuO_2–IrO_2 binary oxide particles prepared by a sol–gel process. *Journal of the Electrochemical Society*. 141:L118–L120.

Murakami Y, Ohkawauchi H et al. 1994b. Preparations of ultrafine IrO_2–SnO_2 binary oxide particles by a sol–gel process. *Electrochimica Acta*. 39:2551–2554.

Murakami Y, Tsuchiya S et al. 1994c. Preparation of ultrafine IrO_2–Ta_2O_5 binary oxide particles by a sol–gel process. *Electrochimica Acta*. 39:651–654.

Murakami Y, Tsuchiya S et al. 1994d. Preparation of ultrafine RuO_2 and IrO_2 particles by a sol–gel process. *Journal of Materials Science Letters*. 13:1773–1774.

Nikiforov AV, Garcia ALT et al. 2011. Preparation and study of IrO_2/SiC–Si supported anode catalyst for high temperature PEM steam electrolysers. *International Journal of Hydrogen Energy*. 36:5797–5805.

Nikkuni FR, Vion-Dury B et al. 2014. The role of water in the degradation of Pt_3Co/C nanoparticles: An identical location transmission electron microscopy study in polymer electrolyte environment. *Applied Catalysis B: Environmental*. 156:301–306.

Nilekar AU, Alayoglu S et al. 2010. Preferential CO oxidation in hydrogen: Reactivity of core-shell nanoparticles. *Journal of the American Chemical Society.* 132:7418–7428.

Nong HN, Gan L et al. 2014. IrO_x core-shell nanocatalysts for cost- and energy-efficient electrochemical water splitting. *Chemical Science.* 5:2955–2963.

Ochal P, de la Fuente JLG et al. 2011. CO stripping as an electrochemical tool for characterization of Ru@Pt core-shell catalysts. *Journal of Electroanalytical Chemistry.* 655:140–146.

Orazem ME and Tribollet B. 2008. *Electrochemical Impedance Spectroscopy.* John Wiley & Sons, Hoboken, NJ.

Ouattara L, Fierro S et al. 2009. Electrochemical comparison of IrO_2 prepared by anodic oxidation of pure iridium and IrO_2 prepared by thermal decomposition of H_2IrCl_6 precursor solution. *Journal of Applied Electrochemistry.* 39:1361–1367.

Owe L-E, Lervik IA et al. 2010. Electrochemical behavior of iridium oxide films in trifluoromethanesulfonic acid. *Journal of the Electrochemical Society.* 157:B1719–B1725.

Owe L-E, Tsypkin M et al. 2011. The effect of phosphate on iridium oxide electrochemistry. *Electrochimica Acta.* 58: 231–237.

Owe L-E, Tsypkin M et al. 2012. Iridium–ruthenium single phase mixed oxides for oxygen evolution: Composition dependence of electrocatalytic activity. *Electrochimica Acta.* 70:158–164.

Parsons R. 2011. Volcano curves in electrochemistry. In: *Catalysis in Electrochemistry: From Fundamentals to Strategies for Fuel Cell Development.* John Wiley & Sons, Hoboken, NJ, pp. 1–15.

Pauli CPD and Trasatti S. 1995. Electrochemical surface characterization of IrO_2 + SnO_2 mixed oxide electrocatalysts. *Journal of Electroanalytical Chemistry.* 369:161–168.

Pauporté T, Aberdam D et al. 1999. X-ray absorption in relation to valency of iridium in sputtered iridium oxide films. *Journal of Electroanalytical Chemistry.* 465:88–95.

Pechini PM. 1967. Method of preparing lead and alkaline earth titanates and niobates and coating method using the same to form a capacitor. US Patent US3330697 A.

Petit MA and Plichon V. 1998. Anodic electrodeposition of iridium oxide films. *Journal of Electroanalytical Chemistry.* 444:247–252.

Petrykin V, Bastl Z et al. 2009. Local structure of nanocrystalline $Ru_{1-x}Ni_xO_2$-δ dioxide and its implications for electrocatalytic behaviors: An XPS and XAS study. *Journal of Physical Chemistry C.* 113:21657–21666.

Petrykin V, Macounova K et al. 2011. Zn-doped RuO_2 electrocatalyts for selective oxygen evolution: Relationship between local structure and electrocatalytic behavior in chloride containing media. *Chemical Materials.* 23:200–207.

Petrykin V, Macounova K et al. 2013. Local structure of Co doped RuO_2 nanocrystalline electrocatalytic materials for chlorine and oxygen evolution. *Catalysis Today.* 202:63–69.

Polonsky J, Mazur P et al. 2014. Performance of a PEM water electrolyser using a TaC-supported iridium oxide electrocatalyst. *International Journal of Hydrogen Energy.* 39:3072–3078.

Polonsky J, Petrushina IM et al. 2012. Tantalum carbide as a novel support material for anode electrocatalysts in polymer electrolyte membrane water electrolysers. *International Journal of Hydrogen Energy.* 37:2173–2181.

Pourbaix MJN, Van Muylder J et al. 1959. Electrochemical properties of the platinum metals. *Platinum Metals Review.* 3:100–106.

Puthiyapura VK, Pasupathi S et al. 2014. Investigation of supported IrO_2 as electrocatalyst for the oxygen evolution reaction in proton exchange membrane water electrolyser. *International Journal of Hydrogen Energy.* 39:1905–1913.

Rabis A, Rodriguez P et al. 2012. Electrocatalysis for polymer electrolyte fuel cells: Recent achievements and future challenges. *ACS Catalysis.* 2:864–890.

Rasiyah P and Tseung ACC. 1984. The role of the lower metal oxide/higher metal oxide couple in oxygen evolution reactions. *Journal of the Electrochemical Society.* 133:803–808.

Rasiyah P, Tseung ACC et al. 1982a. A mechanistic study of oxygen evolution on $NiCO_2O_4$. Part I. Formation of higher oxides. *Journal of the Electrochemical Society.* 129:1724–1727.

Rasiyah P, Tseung ACC et al. 1982b. A mechanistic study of oxygen evolution on $NiCO_2O_4$. Part II. Electrochemical kinetics. *Journal of the Electrochemical Society.* 129:1724–1727.

Rasten E. 2001. Electrocatalysis in water electrolysis with solid polymerelectrolyte. Fakultet for naturvitenskap og teknologi, Norwegian University of Science and Technology, NTNU.

Rasten E, Hagen G et al. 2003. Electrocatalysis in water electrolysis with solid polymer electrolyte. *Electrochimica Acta.* 48:3945–3952.

Rausch S and Wendt H. 1992. Raney nickel activated H2 cathodes. 1. Modeling the current voltage behavior of flat raney-nickel coated microporous electrodes. *Journal of Applied Electrochemistry.* 22:1025–1030.

Reier T, Oezaslan M et al. 2012. Electrocatalytic oxygen evolution reaction (OER) on Ru, Ir, and Pt catalysts: A comparative study of nanoparticles and bulk materials. *ACS Catalysis.* 2:1765–1772.

Ribeiro J, Moats MS et al. 2008. Morphological and electrochemical investigation of RuO(2)-Ta(2)O(5) oxide films prepared by the Pechini-Adams method. *Journal of Applied Electrochemistry.* 38:767–775.

Rodes A and P'erez JM. 2003. Vibrational spectroscopy. In: *Handbook of Fuel Cells – Fundamentals, Technology and Applications.* Wolf Vielstich, Arnold Lamm, Hubert A. Gasteiger (Eds.) John Wiley & Sons, Hoboken, NJ, pp. 191–219.

Roller JM, Arellano-Jimenez MJ et al. 2013. Oxygen evolution during water electrolysis from thin films using bimetallic oxides of Ir–Pt and Ir–Ru. *Journal of the Electrochemical Society.* 160:F716–F730.

Rossmeisl J, Dimitrievski K et al. 2007a. Comparing electrochemical and biological water splitting. *Journal of Physical Chemistry C.* 111:18821–18823.

Rossmeisl J, Logadottir A et al. 2005. Electrolysis of water on (oxidized) metal surfaces. *Chemical Physics.* 319:178–184.

Rossmeisl J, Qu Z-W et al. 2007b. Electrolysis of water on oxide surfaces. *Journal of Electroanalytical Chemistry.* 607:83–89.

Ruban A, Hammer B et al. 1997. *Journal of Molecular Catalysis A: Chemistry.* 115:421–429.

Rüetschi P and Delahay P. 1955. Influence of electrode material on oxygen overvoltage: A theoretical analysis. *Journal of Chemical Physics.* 23:556–560.

Russell J, Nuttall L et al. 1973. Hydrogen generation by solid polymer electrolyte water electrolysis. *Proceedings of the American Chemical Society Division of Fuel Chemistry Meeting.* 18(3):24–40.

Sanchez Casalongue HG, Ng ML et al. 2014. In situ observation of surface species on iridium oxide nanoparticles during the oxygen evolution reaction. *Angewandte Chemie (International ed in English).* 53:7169–7172.

Sasaki K, Takasaki F et al. 2010. Alternative electrocatalyst support materials for polymer electrolyte fuel cells. *ECS Transactions.* 33:473–482.

Savinell RF, Zeller RL et al. 1990. Electrochemically active surface area – Voltammetric charge correlations for ruthenium and iridium dioxide electrodes. *Journal of the Electrochemical Society.* 137:489–494.

Shao YY, Liu J et al. 2009. Novel catalyst support materials for PEM fuel cells: Current status and future prospects. *Journal of Materials Chemistry.* 19:46–59.

Silva TM, Simoes AMP et al. 1998. Electronic structure of iridium oxide films formed in neutral phosphate buffer solution. *Journal of Electroanalytical Chemistry.* 441:5–12.

Siracusano S, Baglio V et al. 2009. Preparation and characterization of titanium suboxides as conductive supports of IrO_2 electrocatalysts for application in SPE electrolysers. *Electrochimica Acta.* 54:6292–6299.

Sivasankar N, Weare WW et al. 2011. Direct observation of a hydroperoxide surface intermediate upon visible light-driven water oxidation at an Ir oxide nanocluster catalyst by rapid-scan FT-IR spectroscopy. *Journal of the American Chemical Society.* 133:12976–12979.

Smith RDL, Sporinova B et al. 2014. Facile photochemical preparation of amorphous iridium oxide films for water oxidation catalysis. *Chemistry of Materials.* 26:1654–1659.

Song S, Zhang H et al. 2008. Electrochemical investigation of electrocatalysts for the oxygen evolution reaction in PEM water electrolyzers. *International Journal of Hydrogen Energy.* 33:4955–4961.

Steegstra P and Ahlberg E. 2012. Involvement of nanoparticles in the electrodeposition of hydrous iridium oxide films. *Electrochimica Acta.* 68:206–213.

Steegstra P, Busch M et al. 2013. Revisiting the redox properties of hydrous iridium oxide films in the context of oxygen evolution. *Journal of Physical Chemistry C.* 117:20975–20981.

Stoerzinger KA, Qiao L et al. 2014. Orientation-dependent oxygen evolution activities of rutile IrO_2 and RuO_2. *Journal of Physical Chemistry Letters.* 5:1636–1641.

Sui S, Ma L et al. 2009. Investigation on the proton exchange membrane water electrolyzer using supported anode catalyst. *Asia-Pacific Journal of Chemical Engineering.* 4:8–11.

Sui S, Ma LR et al. 2011. TIC supported Pt-Ir electrocatalyst prepared by a plasma process for the oxygen electrode in unitized regenerative fuel cells. *Journal of Power Sources.* 196:5416–5422.

Sun S, Shao Z et al. 2014. Investigations on degradation of the long-term proton exchange membrane water electrolysis stack. *Journal of Power Sources.* 267:515–520.

Sunde S, Lervik IA et al. 2009. An impedance model for a porous intercalation electrode with mixed conductivity. *Journal of the Electrochemical Society.* 156:B927–B937.

Sunde S, Lervik IA et al. 2010. Impedance analysis of nanostructured iridium oxide electrocatalysts. *Electrochimica Acta.* 55:7751–7760.

Suntivich J, May KJ et al. 2011. A perovskite oxide optimized for oxygen evolution catalysis from molecular orbital principles. *Science.* 334:1383–1385.

Takasu Y, Onoue S et al. 1994. Preparation of ultrafine RuO_2–IrO_2–TiO_2 oxide particles by a sol–gel process. *Electrochimica Acta.* 39:1993–1997.

Terezo AJ, Bisquert J et al. 2001. Separation of transport, charge storage and reaction processes of porous electrocatalytic IrO_2 and IrO_2/Nb_2O_5 electrodes. *Journal of Electroanalytical Chemistry.* 508:59–69.

Terezo AJ and Pereira EC. 1999. Preparation and characterization of Ti/RuO_2–Nb_2O_5 electrodes obtained by polymeric precursor method. *Electrochimica Acta.* 44:4507–4513.

Terezo AJ and Pereira EC. 2002. Preparation and characterisation of Ti/RuO_2 anodes obtained by sol-gel and conventional routes. *Materials Letters.* 53:339–345.

Thomassen M, Mokkelbost T et al. 2011. Supported nanostructured Ir and IrRu electrocatalysts for oxygen evolution in PEM electrolysers. *Nanostructured Materials for Energy Storage and Conversion.* 35:271–279.

Trasatti S. 1980. Electrocatalysis by oxides – Attempt at a unifying approach. *Journal of Electroanalytical Chemistry.* 111:125–131.

Trasatti S. 1984. Electrocatalysis in the anodic evolution of oxygen and chlorine. *Electrochimica Acta.* 29:1503–1512.

Trasatti S. 1987. Oxide/aqueous solution interfaces, interplay of surface chemistry and electrocatalysis. *Materials Chemistry and Physics.* 16:157–174.

Trasatti S. 1990a. The "absolute" electrode potential – The end of the story. *Electrochimica Acta.* 35:269–271.

Trasatti S. 1990b. Surface chemistry of oxides and electrocatalysis. *Croatica Chemica Acta.* 63:313–329.

Trasatti S. 1991. Physical electrochemistry of ceramic oxides. *Electrochimica Acta.* 36:225–241.

Trasatti S. 1999. Interfacial electrochemistry of conductive oxides for electrocatalysis. In: *Interfacial Electrochemistry.* A Wieckowski (Eds.), Marcel Dekker, Inc., New York, pp. 769–792.

Trasatti S. 2000. Electrocatalysis: Understanding the success of DSA®. *Electrochimica Acta.* 45:2377–2385.

Tseung ACC and Jasem S. 1977. Oxygen evolution on semiconducting oxides. *Electrochimica Acta.* 22:31–34.

Tsuda M and Kasai H. 2006. Ab initio study of alloying and straining effects on CO interaction with Pt. *Physical Review B.* 73:155405-155401–155405-155406.

Tsypkin M, de la Fuente JeLGe et al. 2013. Effect of heat treatment on the electrocatalytic properties of nano-structured Ru cores with Pt shells. *Journal of Electroanalytical Chemistry.* 704:57–66.

Tunold R, Marshall AT et al. 2010. Materials for electrocatalysis of oxygen evolution process in PEM water electrolysis cells. *Electrochemistry: Symposium on Interfacial Electrochemistry in Honor of Brian E Conway.* 25:103–117.

Vojvodic A and Nørskov JK. 2011. Optimizing perovskites for the water-splitting reaction. *Science.* 334:1355–1356.

Wainright JS, Wang J-T et al. 1995. Acid-doped polybenzimidazoles: A new polymer electrolyte. *Journal of the Electrochemical Society.* 142:L121–L123.

Wang J-T, Savinell RF et al. 1996. A H_2/O_2 fuel cell using acid doped polybenzimidazole as polymer electrolyte. *Electrochimica Acta.* 41:193–197.

Wang XP, Kumar R et al. 2006. Effect of voltage on platinum dissolution relevance to polymer electrolyte fuel cells. *Electrochemical and Solid State Letters.* 9:A225–A227.

Wendt H. 1994. Preparation, morphology and effective electrocatalytic activity of gas evolving and gas consuming electrodes. *Electrochimica Acta.* 39:1749–1756.

Williams DB and Carter CB. 2009. *Transmission Electron Microscopy: A Textbook for Materials Science.* New York: Springer.

Wright JD and Sommerdijk NAJM. 2003. *Sol-Gel Materials: Chemistry and Applications.* London, U.K.: Taylor & Francis.

Wu X and Scott K. 2011. RuO_2 supported on Sb-doped SnO_2 nanoparticles for polymer electrolyte membrane water electrolysers. *International Journal of Hydrogen Energy.* 36:5806–5810.

Xu JY, Li QF et al. 2012. Antimony doped tin oxides and their composites with tin pyrophosphates as catalyst supports for oxygen evolution reaction in proton exchange membrane water electrolysis. *International Journal of Hydrogen Energy.* 37:18629–18640.

Yamaguchi M, Shinohara T et al. 2000. *Development of 2500 cm² Five Cell Stack Water Electrolyzer in WE-NET.* Pennington, NJ: Electrochemical Society Inc.

Yuan XZ, Li H et al. 2011. A review of polymer electrolyte membrane fuel cell durability test protocols. *Journal of Power Sources.* 196:9107–9116.

Yuen MF, Lauks I et al. 1983. pH dependent voltammetry of iridium oxide films. *Solid State Ionics.* 11:19–29.

Zhang G, Shao ZG et al. 2012. One-pot synthesis of Ir@Pt nanodendrites as highly active bifunctional electrocatalysts for oxygen reduction and oxygen evolution in acidic medium. *Electrochemical Communications.* 22:145–148.

4

Electrocatalysts for the Hydrogen Evolution Reaction

Marcelo Carmo, Wiebke Lüke, and Detlef Stolten

CONTENTS

4.1 Introduction

The hydrogen evolution reaction (HER) is one of the most investigated and fundamentally important electrochemical reactions in heterogeneous catalysis, whereby hydrogen protons combine with electrons at an electrode surface to form hydrogen chemisorbed species and then desorb free hydrogen gas molecules (see Figure 4.1).

As early as 1952, Bockris attempted to clarify the outstanding problems of the concept and mechanism in the field of the cathodic HER (Bockris et al., 1950; Bockris and Azzam, 1952; Bockris and Mauser, 1959). At that time, the electrolytic evolution of hydrogen was already the main focus of experimental and theoretical research in electrode kinetics because the HER was thought to be the simplest of electrode reactions. Studies on understanding the concepts, aspects, and mechanisms related to the HER for water electrolysis were also supported by the fact that in the early 1900s, more than 400 industrial water electrolyzers were in operation. Likewise in 1939, the first large water electrolysis plant with a capacity of 10,000 N m³ of hydrogen per hour went into operation (Kreuter and Hofmann, 1998). The HER is also very important for the chloralkali process, in which sodium chloride is electrolyzed into chlorine, hydrogen, and sodium hydroxide (caustic soda), which are important chemical commodities required by industry.

Later in the 1960s, with the development of an acidic solid polymer electrolyte (SPE) membrane by Grubb and Niedrach at General Electric (GE), polymer electrolyte membrane or proton exchange membrane (PEM)

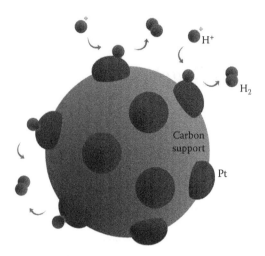

FIGURE 4.1
Representation of the hydrogen evolution on platinum nanoparticles supported on carbon particles (biggest ball: carbon particle, medium balls: platinum nanoparticle, smallest balls: hydrogen).

water electrolysis became an alternative for the electrolytic production of hydrogen. PEM water electrolysis was especially important when considering, for example, the challenges involved in sealing, circulating, and working with liquid alkaline electrolytes (usually 25%–35% KOH). Another important fact was that in the acidic regime, the kinetics of the HER on platinum (Pt) is so fast that the overpotential at the cathode side is rather insignificant compared with the HER in an alkaline regime (see Figure 4.2). However, with faster kinetics for the cathode side when operating in an acid regime, a higher anode overpotential for the oxygen evolution reaction (OER) is obtained. It turns out that the anode overpotential becomes the major factor limiting the overall cell performance in PEM water electrolysis. Hence, due to the fast kinetics of the HER

for PEM water electrolysis, low Pt loading is required, and cell voltage losses even for very low Pt loadings (<200 μg_{Pt} cm^{-2}) could be achieved.

In most of the early studies and developments of PEM electrolysis, researchers used Pt black as a standard catalyst on the cathode side. Later, from the experience gained in the development of catalysts for PEM fuel cells, researchers started to use commercially available Pt nanoparticles supported on carbon black (Pt/C) (supplied by various manufacturers, e.g., ETEK/BASF, Tanaka, Umicore, and Johnson & Matthey) as their standard catalysts for the HER in PEM electrolysis. However, despite the lower Pt loadings, the cathode catalyst still represents a considerable portion of the total stack cost, especially if degradation of the Pt/C catalyst occurs. Today, for commercial units, the loadings for the cathode side in PEM electrolysis range between 0.5 and 2 mg$_{Pt}$ cm^{-2}.

Extensive research and development of HER catalysts has been performed. The great majority of these studies concentrated on further reductions of Pt loadings and potential Pt substitution (creating the so-called Pt-free catalysts). When using high Pt loadings, today lifetimes greater than 60,000 h can be achieved (Anderson and Capuano, 2013). Studies on Pt loading reduction or Pt substitution now have to be pursued with major focus on the durability of these new HER catalyst systems (Ayers et al., 2010). If coupling PEM water electrolysis to renewable energy sources like wind energy and photovoltaics, the catalyst system must be able to withstand the seasonal characteristics of the power input (on/off, partial load operation, and overload operation) over thousands of hours. However, it is still questionable whether the intermittent power input condition will indeed affect the catalyst system durability.

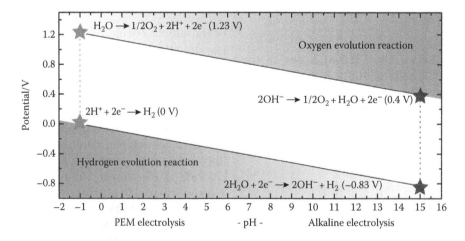

FIGURE 4.2
Variation of the potential for the different half-reactions in water electrolysis in relation to the pH of the electrolyte. (Stolten, D. S. V. 2013. *Transition to renewable energy systems energy process engineering* [Online]. Available: http://search.ebscohost.com/login.aspx?direct=true& scope=site&db=nlebk&db=nlabk&AN=581898. Figure at the bottom reprinted with permission.)

4.2 Reaction Mechanisms

Since the first studies by Tafel around 1900 (Tafel, 1905), the mechanisms of the electrode reaction for the cathodic hydrogen evolution (Equation 4.1) have been extensively studied and discussed.

$$2H^+ + 2e^- \leftrightarrow H_2 \qquad (4.1)$$

The same also applies for the reverse reaction, where the hydrogen oxidation occurs, a reaction with major importance for the development of electrode materials for PEM fuel cells. The HER is considered to be one of the simplest electrode reactions. Essentially, it has the following characteristics (Bagotskii, 2006):

1. The electron transfer step is the only reaction step, which means that other parallel or consecutive steps are absent.
2. Neither the starting material nor the reaction product, nor any intermediates, is adsorbed on the electrode.
3. During the reaction, chemical bonds are not broken, new chemical bonds are not formed, and the geometry of the reacting species remains unchanged.

However, to date, the reaction mechanisms for the HER are not entirely understood, as there is a strong dependence on the operating conditions and electrode materials. Nevertheless, a simple general mechanism that can be separated into a complex two-electron reaction, occurring through several steps, is commonly accepted. It involves the following:

1. Transport of water molecules to the phase boundary and adsorption
2. Charge transfer to the catalyst, following three alternatives:
   a. Formation of adsorbed hydrogen atoms (discharge): reaction on free atom sites that are not occupied by H-adsorbed atoms

   Volmer step: $H_3O^+ + M + e^- \leftrightarrow M-H + H_2O$ (4.2)

   b. Formation of H_2 molecules (electrochemical desorption): reaction on atom sites that are occupied by H-adsorbed atoms

   Heyrovski step: $H_3O^+ + M-H + e^- \leftrightarrow M + H_2 + H_2O$ (4.3)

   c. Recombination of adsorbed H atoms formed in Equation 4.2

   Tafel step: $M-H + M-H \leftrightarrow 2M + H_2$ (4.4)

3. Desorption of H_2 gas molecules from the metal surface.
4. Migration of H_2 by
   a. Diffusion and convection
   b. Gas bubble formation

Taking the reaction steps mentioned earlier into consideration, the mechanism will follow three elementary steps, with the formation of only one intermediate species. The three steps described can then be combined to provide three possible ways for completion of the HER:

1. Volmer–Tafel (Equations 4.2 and 4.4)
2. Volmer–Heyrovsky (Equations 4.2 and 4.3)
3. Reversed Heyrovsky–Tafel (Equations 4.3 and 4.4)

It is now important to obtain information on which of the two consecutive steps for every pathway constitute the rate-determining step. In 1905, Tafel proposed that the recombination of adsorbed hydrogen is the rate-determining step, since the recombination of two atoms should not occur instantaneously, but at some very finite and slow rate (Tafel, 1905). On the other hand, the discharge reaction was also considered to provide the slowest rate. This was attributed to the appreciable energy required to break up the complexes formed by the ions in solution. Moreover, according to Bagotsky, another contribution to the activation energy arises from the need for reorganization of the solvent molecules close to the ions undergoing discharge (Bagotskii, 2006).

After decades of research on the mechanisms and kinetics of the HER, and despite being one of the simplest of such reactions, its full understanding is still incomplete. Unfortunately, no general rate expression exists that can simultaneously account for the alternate pathways in terms of the accepted three steps presented earlier. This becomes even more complicated if one considers resolving the mechanisms for metal alloys or non-PGM metals. More research is required to obtain a more complete picture of this important reaction system, potentially including the elucidation of parallel pathways and dominant steps.

4.3 Kinetics

Electrocatalysis and heterogeneous catalysis are closely related, and the difference essentially relies on the net charge transfer that occurs from a given

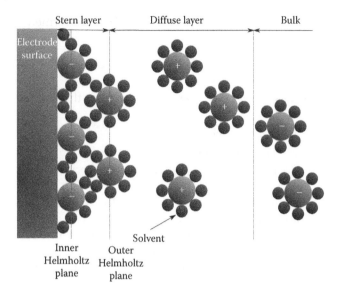

| Stern layer | Diffuse layer | Bulk |

Inner Helmholtz plane

Outer Helmholtz plane

Solvent

FIGURE 4.3
Schematic representation of the electric double layer (EDL). The surface due to the presence of charge has a potential E^0. The outer Helmholtz plane (OHP) marks the closest distance that counter ions can come to the surface.

electrochemical reaction. In other words, the reaction rate for an electrochemical reaction (in our case, the HER) will depend on the electrostatic potential drop, which develops at the interface between the electrode and electrolyte. The existence of a potential drop between two different conducting phases is related to the formation of an electric double layer (see Figure 4.3) at the phase boundary (i.e., formation of two parallel layers of charges with opposite signs, each on the surface of one of the contacting phases). It is a special feature of such an electric double layer that the two layers forming the double layer are a very small (molecular) distance apart, between 0.1 and 0.4 nm. For this reason, electric double-layer capacitances are very high (i.e., tenths of μFcm^{-2}). It is, however, important to mention that the double-layer capacitance will be strongly influenced by the type and concentration of electrolyte used. It emerges that the electrode potential is the main parameter (but not the only one) that one uses to influence the kinetics of electrochemical processes.

Unfortunately, the real potential between two electrodes of different types cannot be measured by any means. Methods in which the force acting on a test charge is measured cannot actually be used here because any values that could be measured would be distorted by the chemical forces. The same holds true for determinations of the work of transfer. At least one more interface is formed when a measuring device such as a voltmeter or potentiometer is connected, and the Galvani potential of that interface will be contained in the quantity being measured. Moreover, these potentials cannot be calculated from indirect experimental data in any rigorous thermodynamic way. Thus, potential differences

can only be measured between points located within phases of the same nature. Hence, electrode potential values always correspond to the differences between two electrodes, or between the working electrode and a reference one. Since the equilibrium potentials (ΔE^0) for electrodes in real electrolytic cells correspond to the difference in electronic energies between the two electrodes (anode and cathode), where the driving force for the current flow is the Gibbs energy of the overall cell reaction, we obtain

$$\Delta E^0 = -nF\Delta_r G \qquad (4.5)$$

where
    ΔE^0 is evaluated under the standard conditions of unit activities of all ions in solution and atmospheric pressure for all gases
    n is the number of electrons
    F is the Faraday constant
    ΔG is the variation of the Gibbs energy

It follows from Equation 4.5 that the potential equilibrium depends only on the nature of the two phases (bulk properties, which are decisive for the values of ΔE), and not on the state of the interphase (i.e., its size, contamination, etc.). In this case, the cell voltage E is then given for arbitrary ionic activities and gas pressure by $E = -\Delta_r G/nF$, from which follows

$$\Delta_r G = \sum_i v_i \mu_i = \sum_i v_i \mu_i^0 + \sum_i v_i RT \ln a_i$$

$$= \Delta_r G^0 + \sum_i v_i RT \ln a_i \qquad (4.6)$$

Rearranging Equation 4.6 by expressing the potential difference between the electrode potential and standard electrode potential $E_0 = -\Delta_r G^0/nF$, we can obtain the well-known Nernst equation that describes the concentration dependence of E vs. SHE (standard hydrogen electrode). For a simple case, for a redox reaction $O + ne \leftrightarrow R$,

$$E = E^0 - \frac{RT}{nF} \sum_i v_i \ln a_i \qquad (4.7)$$

where
    a_i represents the activities of the oxidized and reduced species
    n is the number of electrons
    F is the Faraday constant
    T is the temperature
    R is the molar gas constant
    E^0 is the standard electrode potential (for all substances when activity is equal to one)

It is also important to mention the convention that v_i is positive for products and negative for reactants. If a product or reactant is present in the electrolyte as a dissolved gas in equilibrium with the corresponding gas in the vapor phase with partial pressure p_i, then the activity a_i must be replaced by p_i/p_0. Therefore, we obtain for the following equation:

$$H_2O \leftrightarrow H_2 + \frac{1}{2}O_2 \qquad (4.8)$$

the following:

$$E = E^0 + \frac{RT}{2F}\ln a_{H_2O} - \frac{RT}{2F}\ln\left\{\left(\frac{p_{H_2}}{p^0}\right)\cdot\left(\frac{p_{O_2}}{p^0}\right)\right\} \qquad (4.9)$$

From the expression $\Delta_r G^0 = \Delta_r H^0 - T\Delta_r S^0$, the value of $\Delta_r G^0$ can be simply obtained from the standard enthalpies of product formation and from the standard entropies of the reactants and products. We can now calculate the free energy for Equation 4.8 using the data from the literature to obtain the following:

$$\Delta_r H^0 = +\left(\Delta_f H^0_{H_2}\right) + \left(1/2\Delta_f H^0_{O_2}\right) - \left(\Delta_f H^0_{H_2O}\right)$$

$$= (0+0-285.25) = 285.25 \ \text{kJ mol}^{-1}$$

$$\Delta_r S^0 = +\left(\Delta_f S^0_{H_2}\right) + \left(\frac{1}{2\Delta_f} S^0_{O_2}\right) - \left(\Delta_f S^0_{H_2O}\right)$$

$$= \left(130.74 + \frac{1}{2}\cdot 205.25 - 70.12\right)$$

$$= 163.275 \ \text{J mol}^{-1} = 0.163275 \ \text{kJ mol}^{-1}$$

$$\Delta_r G^0 = \Delta_r H^0 - T\Delta_r S^0$$

$$= 285.25 - 298 \cdot 0.163275 = 236.60 \ \text{kJ mol}^{-1}$$

$$E^0 = \frac{\Delta_r G^0}{nF} = \frac{236.60}{2\cdot 96485} = 1.226 \ V$$

The potential value of 1.226 V assumes that $T\Delta S_R$ is integrated in the electrolysis process in the form of heat. Here, this is also termed as the lower heating value (LHV), which corresponds to an energy content for gaseous hydrogen of 3.0 kWh N m^{-3} that can be obtained using the second Faraday law of electrolysis:

$$m = \frac{Q}{F}\cdot\frac{M}{z} = \frac{It}{F}\cdot\frac{M}{z} \qquad (4.10)$$

where
  m is the mass of produced gas
  Q is the charge in Coulomb
  F is the Faraday constant
  M is the molar mass
  I is the current in amperes
  z is the number of electrons
  t is the time in seconds

If the thermal energy is introduced in the form of electrical energy (e.g., industrial electrolyzers) then the thermoneutral expression must be used, giving

$$E^0 = \frac{\Delta_r H^0}{nF} = \frac{285.25}{2\cdot 96485} = 1.478 \ V \qquad (4.11)$$

This case is, therefore, defined as the higher heating value (HHV), which corresponds to the energy content for gaseous hydrogen of about 3.54 kWh N m^{-3}, which can also be calculated using the second Faraday law of electrolysis (Equation 4.10). Figure 4.4 shows the standard cell potential variation with temperature. The dashed lines show the limits where water electrolysis assumes an endothermic and exothermic character, limited by the values of LHV and HHV at a given temperature. At this cell voltage, the electrical energy is equal to the total reaction enthalpy of the decomposition of water.

The pressure dependence of the cell voltage for water electrolysis is of special importance for PEM electrolyzers where hydrogen can be produced with pressures up

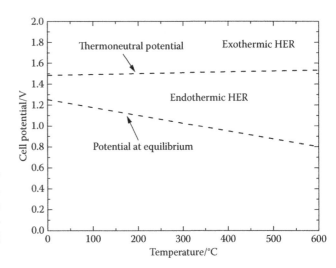

FIGURE 4.4
Variation of the standard cell potential with the temperature with respect to the lower heating values and higher heating value for water electrolysis.

to 80 bar (Carmo et al., 2013). Hence, we can also obtain the equation for pressure dependence as follows:

$$\left(\frac{\partial E}{\partial p}\right)_T = -\frac{1}{nF}\cdot\left(\frac{\partial \Delta_r G}{\partial p}\right)_T = -\frac{\Delta_r V}{nF}$$

$$= -\sum_j v_j \frac{RT}{nFp_j} \equiv \sum_j v_j \left(\frac{\partial E}{\partial p_j}\right)_T \qquad (4.12)$$

$$E = E^0 - \sum_j v_j \left(\frac{RT}{nF}\right)\ln\frac{p_j}{p^0}$$

$$= E^0 - \sum_j v_j \left(\frac{0.059}{n}\right)\log_{10}\frac{p_j}{p^0}\,(\text{for } T = 298 \text{ K}) \qquad (4.13)$$

For our specific example, we see that, due to a simultaneous increase in operational pressure (for H_2 and O_2), from 1 to 10 bar, an increase in the cell voltage to 1.319 V is obtained.

According to the Nernst equation, the potential difference (E) between two electrodes can be expressed by the activities of the reactants and products when in equilibrium ($I = 0$). However, it is very important to understand the cases where the electrochemical system is out of its equilibrium state, or when we observe the flow of electrical current. Out of its equilibrium ($I \neq 0$), the achievable or practical electrode potential is much higher than the theoretical reversible cell voltage (i.e., $E(I \neq 0) \neq E^0 = 1.226$ V). Moreover, the electrode potential difference will increase when the current is increased $\left(E = f(I)\right)$. In batteries or fuel cells, the negative electrode is the anode

and the positive electrode is the cathode. For electrolyzers, the opposite convention applies—the negative electrode is the cathode and the positive electrode is the anode. It is, therefore, crucial to consider the fact that the concepts of "anode" and "cathode" are related only to the direction of current flow, and not to the polarity of the electrodes. Consequently, owing to polarization of an electrolyzer, the potential of the cathode moves in the negative direction and that of the anode moves in the positive direction, that is, the potentials of the two electrodes move farther from each other, and the voltage increases (see Figure 4.5). The deviations from the standard electrode potential can be called overpotentials, potential losses, or electrode polarizations. Three types of potential losses are normally observed and defined:

1. Activation losses that are caused by irreversible processes at the electrode surface; meaning that the rate-determining step is the charge-transfer reaction at the electrode surface.

2. Ohmic losses that are caused by the ohmic resistances (electrolytes, separator, and electrodes).

3. Mass transport losses that are caused when the reactants and products cannot overcome the diffusion barriers across the electrochemical system, and/or their concentration at the electrode surface is diminished.

The HER system can be optimized to provide very low activation, ohmic, and mass transport losses. By using Pt-based electrodes, activation losses at the cathodes of PEM water electrolysis are very small. By using very

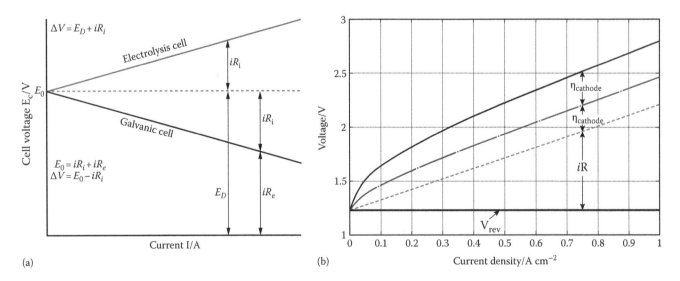

FIGURE 4.5
Representation of the cell voltage variation when current is applied to the system. (a) Comparison between a galvanic cell (fuel cells or batteries) with electrolysis. (b) Water electrolysis showing the different losses inside the system associated with an increase in the current density. (Stolten, D. and Scherer, V.: *Transition to Renewable Energy Systems*. 2013. Copyright Wiley-VCH. Figure at the bottom reprinted with permission.)

thin polymer electrolyte membranes, the ohmic losses can be drastically reduced. However, when current starts to flow, the concentrations of the different species will be different (higher or lower) at the electrode surface in comparison to its concentration in the bulk. This causes the formation of concentration profiles due to the effects of simple species diffusion. In other words, one can neither supply sufficient protons to the surface nor remove hydrogen that evolves from the electrode that would satisfy the flow of electrons. This is especially important when considering the triple-phase boundary, which, to date, has not been optimized for PEM water electrolysis.

All losses mentioned will produce an increase in the cell voltage relative to the standard cell voltage or, in the practical case, the open circuit voltage. For a current density i (where $i > 0$), the real cell voltage E_{cell} is composed of the sum of the reversible cell voltage E^0, the losses related to the ohmic drop E_{ohm}, the activation losses related to the anode E_{an} and to the cathode E_{cath}, and the mass transport losses E_{transp}.

$$E_{cell} = E^0 + E_{an} + E_{cath} + E_{ohm} + E_{transp} \qquad (4.14)$$

Figure 4.5 shows a typical current–voltage characteristic for water electrolysis and its breaking into the main polarization losses. As in PEM water electrolysis, the water is fed to the system with very high lambda value, and the cell and its components are always flooded with water, and if well-designed/structured porous transport layers are used, mass transport losses can be almost neglected.

The interfaces inside an electrochemical system are very dynamic; even in a situation of equilibrium, with species crossing from one direction to the other and vice versa, creating a counter flux. This provides a constant exchange of charged species over time. At the interface between electrode and electrolyte, the exchange of species and charges will be associated with the respective partial reactions. Therefore, when partial current densities (cathode and anode) are located in an equilibrium condition, we obtain

$$\vec{\imath} - \overset{\leftarrow}{\imath} = 0 \quad \text{or} \quad \vec{\imath} - \overset{\leftarrow}{\imath} = i^0 \qquad (4.15)$$

where

$\vec{\imath}$ is the partial current density in the cathodic direction
$\overset{\leftarrow}{\imath}$ is the partial current density in the anodic direction
i^0 is the exchange current density

The values of exchange current density observed for different electrodes (or reactions) vary within wide limits. The higher the exchange current density value (or the more readily charges cross the interface), the

more readily the equilibrium potential will be established, and the higher will be the stability of this potential against external effects. Electrode reactions for which equilibrium is readily established are called thermodynamically reversible reactions. But low values of the exchange current indicate that the electrode reaction is slow (kinetically limited). The HER on Pt using acidic solutions is one of the fastest known electrochemical reactions, and very high exchange current densities ($i^0 > 10^{-3}$A cm^{-2}) in acidic electrolytes are obtained. However, the value chosen for the exchange current densities tends to vary over several orders of magnitude throughout the literature. It is, therefore, very difficult to accurately determine the HER kinetics, especially in the acidic regime, where it is experimentally problematic to eliminate the mass transport losses inside the electrochemical cell. Hence, it is a challenge to understand or elucidate the kinetics of the HER for different materials and different electrode structures. One way to circumvent this is by reducing the temperature of operation and/or by performing the electrochemical characterization in alkaline solutions, since the kinetics of the HER is significantly hindered under these conditions.

4.4 Techniques Used to Study the Hydrogen Evolution Reaction

Various physicochemical and electrochemical techniques have been used to characterize the HER. The most widely used are cyclic voltammetry (CV), chronoamperometry (CA), and linear sweep voltammetry (LSV). Thousands of articles, research reports, and dissertations report the use of these techniques to characterize different catalysts and components for the HER. It is important to provide more detailed descriptions of these three powerful techniques and also of high relevance to provide an overview of the physicochemical methods used for the characterization of catalysts and electrode systems for the HER.

4.4.1 Physicochemical Characterization Techniques

A study of interfacial regions between electrode and electrolyte is essential for the development of PEM water electrolysis. The interfacial regions contain the electrochemically active sites, that is, the three-phase boundary where reactants, electrode, and electrolyte are combined. However, due to the different potential losses, even well-designed interfaces contribute to specific losses and will limit the overall cell performance. Moreover, these interfacial regions are vulnerable to microstructural changes during cell operation. These regions are also active sites for mass transport, diffusion, mechanical and chemical stress, and material segregation under the operating conditions. Microstructural changes in the interfacial regions inevitably affect the cell performance and, in many cases, cause degradation of the components. Thus, there is a demand for more precise analysis of the interfacial microstructure on a nanoscale. In this context, scanning electron microscopy (SEM) and transmission electron microscopy (TEM) are the preferred analytical techniques. Difficulties associated with determining the nanoscale structure and chemistry have led to disagreement over the nature of the best catalyst for the HER. The use of TEM with subnanometer analytical probe sizes now permits the direct analysis of the chemical state and structure of individual nanoparticles by energy dispersive x-ray spectroscopy (EDX) and electron energy loss spectroscopy (EELS), enabling the identification of the best Pt catalyst morphology.

4.4.1.1 Identical Location Transmission Electron Microscopy (IL-TEM)

Fundamental investigations of electrocatalysts are essential for the further development of electrode systems for PEM water electrolysis. Besides the activity, and selectivity for certain electrochemical reactions, the long-term stability of the catalyst is of major interest. Whereas many advances in the performance of catalysts for the HER have recently been reported, improvements in catalyst durability are scarce (Sethuraman et al., 2008; Zhang et al., 2009; Debe et al., 2012; Carmo et al., 2013). This is partially due to the lack of investigative techniques that enable an effective analysis of degradation processes in electrolyte solutions. Generally, the loss of electrochemical active surface area is determined in situ by electrochemical methods (Schmidt et al., 1998). In addition, x-ray diffraction (XRD) and TEM are applied to obtain the average crystallite size and complete size distributions, respectively. However, these techniques are considered to be destructive for the *in situ* study of electrocatalysts since the catalyst has to be removed from the working electrode. Information gathered on the surface area loss and/or particle growth is often not sufficient for a detailed description of occurrences on the catalyst particles on the nanometer scale. As a consequence, several theories have been proposed to explain the loss in the active surface area of catalysts. There are four primary mechanisms believed to be of relevance to low-temperature fuel cells and which can be potentially transferred to PEM water electrolyzer catalysts, since similar micro/nanostructures for the electrodes are found.

The four mechanisms can be described as follows:

1. Ostwald ripening: metal ions dissolve from smaller particles, diffuse, and redeposit onto larger particles, resulting in reduced metal surface area via a minimization of the surface energy.

2. Reprecipitation: Pt dissolves into the ionomer phase within the cathode and then precipitates again as newly formed Pt particles via the reduction of hydrogen.

3. Particle coalescence: Pt particles that are in close proximity sinter together to form larger particles.

4. Corrosion of the carbon support that anchors the Pt particles and provides electrical contact.

These particle growth mechanisms and their rates may vary as a function of electrode potential, intermittent power input, current density, particle size and shape, the hydration state of the membrane, and many other operating conditions (temperature and pressure). To obtain an improved understanding of the degradation mechanism of fuel cell catalysts, a new technique identical location transmission electron microscopy (IL-TEM) was developed by Mayrhofer et al. 2008). It is a nondestructive method, based on TEM analysis, which enables the investigation of identical locations on a catalyst before and after electrochemical treatments (Mayrhofer et al., 2008). The results from studies on a standard, commercially available fuel cell Pt catalyst demonstrate the high potential of this method for PEM water electrolysis.

4.4.1.2 X-Ray Spectroscopy

XRD methods are almost exclusively used to identify the phases of crystal growth. For the most part, the diffraction measurements are performed on catalyst samples so as to provide very accurate phase determinations, crystallinity and/or amorphous degree, and particle size.

4.4.1.3 Thermal Analysis

Determination of the behavior of compounds with increasing temperature can be accomplished by using two main techniques: differential scanning calorimetry (DSC) and thermal gravimetric analysis (TGA). The presence and characterization of phase transitions both above and below room temperature, metal loadings, and T_g value for electrocatalysts and catalyst/electrolyte systems can be accomplished by using TGA and DSC measurements.

4.4.2 Electrochemical Characterization Techniques

Most electrochemical studies are carried out using a three-electrode setup comprising a working electrode, a reference electrode, and a counter electrode. The majority of electrochemical experiments are carried out using a two-compartment electrochemical jacketed glass cell. Figure 4.6 shows the picture of a three-electrode setup in an electrochemical cell. The working electrode is the electrode where the HER is studied. The counter electrode serves to complete the electrochemical circuit as the current flows between the working electrode and the counter electrode. A problem that often occurs when using CV is the dissolution of the counter electrode material and redeposition at the working electrode. In order to avoid this phenomenon, a counter electrode of the same material as the working electrode is usually employed and recommended. When working in potentiostatic mode, or when a potential difference is applied to the working electrode, the counter electrode will adjust the potential so that current spontaneously flows between the two electrodes. However, the applied potential is always measured to a reference value or to a reference electrode where a well-defined redox reaction at the reference electrode occurs. The most commonly used reference electrode is the SHE. The SHE is a redox electrode. It forms the basis of the thermodynamic scale of oxidation–reduction potentials and a basis for comparison with all other electrode reactions. It is declared to be zero at all temperatures. The potential differences of any other electrodes should, nonetheless, be compared with that of the SHE at the same temperature.

The SHE is based on the redox half-cell at a platinized Pt electrode:

$$2H^+_{(aq)} + 2e^- \rightarrow H_{2(g)} \tag{4.16}$$

A common and simple way to fabricate SHE is by dipping a Pt wire in an acidic solution through which pure hydrogen gas is bubbled. It is important to ensure that the concentration of both the reduced and oxidized hydrogen remains at unity, implying that the pressure of hydrogen gas is at 1 bar and the activity of hydrogen ions in the solution is one. The activity of hydrogen ions is their effective concentration, which is equal to the formal concentration multiplied by the activity coefficient. The unit-less activity coefficient is close to one for very dilute aqueous solutions, but it is usually lower for more concentrated solutions. It is, therefore, necessary for the potential of the reference electrode to remain practically constant in order to have a precise value for the potential at the working electrode. This is accomplished by using very pure and clean components and acidic solutions and by maintaining stable activities of the species

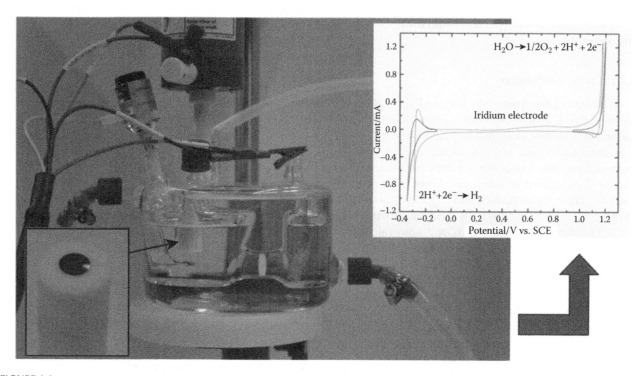

FIGURE 4.6
Schematic representation of a two-compartment jacket electrochemical cell with a Pt working electrode (WE), a Pt mesh counter electrode (CE), and a Gaskatel standard hydrogen reference electrode (SHE).

in the half-cell reaction quotient. Pt is the best choice as an electrode for the SHE. Reasons for this include the following: Pt is inert, it will not corrode, and it will catalyze both oxidation and reduction reactions. Pt has a high intrinsic exchange current density for proton reduction, and excellent reproducibility of the potential (potential difference < 10 μV when two well-made hydrogen electrodes are compared). It is preferable to use platinized Pt wire (i.e., Pt covered with electrodeposited Pt black) so that the total surface area is increased. This will improve the reaction kinetics and maximize the current density for the redox reaction. The increased surface area will also better adsorb hydrogen at its interface. It is also important to protect the reference electrode from contamination, and temperature or concentration gradients during the electrochemical analysis, so that potential shifts or errors during the experiments are avoided.

For most electrochemical experiments regarding the HER, use of the SHE is very practical and convenient as it will be used in the same electrolyte as the working electrode compartment of the electrochemical cell, ensuring no contamination through the diffusion of ions from SHE. The electrochemical characterization can be simply performed by using the SHE in the same compartment as the main electrode, where the working electrode and the SHE are in electrolytic contact via a Luggin capillary that is filled with the same electrolyte. A glass frit or a Nafion membrane can also be used to avoid mixed potentials and contamination through the evolution of oxygen at the counter electrode. However, special attention must be given to avoid undesired extra ohmic resistances.

4.4.2.1 Cyclic Voltammetry

CV is a powerful electroanalytical technique that essentially relies on the imposition of a triangular potential wave on the working electrode that one wants to study.

It is used to study a variety of different processes, demonstrating the versatility of this technique. The response is measured in terms of current (I) or current density (i). The cycling turns (from $E_{initial}$ to E_{final}) usually lie between the HER and the OER. The rate at which the potential is scanned is referred to as a scan rate (v). In practice, v values of 1 mV s^{-1} to several hundred V s^{-1} are common, and any adsorbed impurities at the working electrode surface can be removed by either oxidation or reduction. The resulting profile of a given studied material can be thought of as an electrochemical profile, indicating at which potential value or window the specific redox electrochemical process takes place. By scanning a given electrode material, very reproducible profiles can be obtained, especially for metals like Pt, Au, and Pd. This reproducibility will always depend on the potential window that is chosen, the electrolyte

purity, the cleanliness of the cell and its components, the condition and composition of the working electrode, the scan rate, and other factors.

The current response of electrodes with different sizes and/or surface morphology is often given as current density, where the absolute obtained current can be divided by the electrode geometric area, electrode weight (mass in grams)—mass activity, or divided by the electrochemically active surface area (A_{ecsa})—specific activity. The electrochemically active surface area (ECSA) of Pt-based catalysts can be measured by the electrochemical hydrogen adsorption/desorption method. This method is based on the formation of a hydrogen monolayer electrochemically adsorbed on the catalyst's surface. In the case of Pt catalysts, two or sometimes three well-resolved peaks on the anodic sweep at the low potential area correspond to the hydrogen desorption on the electrode surface (see Figure 4.7). This technique is ubiquitous in catalysis laboratories working with supported metals. It gives a precise measure of the dispersion of the metal as each surface atom chemisorbs (in most cases) a single hydrogen atom. It is well accepted that the characteristic value of charge density associated with a monolayer of hydrogen adsorbed on polycrystalline Pt is 210 μC cm^{-2}, and the ratio of the ECSA to the geometric area can also provide the roughness factor. As an example, for catalysts like Pt/C, the roughness factor can be very high (1–10^5), depending on the dispersion of the active components and the microstructure of the nanoparticles and support material (Zhang, 2008). The measured current is also a combination of faradic current coming from the electrochemical processes at the electrode surface and non-faradic current (capacitive current) coming from the charging process of the electrochemical double layer at the interface of the electrode and the electrolyte. It is consequently crucial to extract the charges ascribed to double-layer capacitive currents when calculating the ECSA.

Figure 4.7(a) shows a schematic representation of the potential versus time behavior at the working electrode following imposition of a triangular waveform typical of CV, and Figure 4.7(b) shows an example of a CV profile that corresponds to a polycrystalline Pt electrode in aqueous sulfuric acid electrolyte. First, the non-faradic current associated with the charging of the electrochemical double layer increases when the scan rate is increased. Since a combination of faradic and non-faradic current is measured, the entire curve is also shifted upward due to this increase in double-layer current upon increasing the scan rate.

4.4.2.2 Chronoamperometry and Chronopotentiometry

Chronoamperometry (CA) is the measurement of the current or current density response of an electrochemical

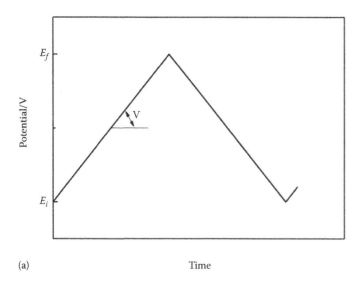
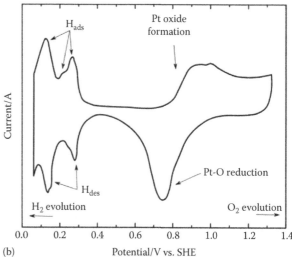

(a) Time

(b) Potential/V vs. SHE

FIGURE 4.7

(a) Program for cell potential variation applied to the working electrode for cyclic voltammetric analysis. (b) Typical cyclic voltammogram profile for polycrystalline Pt in sulfuric acid.

reaction by applying a fixed potential, whereas chronopotentiometry is the measurement of the potential response obtained by applying a fixed current. The response is, therefore, recorded as a function of time. Ideally, an equilibrium rest potential step is chosen to provide adequate stabilization of the electrochemical system, that is, a potential value where no faradaic electrochemical reaction takes place. A second potential step should follow at a potential region where the reaction, process, or condition of interest exists. In this case, the current response does not correspond only to the faradic electrochemical process, but partially to the charging of the double layer that gives rise to a sharp peak, especially at the beginning of the CA profile, where the faradic current is completely masked by the contribution of the electrochemical double layer (Hamann, 1998). Essentially, analysis of the *I vs. t* or CA profiles provides qualitative information about how the reaction that is taking place behaves over time. For most HER systems, we observe a mixed behavior, where the rate of diffusion and the rate of electron transfer occur equally or simultaneously. In order to obtain the response from only the kinetic control process, the rotation of the electrode at a given speed is used.

Quantitative information can also be obtained from CA measurements. The *I* response may be integrated and converted to charge (*Q*), and plotted as a function of time. This curve can then be integrated to determine the total charge transferred, which can be compared to product yields or other analytical information obtained from the system. Alternatively, many *I/E* instruments have the ability to run chronocoulometry: *Q* is directly measured using a built-in capacitor for real-time charge integration.

4.4.2.3 Rotating Disk Electrode Technique

Over the last century, research and development efforts have been invested to find Pt electrocatalysts' alternatives for PEM water electrolysis. However, nanoparticles of Pt supported on high surface area carbon black still remain as the catalyst benchmark for the HER. Whereas the cathodic evolution of hydrogen or its oxidation on Pt is facile and well understood, the Pt substitution is not so trivial. Consequently, the development of an improved cathode catalyst would provide a major boost for the technology on the road toward its aspired commercialization.

Since testing catalysts in the form of a membrane electrode assembly is quite complex and time consuming, a more straightforward alternative based on the rotating disk electrode (RDE) method is commonly used. The supported catalyst is usually deposited onto a glassy carbon disc and can be readily tested in an ordinary electrochemical glass cell (see Figure 4.6). As a consequence of a relatively thin or thick catalyst layer combined with a high Nafion content, the early stages of this method's development faced complications due to a strong diffusion resistance of the reactants through the catalyst film. By attaching the catalyst layer to the glassy carbon support via an only submicrometer thick Nafion film placed on top of the dried catalyst layer, Schmidt et al. (1998) were able to minimize the film diffusion resistance.

Thus, it became possible to determine the kinetic current densities directly from the RDE data by using the general mass transport correlations of the RDE for flat electrodes (Vielstich et al., 2003; Gasteiger et al., 2005). In addition to this, the significantly reduced catalyst loading extends the potential region in which the kinetics can be studied, so that a comparison with HER performance data from the literature can be established. The direct correlation of kinetic activities obtained using RDE was recently verified by Gasteiger et al. (Sheng et al., 2010). However, despite these advancements in testing catalysts using the RDE methodology, quite contradictory reports on the catalytic activity of the HER, even for simple systems such as polycrystalline Pt, can still be found in the literature (Carmo et al., 2013). For high surface area Pt catalysts, the analysis of the kinetic activity is even more complex and the literature data more ambiguous. It is beyond doubt that, for meaningful studies, both fundamental in nature and for practical applications, unambiguous activity benchmarks are a prerequisite. Only then can the influence of factors such as the catalyst composition and structure on the catalytic activity be examined.

4.4.2.4 Electrochemical Impedance Spectroscopy

Over the last 30 years, great advances have been made in electrochemical impedance spectroscopy (EIS) (Macdonald, 2006). It is a powerful analytical technique that can provide a wealth of information on the cell resistivity, charge transport, mass transport, electrical charge transfer for fuel cells (Gomadam and Weidner, 2005, Wippermann et al., 2014), and for PEM water electrolysis (Dedigama et al., 2014). It offers a number of advantages over conventional electrochemical polarization techniques. However, the fitting and interpretation of EIS spectra is the key to extracting useful information with regard to the details of the structure and the electrochemical processes of HER electrodes. EIS fitting is usually carried out through an equivalent circuit, which is an assembly of circuit elements, representing the physical and electrical characteristics of the electrochemical interface.

4.5 Electrocatalysts for the Hydrogen Evolution Reaction

4.5.1 State of the Art

Since the electrolysis phenomenon was discovered by Troostwijk and Diemann in 1789 (Trasatti, 1999), Pt was always recognized as the best and therefore standard catalyst for the HER. Over the years, much research and development has been carried out,

aimed at reducing the amount of Pt to very low loadings, even to the point of substituting it. For classic alkaline electrolysis, electrolysis companies rely on Ni-based catalysts. Whenever used, Pt with very low loadings was mixed or added to Ni-based catalysts, aiming to provide more stable catalysts against the nickel hydrate formation on the HER of classic alkaline electrolysis cathodes (Zeng, 2010). However, in the 1960s, due to the three major issues normally associated with alkaline electrolyzers, namely, a constrained low partial load range, limited current density, and low operating pressure, GE started to develop water electrolysis based on an SPE membrane (Russell et al., 1973). PEM electrolyzers can operate at much higher current densities and values greater than 2 A cm^{-2} can be easily achieved (Carmo et al., 2013). Due to the very low ohmic losses (thin electrolyte membrane capable of providing good proton conductivity: 0.1 S cm^{-1}), higher current densities can be achieved. The low gas crossover rate of the polymer electrolyte membrane (yielding hydrogen with high purity), enables the PEM electrolyzer to function over a wide power input range (an important economical aspect). As discussed earlier, when alkaline electrolyzers operate with low load, the rate of hydrogen and oxygen production decreases, while the hydrogen permeability through the diaphragm remains constant, yielding a larger concentration of hydrogen on the anode (oxygen) side, and thus creating hazardous and less efficient conditions (Barbir, 2005). In contrast to alkaline electrolysis, PEM electrolysis covers a much larger power density range. This is due to the high proton conductivity of Nafion membranes and the low permeability of hydrogen through Nafion (<1.25 10^{-4} cm^3 s^{-1} cm^{-2} for Nafion 117, standard pressure, 80°C, 2 mA cm^2) (Barbir, 2005). A solid electrolyte also allows for a compact system design with strong/resistant structural properties, in which high operational pressures (equal or differential across the electrolyte) are achievable (Medina and Santarelli, 2010).

The corrosive acidic regime provided by the proton exchange membrane requires the use of distinct materials. These materials must not only resist the harsh corrosive low pH conditions, but also sustain the high applied overvoltage, especially at high current densities. This demands the use of scarce, expensive materials and components based on platinum group metals (e.g., Pt, Ir, and Ru), and Ti-based current collectors and separator plates.

Catalyst development for water electrolysis using liquid acid electrolytes, for both HERs and OERs, was reported several decades ago. A series of studies on the HER have been published by Furuya and Motoo (Furuya and Motoo, 1976, 1977, 1978, 1979a,b,c).

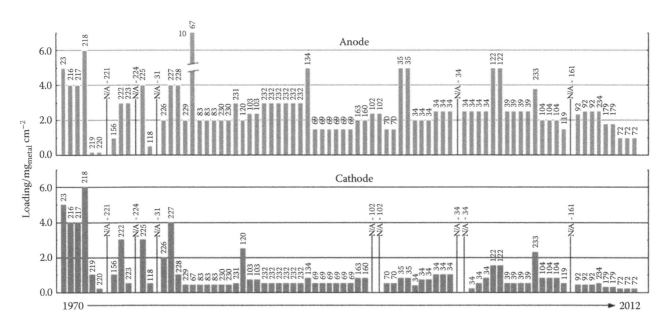

FIGURE 4.8
Representation of the catalysts loadings (anode and cathode) of PEM electrolysis experiments. Numbers above the bars designate the references related to the individual results. (Reprinted from *International Journal of Hydrogen Energy*, 38, Carmo, M., Fritz, D.L., Mergel, J., and Stolten, D., A comprehensive review on PEM water electrolysis, 4901–4934, Copyright (2013), with permission from Elsevier.)

Miles and Thomason have reported a summary of activities for a variety of elements for the HER and the OER using cyclic voltammetric techniques (Miles and Thomason, 1976). When using single elements alone, the HER will be essentially dependent on noble elements. For the HER, in 0.1 mol L^{-1} H$_2$SO$_4$ at 80°C, it was found that the catalytic activity order was Pd > Pt > Rh > Ir > Re > Os > Ru > Ni. Figure 4.8 gives the values of catalyst loading as reported in several publications. Initially, the catalyst loadings for the cathode were very high, but they soon decreased since high amounts of catalysts for the HER were not necessary (OER is the limiting reaction). To date, the goal for catalyst loading reduction for the cathode side ranges between 0.05 and 0.2 mg cm^{-2}.

4.5.2 Commercial Catalysts

Scientists are required to solve a fundamental problem in order to promote renewable energy generation and promote the commercialization of PEM water electrolysis components. This, in turn, requires the development of high-performance and durable electrocatalysts. These catalysts should also be of low cost and have an economical and scalable production. Fortunately, for the cathode side, PEM water electrolysis companies can rely on the well-known Pt/C catalysts already developed for fuel cell applications. The cost of Pt is currently (2015) around U.S. $1200 per troy ounce. For research purposes, scientists have for a long time based their standards on

Pt/C catalysts from Johnson Matthey, BASF, Umicore, Tanaka, and 3M, among others.

4.5.3 Research Directions

The challenges related to PEM electrolysis have mainly been associated with the development of electrocatalyst systems, porous transport layers, and bipolar plates for the anode side. In most of the early studies, researchers used Pt black as a standard catalyst for the HER. Later, due to experience gained in the development of catalysts for PEM fuel cells, researchers diverted to use Pt/C from different manufacturers as their standard catalysts. However, despite the lower Pt loadings compared to the anode side, the cathode catalyst still represents a considerable portion of the total system cost. Today, loadings for the cathode side range between 0.5 and 2 mg cm^{-2} and further reductions will be always desired, with potentials for reaching values lower than 0.2 mg cm^{-2}. After 2005, a few reports appeared attempting to reduce the Pt loadings, improve catalyst utilization (homogeneity and particle size), and potentially apply the substitution of Pt-based catalysts (creating the so-called Pt-free catalysts) (Ayers et al., 2010).

When studying the cathode catalyst, due to the dominance of the OER, experimentation is usually carried out in a liquid electrolyte half-cell or three-electrode apparatus to isolate the contribution of the cathode. Research groups have experimentally studied MoS$_2$ as an alternative catalyst for the HER (Hinnemann et al., 2005a,b, Li et al., 2011, Ge et al., 2012, Vrubel et al., 2012, Chung et al., 2014, Lu et al., 2014, Ma et al., 2014, Wang et al., 2014a,b, Xie et al., 2014, Yu et al., 2014). However, most of these experiments showed that MoS_2 is a reasonable material for the HER, but current densities are significantly lower compared to conventional Pt cathodes. The same observation applies for the study of other alternative materials for the HER, like $CuNiWO_4$ (Selvan and Gedanken, 2009), heteropolyanions of tungstophosphoric acid hybridized with carbon nanotubes (CNTs) (Xu et al., 2007), and WO_3 nanorods (Rajeswari et al., 2007, Zheng and Mathe, 2011). These catalysts are presented as good alternatives for the HER, but unfortunately proper comparisons to standard Pt catalysts are generally lacking. A family of macrocycles has also recently received attention for the HER because of potential use as molecular electrocatalysts for many electrochemical reactions. Co and Ni glyoximes were evaluated by Pantani et al. (2007); the possibility of using these immobilized metals (Co and Ni) in the HER was investigated. It was reported that their activities remained stable, but the results were not comparable to what could be achieved with Pt. The redox potentials need to be shifted to higher voltages, closer to 0 V vs. the SHE, by chemically modifying the ligands bonded to the metal active sites. It was also suggested that these compounds must be dispersed onto an appropriate electronic conductor to increase the contact area between the catalyst and the electrolyte. Before glyoximes can be seen as a viable Pt replacement, the reaction rate and stability must be improved. In order to achieve this, a better understanding of the HER mechanism is needed (Millet et al., 2009). Pd/CNTs have also been tested for the HER, but no consistent difference was obtained compared to Pt/CNTs (Grigoriev et al., 2008, 2011). CNTs are commonly used as supports because they are generally recognized as having higher electron conductivities and corrosion resistance compared to conventional carbon black (Carmo et al., 2005; Taylor et al., 2008).

4.5.4 Advanced Materials and Catalyst Structures

Based on the positive results of electrocatalysts recently developed for PEM fuel cells, analogous catalyst materials and methods were studied for possible employment for PEM electrolysis. Much knowledge is available in many publications related to PEM fuel cells and electrochemical systems, such as batteries and solar cells. This knowledge can be potentially assigned to PEM water electrolysis. In catalysis, it is generally recognized that high-surface area multicomponent nanowire alloys are

particularly promising for increasing the activity and utilization of precious metal catalysts. These improvements are due to ensemble effects that arise when dissimilar surface atoms induce electronic charge transfer between themselves, thus improving their electronic band structure (Strasser et al., 2010). However, the common strategies or processes to produce advanced catalyst systems with these characteristics involve complex synthesis methods. This is due to difficulties in fabricating these advanced nanostructures with high dispersion and noble metal utilization.

Core-shell catalysts essentially consist of atomic metallic monolayers (such as Pt) supported on a non-noble metallic core substrate (such as Cu) (Alayoglu et al., 2008). They provide beneficial effects of bimetallic catalysis, tuning the surface, and catalytic reactivity of these catalysts for different electrochemical reactions. Core-shell systems are prepared by preferential dissolution (removal) of the electrochemically more reactive component from a bimetallic alloy. In other words, in a Pt–Cu configuration, Cu will be removed from the surface and concentrated on the core of the catalysts' structure, with Pt being concentrated on the outer shell of the catalyst. According to Strasser et al., the dealloyed Pt–Cu nanoparticles have presented uniquely high catalytic reactivity for the oxygen reduction reaction in fuel cell electrodes, with pure Pt as the preferred catalyst (Strasser, 2009, Strasser et al., 2010, Marcu et al., 2012). It has also been pointed out that the core-shell catalysts meet and exceed the technological activity targets in realistic fuel cells and can reduce the required amount of Pt by more than 80%. In order to decrease the metal loadings used in PEM electrolysis, core-shell structures could be a promising alternative. Pt–Cu core shell catalysts could be used on the cathode side, drastically reducing the loading to values lower than 0.2 $mgcm^{-2}$, and still improving the efficiency.

Bulk metallic glasses (BMGs) were recently demonstrated as a new catalyst platform for fuel cells (Carmo et al., 2011b, Mukherjee et al., 2012, Mukherjee et al., 2013, Sekol et al., 2012, 2013). BMGs exist in a wide range of compositions and can be thermoplastically formed into complex geometries over a length scale ranging from 10 nm to a few centimeters (Schroers, 2010). The absence of grain boundaries and dislocations in the BMG amorphous structure result in a homogeneous and isotropic material down to the atomic scale, which displays very high strength and elasticity combined with good corrosion resistance. These studies also showed that Pt-BMG properties, composition, and geometry are possibly suitable for high-performance electrocatalysts. Furthermore, the high level of controllability of Pt-BMG during thermoplastic formation in the highly viscous supercooled liquid region results in high catalyst dispersion without the need for a high surface area conductive support (e.g., carbon black).

Nanostructured thin films (NSTFs) from 3M are an another example of advanced structures developed for PEM fuel cells and PEM electrolysis. According to Vielstich et al. and Debe et al., the use of NSTF catalysts eliminates or significantly reduces many of the performance, cost, and durability barriers standing in the way of cathodes and anodes for H_2/air PEM fuel cells (Vielstich et al., 2003). The specific activities and durability of NSTF catalysts seem to be significantly higher than those of conventional carbon-supported Pt catalysts, and mass activities closely approach the 2015 DOE targets of 0.44 A mg^{-1} at 900 mV (Debe et al., 2012). The NSTF catalysts are formed by vacuum sputter deposition of catalyst alloys onto a supported monolayer of oriented crystalline organic pigment whiskers. Whiskers are corrosion resistant, therefore eliminating the high voltage corrosion issue affecting carbon blacks. According to a study by Debe et al., the NSTF whisker is resistant to chemical or electrochemical dissolution/corrosion, because the perylene dicarboximide pigment is insoluble in typical acids, bases, and solvents, while its single crystalline nature confers it with a "band gap" that does not support electrochemical corrosion currents greater than 2 V. NSTF catalysts with low catalyst loadings were assembled in a single cell and short stack PEM electrolyzer. The catalysts were tested for performance and durability at Pt loadings of 0.10–0.15 mg cm^2. The NSTF catalyst alloys were reported to show good OER and HER activities. The cathode performance of NSTF $Pt_{68}Co_{32}Mn_3$ and sputtered NSTF $Pt_{50}Ir_{50}$ catalysts was claimed to be equivalent to that of standard Pt blacks, having loadings of an order of magnitude higher and with durability greater than 4000 h.

4.6 Catalyst Supports for the Hydrogen Evolution Reaction

4.6.1 Standard Concept/Approach

By definition, a catalyst support is the material, usually a solid with a high surface area, to which the catalyst nanoparticles are anchored. The reactivity of heterogeneous catalysts and nanomaterial-based catalysts occurs at the surface atoms. Consequently, great efforts are made to maximize the surface area of a catalyst by distributing it over the support. The support may be inert or participate in the catalytic reactions. Typical supports include various kinds of carbon and transition metal oxides. Carbon black supports typically have surface areas greater than 200 m^2 g^{-1} (Antolini, 2009). Thus, if a precursor salt of the metal can be effectively distributed over this area, metallic particles smaller than 10 Å can be obtained. In the case of costly materials, such as noble metals, obtaining smaller particles becomes a necessity as they offer a larger surface-to-volume ratio and thus greater catalytic utilization of the metal. Maintenance of this dispersion under severe processing conditions is another function of the support because separation of the metal particles inhibits their agglomeration. Clearly, the primary function of the support is physical in nature, and effective dispersion is not dependent on any degree of metal. There are a few processes that can be used to activate the support before anchoring the nanoparticles (Carmo et al., 2008, 2009a,b, 2011a). For example, many catalyst supports are activated by exposure to oxidizing agents to promote the formation of oxygenated species on the catalyst-support surface. Pt/C is normally used as the standard cathode catalyst in PEM water electrolysis. Because the activity of a catalyst increases as the reaction surface area of the catalyst increases, catalyst particles should be reduced in the diameter to increase the active surface. However, one must consider that the specific activity of the metal nanoparticles can decrease with decreasing particle size (particle size effect) (Antolini, 2009). In contrast to the anode catalyst, the structure and proper dispersal of the Pt metal particles make low loading catalyst feasible for PEM electrolysis operation. In addition to a high surface area, which may be obtained through high porosity, a support must also have sufficient electrical conductivity so that the support can act as a path for the transfer of electrons. Moreover, carbon supports should have a high percentage of mesopores in the 20–40 nm region to provide a high accessible surface area to the catalyst and to monomeric units of the Nafion ionomer and to enhance the diffusion of chemical species. Besides the dispersion effect of the support material, an interaction effect between the support material and the metal catalysts exists.

Since the catalysts are bonded to the support, the support material can potentially influence the activity of the catalyst. This interaction effect can be explained in two distinct ways. First, the support material could modify the electronic character of the catalyst particles. This electronic effect could affect the reaction characteristics of the active sites present on the catalyst surface. The second is a geometric effect. The support material could also modify the shape of the catalyst particles. Those effects could change the activity of the catalytic sites on the metal surface and modify the number of active sites present (Antolini, 2009).

On this basis, an important issue of the research is addressed: the development of new carbon and noncarbon supports, which could improve the electrochemical activity of the catalysts (Carmo et al., 2005, 2013). The stability of the catalyst support in the PEM electrolysis environment is of great importance when developing new substrates. In addition to high surface area, porosity, and electrical conductivity, corrosion resistance is also an important factor in the choice of a good catalyst support. If the catalyst particles cannot maintain their structure over their operational lifetime, change in the morphology of the catalyst layer from the initial state will result in a loss of electrochemical activity. For these supported catalysts, several requirements have to be met to achieve the required long-term stability of 40,000–60,000 h. During the development of these catalysts for fuel cell systems, it has been found that the carbon catalyst support degrades over time. It was found that carbon is lost from the system through oxidation, leading to significant losses of carbon over a short period of time. With respect to carbon blacks, these new carbon materials are different both at the nanoscopic level in terms of their structural conformation (e.g., when compared to carbon nanotubes) and pore texture (e.g., mesopore carbons) and/or at the macroscopic level in terms of their form (e.g., microspheres).

4.6.2 Triple-Phase Boundary

The basic definition holds that the HER can only occur at confined spatial sites, called "triple-phase boundaries," where electrolyte, gas, and electrically connected catalyst regions contact. A simplified schematic of the triple-phase boundary is shown in Figure 4.9. The reaction kinetics depends very much on the losses inside the cell, and the triple-phase boundary, if not well optimized, will consequently increase the electrode losses. Therefore, understanding, characterizing, and optimizing the triple-phase boundary for PEM water electrolysis is essential for obtaining performance enhancement.

To date, efforts to optimize the triple-phase boundary in PEM water electrolysis have been rather scarce. However, when using advanced nanostructured catalyst layers, optimization of the triple-phase boundary could be achieved. By employing nanoscale composites of a catalyst material, a conductive support, solid electrolyte, and gas pore space, it is possible to significantly increase the characteristics of the triple-phase boundary, thus improving overall performance. For PEM fuel cells, recent efforts have been made to clearly delineate the nature and properties of the triple-phase boundary (Vielstich et al., 2003). In many of these studies, there is a growing realization that the simple concept of the triple-phase boundary as a singularity is unrealistic; rather, it should be thought of as a "zone," whose width, properties, and behavior depend on a complex interplay between coupled

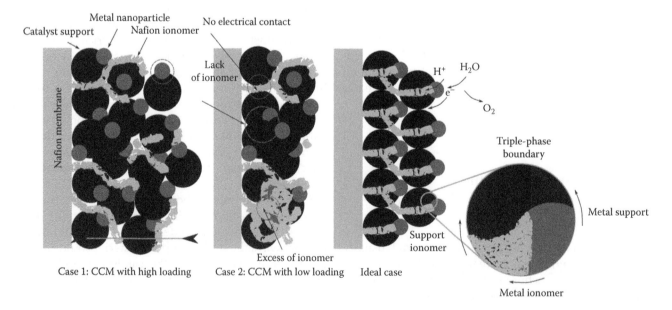

FIGURE 4.9
Representation of the catalyst-coated membrane (CCM) micro/nanostructure on three different cases.

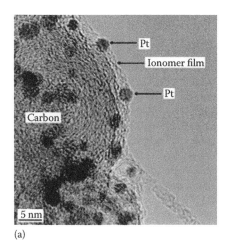
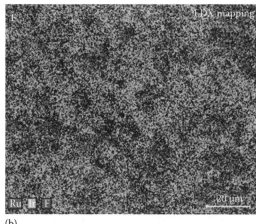

(a) (b)

FIGURE 4.10
(a) TEM analysis of the triple phase boundary of fuel cell electrodes. (b) Catalyst coated membrane (anode side) characterized by EDX mapping showing the inhomogeneous distribution of ruthenium, iridium, and nafion ionomer. In order to distinguish between colors, please check the online version for the references related to each picture. (Reprinted from More, K. et al., 2006, *ECS Transac.*, 3, 733; Xu and Scott, 2010, *Int. J. Hydrogen Ener.*, 35, 12037. With permission.)

reaction and diffusion processes. Figure 4.10 shows the results of a study (Xu and Scott, 2010) in which the physical characterization of the catalyst-coated membranes CCMs for PEM water electrolysis was used to demonstrate how inhomogeneous the triple-phase boundary of conventional electrodes is.

Ideally, the triple phase must be homogenously distributed across the catalyst-coated membrane structure. Unfortunately, due to the use of nonoptimized methods of fabrication, inhomogeneous catalyst layers are obtained, ultimately affecting the performance and durability of the electrodes.

4.6.3 Catalyst Utilization

Studies on reaction kinetics have related the electrode performance to the amount of Pt surface area observable using CV. However, these studies assume 100% catalyst utilization, or that there is negligible potential drop across the electrode (an assumption that will not always be valid). Because of both the resistance to proton transport in the cathode and the charge-transfer resistance change as a function of electrolyzer operating conditions, protons migrating through the ionomer phase of the cathode electrode may react closer to the membrane, in cases where resistance to proton transport is much greater than the charge-transfer resistance of the HER, or react throughout the entire layer, in cases where the resistance to proton transport is much less than the charge-transfer resistance of the reaction.

During the first studies on the use of ionomers inside the catalyst layer of a PEM system, Wilson and

Gottesfeld suggested that the impregnation of Nafion into the catalyst layer was not very uniform—partial or whole areas of the catalyst layer were impregnated with Nafion (Vielstich et al., 2003). Further addition of Nafion resulted in the formation of a film on the external surface of the electrode. To circumvent these problems with the gas diffusion electrode, a thin-film catalyst layer was prepared by mixing Nafion solution with Pt/C catalyst. Nevertheless, Pt utilization in the thin-film catalyst layer is still unknown. As Nafion is an electron insulator, it is considered possible for the Nafion solids to cover the carbon surface and block the conduction of electrons by the carbon particles supported with Pt, subsequently causing the decrease in Pt utilization (see Figure 4.9). The low Pt utilization in the electrodes not only causes a waste of Pt but also restricts further improvement of the electrode performance. To create electrodes with high Pt utilization and better performance, it is necessary to investigate Pt utilization in currently available PEM electrolysis electrodes and determine how it is influenced by other factors such as catalyst-support composition, Pt loading, ionomer loading, and other additives. To date, these objectives have not received much attention and little investigation has been carried out on the influence of Nafion on the Pt in the catalyst-coated membranes or catalyst-coated substrates. Sometimes, polytetrafluoroethylene (PTFE) is also used to mechanically stabilize the catalyst layer; however, no particular advantage is expected when PTFE exists in the immediate vicinity of the catalyst sites. For PEM fuel cells, PTFE has generally been used on the cathode side to facilitate the removal of the water produced. In PEM electrolysis, water is the reactant, and therefore, flooded electrodes are not an issue. In the preparation process of the thin-film catalyst layer, although Pt/C is blended with Nafion solution ultrasonically, it is still unknown whether Nafion will spread uniformly on the carbon surface or form clumps in the catalyst layer after the solvent had evaporated. Additives are sometimes also included in the catalyst paste/ink preparation and will, to some extent, affect the structure of the CCM. This kind of structural difference will greatly influence proton and electron conduction in the catalyst layer. To date, although many efforts have been made to investigate the morphology of the catalyst layer by observing the cross-sectional surface of the electrode using SEM/EDX, the morphology of Nafion in the catalyst layer for fuel cells or electrolyzers is still unclear. Another problem is the difficulty when performing physicochemical characterization. Due to the differences of the components inside the catalyst layer, it is very difficult to clearly distinguish the Pt/C catalyst, the Nafion on the uneven cross-sectional surface, and to research the morphology of Nafion in the catalyst layer.

4.7 Summary

The kinetics for the HER in PEM water electrolyzers is favorable because it occurs in an acidic regime. However, the acidic media provided by the polymer electrolyte membrane and ionomer phase requires the use of Pt-based catalysts. After decades of research, nanoparticles of Pt supported on high surface area carbon black still remain the benchmark catalysts for the cathodic evolution of hydrogen in PEM water electrolyzers. The use of Pt-based catalysts will, however, have a negative impact on the investment costs, and the reduction of Pt loading or its substitution still receives great focus for research and development on PEM water electrolysis. The Pt loadings on the cathode side are much lower in comparison to the anode electrode where IrO_2 is generally used. Today, for Pt, it ranges between 0.5 and 2 mg_{Pt} cm^{-2}. Nowadays, the goal for catalyst loading reduction for the cathode side ranges between 0.05 and 0.2 mg cm^{-2}. In order to reduce the amount of Pt and/or reduce the overall costs, focus on catalyst utilization is required. It is also necessary to optimize the triple-phase boundary for the cathodes of PEM water electrolyzers. A few research groups and companies have concentrated their effort on finding alternative catalysts to Pt. Advanced alternative catalysts such as core-shell catalysts and NSTFs have good potential, but their applications will depend on the ability to scale up its fabrication.

References

Alayoglu, S., Nilekar, A. U., Mavrikakis, M., & Eichhorn, B. 2008. Ru–Pt core–shell nanoparticles for preferential oxidation of carbon monoxide in hydrogen. *Nature Materials*, 7, 333–338.

Anderson, E, Ayers, K., & Capuano, C. 2013. R&D Focus Areas Based on 60,000 hr Life PEM Water Electrolysis Stack Experience. *First International Workshop on Durability and Degradation Issues in PEM Electrolysis Cells and its Components*. Freiburg, Germany.

Antolini, E. 2009. Carbon supports for low-temperature fuel cell catalysts. *Applied Catalysis B: Environmental*, 88, 1–24.

Ayers, K. E., Anderson, E. B., Capuano, C., Carter, B., Dalton, L., Hanlon, G., Manco, J., & Niedzwiecki, M. 2010. Research advances towards low cost, high efficiency PEM electrolysis. *ECS Transactions*, 33, 3–15.

Bagotskii, V. S. E. C. 2006. *Fundamentals of electrochemistry*, Hoboken, NJ, Wiley-Interscience.

Barbir, F. 2005. PEM electrolysis for production of hydrogen from renewable energy sources. *Solar Energy*, 78, 661–669.

Bockris, J. O. & Azzam, A. M. 1952. The kinetics of the hydrogen evolution reaction at high current densities. *Transactions of the Faraday Society*, 48, 145–160.

Bockris, J. O. & Mauser, H. 1959. The kinetics of the evolution and dissolution of hydrogen at electrodes. *Canadian Journal of Chemistry—Revue Canadienne De Chimie*, 37, 475–488.

Bockris, J. O. M., Parsons, R., & Rosenberg, H. 1950. The kinetics of hydrogen evolution. *Journal of Chemical Physics*, 18, 762–763.

Carmo, M., Brandalise, M., Neto, A. O., Spinace, E. V., Taylor, A. D., Linardi, M., & Poco, J. G. R. 2011a. Enhanced activity observed for sulfuric acid and chlorosulfuric acid functionalized carbon black as PtRu and PtSn electrocatalyst support for DMFC and DEFC applications. *International Journal of Hydrogen Energy*, 36, 14659–14667.

Carmo, M., Fritz, D. L., Mergel, J., & Stolten, D. 2013. A comprehensive review on PEM water electrolysis. *International Journal of Hydrogen Energy*, 38, 4901–4934.

Carmo, M., Linardi, M., & Poco, J. G. R. 2008. H(2)O(2) treated carbon black as electrocatalyst support for polymer electrolyte membrane fuel cell applications. *International Journal of Hydrogen Energy*, 33, 6289–6297.

Carmo, M., Linardi, M., & Poco, J. G. R. 2009a. Characterization of nitric acid functionalized carbon black and its evaluation as electrocatalyst support for direct methanol fuel cell applications. *Applied Catalysis A: General*, 355, 132–138.

Carmo, M., Paganin, V. A., Rosolen, J. M., & Gonzalez, E. R. 2005. Alternative supports for the preparation of catalysts for low-temperature fuel cells: The use of carbon nanotubes. *Journal of Power Sources*, 142, 169–176.

Carmo, M., Roepke, T., Roth, C., Dos Santos, A. M., Poco, J. G. R., & Linardi, M. 2009b. A novel electrocatalyst support with proton conductive properties for polymer electrolyte membrane fuel cell applications. *Journal of Power Sources*, 191, 330–337.

Carmo, M., Sekol, R. C., Ding, S. Y., Kumar, G., Schroers, J., & Taylor, A. D. 2011b. Bulk metallic glass nanowire architecture for electrochemical applications. *ACS Nano*, 5, 2979–2983.

Chung, D. Y., Park, S.-K., Chung, Y.-H., Yu, S.-H., Lim, D.-H., Jung, N., Ham, H. C., Park, H.-Y., Piao, Y., Yoo, S. J., & Sung, Y.-E. 2014. Edge-exposed MoS$_2$ nano-assembled structures as efficient electrocatalysts for hydrogen evolution reaction. *Nanoscale*, 6, 2131–2136.

Debe, M. K., Hendricks, S. M., Vernstrom, G. D., Meyers, M., Brostrom, M., Stephens, M., Chan, Q., Willey, J., Hamden, M., Mittelsteadt, C. K., Capuano, C. B., Ayers, K. E., & Anderson, E. B. 2012. Initial performance and durability of ultra-low loaded NSTF electrodes for PEM electrolyzers. *Journal of the Electrochemical Society*, 159, K165–K176.

Dedigama, I., Angeli, P., Ayers, K., Robinson, J. B., Shearing, P. R., Tsaoulidis, D., & Brett, D. J. L. 2014. In situ diagnostic techniques for characterisation of polymer electrolyte membrane water electrolysers – Flow visualisation and electrochemical impedance spectroscopy. *International Journal of Hydrogen Energy*, 39, 4468–4482.

Furuya, N. & Motoo, S. 1976. Electrochemical behavior of Ad-atoms and their effect on hydrogen evolution. 1. Order–disorder rearrangement of copper Ad-atoms on platinum. *Journal of Electroanalytical Chemistry*, 72, 165–175.

Furuya, N. & Motoo, S. 1977. Electrochemical behavior of Ad-atoms and their effect on hydrogen evolution. 2. Arsenic Ad-atoms on platinum. *Journal of Electroanalytical Chemistry*, 78, 243–256.

Furuya, N. & Motoo, S. 1978. Electrochemical behavior of Ad-atoms and their effect on hydrogen evolution. 3. Platinum Ad-atoms on gold, and gold Ad-atoms on platinum. *Journal of Electroanalytical Chemistry*, 88, 151–160.

Furuya, N. & Motoo, S. 1979a. Electrochemical behavior of Ad-atoms and their effect on hydrogen evolution. 4. Tin and lead Ad-atoms on platinum. *Journal of Electroanalytical Chemistry*, 98, 195–202.

Furuya, N. & Motoo, S. 1979b. Electrochemical behavior of Ad-atoms and their effect on hydrogen evolution. 5. Selenium Ad-atoms on gold. *Journal of Electroanalytical Chemistry*, 102, 155–163.

Furuya, N. & Motoo, S. 1979c. Electrochemical behavior of Ad-atoms and their effect on hydrogen evolution. 6. Germanium Ad-atoms on platinum. *Journal of Electroanalytical Chemistry*, 99, 19–28.

Gasteiger, H. A., Kocha, S. S., Sompalli, B., & Wagner, F. T. 2005. Activity benchmarks and requirements for Pt, Pt-alloy, and non-Pt oxygen reduction catalysts for PEMFCs. *Applied Catalysis B: Environmental*, 56, 9–35.

Ge, P., Scanlon, M. D., Peljo, P., Bian, X., Vubrel, H., O'neill, A., Coleman, J. N., Cantoni, M., Hu, X., Kontturi, K., Liu, B., & Girault, H. H. 2012. Hydrogen evolution across nano-Schottky junctions at carbon supported MoS$_2$ catalysts in biphasic liquid systems. *Chemical Communications*, 48, 6484–6486.

Gomadam, P. M. & Weidner, J. W. 2005. Analysis of electrochemical impedance spectroscopy in proton exchange membrane fuel cells. *International Journal of Energy Research*, 29, 1133–1151.

Grigoriev, S. A., Mamat, M. S., Dzhus, K. A., Walker, G. S., & Millet, P. 2011. Platinum and palladium nano-particles supported by graphitic nano-fibers as catalysts for PEM water electrolysis. *International Journal of Hydrogen Energy*, 36, 4143–4147.

Grigoriev, S. A., Millet, P., & Fateev, V. N. 2008. Evaluation of carbon-supported Pt and Pd nanoparticles for the hydrogen evolution reaction in PEM water electrolysers. *Journal of Power Sources*, 177, 281–285.

Hamann, C. H., Hamnett, A., & Vielstich, W. 1998. *Electrochemistry*, Weinheim; New York, Wiley-VCH.

Hinnemann, B., Moses, P. G., Bonde, J., Jorgensen, K. P., Nielsen, J. H., Horch, S., Chorkendorff, I., & Norskov, J. K. 2005a. Biomimetic hydrogen evolution: MoS$_2$ nanoparticles as catalyst for hydrogen evolution. *Journal of the American Chemical Society*, 127, 5308–5309.

Hinnemann, B., Moses, P. G., Bonde, J., Jorgensen, K. P., Nielsen, J. H., Horch, S., Chorkendorff, I., & Norskov, J. K. 2005b. Biomimetic hydrogen evolution: MoS(2) nanoparticles as catalyst for hydrogen evolution. *Journal of the American Chemical Society*, 127, 5308–5309.

Kreuter, W. & Hofmann, H. 1998. Electrolysis: The important energy transformer in a world of sustainable energy. *International Journal of Hydrogen Energy*, 23, 661–666.

Li, Y. G., Wang, H. L., Xie, L. M., Liang, Y. Y., Hong, G. S., & Dai, H. J. 2011. MoS$_2$ nanoparticles grown on graphene: An advanced catalyst for the hydrogen evolution reaction. *Journal of the American Chemical Society*, 133, 7296–7299.

Lu, Z., Zhu, W., Yu, X., Zhang, H., Li, Y., Sun, X., Wang, X., Wang, H., Wang, J., Luo, J., Lei, X., & Jiang, L. 2014. Ultrahigh hydrogen evolution performance of underwater "Superaerophobic" MoS$_2$ nanostructured electrodes. *Advanced Materials*, 26, 2683–2687.

Ma, C.-B., Qi, X., Chen, B., Bao, S., Yin, Z., Wu, X.-J., Luo, Z., Wei, J., Zhang, H.-L., & Zhang, H. 2014. MoS$_2$ nanoflower-decorated reduced graphene oxide paper for high-performance hydrogen evolution reaction. *Nanoscale*, 6, 5624–5629.

Macdonald, D. D. 2006. Reflections on the history of electrochemical impedance spectroscopy. *Electrochimica Acta*, 51, 1376–1388.

Marcu, A., Toth, G., Srivastava, R., & Strasser, P. 2012. Preparation, characterization and degradation mechanisms of PtCu alloy nanoparticles for automotive fuel cells. *Journal of Power Sources*, 208, 288–295.

Mayrhofer, K. J. J., Ashton, S. J., Meier, J. C., Wiberg, G. K. H., Hanzlik, M., & Arenz, M. 2008. Non-destructive transmission electron microscopy study of catalyst degradation under electrochemical treatment. *Journal of Power Sources*, 185, 734–739.

Medina, P. & Santarelli, M. 2010. Analysis of water transport in a high pressure PEM electrolyzer. *International Journal of Hydrogen Energy*, 35, 5173–5186.

Miles, M. H. & Thomason, M. A. 1976. Periodic variations of overvoltages for water electrolysis in acid solutions from cyclic voltammetric studies. *Journal of the Electrochemical Society*, 123, 1459–1461.

Millet, P., Dragoe, D., Grigoriev, S., Fateev, V., & Etievant, C. 2009. GenHyPEM: A research program on PEM water electrolysis supported by the European Commission. *International Journal of Hydrogen Energy*, 34, 4974–4982.

More, K., Borup, R., & Reeves, K. 2006. Identifying Contributing Degradation Phenomena in PEM Fuel Cell Membrane Electride Assemblies Via Electron Microscopy. *ECS Transactions*, 3, 717–733.

Mukherjee, S., Carmo, M., Kumar, G., Sekol, R. C., Taylor, A. D., & Schroers, J. 2012. Palladium nanostructures from multi-component metallic glass. *Electrochimica Acta*, 74, 145–150.

Mukherjee, S., Sekol, R. C., Carmo, M., Altman, E. I., Taylor, A. D., & Schroers, J. 2013. Tunable hierarchical metallic-glass nanostructures. *Advanced Functional Materials*, 23, 2708–2713.

Pantani, O., Anxolabehere-Mallart, E., Aukauloo, A., & Millet, P. 2007. Electroactivity of cobalt and nickel glyoximes with regard to the electro-reduction of protons into molecular hydrogen in acidic media. *Electrochemistry Communications*, 9, 54–58.

Rajeswari, J., Kishore, P. S., Viswanathan, B., & Varadarajan, T. K. 2007. Facile hydrogen evolution reaction on WO_3 nanorods. *Nanoscale Research Letters*, 2, 496–503.

Russell, J. H., Nuttall, L. J., & Fickett, A. P. 1973. Hydrogen generation by solid polymer electrolyte water electrolysis. *Abstracts of Papers of the American Chemical Society*, 24–40.

Schmidt, T. J., Gasteiger, H. A., Stab, G. D., Urban, P. M., Kolb, D. M., & Behm, R. J. 1998. Characterization of high-surface area electrocatalysts using a rotating disk electrode configuration. *Journal of the Electrochemical Society*, 145, 2354–2358.

Schroers, J. 2010. Processing of bulk metallic glass. *Advanced Materials*, 22, 1566–1597.

Sekol, R. C., Carmo, M., Kumar, G., Gittleson, F., Doubek, G., Sun, K., Schroers, J., & Taylor, A. D. 2013. Pd–Ni–Cu–P metallic glass nanowires for methanol and ethanol oxidation in alkaline media. *International Journal of Hydrogen Energy*, 38, 11248–11255.

Sekol, R. C., Kumar, G., Carmo, M., Gittleson, F., Hardesty-Dyck, N., Mukherjee, S., Schroers, J., & Taylor, A. D. 2012. Bulk metallic glass micro fuel cell. *Small*, 9, 2081–2085.

Selvan, R. K. & Gedanken, A. 2009. The sonochemical synthesis and characterization of Cu(1–x)Ni(x)WO(4) nanoparticles/nanorods and their application in electrocatalytic hydrogen evolution. *Nanotechnology*, 20, 105602.

Sethuraman, V. A., Weidner, J. W., Haug, A. T., & Protsailo, L. V. 2008. Durability of perfluorosulfonic acid and hydrocarbon membranes: Effect of humidity and temperature. *Journal of the Electrochemical Society*, 155, B119–B124.

Sheng, W., Gasteiger, H. A., & Shao-Horn, Y. 2010. Hydrogen oxidation and evolution reaction kinetics on platinum: Acid vs alkaline electrolytes. *Journal of the Electrochemical Society*, 157, B1529–B1536.

Stolten, D. S. V. 2013. *Transition to Renewable Energy Systems energy process engineering* [Online]., Available: http://search.ebscohost.com/login.aspx?direct=true&scope=site&db=nlebk&db=nlabk&AN=581898.

Strasser, P. 2009. Dealloyed core–shell fuel cell electrocatalysts. *Reviews in Chemical Engineering*, 25, 255–295.

Strasser, P., Koh, S., Anniyev, T., Greeley, J., More, K., Yu, C. F., Liu, Z. C., Kaya, S., Nordlund, D., Ogasawara, H., Toney, M. F., & Nilsson, A. 2010. Lattice-strain control of the activity in dealloyed core–shell fuel cell catalysts. *Nature Chemistry*, 2, 454–460.

Tafel, J. 1905. The polarisation of cathodic hydrogen development. *Z Phys Chem Stoch Verwandt*, 50, 641–712.

Taylor, A. D., Sekol, R. C., Kizuka, J. M., D'Cunha, S., & Comisar, C. M. 2008. Fuel cell performance and characterization of 1-D carbon-supported platinum nanocomposites synthesized in supercritical fluids. *Journal of Catalysis*, 259, 5–16.

Trasatti, S. 1999. Water electrolysis: Who first? *Journal of Electroanalytical Chemistry*, 476, 90–91.

Vielstich, W., Lamm, A., & Gasteiger, H. A. 2003. *Handbook of Fuel Cells: Fundamentals, Technology, and Applications*, Chichester, U.K.; New York, Wiley.

Vrubel, H., Merki, D., & Hu, X. 2012. Hydrogen evolution catalyzed by MoS_3 and MoS_2 particles. *Energy & Environmental Science*, 5, 6136–6144.

Wang, D., Pan, Z., Wu, Z., Wang, Z., & Liu, Z. 2014a. Hydrothermal synthesis of MoS_2 nanoflowers as highly efficient hydrogen evolution reaction catalysts. *Journal of Power Sources*, 264, 229–234.

Wang, H., Lu, Z., Kong, D., Sun, J., Hymel, T. M., & Cui, Y. 2014b. Electrochemical tuning of MoS_2 nanoparticles on three-dimensional substrate for efficient hydrogen evolution. *ACS Nano*, 8, 4940–4947.

Wippermann, K., Lohmer, A., Everwand, A., Muller, M., Korte, C., & Stolten, D. 2014. Study of complete methanol depletion in direct methanol fuel cells. *Journal of the Electrochemical Society*, 161, F525–F534.

Xie, J., Zhang, J., Li, S., Grote, F., Zhang, X., Zhang, H., Wang, R., Lei, Y., Pan, B., & Xie, Y. 2014. Controllable disorder engineering in oxygen-incorporated MoS_2 ultrathin nanosheets for efficient hydrogen evolution (vol 135, pg 17881, 2013). *Journal of the American Chemical Society*, 136, 1680–1680.

Xu, W., Liu, C., Xing, W., & Lu, T. 2007. A novel hybrid based on carbon nanotubes and heteropolyanions as effective catalyst for hydrogen evolution. *Electrochemistry Communications*, 9, 180–184.

Xu, W. & Scott, K. 2010. The effects of ionomer content on PEM water electrolyser membrane electrode assembly performance. *International Journal of Hydrogen Energy*, 35, 12029–12037.

Yu, Y., Huang, S.-Y., Li, Y., Steinmann, S. N., Yang, W., & Cao, L. 2014. Layer-dependent electrocatalysis of MoS$_2$ for hydrogen evolution. *Nano Letters*, 14, 553–558.

Zeng, K. Z. D. 2010. Recent progress in alkaline water electrolysis for hydrogen production and applications. *Progress in Energy and Combustion Science*, 36, 307–326.

Zhang, J. 2008. *PEM Fuel Cell Electrocatalysts and Catalyst Layers: Fundamentals and Applications*, London, U.K., Springer.

Zhang, S. S., Yuan, X. Z., Wang, H. J., Merida, W., Zhu, H., Shen, J., Wu, S. H., & Zhang, J. J. 2009. A review of accelerated stress tests of MEA durability in PEM fuel cells. *International Journal of Hydrogen Energy*, 34, 388–404.

Zheng, H. & Mathe, M. 2011. Hydrogen evolution reaction on single crystal WO(3)/C nanoparticles supported on carbon in acid and alkaline solution. *International Journal of Hydrogen Energy*, 36, 1960–1964.

5

3M NSTF for PEM Water Electrolysis

Krzysztof Lewinski

CONTENTS

5.1 Introduction

Ever since its discovery, hydrogen has been a source of fascination for the scientific community. As early as the sixteenth century, scholars reported a gas evolution reaction resulting from adding metal to acidic solutions. In 1766, Henry Cavendish recognized that the evolved gas was in fact hydrogen (Cavendish 1766; Al-Khalili 2010). Hydrogen is, after all, the most common element in the universe (Palmer 1997) and is responsible for the majority of the mass-to-energy conversion happening in the stars. It also forms the simplest atomic entity H, a combination of a single proton and a single electron, the simplest example of multi-atomic molecule H_2, the simplest of all ionic species H^+, and, of interest to electrochemists worldwide, presents the simplest electrochemical reaction

$$2H^+ + 2\overline{e} \leftrightarrow H_2 \qquad (5.1)$$

On a (much) smaller scale, our planetary scale, it can also be a foundation of the carbon-free energy conversion scheme, the so-called "hydrogen economy," a term coined and a concept formed in the early seventies by one of electrochemistry's greatest minds, Dr. Bernhardt Patrick John O'Mara Bockris (Bockris and Appleby 1972), a big proponent of a global conversion from carbon-based energy carriers to the hydrogen-based ones.* Bockris' dreams of hydrogen economy were perhaps well ahead of their time in the seventies and were not getting much of the traction back then. Today, however, the situation seems to be changing dramatically, and the vision of hydrogen economy may not be such a pipe dream anymore.

---

* As a side note, the author of this chapter had a truly great pleasure to work with Dr. A. John Appleby and briefly interact with Dr. John O'Mara Bockris himself while pursuing industrial postdoctoral work in the area of fuel cells in College Station, TX, where both scientists resided and did their research work in the mid-1990s.

The reasons for changing fortunes of the hydrogen economy concept have to do with the global realization that the continued reliance on carbon-based fossil fuels is not sustainable. One immediately obvious problem with the hydrogen economy is that hydrogen practically does not exist in nature (at least on our planet) in its molecular form—it has to be made. This is an energy-intensive process. While it is not likely that the cost of producing hydrogen for use as a fuel is ever going to be competitive with the cost of pumping hydrocarbon fuels from the ground, where they had been deposited over millennia, there are, however, additional indirect costs associated with hydrocarbon extraction from the earth. Some of them are measurable (the cost of permits, cost of exploration, drilling, transporting and refining, depreciation of the equipment, and the cost of capital), but some are more elusive (the social cost of exploitation in the undeveloped and underdeveloped places, and risks associated with the unrest in those regions, the progressively more technically challenging conditions of exploitation of oil and gas resources, the cost of exploration for new reserves, the ever-increasing magnitude and rate of fossil fuel extraction, and last but not least, the health and environmental costs, that is, the environmental damage associated with such activities). The related problem, namely of CO_2, an acknowledged greenhouse gas, and emissions associated with the use of fossil fuels for transportation and energy generation, is particularly very contentious recently. The emissions associated with energy production based on combustion of hydrocarbon fuel use are blamed for the sudden and, thus far, seemingly unstoppable rise in the level of carbon dioxide in our atmosphere. Dire consequences to the earth's climate and habitability of our planet, resulting from such un-countered raise in CO_2 levels in the earth's atmosphere, are being predicted by virtually every climate model out there.

So, as discussed earlier, it stands to reason that we should at least give hydrogen economy a serious thought as one of the possible (perhaps, at present the best possible) alternatives for energy storage and carrier medium that would not further increase the danger of causing an irreversible environmental disaster. To do this, however, we must entertain the question of how exactly are we going to obtain that hydrogen on the required scale. Today, the most common source of industrial hydrogen is steam reforming of hydrocarbon fuels (Staff 2007). It is perhaps the most cost-efficient method today, but it does produce CO_2 emissions and is also subject to all the pitfalls associated with the hydrocarbon fuel extraction from the ground. The other sources of hydrogen are by-products of other industrial activities, such as brine electrolysis in the chlor–alkali industry. Hydrogen sourced from these supply streams is typically not very pure and would need substantial postprocessing for use

as fuel in electrochemical energy conversion devices (such as fuel cells and fuel cell/flow battery hybrids); however, it could be used with less purification in the traditional heat engines (turbines, internal combustion engines, etc.). Such an application can be envisioned as an intermediary step before a full conversion to hydrogen economy can take place, although heat engines do not have anywhere near efficiencies typically associated with electrochemical energy conversion devices (that are not subject to Carnot cycle limitations like the heat engines are [Giordano 2009]). Needless to say, one of the best sources of hydrogen for the downstream use in the electrochemical energy conversion is water electrolysis. The process is very simple in its principle: an application of a voltage in excess of the thermodynamical stability window for water across two suitable electrodes in aqueous electrolyte should lead to decomposition of water according to the following reaction:

$$2H_2O \leftrightarrow 2H_2 + O_2 \qquad (5.2)$$

It produces very pure hydrogen that is already fully humidified—a big advantage, if the need for reusing it in a fuel cell is immediate. Otherwise, a simple dehumidification process will allow for both extended storage and/or transport of the electrolytic hydrogen. It should not be a surprise that devices performing this reaction do indeed exist and are in fact a rather mature technology. What might be a surprise to some readers is that the process of electrolysis of water was invented quite early in the history of modern science—in AD 1789 by Dutchmen Jan Rudolph Deiman and Adriaan Paets van Troostwijk (Levie 1999), which means it predates both the internal combustion engine (that oddly enough used combustion of hydrogen and oxygen) (web reference—about.com) of AD 1807 and fuel cells (Grove 1839), and invented by William Robert Grove in AD 1838 (Grove 1838).

A recent societal recognition of the greenhouse gases problem that triggered the global warming and the resulting shift into governmental regulations aiming to curb greenhouse gas emissions as well as promotion of more environmentally friendly technologies, which encourage the deployment and use of so-called renewable energy sources (such as solar, wind, tide, wave, and geothermal) can be the game changer in economics of electrochemical hydrogen generation and can make the "hydrogen economy" quite possible in the future. The benefits are clear and immediate, that is, generation of power without generation of greenhouse gases, but there is another positive side effect of wide deployments of renewable technologies from the point of view of hydrogen economy in general, and proton exchange membrane (PEM) water electrolysis in particular. The unpredictability and uncontrollability of energy production by renewable sources tends to destabilize electrical energy grid, and at certain penetration levels of renewable energy generation into a large scale, for example, a nation-wide scale electrical energy grid, such destabilization may be hard to counter. While there is still a debate at what degree of penetration of renewables into the energy grid, the grid stability problems start occurring; European Union, which as of today has over 117 GW of wind power installed with Germany @ 33.7 GW, Spain @ 22.9 GW, and United Kingdom @ 10.5 GW being the most intensive adopters where the wind power is even more dominant (Miloradovic et al. 2014), has already started to experience and report the region-wide grid instabilities. For one, this increased variability requires placing an additional load on the grid for stabilization during periods of overproduction and placing a substantial and geographically diverse overcapacity to guarantee the desired overall grid production levels during periods of local low energy production in certain geographical regions. All this bodes well for the hydrogen economy and water electrolysis as electrolyzers in general, some types more than others, are very compatible with unpredictability and variability of the renewable energy production. Therefore, the employment of electrolyzers on the oversized renewable grid can then serve two purposes: grid stabilization and energy capture/storage (in the form of hydrogen) during the periods of energy overproduction. Hydrogen generated during such periods of electrical energy overproduction can be shipped to the end point of use and then converted back to electricity either in conventional heat engines (turbines) or, better yet, in even more efficient fuel cells. Such concepts of remote hydrogen generation, followed by transportation to the user (via pipeline or tanker), had already been considered in the past (Justi 1987; Ogden and Williams 1989; Gretz et al. 1990; Coluccia et al. 1994; Drolet et al. 1996). Today, such concepts start making substantially more economic and practical sense. The fact that such concepts have merits is being proven by the German power-to-gas initiative, where electrolyzers are placed at the output of large wind and solar installations for the purpose of electrolytic hydrogen production, followed by injection of generated hydrogen into the natural gas pipelines. As of now, the intention is to burn that hydrogen as fuel in the internal combustion devices (turbines mostly), a very inefficient use of that hydrogen; but in the future, when fuel cells become common place, such hydrogen may be retrieved back from pipelines and used to fuel them. At some point, a complete switch to carbon-free energy carrier can, therefore, be envisioned. As an additional bonus, there are energy savings associated with reduced transmission losses. Research shows that transporting hydrogen in a gas pipeline is more efficient then transporting equivalent energy levels using high-voltage transmission lines (Oney et al. 1994; Keith and Leighty 2002).

Today, a dominant type of commercial water electrolyzer in industrial use is a type called alkaline water electrolyzer, or sometimes simply alkaline electrolyzer (AEL). These typically are very large devices that use a filter press concept to combine multiple individual electrolysis cells into a stack, where cells are connected in a series. The current, at suitable voltage, is then fed into the extreme ends of the electrolyzer stack and the supply of liquid electrolyte (typically highly concentrated solution of KOH) is fed to individual cells. The electrodes are typically Ni based with the balance of hardware (compression plates, bipolar plates, connecting pipes, and compression hardware being made of stainless steel or Ni-plated stainless steel). One of the best-known early adopters of this technology is the Norwegian company Norsk Hydro ASA. Other well-known companies offering alkaline water electrolyzers include Brown Boveri, Lurgi, Verde LLC, Industrie Haute Technologie, Hydrogenics, Teledyne, and DeNora. A picture of a large modern alkaline water electrolyzer from Verde LLC is shown in Figure 5.1 as an example. This is a V-series electrolyzer whose output can range from 10 to 600 m^3 h^{-1} @ 3.1 MPa output pressure and can be tailored to user's needs as it is a very modern modularized design.

A rather recent emergence of an alternative technology to AEL electrolyzers based on a solid PEM electrolyte brings in a worthy competition to the well-established and mature electrolytic hydrogen market. For a more in-depth review of what PEM electrolyzers are and how they function, the author suggests consulting a great comprehensive review on PEM water electrolysis contained in reference (Carmo et al. 2013) as well as

FIGURE 5.1
KOH electrolyzer, Verde LLC, 1285 kg day^{-1} (or 600 m^3 h^{-1}) of hydrogen.

chapters of this publication. Instead, the remainder of this chapter will concentrate on 3M's proprietary nanostructured thin-film (NSTF) catalyst system and how it fits into the overall scheme of using it for the purpose of efficient, high-power PEM water electrolysis.

5.2 History of 3M NSTF Catalyst Technology

In this section, we will touch on the historical origins of NSTF's Perylene Red (PR) support whiskers, the first attempts at identifying a suitable application for the structure (CO sensor), the initial success as a catalyst support in the fuel cell area, the development of NSTF-based fuel cell catalyst, and finally, the beginnings of the fuel cell program at 3M. Then, we will cover the expansion into nonfuel cell applications of fuel cell–derived technologies and into perhaps the best fitting application for NSTF catalyst, that is, the electrolytic hydrogen production area. We will go back to the roots of the 3M NSTF catalyst technology and describe how it came to being, and why interesting properties of the single-crystal PR support called for an identification of an application that could further advance the technology beyond simply an academic curiosity.

The ultimate shift in thinking about the nanostructured PR support came when the whiskers were vapor coated with metal and the structure's conductivity and high surface area suggested possible applicability to electrochemical systems. CO became the first electrochemical application of what we now know as the NSTF catalyst. However, once the potential of the catalyst had been realized, the switch to fuel cell and forming the

fuel cell research project was almost automatic (a vastly larger market prospect was the main driver). The success of the first tests as the fuel cell catalyst, and the quick progress in performance improvements, ensured the future of the NSTF, the formation of the fuel cell program, and, ultimately, the expansion into the neighboring electrochemical energy conversion–related areas, such as the water electrolysis.

5.2.1 Historical Origins

The NSTF catalyst history starts with the area of vacuum science exploring variations of physical microstructure of vapor-deposited organic films. At some point, it was recognized that some large heterocyclic organic molecules, such as perylenes and phthalocyanines, under some deposition conditions tend to form crystalline filaments growing preponderantly parallel to the surface. After further exploration of different growing conditions, it was realized that at higher supersaturation levels the resulting organic film forms what was described as a "crabgrass-like" structure (thought to be a result of an explosive dendritic growth process). On the other hand, allowing the growth conditions to reach near equilibrium produced unique and highly oriented microstructures. Such a process was the most intensely followed with two particular molecules of interest, namely, *N,N'*-di-(3,5-xylyl)-perylene-3,4,9,10-bis(dicarboximide), commonly known as PR (Figure 5.2) first reported to form ordered nanostructures by Kam et al. (1987) and *N,N'*-bis(2-phenylethyl)-perylene-3,4,9,10-bis(dicarboximide), also known as PEP (Figure 5.3) first reported to form ordered nanostructures in reference by Debe et al. (1988). Some phthalocyanines, and more specifically the

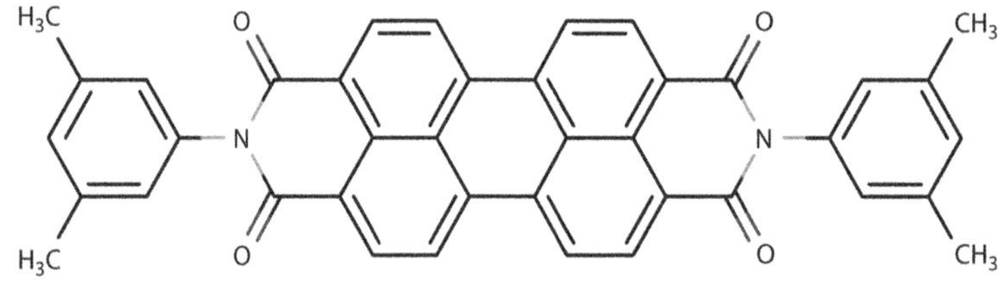

FIGURE 5.2
N,N'-di-(3,5-xylyl)-perylene-3,4,9,10-bis(dicarboximide), commonly known as Perylene Red (PR).

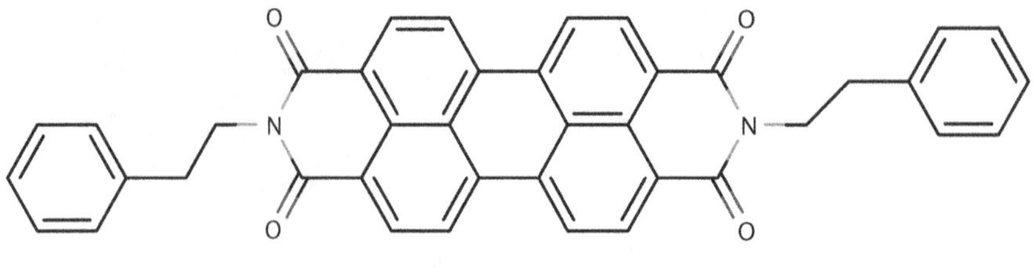

FIGURE 5.3
N,N'-bis(2-phenylethyl)-perylene-3,4,9,10-bis(dicarboximide), also known as PEP.

films they form, were even subjected to the microgravity growth experiments performed in space (e.g., PEP) (Debe and Poirier 1990). All of the earlier-mentioned compounds are organic pigments, and such highly oriented microstructures grown from them in vacuo using a simple dry sublimation process were thought to be a perfect basis for further modification. Therefore, a hunt for possible practical applications had started.

In the late 1980s and early 1990s, scientists realized that PR highly oriented microstructures can be easily coated with various metals, oxides, suboxides, nitrides, carbides, and their mixtures via any practical physical vapor deposition (PVD) methods, thus forming a great basis for the development of high surface catalysts. One of the very first applications in which such catalysts were tested in was a CO sensor, an electrochemical device that bases its signal on the measurements of current resulting from the diffusion-limited oxidation of carbon monoxide.

$$CO + H_2O \rightarrow CO_2 + 2H^+ + 2\overline{e} \qquad (5.3)$$

CO sensors, despite being a very promising application for this catalyst, are rather small devices and do not hold promise for large volume production. It did not take long to realize similarities between the principle of operation of the CO sensor and that of the fuel cell. In around 1995, contacts were made with Los Alamos National Laboratory (ultimately resulting in 1996 in CRADA being signed between 3M and Los Alamos National Laboratory); the first samples of Pt-coated PR highly oriented microstructures were made into membrane electrode assemblies (MEAs), tested for performance in the fuel cell application—and the rest, as we say it, is history. These very first experiments performed at Los Alamos National Laboratory and later also at 3M (the very first fuel cell test station to study NSTF and its application in fuel cells was purchased in 1995) were the humble beginnings of what we know today as the NSTF catalyst system. Since 1997, a string of successful proposals has resulted in fuel cell funding from government agencies (mostly, the U.S. Department of Energy) to the amount of more than $70M and a series of very successful developmental projects. In recognition of the increased interest in NSTF, in the year 2000 the 3M Fuel Cell Components Program was officially launched with the aim of commercialization of the fuel cell components, NSTF catalyst-based catalyst-coated membranes (CCMs), and MEAs being one of its main products. Finally, in 2011, a decision to broaden the scope of applications of materials and processes developed for fuel cells was made and the Electrochemical Energy Conversion and Storage Laboratory was born, with the purpose to conduct research and developmental activities in broadly defined areas of, as the name implies, electrochemical energy conversion and storage.

These were in time to include PEM water electrolysis, flow batteries, hybrid systems, and electrochemical reactors. In 2014, in recognition of successes achieved in this newly broadened application area as well as in order to better reflect its scope, the name of the program was changed from 3M Fuel Cell Components Program to 3M Energy Components Program.

5.2.2　Uniqueness of NSTF Catalysts in Comparison with Nanodispersed Carbon-Supported PGM Catalysts

The first interesting characteristics of the NSTF catalyst, or more precisely PR highly oriented microstructures that form the basis for an NSTF catalyst (which, for the sake of simplicity, is called PR whiskers that they resemble) is the fact that they are composed of a pure, organic molecular solid in a crystalline form. As a matter of fact, all of the lath-shaped PR whiskers themselves are single crystals of nanometer dimensions, with very reproducible surface structures. Properly grown on any suitable substrate by a surface sublimation mechanism, they have a high aspect ratio and number densities in the 3–5 billion per cm^2 range. Thanks to the molecular crystallinity, both the surface properties of whiskers and any coatings composed of them are very reproducible and uniform. A general consensus among many of my colleagues at 3M strongly reinforced by review publications in prestigious scientific literature such as *Nature* (Debe 2012) or *Handbook of Fuel Cells* (Debe 2010) would lay a claim that the PR in NSTF catalyst, an underlying organic pigment, is both thermally and chemically stable, nontoxic, insoluble, and nonreactive in most known solvents (water included), acid and base resistant, UV resistant, and, most importantly, not electroactive or subject to electrochemical corrosion. This of course would make it essentially an ideal support and basis for an active catalyst. We learned since, however, that despite its general unreactiveness and resistance to solvents PR can be dissolved in very strong acids such as sulfuric acid (interestingly forming a deep blue solution) as well as in many common ionic liquids for instance, those based on alkyl imidazolia (MIM, EMIM, BMIM, HMIM, etc.) when paired with suitable anions (BF_4^-, Cl^-, HCO_3^-, HSO_4^-, triflate, TFSI, and many others). Even more surprisingly PR is actually electroactive. An interesting (and unexpected) voltammetric response of PR is shown in Figure 5.4. A very well-defined pair of red-ox peaks is found at potentials close to RHE (an Ag wire reference was used since the experiment was conducted in ionic liquid environment). The experiment was performed at room temperature with PR reduced (and then reoxidized) on Ag working electrode (no electrochemistry of Ag in this potential window) in $EMIM^+Cl^-/H_2O$ solution with $H_2/0.5$ M H_2SO_4 system

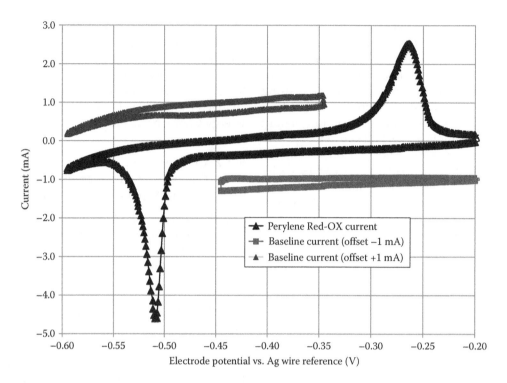

FIGURE 5.4
An interesting (and unexpected) voltammetric response of Perylene Red in ionic liquid.

used on Pt counter electrode with perfluorosulfonic acid (PFSA) separator. Baseline currents are offset by 1 mA to enhance clarity. This experiment clearly shows the electroactivity of PR observed at potentials just below that of reversible hydrogen electrode (Lewinski 2014).

The electroactivity of PR has not been previously observed due to a strong R&D focus on developing it as a basis for the NSTF catalyst system for use in fuel cells and therefore, a strong focus on relatively narrow potential window accessed during the fuel cell operation. As such, the electroactivity of PR on the negative side of hydrogen electrode would not have interfered with that intended fuel cell application. The matter is quite different if one is to consider utilizing the NSTF and NSTF-derived catalysts in applications, such as water electrolysis, where cathode potentials may reach the negative enough values, where PR is electrochemically active. The identification of this phenomenon opened a new area of research and created a definite need to research the compatibility of NSTF with the use in electrolyzer applications and determine the impact on the durability of the catalyst (or at least the support) in such a use, if any.

Another interesting characteristic of the NSTF catalyst is the nature of the catalyst structure itself. As mentioned in a more detail in the next section (Section 5.3), the NSTF catalyst is an extended surface area (ESA) catalyst and as such it behaves as a polycrystalline film rather than as an assembly of individual discrete particles. We believe this property allows it to behave more like a bulk metal and

is responsible for high reported specific activities. This is quite fortuitous for NSTF, as the basic geometric surface enhancement factor (SEF), a measure of the true geometric surface area ratio to the nominal geometric surface area, depends for the most part on the aspect ratio and the number (area density) of the whiskers. With typical whiskers being about 30 nm × 55 nm × up to a 1 μm and the earlier-mentioned 3–5 billion whiskers per cm², the SEF values hover around 10, which is a surprisingly low number. In order to compete with the typical dispersed carbon-supported fuel cell catalysts with surface areas in high 10 s to low 100 s, the NSTF catalyst must provide a substantial specific activity gain. Luckily, we have found this condition generally to be met. A lot of research over the years went into maximizing that SEF value. Due to the ESA nature of the NSTF catalyst and its typically relatively large crystallite sizes, we found it to be substantially more resistant to agglomeration, corrosion, and Oswald ripening than the compositionally similar classical dispersed catalysts. Also, at least in fuel cell applications, production of hydrogen peroxide, a precursor to very damaging peroxo-radicals, is orders of magnitude lower on NSTF than on dispersed catalysts. It has tremendous benefit of increasing the longevity of the PEM (again, in fuel cell applications—it has not been yet determined if the same is true for electrolyzer applications or if it is even a concern).

A third interesting and very appealing characteristic of the NSTF catalyst is that it offers a multitude of compositional and structural possibilities, including alloying, layering, a control of the degree of intermixing and layer thickness, the ability to mix metals with nonmetals, a control over dominant surface facets, the ability to introduce controlled amounts of oxygen, nitrogen, carbon, boron, silicon, etc., and hence tremendous opportunities to tailor the catalyst to the intended application. The types and parameters of depositions can be changed giving opportunities to play with the energy of incoming species and the degree of crystallinity of deposits—a completely amorphous catalyst, for example, can be (and have been) deposited on top of crystalline whiskers. And this brings us to the actual NSTF catalyst preparation process.

5.3 NSTF Catalysts: Preparation and Physical Properties

The NSTF catalyst belongs to the class of catalysts we now call ESA catalysts. They are characterized by the extension of their large surface areas principally in two dimensions. In general, it is thought that the fact these catalysts' primary nanometer-sized particles have a large number of immediate neighbors, which themselves have immediate neighbors, and *ad infinitum*, makes them behave electrochemically more like bulk catalysts, helps diffuse, and therefore lessens the forces promoting

agglomeration, surface rearrangements, and dissolution. Most of all, their electronic structure properties endow them with typically very high specific activities (similar to that of metallic fines or blacks). Other types of ESA catalysts, at least those known today, are single crystalline bulk surfaces (Paulus et al. 2002; Stamenkovic et al. 2002, 2007a,b), porous catalysts (films, leaves, membranes) (Erlebacher et al. 2001; Erlebacher and Snyder 2009; Moffat et al. 2009; Imbeault et al. 2010; Yang et al. 2010), and NSTF-like modified nano-wires, nano-fibers, and nano-tubes (Chen et al. 2007; Zhou et al. 2009; Wang et al. 2010; Adzic 2011; Park et al. 2011; Van der Vliet et al. 2012; Alia et al. 2013; Alia et al. 2014a,b).

5.3.1 Preparation of NSTF Catalyst

The NSTF catalyst manufacturing process starts with the preparation of a temporary substrate, on which the PR whiskers are later grown. Any suitable substrate can, in principle, be used for the process, but there are some particular requirements related to the mechanical stability and temperature resistance. The substrate specifically used at 3M, called microstructured catalyst transfer substrate (MCTS), is rather unique and is manufactured by 3M solely for its energy components program. The MCTS has a 3D microstructure built on its top surface, serving as a facilitator in the manufacturing process, with feature sizes many times larger than that of the PR whiskers. Figure 5.5 shows a top

FIGURE 5.5
150× magnification of the 90/6 MCTS, top view.

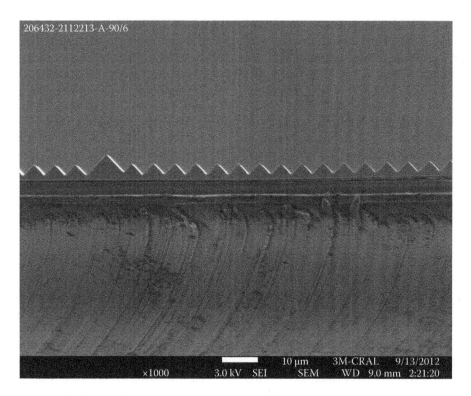

FIGURE 5.6
1000× magnification of the 90/6 MCTS, cross section.

view of the so-called 90/6 MCTS at 150× magnification. Figure 5.6 shows a cross section of the same 90/6 MCTS this time at 1000× magnification. Figure 5.7 shows the same cross section of the MCTS at still increased magnification of 5000× to show fine details of the microstructure. And finally, Figure 5.8 shows a picture of a roll of the so-called 90/6 MCTS as used for the production of NSTF catalyst.

The side effect of the presence of this microstructure is that it allows for microscale increases in the available catalyst loading (by a factor of $\sqrt{2}$) when compared to the flat growth surface. The PR itself is then deposited onto the MCTS and cultured into growing whiskers in vacuum ovens. The process is completely dry (no solvents of any kind are involved as the PR transfer to the MCTS is via direct sublimation), roll-to-roll high-volume compatible, and the PR deposition and whisker growth are done in a single continuous process. The PR (PR-149 to be more specific) is itself a volume product (pigment) used to color polyolefins and is widely available from multiple sources.

The next step in the NSTF catalyst making process is a deposition of catalytic species on top of the MCTS microstructure covered with nanostructure of PR whiskers. Any material capable of being deposited in vacuo can in principle be applied and used. In our case, we principally use PVD methods (such as e-beam and magnetron sputtering) to deposit layer by layer, in an almost atomically controlled deposition process, PGM group metals as catalysts in our electrochemical energy conversion devices. Such processes are very efficient methods to finely disperse materials that start their lives as bulk metals. Prepared in this fashion, deposits of catalytic species are very uniform. Platinum and its alloys as of today dominate fuel cell–targeted catalysts, while alloys of iridium with other platinum group metals are typically being used for water electrolysis, but other catalyst compositions have been deposited as well. Figure 5.9 shows an example of the resulting catalyst, in this case a view of 10,000× magnified the so-called NSTF catalyst as grown onto the MCTS film and catalyzed, top view (peak of the ridge being in focus). Figure 5.10 shows 10,000× magnification of the cross section of the NSTF catalyst as grown onto the MCTS film and catalyzed. Figure 5.11 shows the same 10,000× magnification of the cross section of the NSTF catalyst as grown onto the MCTS film, but with finer pitch, while Figure 5.12 shows much increased 50,000× magnification of the cross section of the NSTF catalyst as grown onto the MCTS film, to show fine details of whiskers and the substrate. Figure 5.13, similar to Figure 5.8 for MCTS, shows a picture of a roll of the finished NSTF catalyst—an example of the output of high-speed, roll-to-roll production process of NSTF catalyst.

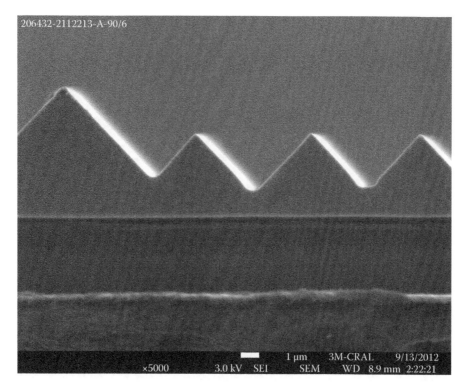

FIGURE 5.7
5000× magnification of the 90/6 MCTS, cross section, increased magnification to show details of the microstructure.

FIGURE 5.8
Picture of a roll of the 90/6 MCTS as used for the production of NSTF catalyst.

What is perhaps more important is the fact that not only pure metals can be deposited, but also metal alloys, metal–nonmetal compositions, metal oxides (stoichiometric or nonstoichiometric), nitrides, carbides, non-metal oxides, etc., layered structures (with very good control of layer thickness and order), top-surface modifiers—in this sense, the NSTF was probably the original core-shell catalyst as the concept of different top layers vs. the underlying bulk of the whisker was applied almost at the beginning of the history of NSTF catalyst at 3M.

Perhaps even more interestingly, one can think of the entire NSTF catalyst concept as being a core-shell catalyst since the PR de facto served as the core to the Pt catalyst shell in even the very first implementations of NSTF. One can argue that the organic core cannot act as a modifier of the properties of metallic shell on top of it, but at least in one case the evidence of a specific interaction of the PR with ruthenium deposited on whisker was documented (Atanasoska et al. 2011, 2012; Atanasoski et al. 2013a,b,c; Cullen et al. 2014). In addition to layers, discrete intermixtures can be obtained, down to almost alloys (in addition to true alloys in case where the sputtering target was an alloy target). The degree of crystallinity and facets sizes can be controlled by the choice of deposition conditions and/or posttreatment processes (such as annealing, chemical or electrochemical etching, or other chemical reactions), as well as the native surface energy interactions of species forming the catalyst film. Again, all the described catalyzation processes can be in principle applied in the very same vacuum system used to deposit and grow PR whiskers, and are roll-to-roll, high-volume manufacturing capable.

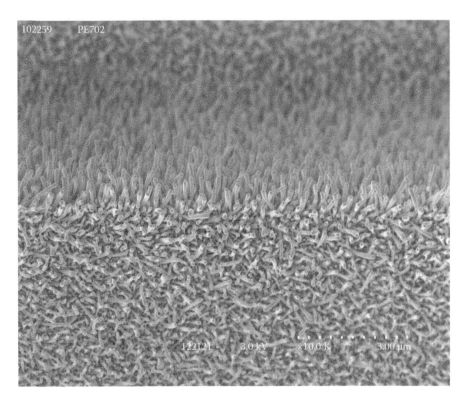

FIGURE 5.9
10,000× magnification of the NSTF catalyst as grown onto the MCTS film and catalyzed, top view (peak of the ridge being in focus).

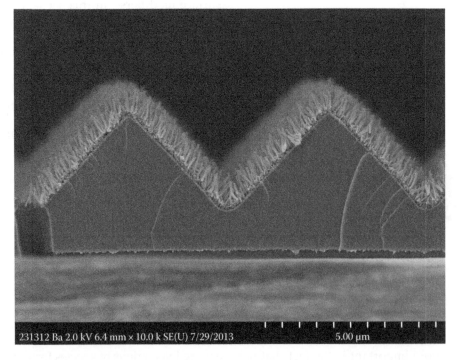

FIGURE 5.10
10,000× magnification of the NSTF catalyst as grown onto the MCTS film and catalyzed, cross section.

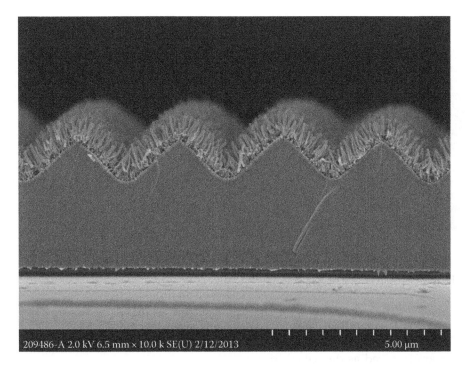

FIGURE 5.11
10,000× magnification of the NSTF catalyst as grown onto the MCTS film with finer pitch and catalyzed, cross section.

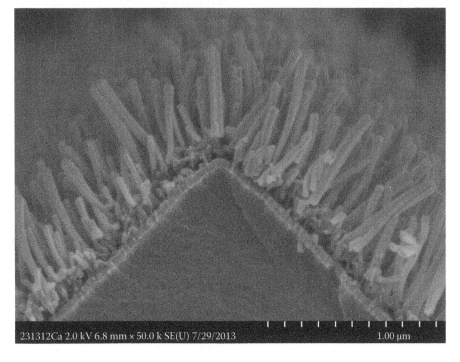

FIGURE 5.12
50,000× magnification of the so-called NSTF catalyst as grown onto the MCTS film and catalyzed, cross section, increased magnification to show fine details of whiskers and the substrate.

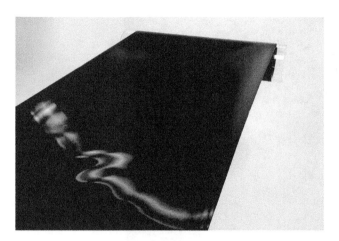

FIGURE 5.13
Picture of a roll of the NSTF catalyst—an example of the output of high-speed, roll-to-roll production process of NSTF making catalyst.

5.3.2 Application of NSTF Catalyst for Making MEAs

The NSTF catalyst, manufactured as described in the previous sections, has to be incorporated into the MEA structure in order to be used in membrane-based electrochemical devices. This incorporation is an example of yet another all dry, roll-to-roll, high-volume manufacturing capable process of laminating the output of the catalyzation process (a catalyzed MCTS) to the suitable ion exchange membrane. Typically, for the fuel cell and electrolyzer use today, that membrane is a variant of the PFSA PEM. During the lamination process, the NSTF catalyst is transferred from the temporary MCTS to the surface of PFSA membrane in a somewhat traditional albeit substantially modified to 3M specifications laminator, and under a suitable set of transfer conditions of temperature, pressure, and feed rate. The process itself is not unlike a decal transfer process, but it has much less sensitivity to the application of correct transfer conditions, which in case of decal transfer could have resulted in either poor transfer or the collapsed internal structure of the catalyst film being transferred. The used MCTS is currently discarded, but plans for its multiple reuses are already being implemented. The output of the lamination process is called CCM and is the heart of both the fuel cell and the electrolyzer MEA as well. Many linear kilometers of CCM for fuel cell application had been successfully made to date. An example of such CCM can be seen in pictures in the following.

Figure 5.14 shows a composite image of CCM where two layers of NSTF catalyst (an anode and a cathode) are shown to be laminated to PFSA membrane. Cross sections of progressively increasing magnifications (1500×, 4000×, and 20000×) and a top view (4000× magnification) are shown. Figure 5.15 shows a 2000× magnification of one of the electrolyzer-specific NSTF catalyst compositions (PtRuO$_x$ in this case) as embedded into the CCM film, again cross section is shown. Figure 5.16 shows 8000× magnification of one of the electrolyzer-specific NSTF catalyst compositions (the same PtRuO$_x$ as in Figure 5.15) embedded into the CCM film, cross section, increased magnification to show intimate bonding of NSTF catalyst whiskers, and the PFSA membrane. Finally, Figure 5.17 shows picture of a roll of the so-called catalyst-coated membrane (CCM)—an example of the output of high-speed, roll-to-roll production process of CCM making.

All that one needs to make an MEA from a CCM is to add suitable gas diffusion layers (GDLs), seals and gaskets, and edge protection if any. Such an MEA, sandwiched between a pair of bipolar plates stacked up to form multicell stacks surrounded by compression hardware and fluid distribution manifolds, forms our energy conversion devices.

5.4 Manufacturing of NSTF Catalysts

As mentioned earlier, the processes used to manufacture the NSTF catalyst are all dry, high-speed roll-to-roll capable. This is important for two reasons mainly: cost and the ability to provide sufficient supply to the world manufacturers once the conversion to the hydrogen economy begins. Since the cost and volume are so intrinsically related we will not separate them here, but instead in this chapter we will discuss the manufacturing advantages the NSTF catalyst can offer over the alternative wet manufacturing processes as we see them.

5.4.1 Fabrication Advantages of NSTF

First of all, we must address the question of whether all these vacuum processes mentioned in the previous sections and required to generate the NSTF catalyst are common and themselves capable of high-volume low-cost manufacturing. The answer is the unequivocal "yes." As we have mentioned, the PR is already an industrial, high-volume product, and so are bulk PGM metals required to make the NSTF catalyst. The MCTS-generating process, although proprietary to 3M and unique to the NSTF product line, is related to the Double Brightness Enhancement Film (DBEF) process used to make light management surfaces for the electronic industry. It is used in virtually all cell phones, tablets, laptops, PCs, and flat panel TVs. It is, without a doubt, a large-scale manufacturing process, and 3M is a major player in the DBEF market worldwide. Vacuum deposition itself is also a well-established, mature, and

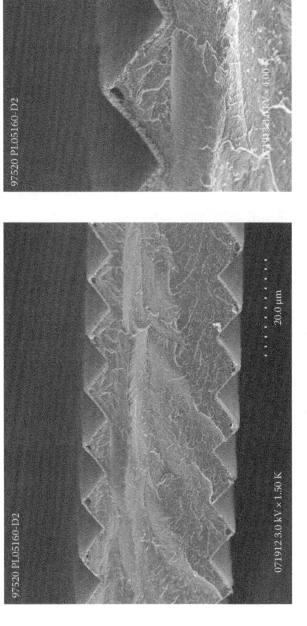

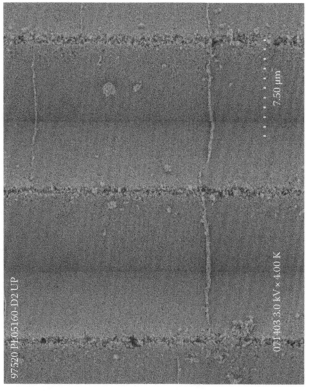

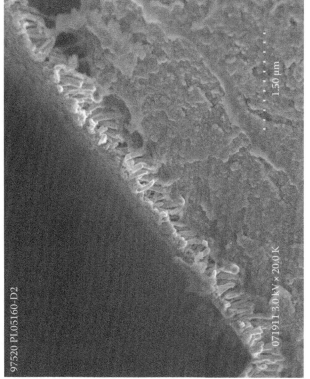

FIGURE 5.14

A composite image of catalyst-coated membrane (CCM) where two layers of NSTF catalyst (an anode and a cathode) are shown to be laminated to PFSA membrane, cross sections shown.

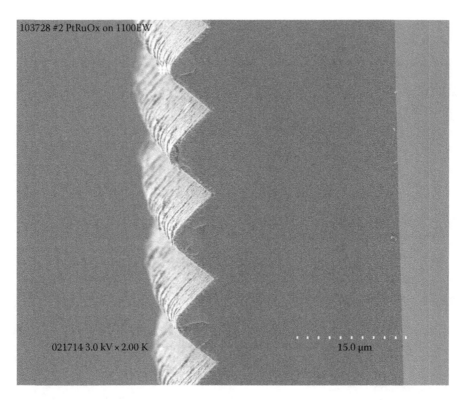

FIGURE 5.15
2000× magnification of one of the electrolyzer-specific NSTF catalyst compositions (PtRuO$_x$ in this case) as embedded into the CCM film, cross section.

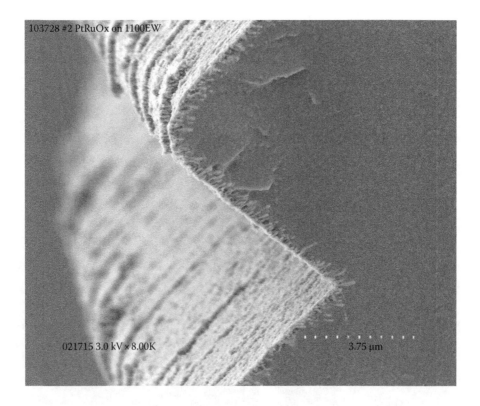

FIGURE 5.16
8000× magnification of one of the electrolyzer-specific NSTF catalyst compositions (the same PtRuO$_x$ as in Figure 5.15) embedded into the CCM film, cross section, increased magnification.

FIGURE 5.17
Picture of a roll of the catalyst-coated membrane (CCM)—an example of the output of high-speed, roll-to-roll production process of CCM making.

high-volume industrial process. An example of this may be aluminum coatings on the now ubiquitous potato chip bags (a simple e-beam process) or a more complex multilayer electronic data storage industry (hard drives), and low-emissivity glass manufacturing. The latter, based on the magnetron sputtering process, has about the same complexity as the NSTF catalyst deposition process, where multilayer constructions are being deposited with high precision, reproducibility, quality, yields, and speed needed to meet even the most optimistic predictions for future fuel cell/electrolyzer-based device penetration rates into energy and transportation markets. To offer a frame of reference, some 250 million square meters of low-emissivity glass had been manufactured in 2005 (ResearchInChina 2010). One can easily envision lamination as being simple enough to be compatible with high-speed manufacturing on the required scale.

It is worth noting that the projections for the future scale of manufacturing required to meet emerging fuel cell market requirements are quite staggering. The production rates required for true commercialization call for the ability to assemble tens to hundreds of thousands of fuel cell vehicles per year. This means potentially tens of millions of MEAs per year. Most automotive OEMs are targeting 2017–2020 as the timeframe for the introduction of fuel cell vehicles en mass. The process has already started with Hyundai's introduction of their fuel cell–based vehicle to North America earlier this year (2014) and Honda and Toyota planning an introduction of their respective fuel cell automotive offerings in 2015. This, with the circa five-year insertion schedule for qualified MEAs into vehicular stack, means that the time to start seriously considering volume manufacturing is now, and the scalability of any considered manufacturing process has to be considered from the outset.

Related to that is the expectation that in order to feed all these fuel cell vehicles, one has to consider the source of hydrogen they use and the best environmentally friendly way to make it is via electrolysis coupled to renewable (sun, wind, tide, geothermal, etc.) or nuclear prime energy sources. What this means for our field is that there will be as large of a need for electrolyzer MEAs as there will be for fuel cell MEAs—a very nice prospect indeed.

5.4.2 3M Manufacturing Capabilities

3M, as mentioned earlier, is already a major player in DBEF manufacturing worldwide, with film fabrication and roll-to-roll processing pretty much being the proverbial "bread and butter" for the company. 3M has been extensively using vacuum technologies in many lines of our products, so all the basic technological ingredients for implementation of the vacuum-based thin-film catalyst manufacturing were already present in the company's portfolio of technology platforms, when we started to develop catalysts for PEM fuel cells back in the mid-1990s. The very same technologies, processes and equipment can be and, as a matter of fact, have already been used for the development and not-too-distant future commercialization of PEM electrolyzer-specific catalysts as well. Of interest specifically to PEM water electrolysis, we can bring to bear a series of vacuum deposition coaters, from the lab Mark 50 and ML2 units used mostly for experimental laboratory work to what we call P-coaters, a semi-pilot scale production PVD deposition units. Figure 5.18 shows picture of one of the NSTF production P-coaters in Menomonie factory. Both whisker growth and catalyzation can be done in this machine.

FIGURE 5.18
Picture of one of the NSTF production P-coaters in Menomonie, WI factory. Both whisker growth and catalyzation can be done in this machine.

PEM for use in both PEM fuel cell MEAs and PEM electrolyzer MEAs is in very simplified terms a plastic film and can be manufactured by either extrusion of a thermoplastic form of the ionomer (as it is done for the thicker, chloroalkaline variants of PEMs) or, more preferably, by solvent casting. There is a general consensus that the former manufacturing method is more suitable to film thicknesses of 50 μm and above and the latter—for film thicknesses of 50 μm and below. There is, however, a substantial overlap, where both methods can compete. In addition, the fact that some membranes are reinforced and some have additives mixed in that may be heat sensitive may shift the sweet spot for any membrane in one way or another. 3M, at least up to this point, has been focused on the development of casting membranes, a decision mainly driven by the fuel cell market and its continuous drive toward the ever thinner electrolyte layer. An example of one of our PFSA-based PEM casting lines can be seen in Figure 5.19, which shows a picture of the 3M PFSA membrane production line in St. Paul, MN. The line is being used for both fuel cell and electrolyzer membrane production.

3M is one of the largest fuel cell membrane makers in the world and, akin to the case of producing CCMs, many linear kilometers of PFSA membranes had been made for fuel cell use. We are at the moment in the process of developing electrolyzer-specific PEMs and have realized that the operational requirement differences between the fuel cell and electrolyzer are substantial enough to make it unlikely for one type of membrane to be usable in both applications. More specifically, electrolyzer's mechanical, conductivity, and gas permeability requirements are much more stringent than that of the corresponding fuel cell. The electrolyzer

PEM is also not, in principle, exposed to humidity variations, nor is it subject to many start–stop cycles. It is, however, operating at substantially higher operating potentials and cell currents, and often at high-pressure differentials. So, while some requirements for the desired electrolyzer PEM can be relieved compared to the fuel cell membrane, most are actually much more stringent.

On the subject of CCM and MEA assembly, for any sort of activity resembling manufacturing, low cost, reproducibility, and controllability of the process have to be essential. Such requirement simply cannot be met by employing manual labor for the CCM or MEA assembly functions. Aside from the cost aspect of such enterprise, human make mistakes, and the errors in heavily serialized devices, such as fuel cell or electrolyzer stacks, would be unforgivable. It is estimated that for the economical assembly of electrochemical devices, the manufacturing process would have to be done with better than 6-sigma (actually in fact a much better than 6-sigma) precision. It is, therefore, rather obvious that the only way to implement such quality requirements for CCM and MEA assembly is for them to be done on a high speed, fully automated production line. An example of such production line that was developed by 3M to make fuel cell MEAs can be seen in Figure 5.20. What is shown here is 3M fuel cell MEA production line in St. Paul, MN that is presently being used for the production of fuel cell MEAs for both stationary and automotive markets. We anticipate it will also be used for the production of electrolyzer-specific MEAs in not-too-distant future.

It is worth stressing that it may take some modification before this line can be used to manufacture MEAs

FIGURE 5.19
A picture of the 3M PFSA membrane production line in St. Paul, MN. The line is being used for both fuel cell and electrolyzer membrane production.

FIGURE 5.20
A picture of the 3M fuel cell MEA production line in Menomonie, WI. We anticipate it will also be used for the production of electrolyzer-specific MEAs in the not-too-distant future.

for PEM electrolyzers, since the properties of GDL materials, as they are today—porous, rigid Ti sinters, do not make them directly amenable to assembly using the earlier pictured equipment for the manufacturing of electrolyzer MEAs.

This is perhaps a side note, but taking into account the anticipated future production volumes of both fuel cell and electrolyzer catalysts, an impact of mass-production on environment comes to mind. It should be stressed that such production has to have an aspect of sustainability with a minimal effect on environment. We can safely say that our envisioned solvent-free vacuum-based production process does seem to fit nicely into such an environmentally friendly requirement.

5.5 Application of NSTF Catalysts in PEM Electrolysis

Why did we at 3M engage in R&D in the electrolyzer area? Well, 3M already has a well-established fuel cell program. To date, almost 20 years of research and development went into inventing and refining NSTF for that application and a lot of techniques, procedures, and processes have been developed for making and testing NSTF-based fuel cell MEAs—something that is readily adaptable to the electrolyzer MEAs as well. 3M has also made an investment in facilities and equipment for making fuel cell–related materials—all again readily adaptable to electrolyzer materials. 3M's proprietary NSTF catalyst seems to be an extremely good fit to water electrolysis (carbonless—for durability, thin layer—for low mass transport losses, high inherent conductivity (demonstrated by device performance), good heat dissipation, and strongly hydrophilic—perhaps a weakness in the fuel cell application, but a tremendous strength in water electrolysis. And, on the business side, a source of H_2 is necessary for the anticipated deployment of fuel cells with water electrolysis being one of the most promising sources (environmentally clean, can use renewables, meets energy independence goals, etc.). No wonder then that it was just a matter of time before the NSTF catalyst was tried in this new promising application.

5.5.1 NSTF Catalyst Concept in PEM Electrolysis

In this section, we will focus on similarities between PEM fuel cell operation and PEM electrolysis and how we adapted the NSTF catalyst concept for use in electrolysis, and why we think this application will be one of the most successful ones for NSTF catalysts.

We will start with the brief historical overview of the events that led to 3M and the NSTF catalyst entering the electrolysis area, including why the area is of interest to 3M and why we believe there is a good fit between the basic NSTF technology and its application in electrolysis. We will then show the progress of our work from the initial attempts to use fuel cell catalysts in the electrolysis mode to the performance of the best 3M catalyst compositions today.

As we have shown in previous sections, the NSTF catalyst is a very versatile platform, adaptable to many possible applications in the emerging electrochemical energy conversion and storage field. Fuel cells are the first large-scale application area for the NSTF catalyst, which for years had set performance benchmarks in low loading applications, and later quickly became a well-known and the only serious rational alternative to the ubiquitous carbon-supported catalysts. A less known fact perhaps is that the NSTF catalysts were also at the forefront of high-performance fuel cell research, particularly back in the days when testing was done under high-pressure hydrogen and pure oxygen with the goal of getting an insight into the operation of fuel cells, understanding of the performance limitations and mechanisms governing them, and exploring the limits of what was possible, rather than concentrating on practical and commercial aspects that dominate fuel cell research today. It was none other than an NSTF catalyst–based fuel cell in our laboratories back in the late 1990s that set the fuel cell current density benchmark at 10 A cm^{-2}, a result that, to the best of our knowledge, has not been outdone to date (Lewinski and Mao 1998). Most of that research was performed with today no longer existing Dow PFSA low equivalent weight polymer–based membranes, radical experimental treatments of these membranes, as well as some rather extreme operating conditions. Some of the less radical results obtained in our laboratories were published in the *Handbook of Fuel Cells—Fundamentals, Technology, and Applications* (Debe 2010).

Around that time, we realized there might be a good fit between the NSTF technology and another electrochemical application where qualities of the NSTF and its high current density potential could be game changers to PEM water electrolysis. The ability to easily adapt and tailor catalyst compositions, and to mate the NSTF to pretty much any PFSA membrane known at that time, a thin, dense catalyst layer (meaning low transport limitations), chemically and electrochemically stable catalyst support, high catalyst layer conductivity that does not rely on the conductivity of the ionomer, and its hydrophilicity and propensity to flood, a handicap in fuel cell application, but a gigantic advantage for electrolyzer use, all meant the NSTF could be a very strong contender indeed. The fact that fuel cells were at that time very much a resurgent area of research popular with both investors and funding agencies alike, combined with the fact that water electrolysis was not high on anybody's agenda at that time, meant that quite a few years had to have passed before the idea to adapt the 3M NSTF catalyst to PEM water electrolysis was actually acted upon.

The beginnings were quite humble. Since 3M had no prior experience with solid-state water electrolysis and no hardware to test PEM electrolyzers, the very first attempts to use the NSTF catalysts in electrolyzers were done at our partners, Giner Corporation, Proton OnSite, and HySA. These interactions started around 2009 and took mainly a form of shipping samples to our partners and prospective customers for testing (there was no such capability at 3M at that time yet). Multiple low loading CCMs and ½ CCMs with fuel cell–derived cathodes (such as $Pt_{68}Co_{29}Mn_3$ with 0.1–0.15 mg cm^{-2} catalyst loading on a Pt basis), and later—various combinations of $Pt_xIr_yRu_z$ anodes were tested for performance and durability. Typical tests involved testing NSTF materials on one side of the cell (½ CCM) with a matching electrode being a standard PGM black electrolyzer electrode. Performance comparisons to the then state-of-the-art were made. A very positive finding was that the durability of the NSTF catalyst in those early samples, in both anode and cathode applications, was very good (e.g., 2000 h anode, 4000 h cathode as tested at Proton OnSite). The NSTF cathode catalyst activity was even claimed to be superior compared to the Pt black–based reference cathode, but the actual published data were perhaps somewhat less conclusive—about a 20 mV gain over the baseline was reported in one experiment, and between 0 and 100 mV loss in other tests, although the membrane was not the same in all tests, making direct comparison difficult (Debe et al. 2012). The NSTF anode catalyst activity was claimed to be "adequate," but it was true only for the Pt-based, iridium, and ruthenium containing compositions, and there was clearly a room for improvement. 3M MEA looked best at highest current densities tested (up to 4.0 A cm^{-2}), indicating a low IR drop and therefore good catalyst layer conductivity. All this indicated a solid foundation upon which our electrolyzer program could base a further OER-specific NSTF catalyst development.

In 2011, 3M decided to fully engage in research on non-fuel cell electrochemical applications of fuel cell–derived technologies, and the Electrochemical Energy Conversion and Storage Laboratory was set up. As a result of this new direction in research and new testing capabilities, we applied for government funding to help accelerate the development and best leverage 3M

investment dollars. In short succession, our efforts in the electrolyzer area resulted in participation in the following PEM water electrolyzer projects (sponsored by both 3M and government agencies):

- Giner/NASA SBIR Phase I subcontract (2011–2012, $Pt_1Ru_1Ir_x$ compositions tested, project is now completed)
- Proton OnSite DOE SBIR Phase II subcontract (2011–2013), electrolyzer membrane, $Pt_xRu_yIr_z$ compositions tested (project is now completed)
- Giner DOE SBIR Phase II Subcontract (2013–ongoing)
- Continued interactions with Hydrogen South Africa (HySA) with over 200 NSTF catalyst samples supplied to date (even closer collaboration desirable)

To perhaps best summarize the progress we made in those past few years at 3M in the PEM electrolyzer area, a graphic representation is worth a thousand words. The graph as in Figure 5.21 shows representative performance of our catalysts in time periods indicated.

Figure 5.21 illustrates graphically the progress of 3M NSTF electrolyzer catalyst development in recent years, from the first attempts to use fuel cell–derived catalysts such as $Pt_xCo_yMn_z$ and Pt_xNi_y to do OER in PEM electrolyzers to the first and second generation of electrolyzer-specific NSTF catalyst compositions. Data shown (polarization curves) were collected in potential controlled mode @ 80°C. Each successive generation shows improvement in turnover rates over 1000× that of the previous generation.

It is also to be understood that the performance improvements shown are a composite of both the improvements in the intrinsic catalytic activity of the NSTF catalyst and the improvements in the conductivity of the electrolyzer membrane. Together, they represent several million fold increase in performance (a turnover rate at given voltage) compared to our earliest attempts.

5.5.2 Exemplary Applications of NSTF Catalysts in PEM Electrolysis

In this section, we would like to show a few examples of successful applications of the NSTF catalyst system to the PEM electrolysis, as done in a laboratory scale at 3M. We will then demonstrate the performance of selected catalyst compositions, which will show how well the NSTF catalyst as a platform fits this application. Finally, we will discuss performance possibilities in PEM water electrolysis and the resulting practical implications.

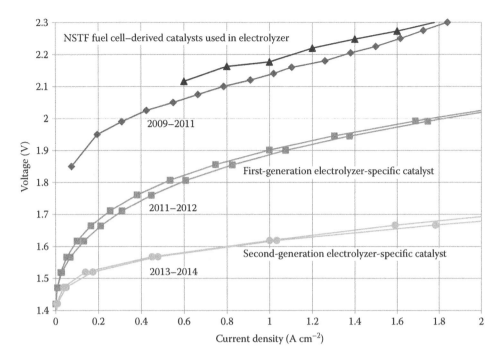

FIGURE 5.21
Progress of 3M NSTF electrolyzer catalyst development in recent years. Each successive generation shows improvement in turnover rates over 1000× that of the previous generation.

5.5.2.1 Catalyst Effects

To talk about the performance of the electrolyzer, we must first address the electrolyzer catalyst itself. As mentioned earlier, highly dispersed metal fines or blacks are normally used with very large mass loadings (as much as tens of milligrams of PGM materials per cm² of the electrode area have been used). As a general rule, such catalysts are unsupported and no ionomer is typically used in the catalyst layer. The NSTF system in this application seems to be a little bit handicapped, as the very nature of the structured whisker surface limits the practical amount of the catalytic material that can be applied to it before the effective surface area of such catalyst (the ElectroChemical Surface Area (ECSA)) starts to actually decrease with the increased loading due to the overfilling of spaces between the whiskers. Such an effect was found to be very detrimental to the performance of the fuel cell catalysts and it reduced the practical upper catalyst loading limits in that application to about 0.25 mg cm⁻² on a Pt basis. This had no impact on the development of the NSTF catalyst for the fuel cell application, however, as from the very early stage, fuel cells tended to lean toward favoring a low catalyst loading for cost-related reasons. Such cost pressures are much lower in the electrolyzer area, as catalyst performance decisively dominates any cost models, and a justification case for higher PGM loadings in exchange

for the performance boost can be made rather easily. So, with one of the key advantages of the NSTF catalyst it enjoys in the fuel cell area eliminated, we had to concentrate on the other key advantages, namely, typically a very high intrinsic catalytic activity of the NSTF, a characteristic seemingly common to all ESA types of catalysts. In addition, in order to bring the NSTF performance to par with the state-of-the-art electrolyzer-specific catalysts we needed to be able to grow the ECSA of the NSTF catalyst despite ever-increasing catalyst loadings—a task made even more difficult as this was never a focus area in the history of NSTF catalyst development.

There were quite a few surprises we ran into during the course of the catalyst development work. For one, in fuel cells, the conductivity of the NSTF catalyst layer has never been put into question. However, due to the high operating potentials on the electrolyzer anode, most of all the known elements are, at least in equilibrium, in their oxidized state. Therefore, there was a very serious concern about the conductivity along what was essentially an electrical isolator whisker. To compound the potential problem, one would have to deal with substantially higher current densities than those encountered in fuel cell applications. Our research showed, to our delight, that most of the suitable NSTF compositions did offer conductivities sufficient to meet the demands

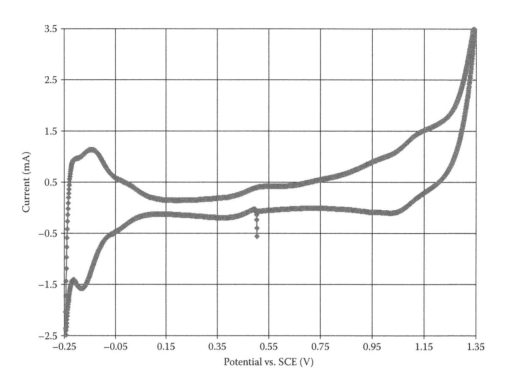

FIGURE 5.22
Cyclic voltammogram of polycrystalline Ir electrode. Please note general similarities to voltammograms of polycrystalline Pt.

of the intended application, even despite these very high operating anode potentials and essentially full oxidation of the catalyst surfaces. We were also able to retain or even increase the available electrochemical surface area, despite a substantial increase in the electrolyzer catalyst loadings as compared to its fuel cell relatives.

Our most successful NSTF electrolyzer catalysts are currently based on iridium and its alloys. It is, however, a nontrivial task to measure the ECSA of the Ir-based catalyst. Platinum has a very well-developed hydrogen UPD zone and has been studied by hundreds of electrochemists for many years and has a well-documented relationship between the H_2-UPD and the surface on which hydrogen adsorbs. In contrast, iridium, being more oxophilic, suffers from the late onset of reduction of its oxides. This makes measurements of H_2-UPD on Ir rather irreproducible and highly dependent on the history of the tested specimen. Also, the relationship between the H_2-UPD and the surface on which hydrogen adsorbs for Ir is not that well established (although it can reasonably be assumed to be close to that of Pt). It is, therefore, the author's opinion that, at least until a suitable measurement procedure is identified and vetted, only the relative measurements of Ir-based catalyst surfaces having a similar history and method of preparation should be compared. Also, since it was not trivial to find a quality voltammogram of the Ir electrode in recent literature, the one recorded in our laboratories

(non-NSTF) can perhaps be used as an illustration. Figure 5.22 is an example of a cyclic voltammogram of polycrystalline Ir electrode recorded in 0.5 M H_2SO_4 at room temperature and at 25 mV s^{-1} scan rate. Working electrode potentials were referenced to SCE. Please note similarity to voltammetry of polycrystalline Pt with the notable exception of missing (retarded) oxide reduction peak. Scan recorded this way (negative going from OCV) allows for the estimation of H_2-UPD charge and hence the electrode ECSA area.

Another unforeseen surprise we encountered was an unusual relationship between the electrolyzer performance and the catalyst loading. What we found was a behavior resembling that of the threshold loading type, where the performance of the catalyst stayed low until a certain critical loading was reached. The performance then increased with the increase in loading and leveled off for even higher catalyst loadings. An example of this behavior can be seen in the loading study presented in Figures 5.23 and 5.24. Figure 5.23 shows loading study of Ir-based NSTF catalyst. The graph plots cell voltage as a function of catalyst loading (normalized to the highest loading tested) at three different current densities: 0.5, 2.0, and 5.0 A cm^{-2}. The lowest loading electrodes were incapable of running at the two highest current densities chosen for this test. Tests were done at 80°C and ambient pressure. Figure 5.24 shows loading study of Ir-based NSTF catalyst. The same data as in Figure 5.23

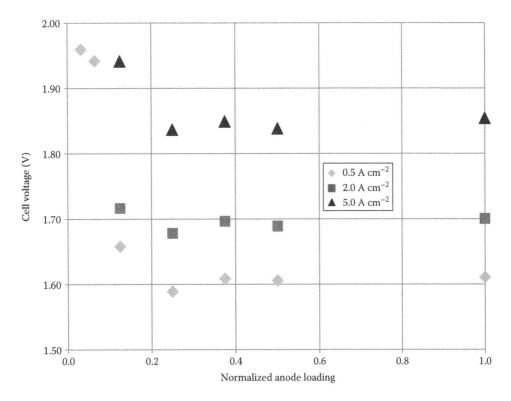

FIGURE 5.23
Loading study of Ir-based NSTF catalyst. Graph shows the cell voltage as a function of catalyst loading at three different current densities: 0.5, 2.0, and 5.0 A cm^{-2}.

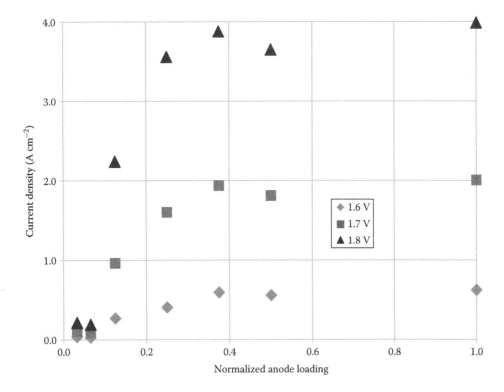

FIGURE 5.24
Loading study of Ir-based NSTF catalyst. This graph shows cell current density as a function of catalyst loading at three different cell voltages of 1.6, 1.7, and 1.8 V.

are plotted differently. This graph shows cell current density as a function of catalyst loading (likewise normalized to the highest loading tested) at three different cell voltages of 1.6, 1.7, and 1.8 V. Tests were done at 80°C and ambient pressure. This experiment demonstrates that there are practically no further performance gains past some optimum catalyst loading (at least for Ir-based NSTF catalyst tested here).

This behavior could be related to the catalytically active surface area formation on the surface of the whisker itself, combined perhaps with the particle-to-particle electrical contact requirements. Once those particle-to-particle contacts are properly established, the rapid growth of the catalyst surface that can be electrically accessed commences and the performance of the catalyst goes up dramatically. At the higher end of the loading spectrum, the performance still seems to be going up (an indication that the additional catalyst continues to be accessible), but the law of diminishing returns seems to be kicking in. And, while a case can be made that for the durability sake, the catalyst loadings should be kept at a high level, there seems to be an optimum catalyst loading that would seem to be best from the performance-per-dollar point of view. Anything above that optimum catalyst loading does not substantially increase the performance anymore, but, as mentioned earlier, may be beneficial from the durability point of view.

5.5.2.2 *Membrane Effect*

Another rather under-appreciated, but very important contributor (one can even say a dominant contributor, at least at higher current density ranges) to the electrolyzer performance, is the membrane. While the commercial PEM electrolyzer world had been relying on the N117 and N1110 membranes, there is a work going on aiming at improving the performance of the electrolyzer via improvements in performance of the membranes. There are, however, legitimate reasons for not dismissing N117 and N1110 as the performance benchmark just yet. They had been extensively used in many long-term trials and were found to be very durable. They also react well to high differential pressures and have very sought-after low gas permeabilities. All new contenders will have to show that they can last as long in this very demanding (and downright atrocious) application, can take mechanical stresses, and that any permeability losses can be more than offset by performance increases.

Having said that, we believe our Energy Components Program is very well positioned to successfully compete in this area. 3M has a very well-established and regarded PFSA membrane program, with the ability to tailor make the membranes with a choice of chemistries, equivalent weights, and reinforcement strategies, should these be required or desired. In order to demonstrate the effect of the membrane on PEM electrolyzer performance, Figure 5.25 shows how different

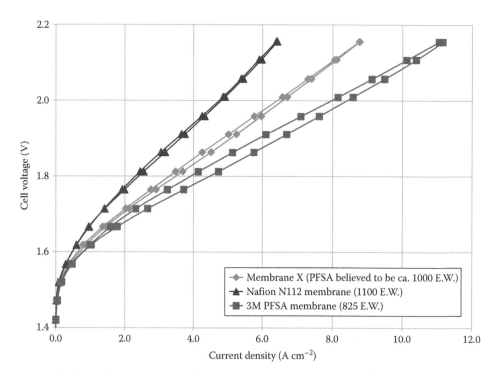

FIGURE 5.25
Membrane E.W. impact study—an example of the effect of equivalent weight of membrane ionomer and its effect on performance.

membranes of the same thickness are tested using the same hardware, assembly methods, catalyst, GDLs, and the identical testing protocol. What is different between these membranes is the equivalent weight of the PFSA polymer, more specifically. Figure 5.25 presents membrane E.W. impact study. Three PFSA membranes of the same 50 μm thickness were tested with the same catalysts and GDL in the same cell on the same test station. The main difference was the equivalent weight of the PFSA polymer with 3M membrane being 825 E.W. Membrane X is believed to be ca. 1000 E.W. and N112 has E.W. of 1100. This example shows direct and dramatic impact of membrane E.W. (and hence its conductivity) on performance in virtually the entire current range. Tests were done at 80°C and ambient pressure.

What one can see from Figure 5.25 is that the membranes with a lower equivalent weight offer lower resistance and thus, a lower slope of the polarization curve. Depending on the targeted operating point, this can substantially improve the electrolyzer performance. This is particularly true at higher current densities, at which membrane-related losses dominate. Any improvement of the membrane conductivity has, therefore, an immediate and measurable effect of performance boost; the larger the difference, the more substantial the performance gain.

Another important aspect of the membrane performance is its mechanical integrity. It is an important factor in the fuel cell area as well, but due to the typically much larger operating pressures, and pressure differentials in particular, as well as the use of mostly incompressible metallic GDLs the stresses that the membrane in the PEM electrolyzer application is subjected to are much larger. On top of that, the membrane in the electrolyzer application operates in the essentially 100% water vapor–saturated environment and even in the direct contact with liquid water. This results in a much more substantial degree of swelling than the swelling seen in the fuel cell applications. It is well known that substantially swollen membranes offer a much reduced mechanical strength. That omnipresence of water and high humidity, combined with high operating temperatures and mechanical stresses, also adds to the problem of the propensity of electrolyzer membranes to coldflow or creep. While not too troubling for short-term performance tests, this could be a substantial impediment to successful applications of membranes for longer tests, particularly those membranes that have inherently lower mechanical robustness. It is, therefore, not a surprise that despite their general undesirability on the performance loss grounds, the use of various reinforcement schemes is seriously considered. Since from the point of view of the gas permeability, there are no performance benefits resulting from the membrane reinforcement, one has to carefully weight pros and cons of such schemes. An example of the effect of reinforcement addition to the 50 μm thick 3M 825 E.W. membrane can be seen in Figure 5.26. Figure 5.26 shows a membrane reinforcement impact study—an example of the effect of

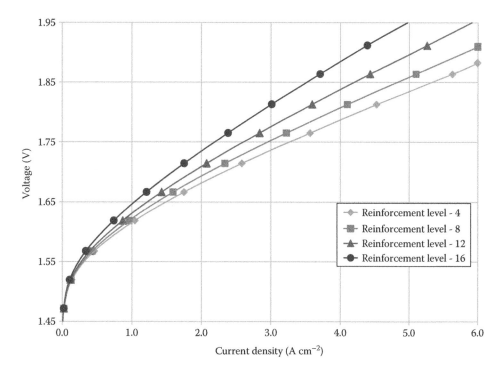

FIGURE 5.26
Membrane reinforcement impact study—an example of the effect of reinforcement addition to the 50 μm thick 3M 825 E.W. membrane and its effect on performance.

reinforcement addition to the 50 μm thick 3M 825 E.W. membrane and its effect on performance. All tests were done at 80°C and ambient pressure.

As we can see, there is a nice correlation between the performance increase and the amount of reinforcement used as exemplified by the series of progressively more downward tilted polarization curves with lower levels of reinforcement in membranes. At essentially any reasonable current density, the impact of the reinforcement is measurable and directly proportional to the amount of reinforcement. Similar to the impact of equivalent weight, the higher the operating current, the more dominant the resistive losses are and the larger the impact on the electrolyzer's performance. This effect can be best summarized in a table format that can be found in Table 5.1.

TABLE 5.1

ASR resistance of 3M 50 μm membranes with various levels of reinforcement.

Support Level	ASR (DC) Resistance (Ω cm^2 @ $i > 3$ A cm^{-2})	ASR Resistance Due to Support (Ω cm^2)	Apparent E.W. (g mol$_{H+}^{-1}$)
16	0.071	0.024	993
12	0.060	0.013	943
8	0.052	0.005	899
4	0.048	0.002	863
0	0.047	0.000	825

The earlier table summarizes the increase in ASR resistance, ASR resistance due to support, and the Apparent E.W. of resulting "composite" electrolyte with the increase in level of support. Due to the similarity in effects, one can start considering the effect of the reinforcement addition as essentially the virtual equivalent weight increase. Figure 5.27 shows a plot of the ASR contribution of the support to the ASR of the base nonreinforced 50 μm 3M 825 E.W. membrane as a function of the reinforcement level (the higher numbers correspond to increase in the level of reinforcement) and its impact on "Apparent Equivalent Weight" of the membrane polymer. Tests were done at 80°C and ambient pressure.

5.5.3 High-Efficiency Operation of PEM Electrolyzer

After briefing our reader on the two most relevant contributions to the performance of PEM electrolyzers, namely, the effect of the catalyst and the effect of the membrane, it is time to say a few words about PEM electrolyzer operating modes. A historically dominant and, as of a few months ago, the only operating mode considered practical was the high-efficiency mode. The driving force for this point of view was an assumption that the hydrogen produced as either a fuel or a chemical will have to compete on the market with the alternatively sourced hydrogen (such as from the steam reforming of natural gas or coal gasification).

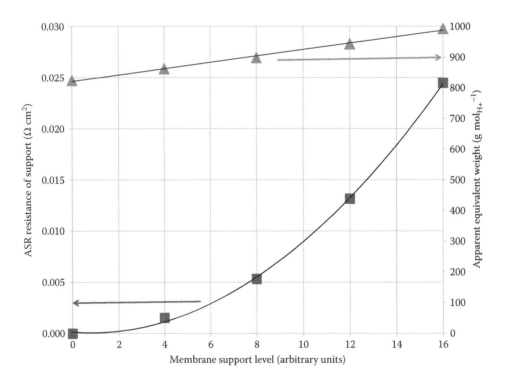

FIGURE 5.27

Plot of the ASR contribution of the support to the ASR of the base nonreinforced 50 μm 3M 825 E.W. membrane as a function of the reinforcement level and its impact on "Apparent E. W."

Such an assumption placed the electrolytic hydrogen at an immediate disadvantage based on the production cost, and demanded that these production costs are minimized as much as possible to allow for competition in applications that needed a higher quality of input gas. This drove the perceived need for operating electrolyzers at highest practical efficiency (meaning typically at 1.0–2.0 A cm^{-2}). Applications requiring a low, but steady flow of hydrogen (e.g., such as gas chromatography), were typically well served by electrolyzers, and in these applications, a pressure to operate at the highest efficiency was somewhat lessened. Therefore, at that time it was not a significant impediment that the available membranes had a relatively large thickness (again, typically N117 and N1110 were used) and, as a result, high resistivity. The required current densities were low, and so was the impact of the membrane—all performance gains were obtained solely via improvements of the performance of the catalyst alone. The catalyst was considered the most relevant performance knob that could be turned to coax the very last bits of efficiency gains that could be had from the PEM electrolyzers. The fact that in this low current density–high-efficiency range, the catalyst performance dominates the effect of the membrane is still, however, very significant. To demonstrate, Figure 5.28 presents examples of the performance of our electrolyzers, where performance of both the catalyst and the membrane had been optimized. The results push the overall PEM electrolyzer efficiency benchmarks even at the catalyst loadings that are substantially smaller than the catalyst loadings in the past. The contribution of the membrane IR to the performance of the PEM electrolyzer, even at these low current density ranges, is still nevertheless quite substantial.

Figure 5.28 is an example of high-performance water electrolysis in a low current density range ($i < 4.0$ A cm^{-2}) and the effect of temperature on performance in this range (performance is strongly influenced by the activity of the catalyst, hence huge temperature impact). Tests were done at 80°C, 90°C, and 100°C, near ambient pressure (just enough pressurization to keep water from boiling), DI water flow was set at 1.25 cm^3 s^{-1}, the membrane under test was a 50 μm 3M 825 E.W. PFSA electrolyzer membrane anode PGM loading was 0.25 mg cm^{-2} and the cell active area was 50 cm^2.

5.5.4 High-Power Operation of PEM Electrolyzer

What is most interesting we believe, however, is the potential for finding new viable applications for PEM electrolyzers in the renewable energy area. As described earlier, we believe in the need for the world to move away from the carbon-based energy sources and to eventually transition into the hydrogen economy. To provide that hydrogen, a tremendous volume of which

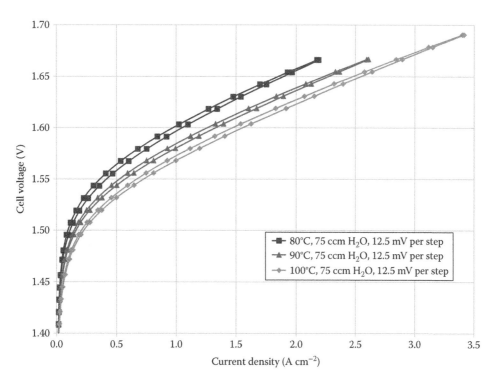

FIGURE 5.28

Example of high-performance water electrolysis in a low current density range ($i < 4.0$ A cm^{-2}). DI water flow was set at 1.25 cm^3 s^{-1}, the membrane under test was a 50 μm 3M 825 E.W. PFSA.

can be envisioned, the world will need a massive electrolyzer capacity with a low cost per installed kW equivalent. At the same time, if one can take Germany as an example, there is going to be an oversupply of energy from the renewable sources at times when generation exceeds demand. This phenomenon has the potential to greatly impact the economics of the hydrogen production. As in this new scenario one can count on low operating expenses, capital expenses will start to dominate the cost of the installation and operation of electrolyzers, and the efficiency of the electrolyzer will become much less important. This, in our opinion, opens a window into operating the electrolyzer at its highest sustainable power, namely, at a very high current density range. To our knowledge, such an operation of PEM electrolyzers at 10, 15, or even 20 A cm^{-2} has not been previously demonstrated, or even considered possible, and it is with a great pleasure that the author and his group present such data in this publication for the first time. The intent is to show what is possible and to open a public discussion on whether such a high-power operating range is useful, whether it can be used outside of laboratory in industrial settings, and whether it can help the world in its transition to hydrogen economy.

Figure 5.29 shows an example of high-performance water electrolysis in a high current density range

($i > 4.0$ A cm^{-2}), and the effect of temperature on performance in this range (performance is strongly influenced by the conductivity of the membrane; catalyst activity carries through gains from low current density range). Tests were done at 80°C and 100°C, near ambient pressure (just enough pressurization to keep water from boiling), DI water flow was set at 1.25 cm^3 s^{-1}, the membrane under test was a 50 µm 3M 825 E.W. PFSA electrolyzer membrane, and the cell active area was 50 cm^2. Figure 5.30 is an example of what is possible in high-performance water electrolysis in a very high current density range (i—10.0–20.0 A cm^{-2}). Tests were done at 80°C, ambient pressure, DI water flow was set at 1.67 cm^3 s^{-1}, the membrane under test was a 50 µm 3M 825 E.W. PFSA electrolyzer membrane, and the cell active area was 25 cm^2. Power density of this electrolyzer is plotted on the secondary axis and exceeded 50 W cm^{-2} mark – this is, as far as the author is aware, the highest power density electrochemical device of any kind ever reported per unit of electrode area. The instabilities of the polarization curve noted at the highest current density range ($i > 15.0$ A cm^{-2}) reflect the extreme rate and violence of gas evolution which the cell we used was not designed to handle as well as stability issues with the power supply itself.

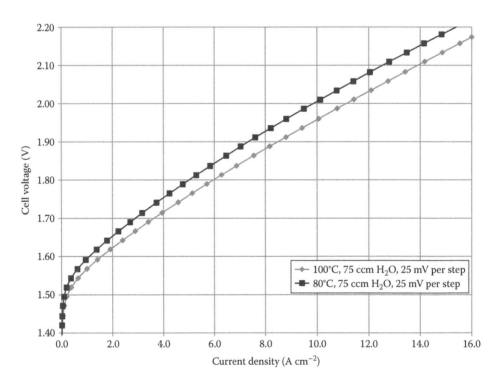

FIGURE 5.29
Example of high-performance water electrolysis in a high current density range ($i > 4.0$ A cm^{-2}), DI water flow was 1.25 cm^3 s^{-1}, the membrane was a 50 µm 3M 825 E.W. PFSA, and the cell active area was 50 cm^2.

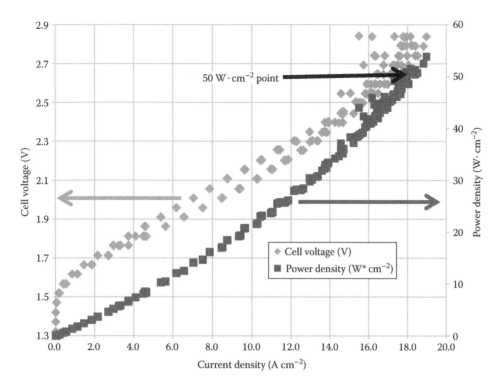

FIGURE 5.30
Example of what is possible in high-performance water electrolysis in a very high current density range (i—10.0–20.0 A cm^{-2}). DI water flow was set at 1.67 cm^3 s^{-1}, the membrane was a 50 μm 3M 825 E.W. PFSA, and the cell active area was 25 cm^2.

5.6 Current Challenges and Future Potential

There are of course business and technical challenges. Business challenges tend to be focused on high capital costs of electrolyzer installations, including the cost of the device itself, cost of control electronics, cost of power conversion, cost of land (lease/buy), cost of buildings and infrastructure, cost of permits, etc. I believe this could be addressed at least to some degree by operating electrolyzers at much higher current densities than it has been typically done to date. This strategy should offer capital-related cost reductions that are directly proportional to current density increases. The other costs, namely, the operational expenses, are related mainly to the cost of electricity used. As mentioned earlier, the increase in penetration levels of green renewable energy technologies in the total energy mix should create opportunities for harvesting the excess electricity for the hydrogen production use. That opportunity, especially when combined with lower capital costs due to operating electrolyzers at higher current densities as proposed (and demonstrated) earlier, should allow one to take advantage of low electricity rates during the periods of overproduction (such as during a mismatch between the production and the demand), or when

generating resources are being stranded due to transmission line congestion, for example.

And then, there are of course technical challenges. We will limit the listing here to the challenges affecting the PEM electrolysis, as all electrolyzer technologies have their own set. Due to a rather fantastic performance the PEM electrolyzers are capable of, the straight performance is not much of the challenge any more. The anode catalyst, even in a relatively low-loaded NSTF form, is quite active for OER (e.g., perhaps a hundred times more so than the relative fuel cell ORR reaction catalyst), although further improvements would of course still be desirable. With the advent of low-impedance membranes and ever-increasing operating current densities of PEM electrolyzers, the impact of the catalyst on the total performance is further reduced as the membrane-related losses dominate at anything past 3–5 A cm^{-2}. Interestingly, the higher the membrane conductivity, the further the importance of the good catalyst performance stretches down the current density range. Eventually though, the resistance always takes over as the dominant source of losses in the PEM electrolyzer cell. Ultimately, it is the ability to maintain that performance for any reasonable amount of time that is the biggest technical challenge for high-performance PEM electrolyzers today. In the past, membranes were thick, perhaps as much as 170–250 µm, and the electrode loadings were very high. The operational conditions were, relatively speaking of course, mild: 50°C operating temperature, full humidification via direct access to liquid water, low current densities (1–2 A cm^{-2} were typical), no need for additional cooling as the temperatures were uniformly distributed, moderate mechanical stresses, etc. It is not too surprising then that electrolyzers were shown to be durable, some demonstrating as high as 60,000 h of continuous operation (Anderson et al. 2013). The only concern was corrosion induced by high operating anode potentials and hydrogen embrittlement issues (to which Ti used as predominant electrolyzer stack building material is known to be prone). Today, these issues remain and are joined by the new set introduced in equal measures by the drive toward improvements in performance (ever thinner membranes of ever lower equivalent weight, more active catalyst layers with lower amounts of catalyst itself—a higher concentration and intensity of electrode reactions, volume wise, larger nonuniformities of stresses exerted on the cell, again due to thinner membranes) as well as the success of those improvements (higher operating current densities, a higher water transfer driven by the electro-osmotic drag, a higher nonuniformity of current and heat flow distributions, much lower catalyst loadings, etc.) on top of the choices of operating conditions (higher operating temperatures and pressures, higher pressure differentials). All these changes require providing new answers to the old durability questions. The answers are particularly needed as the bar for durability has been set very high by the previous generation of electrolyzers and the expectations will be set, respectively. The potential for the performance improvers and aggressively set operating conditions to negatively affect durability is very high. Nothing, for example, is known about the effect of high operating current on the electrolyzer durability. High operating temperatures are likely to be detrimental to durability of both the catalyst and the membrane. In the case of catalyst, lower loadings will very likely diminish the surplus of the catalyst present in the highly loaded variants that would compensate for the catalyst dissolution and migration/redeposition into the membrane. In the membrane case, the lower thicknesses would tolerate only a minimal degree of imperfection in the internal stresses caused by the counteraction of the membrane swelling and the cell compression acting on a pair of now rigid GDLs, a swelling made worse by lower equivalent weights of the electrolyte. To make matters worse, the mechanical integrity of lower equivalent weight membranes is typically lower than the mechanical integrity of higher equivalent weights. Higher electrolyzer operating pressures are also sought, in order to reduce or perhaps even completely eliminate the need for a subsequent mechanical

compression. Likewise, unknown is the question of a crossover, a problem previously manageable, but likely to get worse with the membrane thickness reductions, a problem that can potentially be counteracted by higher gas generation rates (thus offsetting the increased crossover rates).

In our opinion, future R&D priorities in the PEM electrolyzer area will need to be given to the research efforts aiming to answer the previous questions. Without satisfactory answers to durability questions, no new products exploiting the newly available high current density operation window will be built, and no new ways to deploy them will be tested. This is potentially a game-changing opportunity for PEM electrolyzers, and the time, due to the current popularity of renewable energy generation, is right.

5.7 Summary

To very briefly recap the topics discussed earlier, we would like to introduce the NSTF catalyst and present the key differentiating features and advantages it can offer when employed in PEM electrolysis applications:

- Supported catalyst, but no carbon to corrode; support whisker (core) nonconducting, very stable (chemically and electrochemically), conductive catalyst coating (shell)
- Higher area-specific activity (compared to carbon-supported catalysts), high mass-specific activity
- Significantly reduced dissolution and agglomeration (compared to carbon-supported catalysts or nano-sized discrete PGM powders)
- Reduced mass transport over-potential at high current density (thin)
- Strongly hydrophilic—helps water access, enables high current densities in electrolyzer mode
- Very good OER performance at low loadings (low loadings defined as being below 0.5 mg cm^{-2} PGM)
- Cost, high-volume manufacturability derived from existing high-volume thin-film technologies such as low-E glass

Unknowns/areas of potential concern—in H_2O PEM electrolysis application:

- Long-term MEA durability (catalyst, membrane, GDL)
- Performance stability (both short and long term)

- Catalyst dissolution (particularly at high potentials)
- Catalyst deactivation/poisoning
- Membrane decomposition (chemical, thermal, mechanical)
- Membrane poisoning (purity of DI)
- Decomposition of GDLs (oxidation of Ti, oxide buildup, reduction of carbon if any present)
- Durability under very high current densities (are there new modes of failure introduced by running @ very high currents?)
- Effects of extreme heat loads
- Effects of extreme water fluxes
- Effects of extreme gas evolution rates

Can high-power operation be a new paradigm for PEM electrolyzer operation?

- Particularly well suited to applications sensitive to CAPX expenditures
- Current density ~10x times present commercial PEM electrolyzers
- PGM catalyst loading ~1/10th present commercial PEM electrolyzers
- Should equal to substantial cost savings on electrolyzer device (CAPX)
- Only moderate efficiency losses (OPEX) due to higher IR loss (due to large I) could be moderated by the availability of low-cost excess of renewable energy during low demand periods
- Could be used to tap the excess of renewable energy from stranded generating resources lacking sufficient transmission capacities
- Could easily support intermittent renewable energy sources (enough dynamic range to catch peaks/spikes)

We believe that the high current density implementation of PEM H_2O electrolysis introduced here could become a disruptive technology, and if successfully implemented, it could change the nature of competition in energy markets.

References

Adzic, R. 2011. Contiguous platinum monolayer oxygen reduction electrocatalysts on high-stability-low-cost supports. In 2011 DOE Hydrogen Program Annual Merit Review FC-009, http://www.hydrogen.energy.gov/pdfs/review11/fc009_adzic_2011_o.pdf (last accessed on 2015-06-07).

Alia, S.M., Jensen, K., Contreras, C., Garzon, F., Pivovar, B., and Yan, Y. 2013. Platinum coated copper nanowires and platinum nanotubes as oxygen reduction electrocatalysts. *ACS Catalysis*, 3, 358–362.

Alia, S.M., Larsen, B.A., Pylypenko, S., Cullen, D.A., Diercks, D.R., Neyerlin, K.C., Kocha, S.S., and Pivovar, B.S. 2014a. Platinum-coated nickel nanowires as oxygen-reducing electrocatalysts. *ACS Catalysis*, 4, 1114–1119.

Alia, S.M., Pylypenko, S., Neyerlin, K.C., Cullen, D.A., Kocha, S.S., and Pivovar, B.S. 2014b. Platinum-coated cobalt nanowires as oxygen reduction reaction electrocatalysts. *ACS Catalysis*, 4, 2680–2686.

Al-Khalili. 2010. Discovering the Elements. Chemistry: A Volatile History. 25:40 minutes in. BBC Four, Presenter: Jim Al-Khalili, Professor (21 January. 2010) http://www.bbc.co.uk/programmes/b00q2mk5.

Anderson, E., Ayers, K., and Capuano, C. 2013. R&D Focus Areas Based on 60,000 hour Life PEM Water Electrolysis Stack Experience, at *First International Workshop on Durability and Degradation Issues in PEM Electrolysis Cells and its Components*, Freiburg, Germany, 12 March.

Atanasoska, L.L., Cullen, D.A., and Atanasoski, R.T. 2013b. XPS and STEM of the interface formation between ultra-thin Ru and Ir OER catalyst layers and Perylene Red support whiskers. *Journal of the Serbian Chemical Society*, 78, 1993–2005.

Atanasoska, L.L., Vernstrom, G.D., Haugen, G. M., and Atanasoski, R.T. 2011. Catalyst durability for fuel cells under start-up and shutdown conditions: Evaluation of Ru and Ir sputter-deposited films on platinum in PEM environment. *ECS Transactions*, 41, 785–795.

Atanasoski, R.T., Atanasoska, L.L., and Cullen. D.A. 2013a. Chapter 22: Efficient oxygen evolution reaction catalysts for cell reversal and start/stop tolerance, in Shao, M. ed. *Electrocatalysis in Fuel Cells: A Non and Low Platinum Approach*, pp. 637–663. Springer-Verlag, London.

Atanasoski, R.T., Atanasoska, L.L., Cullen, D.A., Haugen, G.M., More, K.L., and Vernstrom, G.D. 2012. Fuel cells catalyst for start-up and shutdown conditions: Electrochemical, XPS, and TEM evaluation of sputter-deposited Ru, Ir, and Ti on Pt-nano-structured thin film (NSTF) support. *Electrocatalysis*, 3, 284–297.

Atanasoski, R.T., Cullen, D.A., Vernstrom, G.D., Haugen, G.M., and Atanasoska, L.L. 2013c. A materials-based mitigation strategy for SU/SD in PEM fuel cells: Properties and performance-specific testing of IrRu OER catalysts. *ECS Electrochemistry Letters*, 2, F25–F28.

Bockris, J.O'M. and Appleby, A.J. 1972. The hydrogen economy: An ultimate economy? *Environment*, 1, 29.

Carmo, M. Fritz, D.L., Mergel, J., and Stolten, D. 2013 A comprehensive review on PEM water electrolysis. *International Journal of Hydrogen Energy*, 38, 4901–4934.

Cavendish, H. 1766. Three papers containing experiments on factitious air, by the hon. Henry Cavendish. *Philosophical Transactions (The University Press)* 56, 141–184. doi:10.1098/rstl.1766.0019.a.

Chen, Z., Waje, M., Li, W., and Yan, Y. 2007. Supportless Pt and PtPd nanotubes as electrocatalysts for oxygen-reduction reactions. *Angewandte Chemie International Edition*, 46, 4060–4063.

Coluccia, M., Gaggio, G., Guarna, S., and Spazzafumo, G., 1994. Pre-feasibility analysis of an energy supply system for Southern Europe: Technical. *International Journal of Hydrogen Energy*, 19, 957–963.

Cullen, D.A., More, K.L., Atanasoska, L.L., and Atanasoski, R.T. 2014. Impact of IrRu oxygen evolution reaction catalysts on Pt nanostructured thin films under start-up/shutdown cycling. *Journal of Power Sources*, 269, 671–681.

Debe, M.K. 2010. Handbook of Fuel Cells – Fundamentals, Technology, and Applications. (Wolf Vielstich, Hubert Gasteiger, Arnold Lamm and Harumi Yokokawa, editors.). John Wiley and Sons, Inc., Hoboken, NJ, ISBN: 9780470974001.

Debe, M.K. 2012. Electrocatalyst approaches and challenges for automotive fuel cells. *Nature*, 486, 43–51.

Debe, M.K., Hendricks, S.M., Vernstrom, G.D., Meyers, M., Brostrom, M., Stephens, M., Chan, Q., Willey, J., Hamden, M., Mittelsteadt, C.K., Capuano, C.B., Ayers, K.E., and Anderson, E.B. 2012. Initial performance and durability of ultra-low loaded NSTF electrodes for PEM electrolyzers. *Journal of The Electrochemical Society*, 159, K165–K176.

Debe, M.K., Kam, K.K., Liu, J.C., and Poirier, R.J. 1988. Vacuum vapor deposited thin films of a perylene dicarboximide derivative: Microstructure versus deposition parameters. *Journal of Vacuum Science & Technology A*, 6, 1907–1911.

Debe, M.K. and Poirier, R.J. 1990. Effect of gravity on copper phthalocyanine thin films. III. Microstructure comparisons of copper phthalocyanine thin films grown in microgravity and unit gravity. *Thin Solid Films*, 186, 327–347.

Drolet, B., Gretz, J., Kluyskens, D., Sandmann, F., and Wurster, R. 1996. The Euro-Quebec hydro-hydrogen pilot project [EQHHPP]: Demonstration phase. *International Journal of Hydrogen Energy*, 21, 305–316.

Erlebacher, J., Aziz, M., Karma, A., Dimitrov, N., and Sieradzki, K. 2001. Evolution of nanoporosity in dealloying. *Nature*, 410, 450–453.

Erlebacher, J. and Snyder, J. 2009. Dealloyed nanoporous metals for PEM fuel cell catalysis. *ECS Transactions*, 25, 603–612.

Giordano, N. 13 February 2009. *College Physics: Reasoning and Relationships*. Cengage Learning. p. 510. Brooks Cole, Belmont, CA. ISBN 0-534-42471-6.

Gretz, G., Baselt, J.P., Ullmann, O., and Wendt, H. 1990. The 100 MW Euro-Quebec hydro-hydrogen pilot project. *International Journal of Hydrogen Energy*, 15, 419–424.

Grove, W.R. 1838. Mr. W. R. Grove on a new voltaic combination. *The London and Edinburgh Philosophical Magazine and Journal of Science*, 430–431.

Grove, W.R. 1839. On voltaic series and the combination of gases by platinum. *Philosophical Magazine and Journal of Science*, XIV, 127–130.

Imbeault, R., Antonio, P., Garbarino, S., and Guay, D. 2010. Oxygen reduction kinetics on PtxNi100-x thin films prepared by pulsed laser deposition. *Journal of The Electrochemical Society*, 157, B1051–B1058.

Justi, E.W. 1987. *A Solar-Hydrogen Energy System*. Plenum Press, New York.

Kam, K.K., Debe, M.K., Poirier, R.J., and Drube, A.R. 1987. Summary abstract: Dramatic variation of the physical microstructure of a vapor deposited organic thin film. *Journal of Vacuum Science & Technology*, 5, 1914–1916.

Keith, G. and Leighty, W. 2002. Transmitting 4,000 MW of new windpower from N. Dakota to Chicago: New HVDC electric lines or hydrogen pipeline. In: *Proceedings of the 14th World Hydrogen Energy Conference*, Montreal, Canada.

Levie, R. de. October 1999. The electrolysis of water. *Journal of Electroanalytical Chemistry*, 476, 92–93. doi:10.1016/S0022-0728(99)00365-4.

Lewinski, K. and Mao, S. 1998. 3M internal unpublished data.

Lewinski, K.A. 2014. 3M internal unpublished data.

Mary Bellis, http://inventors.about.com/library/weekly/aacarsgasa.htm?rd=1 (last accessed on 2015-0607).

Miloradovic, T., Pineda, I., Azau, S., Moccia, J., and Wilkes, J. February 2014. Wind in power, 2013 European statistics. The European Wind Energy Association Publication, http://www.ewea.org/fileadmin/files/library/publications/statistics/EWEA_Annual_Statistics_2013.pdf

Moffat, T.P., Mallett, J.J., and Hwang, S.-M. 2009. Oxygen reduction kinetics on electrodeposited Pt 100-xNix, and Pt 100-xCox. *Journal of The Electrochemical Society*, 156, B238–B251.

Ogden, J.M. and Williams, R.H. 1989. *Solar Hydrogen: Moving Beyond Fossil Fuels*. World Resources Institute, Washington, DC.

Oney, F., Veziroglu, T.N., and Dülger, Z. 1994. Evaluation of pipeline transportation of hydrogen and natural gas mixtures. *International Journal of Hydrogen Energy*, 19, 813–822.

Palmer, D. (13 September 1997), "Hydrogen in the Universe", NASA, http://imagine.gsfc.nasa.gov/docs/ask_astro/answers/971113i.html, (last assessed on 2008-01-23).

Park, S. et al. 2011. Polarization losses under accelerated stress test using multiwalled carbon nanotube supported Pt catalyst in PEM fuel cells. *Journal of The Electrochemical Society*, 158, B297–B302.

Paulus, U.A., Vokaun, A., Scherer, G.G., Schmidt, T.J., Stamenkovic, V., Markovic, N.M., and Ross, P.N. 2002. Oxygen reduction on high surface area Pt-based alloy catalysts in comparison to well defined smooth bulk alloy electrodes. *Electrochimica Acta*, 47, 3787–3798.

ResearchInChina. November 2010. Global and China Low-E Glass Industry Report, at http://pressexposure.com/Global_and_China_Low-E_Glass_Industry_Report,_2010_-_Published_by_ResearchInChina-205310.html

Staff 2007. Hydrogen Basics – Production. Florida Solar Energy Center, University of Central Florida, Cocoa, FL, http://www.fsec.ucf.edu/en/consumer/hydrogen/basics/production.htm, (last accessed on 2015-06-07).

Stamenkovic, V., Schmidt, T.J., Ross, P.N., and Markovic, N.M. 2002. Surface composition effects in electrocatalysis: Kinetics of oxygen reduction on well-defined Pt$_3$Ni and Pt$_3$Co alloy surfaces. *The Journal of Physical Chemistry B*, 106, 11970–11979.

Stamenkovic, V.R., Fowler, B., Mun, B.S., Wang, G. Ross, P.N., Lucas, C.A., and Markovic, N.M. 2007a. Improved oxygen reduction activity on Pt$_3$Ni(111) via increased surface site availability. *Science*, 315, 493–497.

Stamenkovic, V.R., Mun, B.S., Arenz, M., Mayrhofer, K.J.J., Lucas, C.A., Wang, G., Ross, P.N., and Markovic, N.M. 2007b. Trends in electrocatalysis on extended and nanoscale Pt bimetallic alloy surfaces. *Nature Materials*, 6, 241–247.

Van der Vliet, D.F., Wang, C., Tripkovic, D., Strmcnik, D., Zhang, X.F., Debe, M.K., Atanasoski, R.T., Markovic, N.M., and Stamenkovic, V.R. 2012. Mesostructured thin films as electrocatalysts with tunable composition and surface morphology. *Nature Materials*, 11, 1051–1058.

Wang, S., Jiang, S.P., White, T.J., and Wang, X. 2010. Synthesis of Pt and Pd nanosheaths on multi-walled carbon nanotubes as potential electrocatalysts of low temperature fuel cells. *Electrochimica Acta*, 55, 7652–7658.

Yang, R., Leisch, J., Strasser, P., and Toney, M.F. 2010. Structure of dealloyed PtCu$_3$ thin films and catalyst activity for oxygen reduction. *Chemistry of Materials*, 22, 4712–4720.

Zhou, H., Zhou, W.-P., Adzic, R., and Wong, S.S. 2009. Enhanced electrocatalytic performance of one-dimensional metal nanowires and arrays generated via an ambient surfactantless synthesis. *The Journal of Physical Chemistry C*, 113, 5460–5466.

6

Membranes

Hiroshi Ito

CONTENTS

6.1 Introduction

A proton exchange membrane (PEM) electrolyzer offers several advantages over conventional alkaline (liquid electrolyte) water electrolysis including higher energy efficiency, greater hydrogen production rate, and more compact design. These advantages are derived from the solid-state membrane electrolyte compared to a device with free liquid electrolyte and a porous separator (diaphragm). In PEM electrolyzers, liquid electrolytes (KOH solution) used in alkaline electrolyzers are replaced by thin (50–250 μm thick) proton-conducting membranes. The performance and durability of the electrolyzer significantly depend on the membrane properties, which also determine the operation conditions, such as temperature and pressure. In addition, the material used for cell support components depends on the acidity of the membrane (i.e., pH).

The objective of this chapter is to present a comprehensive overview of the membrane electrolyte for PEM electrolyzers. First, functionalities and requirements of membranes are reviewed and examined briefly. Second, various types of alternative membranes are introduced and reviewed. Finally, the several recent research efforts are discussed.

6.2 Functionalities and Requirements of Membranes

The configuration of a PEM electrolyzer is similar to that of a proton exchange membrane fuel cell (PEMFC), consisting of a membrane electrode assembly (MEA), current collectors, bipolar plates with flow channels, bus plates, manifolds, and end plates. The heart of a PEM electrolyzer is the MEA, the same as that of a PEMFC. The electrodes (catalyst layers) are plated on both sides of a PEM, and the MEA serves as the electrolyte as well as the barrier between hydrogen and oxygen. Half-cell reactions during electrolysis are reversible of those during fuel cell operation and can be expressed as follows:

$$H_2O \rightarrow 2H^+ + \frac{1}{2}O_2 + 2e^- \quad \text{at anode} \qquad (6.1)$$

$$2H^+ + 2e^- \rightarrow H_2 \quad \text{at cathode} \qquad (6.2)$$

Liquid water is introduced at the anode, where it dissociates into molecular oxygen, protons, and electrons. Solvated protons formed at the anode migrate through the membrane to the cathode where they are reduced to molecular hydrogen. During migration of protons through the membrane, water molecules accompany the protons through the membrane from the anode to the cathode due to an electric field. Applying the membrane technology, the electrolyzer utilizes pure water as the circulating fluid, which results in a simpler balance of plant (BOP) and a lower leak current. In addition, gaseous products of H_2 and O_2 are produced at the backside of interpolar field, which brings a lower ohmic loss and a higher purity of gases (Ayers et al. 2010; Grigoriev et al. 2009). Considering this configuration of PEM electrolyzer, there are some specific properties required for the membrane electrolyte, such as high ion (proton) conductivity, low gas permeability, chemical and mechanical stability, low cost, and high durability.

6.3 Various Types of Membranes

The archetypal PEM is Nafion® (DuPont), because of its favorable chemical, mechanical, and thermal properties along with high proton conductivity when sufficiently hydrated. First, in this section, the comprehensive review on Nafion is introduced on a wide variety of properties, such as water uptake, conductivity, and gas permeability (Ito et al. 2011). In the case of practical use as an electrolyte of electrolysis, there are some drawbacks of Nafion such as cost, maximum operating temperature (<90°C), and problems associated with the transport of water and gas. This has driven a number of strategies into the design of alternative membranes (Paddison 2003). Subsequently, thus, the recent progress of Nafion-based composite membrane and hydrocarbon-based membrane is introduced and overviewed.

6.3.1 Nafion Membrane

Nafion membrane manufactured by DuPont is the most representative commercial product of perfluorosulfonic acid (PFSA) ionomers that consist of polytetrafluoroethylene (PTFE) backbone and double ether perfluoro side chains terminating in a sulfonic acid group as shown in Figure 6.1. The properties of Nafion have been extensively studied to analyze PEMFC operation (Doyle and Rajendran 2004). However, the hydration state of the membrane differs between fuel cell operation and electrolysis operation. During PEMFC operation, the membrane is humidified by the humidified gases and equilibrated with water vapor, whereas during electrolysis operation, the electrolyte membrane is exposed to the liquid phase of water and fully hydrated during water electrolysis. Therefore, to analyze a PEM electrolyzer, the properties of a Nafion membrane need to be evaluated when the membrane is exposed to and

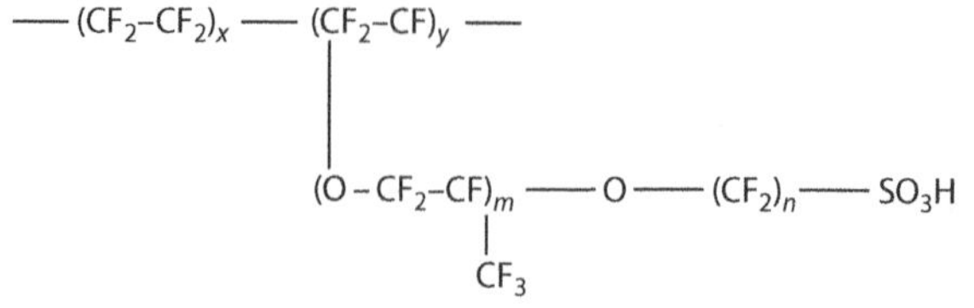

Nafion©: $m \geq 1, n = 2, x = 5-13, y = 1000$
Aquivion™: $m = 0, n = 1$

FIGURE 6.1
Chemical structures of perfluorosulfonated acid polymer electrolyte membranes. (Reprinted from *Progress in Polymer Science*, Rikukawa, M. and Sanui, K. 2000. Proton-conducting polymer electrolyte membranes based on hydrocarbon polymers. 25: 1463–1502, with permission from Elsevier.)

equilibrated with liquid water. This section reviews and examines previous studies on the properties of Nafion membranes such as water uptake, swelling ratio, proton conductivity, and electro-osmotic drag coefficient when the membrane is exposed to and equilibrated with liquid water.

6.3.1.1 Water Uptake and Swelling

PFSA membrane, such as Nafion, needs to be hydrated for proton conduction. The water uptake by a membrane is expressed in two forms (Doyle and Rajendran 2004): weight percent of water (ω) and water content (λ). The water uptake expressed as ω is calculated based on the weight of the wet sample w_{wet} and dry sample w_{dry}:

$$\omega\,[\%] = \frac{w_{\text{wet}} - w_{\text{dry}}}{w_{\text{dry}}} \qquad (6.3)$$

The water uptake expressed as λ represents the number of water molecules per sulfonic acid site (SO_3H^-). The relationship between λ and ω is represented as

$$\lambda = \frac{\omega \times \text{EW}}{M_{H_2O}} \qquad (6.4)$$

where EW is the equivalent weight and is 1100 g/mol for the currently commercial Nafion and M_{H_2O} is the molar weight of water.

The membrane is considered to be saturated with liquid water during water electrolysis. Figure 6.2 shows immersion temperature dependency of λ in a Nafion 117 membrane from liquid water (Hinatsu et al. 1994; Parthasarathy et al. 1992; Yoshitake et al. 1996; Zawodzinski et al. 1993). In the case of equilibration with liquid water, λ strongly depends on the pretreatment of the membrane. As shown in Figure 6.2, Zawodzinski et al. reported that the water uptake (λ) by a membrane dried at room temperature (form 1) is roughly twice that by a membrane dried at an

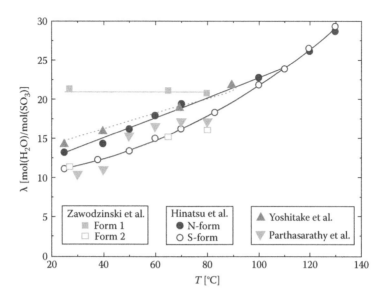

FIGURE 6.2

Water uptake (λ) of a Nafion 117 membrane immersed in liquid water at different temperatures (T): ■ (form-1): dried at room temperature, □ (form-2): dried at room temperature in vacuum followed by 1 h in vacuum at 105°C by Zawodzinski et al. 1993, ● (N-form): dried in vacuum at 80°C for 3 h, ○ (S-form): dried in vacuum at 105°C for 3 h by Hinatsu et al. 1994, ▲: boiled in water for 0.5 h followed by boiling in H_2SO_4 for 0.5 h (1M) by Yoshitake et al. 1996, and ▼: treated with 3% H_2O_2, 9M HNO_3 and water at 80–100°C by Parthasarathy et al. 1992. (Reprinted from Ito, H. et al., *International Journal of Hydrogen Energy*, 35, 10527, 2011. With permission from Hydrogen Energy Publications, LLC.)

elevated temperature of 105°C (form 2). Furthermore, the λ in a membrane dried at room temperature (form 1) is independent of the immersion temperature. Hinatsu et al. observed that the water uptake (λ) of a membrane without a pretreatment of vacuum drying (E-form) was relatively high ($\lambda \approx 23$) and remained constant up to approximately 100°C. Hinatsu et al. also evaluated two other pretreated membranes: membranes pretreated by drying at 80°C (N-form) and membranes pretreated by drying at 105°C (S-form). The λ of an N-form membrane was higher than that of an S-form membrane at an immersion temperature below 110°C. At immersion temperatures higher than the glass transition (100~110°C), the λ was the same for both forms of N and S. Note that the pretreatment of the S-form membrane tested by Hinatsu et al. is nearly identical to that of form-2 by Zawodzinski et al., and both membranes showed similar values of λ. Zawodzinski et al. and Broka and Ekdunge explained these differences of λ between form-1 and form-2 (or N and S forms) by the disintegration of ionic clusters in the polymer membrane during drying at elevated temperatures close to the glass transition temperature (Broka and Ekdunge 1997). The water uptake (λ) data reported by other researchers agree well with the line for either the N-form or the S-form. The electrolyte membrane is always exposed to the liquid phase of water and thus is fully hydrated during electrolysis. However, pretreatment procedures affect the water uptake by the membrane.

Water uptake by Nafion membrane from liquid water is higher than that from saturated water vapor (relative humidity = 100%) (Hinatsu et al. 1994; Zawodzinski et al. 1993). This difference in water uptake can be explained by Schroeder's paradox (Schroeder 1903; Zawodzinski et al. 1993). Based on this paradox, the lower water uptake from the vapor phase might be due to the difficulty in condensing vapor within the pores of a membrane. Furthermore, Weber and Newman pointed out that the water becomes more bulk like in the Nafion membrane equilibrated with liquid water, indicating that a separate liquid phase is not like bound water or cluster water formed in the case of a vapor equilibrium membrane (Weber and Newman 2003).

Due to water uptake, a membrane swells in both directions, namely, in-plane (length/width) and through-plane (thickness). This swelling result is due to a complex interplay between the affinity of the polymer and ionic sites for polar solvents and the resistance of the membrane's structure and crystallinity to volumetric expansion (Doyle and Rajendran 2004). Table 6.1 summarizes the data from previous studies on membrane swelling ratio in the in-plane (δ_i) and through-plane direction (δ_t), which are the swelling ratio of the length (width) and thickness, respectively, and calculated in the same manner as ω (Equation 6.3) (Büchi and Scherer 2001; Sakai et al. 1985; Takenaka et al. 1984). In terms of δ_i, Gabel et al. reported that the ratio of the expansion along the perpendicular and parallel laminating directions is about 1.4, though expansions are identical along the thickness and along the perpendicular direction (Gabel et al. 1993). This

TABLE 6.1

Swelling Ratio of Nafion Membranes Immersed in Liquid Water

References	Membrane	Immersed Water Temperature (°C)	Water Content (λ) mol_H$_2$O/ mol_SO$_3$	Water Content (ω) (wt.%)	Wet Thickness (μm)	Swelling Ratio in In-Plane (δ_i) (%)	Swelling Ratio in Through-Plane (δ_t) (%)
Takenaka et al. (1984)	Nafion 1110	25				10.7	
	Nafion 1110	50				11.5	
	Nafion 1110	75				12.0	
	Nafion 1110	90				12.6	
Sakai et al. (1985)	Nafion 117	Dry	1.0	1.7	185		—
	Nafion 117	100	21.4	35	220		18.9
	Nafion 117	140	30.5	50	237		28.1
	Nafion 117	170	62.3	102	265		43.2
Büchi and Scherer (2001)	Nafion 112	100	21–22	34–36	58–62		13–22
	Nafion 115	100	21–22	34–36	145–150		14–18
	Nafion 117	100	21–22	34–36	200–205		9–12

Source: Reprinted from Ito, H., Maeda, T. et al., *International Journal of Hydrogen Energy*, 35, 10527, 2011. With permission from Hydrogen Energy Publications, LLC.

swelling anisotropy of the membrane can be attributed to the laminating condition.

6.3.1.2 Proton Conductivity

Proton conductivity of a Nafion membrane (κ) depends on the water content and temperature. Zawodzinski et al. measured the conductivity of a Nafion 117 membrane as a function of water content (λ) at 30°C and reported that the conductivity increases linearly with λ when $2 < \lambda < 22$ (Zawodzinski et al. 1993). In addition, they measured the temperature dependence of the conductivity on a Nafion 117 membrane immersed in liquid water in the temperature range from 25°C to 90°C. In their measurement, λ was constant at 22 over this range of temperature. As described earlier, this sample must have been pretreated at room temperature. Based on these measurements data by Zawodzinski et al., Springer et al. (1991) correlated the proton conductivity using the following expression:

$$\kappa \, [\text{S cm}^{-1}] = (0.005139\lambda - 0.00326)\exp\left[1268\left(\frac{1}{303} - \frac{1}{T}\right)\right]$$

(6.5)

where T is the membrane temperature and the activation energy (E_a) used in this expression is 10.5 [kJ mol^{-1}]. Kopitzke et al. and Parthasarathy et al. also measured the proton conductivity as a function of temperature for a Nafion membrane immersed in liquid water (Kopitzke et al. 2000; Parthasarathy et al. 1992). The data reported by Kopitzke et al. were obtained for a membrane pretreated with boiling water and agree well with data

reported by Zawodzinski et al. The correlation of κ and T presented by Kopitzke et al. is

$$\kappa = \kappa_0 \exp\left(-\frac{E_\kappa}{RT}\right)$$

(6.6)

where E_κ = 7.829 [kJ mol^{-1}] and κ_0 = 2.29 [S cm^{-1}]. Figure 6.3 shows the comparison of the data reported by several authors (Doyle et al. 2001; Kopitzke et al. 2000;

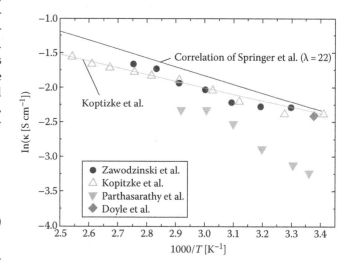

FIGURE 6.3

Arrhenius plots of proton conductivity (κ) of a Nafion 117 membrane immersed in liquid water obtained by several authors (Doyle et al. 2001; Kopitzke et al. 2000; Parthasarathy et al. 1992; Zawodzinski et al. 1993) and correlations presented by Springer et al. (1991) and Kopitzke et al. (2000). (Reprinted from Ito, H. et al., *International Journal of Hydrogen Energy*, 35, 10527, 2011. With permission from Hydrogen Energy Publications, LLC.)

Parthasarathy et al. 1992; Zawodzinski et al. 1993) and lines predicted by Springer et al. (1991) and Kopitzke et al. (2000) The pretreatment condition of the membrane used by Parthasarathy et al. was not reported, although it can be presumed that the membrane was pretreated at the elevated temperature (~105°C) because the conductivity data were obtained for the same membrane used in their measurement of λ, and their λ data were similar to that obtained for a membrane pretreated at the elevated temperature reported by other researchers (Figure 6.2). Therefore, not only the water content but also the proton conductivity of a membrane equilibrated with liquid water might decrease when pretreated by drying at the elevated temperature over about 100°C.

The dependence of water uptake (λ) and conductivity (κ) on membrane pretreatment might have important implications in the use of these membranes not only in fuel cells but also in electrolyzers. The MEA of PEM electrolyzers can be constructed by several methods. Some researchers (Ioroi et al. 2000; Ito et al. 2010; Song et al. 2008; Valdez et al. 2009) applied the hot-pressing process that is used in typical MEA fabrication for fuel cells. During this process, the membrane is fully dried at a temperature higher than the glass transition temperature. Such a high dehydration temperature during pretreatment possibly causes incomplete subsequent rehydration and degradation of proton conductivity even when the membrane is immersed in liquid water. However, we could not find out any reports discussing the relation between the fabrication method of MEA and the proton conductivity.

6.3.1.3 Electro-Osmotic Drag Coefficient

When a proton transverses through a PEM, water molecules accompany the proton (H⁺) when an electric field is applied. This phenomenon is well known as electro-osmosis (Springer et al. 1991; Zawodzinski et al. 1993). The electro-osmotic drag coefficient n_{drag} is defined as the number of water molecules "dragged along" per H⁺. n_{drag} depends on the water content in the membrane as the same as the proton conductivity, and n_{drag} ranges from 2.5 to 2.9 at 30°C when λ = 22, that is, the membrane is fully hydrated with liquid water. Onda et al. (2002) reported that n_{drag} depends only on the membrane temperature under PEM electrolysis conditions where the membrane must also be fully hydrated. They experimentally obtained the following correlation:

$$n_{drag} = 0.0134 \times T + 0.03 \tag{6.7}$$

At 30°C, n_{drag} calculated using this correlation is about 4.1. This drag coefficient (n_{drag}) is an important parameter for designing BOP, such as accumulators and pumps (Ito et al. 2010).

6.3.1.4 Solubility and Diffusivity of Gases

The major impact of high operating pressure in a Nafion membrane is the degradation of current efficiency and gas purity caused by cross-permeation of produced gases (Grigoriev et al. 2011; Millet et al. 2010; Schalenbach et al. 2013). This section reviews and examines previous studies on the gas permeability in Nafion membrane to evaluate the cross-permeation flux during electrolysis operation. The gas permeability (P_m) can be expressed as the product of diffusion coefficient (D) and solubility (S):

$$P_m = DS \tag{6.8}$$

As summarized by Kocha et al., the permeability of gases though a Nafion membrane has been estimated using several measurement techniques, such as the volumetric method, time-lag technique, gas chromatography method, electrochemical monitoring technique, and positron annihilation lifetime spectroscopy (Kocha et al. 2006). In the volumetric method, higher pressure is applied to one side of the membrane and the gas permeation rates are measured at the other side (Sakai et al. 1985). The time-lag method is identical to the volumetric method except that the time to fill a fixed volume downstream of the membrane is measured instead of the flow rate (Chiou and Paul 1988; Sakai et al. 1986). An advantage of the time-lag method is that the diffusion coefficient and solubility can be obtained from the combined permeability. Gas chromatography methods involve measuring the concentration change downstream of the membrane when the same total pressure but different gas concentrations are applied across the membrane (Broka and Ekdunge 1997; Yoshitake et al. 1996). The electrochemical monitoring technique, which monitors current over time for the diffusion-limiting condition through the membrane, has been widely used to characterize the gas permeation rate (Haug and White 2000; Kocha et al. 2006; Lehtinen et al. 1998; Ogumi et al. 1984; Parthasarathy et al. 1992). Positron annihilation lifetime spectroscopy has been used to characterize the free volume in polymers due to the unique property of a positronium atom (Mohamed et al. 2008). The free volume hole size method has been applied to estimate the diffusion coefficient and permeability of gases in a polymer membrane.

Figures 6.4 and 6.5 show Arrhenius plots of the permeability in Nafion membranes reported in the literature for hydrogen and oxygen, respectively, and Table 6.2 summarizes the measurement conditions. The permeability lines for water and PTFE were calculated by the literature data (Pasternak et al. 1970; Wise and Houghton 1966). As shown in Figures 6.4 and 6.5, the permeability of wet membrane was higher than that

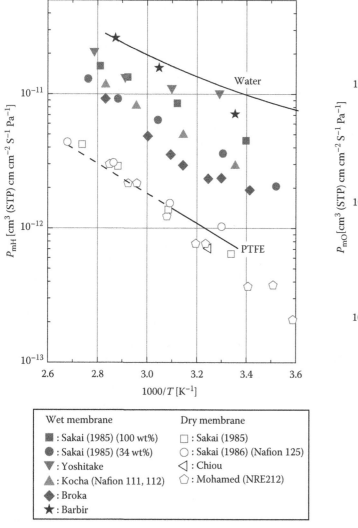

FIGURE 6.4
Arrhenius plots of hydrogen permeability (P_{mH}) in a Nafion 117 membrane at dry and wet conditions obtained from literatures (Barbir 2005; Broka and Ekdunge 1997; Chiou and Paul 1988; Kocha et al. 2006; Mohamed et al. 2008; Sakai et al. 1985, 1986; Yoshitake et al. 1996). Calculated lines for water and PTFE were also determined from literature data (Pasternak et al. 1970; Wise and Houghton 1966). (Reprinted from Ito, H. et al., *International Journal of Hydrogen Energy*, 35, 10527, 2011. With permission from Hydrogen Energy Publications, LLC.)

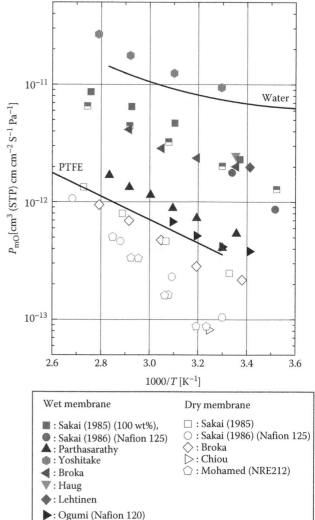

FIGURE 6.5
Arrhenius plots of oxygen permeability (P_{mO}) in Nafion 117 at dry and wet conditions obtained from literatures (Broka and Ekdunge 1997; Chiou and Paul 1988; Haug and White 2000; Parthasarathy et al. 1992; Lehtinen et al. 1998; Mohamed et al. 2008; Ogumi et al. 1984; Sakai et al. 1985, 1986; Yoshitake et al. 1996). Calculated lines for water and PTFE were also determined from literature data (Pasternak et al. 1970; Wise and Houghton 1966). (Reprinted from Ito, H. et al., *International Journal of Hydrogen Energy*, 35, 10527, 2011. With permission from Hydrogen Energy Publications, LLC.)

of dried membrane, and a dried membrane showed similar or lower values than in PTFE. In the case of hydrogen permeability (Figure 6.4), although there are significant differences among the data for the wet membrane, relative good agreement is evident for the data reported by Sakai et al. (34 wt.%), Kocha et al., and Broka and Ekdunge. However, note that the hydrogen permeability data at relatively high-pressure conditions (16–220 atm) observed by Barbir show the highest value and are similar to that in water, though the measurement technique was not reported. Although the literature data for oxygen permeability are scattered

(Figure 6.5), comparable values were reported by Sakai et al., Broka and Ekdunge, Lehtinen et al., and Haug and White. Although the reason for these differences in permeability data for hydrogen and oxygen is not clear at present, one explanation is the difference in pretreatment and hydration state of the membrane. Several researchers have measured the permeability of both hydrogen and oxygen and reported a higher hydrogen permeability than oxygen permeability. Sakai et al. and Broka and Ekdunge reported a ratio

TABLE 6.2

Summary of Measurements for Solubility (S), Diffusion Coefficient (D), and Permeability (P_m) of a Nafion Membrane in the Literature

References	Presented Data	Gas[a]	Media[b]	Technique	Temperature (°C)	Pressure
Ogumi et al. (1984)	D, S	O_2	Nafion 120 (Wet)	Electrochemical	20–50	1 atm
Sakai et al. (1985)	P_m	H_2/O_2	Nafion 117, Nafion 125 (Wet/dry)	Volumetric	20–90	1–30 atm
Sakai et al. (1986)	P_m, D	H_2/O_2	Nafion 117, Nafion 125 (Wet/dry)	Time-lag	−10–100	4–14 atm
Chiou and Paul (1988)	P_m, D, S	H_2/O_2	Nafion 117 (Dry)	Time-lag	35	1 atm
Parthasarathy et al. (1992)	D, S	O_2	Nafion 117 (Wet)	Electrochemical	25–80	5 atm
Yoshitake et al. (1996)	P_m	H_2/O_2	Nafion 117 (Wet)	Gas chromatography	30–85	1 atm
Broka and Ekdunge (1997)	P_m	H_2/O_2	Nafion117 (Wet/dry)	Gas chromatography	20–100	n/a
Lehtinen et al. (1998)	D, S	O_2	Nafion 117 (Wet)	Electrochemical	20	n/a (1 atm)[c]
Haug and White (2000)	D, S	O_2	Nafion 117 (Wet)	Electrochemical	25	n/a (1 atm)[c]
Barbir (2005)	P_m	H_2	Nafion 117 (Wet)	n/a	20–75	16–220 atm
Kocha et al. (2006)	P_m	H_2	Nafion 111, Nafion 112 (Wet)	Electrochemical	20–80	1–2 atm
Mohamed et al. (2008)	P_m	H_2/O_2	NRE212 (Wet/dry)	Positron annihilation lifetime	−30–80	n/a

Source: Reprinted from Ito, H. et al., *International Journal of Hydrogen Energy*, 35, 10527, 2011. With permission from Hydrogen Energy Publications, LLC.

[a] Tested gases without H_2 and O_2 are not listed.
[b] Tested membrane without Nafion series is not listed.
[c] Pressure in parentheses is estimated for calculating permeabilities and solubilities shown in Figures 6.4 and 6.5.

of permeability of hydrogen and oxygen of about 2 for a wet membrane, whereas Yoshitake et al. reported a ratio of about 10.

6.3.1.5 Degradation

In PEM electrolysis, the surrounding hydration condition for the PEM is stable, because the electrolyte membrane is always exposed to the liquid phase of water and thus is fully hydrated. Therefore, it can be noted that membrane degradation during operation is not critical compared to the case of fuel cell applications. For example, a commercial product of Proton OnSite has shown a durability at 165 bar differential pressure to over 20,000 h (Ayers et al. 2010). Nevertheless, we have to pay attention to the state of preservation at standby mode as well as an intermittent power input, when the hydration condition would possibly be changed.

6.3.2 Nafion-Based Composite Membrane

As described earlier, the maximum operating temperature of Nafion is about 90°C. There are some advantages for operating water electrolysis at high temperature, as follows: (1) improving the electrode kinetics and thus reducing the activation overpotential, (2) lowering total thermodynamic energy requirement for the water splitting (ΔH) (284 kJ mol^{-1} at 80°C would reduce to 243 kJ mol^{-1} at 130°C), and (3) lowering the Gibbs free

energy change (ΔG) and thus decreasing the reversible voltage, which should be compensated with an increase in heat adding to the system, and the heat would be supplied from the self-heating due to the ohmic loss (Joule heat) (Hansen 2012; Hansen et al. 2012; Nikiforov et al. 2012). Several attempts have been carried out for operating water electrolysis at high temperature above 100°C using a polymer electrolyte.

The main problem associated with the high-temperature electrolysis operation is the membrane dehydration, because the relative humidity around PEM would be significantly decreased. In terms of PFSA membrane, one solution for this dehydration problem is the inclusion of the hygroscopic oxide particle in the membrane (i.e., Nafion). Antonucci and co-workers executed an electrolysis operation at high temperature above 100°C using a composite Nafion–SiO_2 membrane (Antonucci et al. 2008) and Nafion–TiO_2 membrane (Baglio et al. 2009). Xu et al. also used the PFSA–SiO_2 composite as an electrolyte and tried a high-temperature operation. They could achieve a stable operation until 1.6 A cm^{-2} at 130°C under 4 bar (Xu et al. 2011). According to their observations, the inclusion of inorganic particle should bring the effect of not only improving water retention but also decreasing the gas cross-over due to the presence of inorganic hygroscopic fillers inside the polymeric matrix.

A dimensionally stable membrane (DSM) based on Nafion was proposed by Ginar, Inc (Cropley and

Norman 2008). DSM is a membrane incorporated in a high-strength supporting structure with a definable pattern. Laser micromachining technology was applied to drill holes 30 µm in diameter in the 8 µm thick support structure (polyimide, Kapton® DuPont) as shown in Figure 6.6. To form membrane, the support was sprayed with multiple thin coats of liquid Nafion solution, which is pulled into the pores by capillary action. Eventually, the total thickness of DSM was about 30 µm (Figure 6.6b). The supported membrane shows negligible in-plane dimensional expansion when in contact with liquid water, and this property would alleviate the stress during wet/dry and temperature cycling. The electrolyzer performance of the DSM was significantly better than that of Nafion 117 at 60°C. Performance of a cell with the DSM after 300 h operation was slightly better than the initial performance, indicating that the membrane is stable in short-term testing.

6.3.3 Hydrocarbon-Based Membrane

Until recently, PFSA membrane, such as Nafion, was solely used for a long period of time in water electrolysis application. However, the current industry standards on PFSA polymer have been limited by high cost and loss of membrane stability at elevated temperature above about 90°C. A high-temperature operation is advantageous for PEM electrolysis as described earlier. In addition, the disposal cost of PFSA membrane would also be expensive due to the contained fluorine in the backbone structure. Therefore, the use of hydrocarbon-based membrane, instead of PFSA membrane, has been mainly motivated by cost and stability at elevated temperature above 100°C. Generally, hydrocarbon polymers are easily recycled by conventional methods.

In terms of the electrolysis using hydrocarbon membrane, Masson and Molina (1982) first tried to develop hydrocarbon membrane in the early 1980s. They fabricated the membrane by radiation grafting of styrene groups on a polyethylene matrix, followed by sulfonation of the resulting membrane. The electrolytic cell with this membrane showed a noticeable degradation or increase in the voltage during 3000 h tests. However, the life test was carried out under a low current density of 0.2 A cm^{-2}. According to Wei et al. (2010), Masson et al. reported later that the polyethylene-based membrane was not likely to withstand severe working conditions of 0.5–1.0 A cm^{-2} at 80–100°C.

Sulfonated aromatic polymers, such as polyetheretherketone (PEEK), polysulfone (PSf), and polybenzimidazole (PBI), are expected to have lower production cost as well as satisfactory chemical and mechanical stability. Chemical structures of sulfonated PEEK (S-PEEK) and sulfonated PBI are shown in Figure 6.7. In the case of S-PEEK that is one of representative aromatic

polymers, many studies on electrochemical properties have been executed for the purpose of PEMFC applications (Kreuer 1997, 2001, 2003; Kreuer et al. 2004; Rikukawa and Sanui 2000). The S-PEEK is attractive due to high strength, easy-membrane-forming material, and low cost. However, there is a fatal drawback associated with hydration. Figure 6.8 shows the water uptake of different membranes and implies the swelling behavior as a function of temperature. For Nafion, irreversible swelling starts only at a temperature above 130°C, which corresponds to the glass transition temperature of Nafion (Kreuer 1997, 2001, 2003). Contrary to this, the sulfonated aromatic membrane (S-PEEKK) is morphologically stable in water only up to 80°C. In addition, the high degree of sulfonation causes PEEK polymer to become excessively water swollen or soluble in water, thereby destroying their mechanical structure (Jang et al. 2008a). Subsequently, it must be difficult to use S-PEEK as an electrolyte of water electrolyzer, because the membrane is fully hydrated during water electrolysis.

To achieve high proton conductivity without these drawbacks, chemically cross-linkable ionomeric systems were proposed (Kerres et al. 2004). Jang et al. prepared a composite polymer membrane consisting of covalently cross-linked S-PEEK with tungstophosphoric acid (TPA). The composite membrane showed good electrochemical and mechanical properties. The prepared membrane used in water electrolytic cell and resulting cell voltage was 1.78 V at 1 A cm^{-2} and 80°C, with a platinum loading of 1.28 mg cm^{-2} (Jang et al. 2008a). Jang et al. also developed another composite polymer that consists of copolymer of PSf and S-PEEK blended with TPA. The cell voltage was 1.83 V at 1 A cm^{-2} and 80°C, with a platinum loading of 1.12 mg cm^{-2} (Jang et al. 2008b). Woo et al. and Lee et al. investigated covalently cross-linked S-PEEK with several types of heteropolyacids, such as molybdophosphoric acid (MoPA), tungstosilicic acid (TSiA), and TPA. With membranes of similar thickness (ca. 180 µm), they showed cell voltage around 1.75 V at 1 A cm^{-2}, 80°C and 1 atm (Lee et al. 2011; Woo et al. 2010). Wei et al. used S-PEEK and polyethersulfone (PES) blend membrane for water electrolysis. An electrolytic current of 1.655 A cm^{-2} was obtained at 2 V and 80°C (Wei et al. 2010).

The cell temperature in the studies using S-PEEK as the main polymer backbone has been limited up to 100°C as the same as Nafion, though the application of hydrocarbon membrane is motivated by a high-temperature operation. The experimental work of a high-temperature operation above 100°C can be seen in the studies using phosphoric acid (PA)-doped PBI (Aili et al. 2011; Hansen 2012a). As the same as other hydrocarbon membranes, PA-doped PBI was first proposed as electrolyte for high-temperature fuel cells

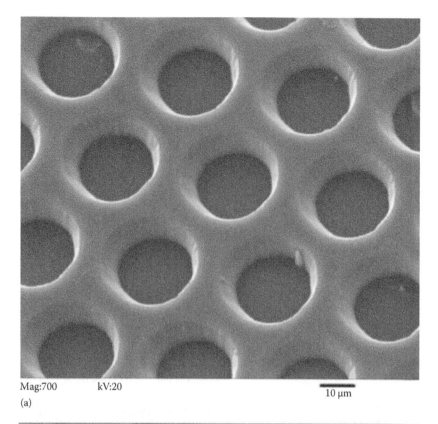

Mag:700   kV:20   10 μm

(a)

HV	Spot	WD	Tilt	Sg	HFW	Mag	20.0 μm
10.0 kV	3.5	9.7 mm	−0.1 °	BSE	85.33 μm	1500×	

(b)

FIGURE 6.6

Scanning electron microscope (SEM) image of the polymer membrane support structure with definable hole pattern (a) and cross section of dimensionally stable membrane (b). (Reprinted from Cropley, A. and Norman, T., A low-cost high-pressure hydrogen generator. DOE Final report, DE-FC36-04GO013029, 2008. With permission from Giner, Inc. and U.S. Department of Energy.)

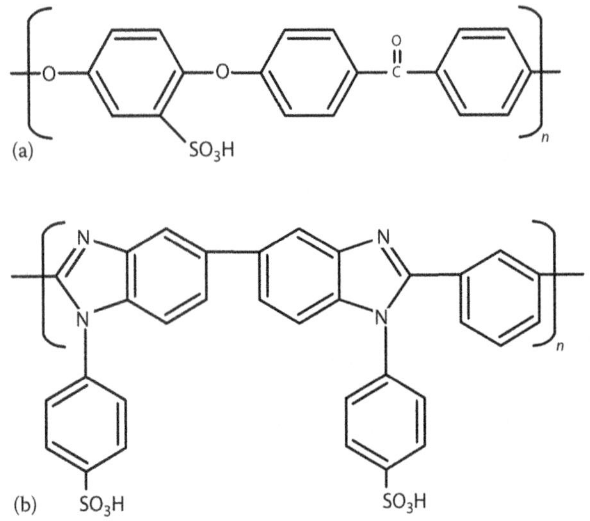

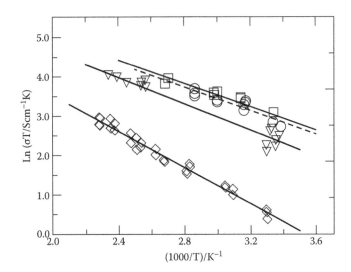

FIGURE 6.7
Chemical structures of polymer electrolyte membranes based on hydrocarbon polymers. (a) Sulfonated PEEK and (b) sulfonated PBI. (Reprinted from *Progress in Polymer Science*, 25, Rikukawa, M. and Sanui, K., Proton-conducting polymer electrolyte membranes based on hydrocarbon polymers, 1463–1502, Copyright (2000), with permission from Elsevier.)

FIGURE 6.9
Arrhenius plots of conductivity of phosphoric acid–doped PBI membranes (————) and Nafion (- - - -). Doping levels of PBI membranes; (□) 1600%, (▽) 1300%, and (◇) 450%. Relative humidity was 92%–98% for Nafion and 80%–85% for PBI membranes. (With kind permission from Springer: *Journal of Applied Electrochemistry*, Phosphoric acid doped polybenzimidazole membranes: Physiochemical characterization and fuel cell applications, 31, 2001, 773–779, Li, Q., He, R. et al.)

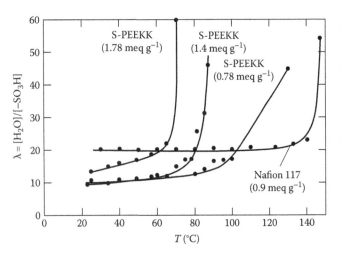

FIGURE 6.8
Water uptake in water of Nafion and sulfonated polyetherketones (PEEKK) with different degrees of sulfonation (ion exchange capacities are given in brackets). (Kreuer, K.D.: *Handbook of Fuel Cells*. 420–435. 2003. Copyright Wiley-VCH Verlag GmbH & Co. KGaA. Reproduced with permission.)

(Wainright et al. 1995). Since then, numerous groups have published high-temperature fuel cell results using PA-doped PBI (e.g., Li et al. 2001, 2003; He et al. 2003, 2006; Ma et al. 2004; Zhai et al. 2007; Mader et al. 2008; Seel et al. 2009). The challenge with an increase in operation temperature was solved by replacing the proton-conducting phase in the membrane with PA, which exhibits high anhydrous proton conductivity

and has very low vapor pressure. PA-doped PBI can be used at operation temperature up to 200°C, because its proton conductivity mechanism is independent of the water presence within the membrane matrix. The proton conductivity of PA-doped PBI membrane depends on the doping level of PA as shown in Figure 6.9. Hansen carried out comprehensive experimental work using PA-doped membrane applied to water (steam) electrolysis and revealed that it is possible to achieve reasonable short-term electrolysis performance with PA-doped PBI membrane at 130°C. However, no long-term durability could be achieved since the membrane failed within hours of electrolysis initiation (Hansen 2012a). The reason for the membrane failure has not been completely understood. A possible explanation could be the lack of chemical stability of the PA-doped PBI membrane at the harsh oxidative conditions during electrolysis at elevated temperature. Linkous discussed the thermodynamic stability problem of PBI especially under oxidative conditions and pointed out that the degradation could be caused by hydrolysis of the imidazole ring (Linkous 1993; Linkous and Anderson 1998). Aili et al. prepared two kinds of membranes doped with PA for high-temperature electrolysis above 100°C, PA-doped Nafion and PA-doped PBI. An MEA based on PA-doped Nafion was operated at 130°C under ambient pressure with a current density of 0.3 A cm^{-2} at 1.75 V, with no membrane degradation observed during a run of 90 h. The PBI-based MEAs showed better polarization curves (0.5 A cm^{-2} at 1.75 V) but poor durability (Aili et al. 2011).

6.4 Recent Research Efforts

This section gives a brief overview of recent efforts on two kinds of alternative membranes.

6.4.1 Short-Side-Chain Perfluorosulfonated Acid Membrane

Several groups have paid attention to a short-side-chain perfluorosulfonated acid (SSC-PFSA) membrane, such as Aquivion™ from Solvay (Figure 6.1). Conventional PFSA polymer, such as Nafion, consists of a PTFE-like backbone and double ether perfluoro side chains terminating in a sulfonic acid group. The Aquivion is a short-side-chain (SSC) version of PFSA with simpler – $OCF_2CF_2SO_3H$ pendant side chains each consisting of two C atoms compared to four C atoms in the case of Nafion. Correspondingly, Nafion is sometimes referred to as the long-side-chain (LSC)–PFSA ionomer. There are two major differences found between LSC– and SSC–PFSA membranes. One is their different behaviors relative to temperature. SSC–PFSA polymer presents a primary glass transition, defined as the "α" transition, at ~160°C, whereas LSC–PFSA polymer shows this transition at ~110°C. This property is favorable to a high-temperature operation of the electrolyzer. Another important difference is found in crystalline characteristic. SSC–PFSA ionomers, as a consequence of their lower-molecular-weight pendant group, show a crystallinity content that is higher than the LSC of the same equivalent weight (EW). Therefore, lower-EW (i.e., higher-ionic-content) membranes can be prepared with the same crystallinity or the same EW membranes with higher crystallinity (i.e., higher mechanical properties). This possibility is important not only for fuel cell but also for electrolysis applications, because higher mechanical properties allow membranes with lower thickness to be attained, which, in turn, means higher membrane conductance (Arcella et al. 2005; Aricò et al. 2010; Ghielmi et al. 2005; Kreuer et al. 2008). Hansen et al. tried to use the Aquivion as an electrolyte of high-temperature steam electrolysis at 130°C, since a dimensional stability at elevated temperature can be expected with SSC–PFSA membrane. Because the proton conductivity of Aquivion at 130°C is not sufficient for the electrolysis operation, the Aquivion membranes were doped with PA for securing sufficient proton conductivity at elevated temperature above 100°C. Steam electrolysis with a PA-doped Aquivion membrane was successfully conducted and current densities was reached up to 0.78 A cm^{-2} at 1.8 V under ambient pressure (Hansen et al. 2012b). The electrolysis performance using this PA-doped Aquivion membrane was compared to that with Nafion-based composite membranes, such as Nafion–SiO_2 and Nafion–TiO_2 membrane (Antonucci et al. 2008; Baglio et al. 2009), and was superior to these two cases due to higher membrane conductivity owing to PA doping. However, the long-term durability of the PA-doped Aquivion membrane still needs to be examined. For example, the leaking of PA from the membrane to the anode under electrolysis conditions is unknown.

6.4.2 Anion Exchange Membrane

The distinct disadvantage of PEM electrolysis is its high capital cost of the cell stack compared to alkaline liquid electrolyte electrolysis. This drawback is mainly attributed to the acidic environment of PEM, which limits the catalyst to noble metals and the cell support structure to anticorrosion metals. On the other hand, water electrolysis using a solid-state polymer electrolyte offers several advantages, such as compact design, easy handling, and simple BOPs. To lower the cost of membrane-based electrolyzers, new materials are needed, which enable less expensive electrolyzer system while maintaining the advantages of the polymer electrolyte architecture. One of the promising new materials is the anion exchange membrane (AEM) that has high internal pH. Because the possibility of nonnoble metal catalyst's use with the AEM is beneficial for the fuel cell, such AEM technology has been developed and demonstrated for fuel cell applications prior to water electrolysis (Dekel 2012; Merle et al. 2011; Piana et al. 2010; Varoe et al. 2006). Today, there is wide variety of commercially available AEMs exhibiting different chemical properties (Merle et al. 2011). Among them, one of the representative commercial AEMs is A201 supplied from Tokuyama Corp. (Watanabe et al. 2010) (Figure 6.10). In the last few years, several groups had successfully demonstrated AEM water electrolysis technologies that were free of noble metal catalyst, offering promising strategies for low-cost water electrolysis (Cao et al. 2012; Faraj et al. 2012; Leng et al. 2012; Wu and Scott 2013; Xiao and Zhang 2012). Xiao et al. executed an AEM water electrolysis test with only pure water feeding for both electrode sides and obtained the cell voltage of about 1.8–1.85 V at 70°C under a current density of 0.4 A cm^{-2} (Xiao and Zhang 2012). Faraj et al. developed a new AEM based on low-density polyethylene (LDPE) backbone and tested the membrane with an electrolytic cell using a Tokuyama membrane as a reference. In these tests, an aqueous solution of 1 wt.% K_2CO_3 was circulated at the anode side, while no liquid water/electrolyte was supplied at the cathode side. The dilute aqueous K_2CO_3 electrolyte plays a role to guarantee ionic conductivity along with pH buffering (ca. 10) (Faraj et al. 2012). Recently, Acta

Structural image

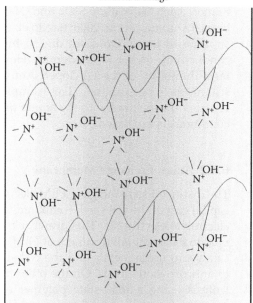

A201

FIGURE 6.10

Chemical structures (left) and photo (right) of commercial anion exchange membrane (A201) and ionomer solution (AS-4) supplied from Tokuyama Corp. (Courtesy of Tokuyama.)

SpA has commercially launched an AEM water electrolyzer (Pavel et al. 2014). The MEA of Acta's electrolyzer was composed of Tokuyama membrane (A201) and non-noble metal catalyst developed by their own. As the same as Faraj et al., liquid water containing electrolyte is supplied only to the anode, which facilitates to produce hydrogen relatively dry and pressurized state (30 bar) at the cathode. Acta also confirmed that the AEM electrolyzer can be operated with intermittent power inputs, indicating the adaptability to unstable renewable sources (Pavel et al. 2014).

6.5 Summary

Applying the membrane architecture, PEM electrolyzers as hydrogen production devices have higher efficiency and higher current density capability. Nafion membrane is still commonly employed as the electrolyte because of its favorable chemical, mechanical, and thermal properties along with high proton conductivity. In order to apply hydrogen to more widespread energy applications, the hydrogen production cost is expected to be cheaper. Since the cost of electricity is the major contributor to the cost of hydrogen produced by electrolysis, low capital cost and efficient operation of the electrolyzer are keys to low-cost hydrogen production for large applications. For the cost reduction of PEM electrolyzer, the cost of membrane is also expected to

be cheaper. Considering the installation of renewable energy source, the electrolyzer would have to be operated under dynamic conditions including part loads and intermittent power supply. These operating conditions are harsh for the membrane, because the hydration state would be changed drastically. In addition, the hydrogen production at high pressure is beneficial for the hydrogen storage. Thus, the required directionality of the membrane development/improvement is the reinforcement of mechanical and chemical stability at a wide range of hydration degrees, as well as the cost reduction. On the other hand, PEM electrolysis at elevated temperature is also expected to achieve higher efficiency. The improvement of thermal stability of the membrane above 100°C is another pathway to achieve efficient operation of PEM electrolyzer.

References

Aili, D., M. K. Hansen et al. 2011. Phosphoric acid doped membranes based on Nafion, PBI and their blends—Membrane preparation, characterization and stem electrolysis testing. *International Journal of Hydrogen Energy* 36: 6895–6993.

Antonucci, V., A. Di Blasi et al. 2008. High temperature operation of a composite membrane-based solid polymer electrolyte water electrolyzer. *Electrochimica Acta* 53: 7350–7256.

Arcella, V., C. Troglia et al. 2005. Hyflon ion membranes for fuel cells. *Industrial and Engineering Chemistry Research* 44: 7646–7651.

Aricò, A. S., A. Di Blasi et al. 2010. High temperature operation of a solid polymer electrolyte fuel cell stack based on a new ionomer membrane. *Fuel Cells* 10: 1013–1023.

Aricò, A. S., V. Gaglio et al. 2006. Proton exchange membranes based on the short-side-chain perfluorinated ionomer for high temperature direct methanol fuel cells. *Desalination* 199: 271–273.

Ayers, K. E., E. B. Anderson et al. 2010. Research advances towards low cost, high efficiency PEM electrolysis. *ECS Transactions* 33(1): 3–15.

Baglio, V., R. Ornelas et al. 2009. Solid polymer electrolyte water electrolyzer based on Nafion-TiO_2 composite membrane for high temperature operation. *Fuel Cells* 9: 247–252.

Barbir, F. 2005. PEM electrolysis for production of hydrogen from renewable energy sources. *Solar Energy* 78: 661–669.

Broka, K., P. Ekdunge. 1997. Oxygen and hydrogen permeation properties and water uptake of Nafion 117 membrane and recast film for PEM fuel cell. *Journal of Applied Electrochemistry* 27: 117–123.

Büchi, F. N., G. G. Scherer. 2001. Investigation of the transversal water profile in Nafion membranes in polymer electrolyte fuel cells. *Journal of the Electrochemical Society* 148: A183–A188.

Cao, Y.-C., X. Wu et al. 2012. A quaternary ammonium grafted poly vinyl benzyl chloride membrane for alkaline anion exchange membrane water electrolysis with no-noble-metal catalyst. *International Journal of Hydrogen Energy* 37: 9524–9528.

Chiou, J. S., D. R. Paul. 1988. Gas permeation in a dry Nafion membranes. *Industrial & Engineering Chemistry Research* 27: 2161–2164.

Cropley, A. T. Norman, 2008. A low-cost high-pressure hydrogen generator. DOE final report (DE-FG36-01GO13029). Available at: http://www.osti.gov/scitech/biblio/926321 OSTI ID: 926321 (DOE/GO/13029-20).

Dekel, D. R. 2012. Alkaline membrane fuel cell (AMFC) materials and system improvement -state of the art-. *ECS Transactions* 50(2): 2051–2052.

Doyle, M., M. E. Lewittes et al. 2001. Relationship between ionic conductivity of perfluorinated ionomeric membranes and nonaqueous solvent properties. *Journal of Membrane Science* 184: 257–273.

Doyle, M., G. Rajendran. 2004. Perfluorinated membranes. In: Vielstich, W., Lamm, A., Gasteiger, H.A., eds. *Handbook of fuel cells*, vol. 3, Chichester, U.K.: John Wiley & Sons, Chapter 30.

Faraj, M., M. Boccia et al. 2012. New LDPE based anion-exchange membranes for alkaline solid polymeric water electrolysis. *International Journal of Hydrogen Energy* 37: 14992–15002.

Gabel, G., P. Aldebert et al. 1993. Swelling study of perfluoro-sulphonated ionomer membranes. *Polymer* 34: 333–339.

Ghielmi, A., P. Vaccarono et al. 2005. Proton exchange membranes based on the short-side-chain perfluorinated ionomer. *Journal of Power Sources* 145: 108–115.

Grigoriev, S. A., P. Millet et al. 2009. Hydrogen safety aspects related to high-pressure polymer electrolyte membrane water electrolysis. *International Journal of Hydrogen Energy* 34: 5986–5991.

Grigoriev, S. A., V. I. Porembskiy et al. 2011. High-pressure PEM water electrolysis and corresponding safety issues. *International Journal of Hydrogen Energy* 36: 2721–2728.

Hansen M. K. 2012. PEM water electrolysis at elevated temperature. PhD thesis. Technical University of Denmark, Denmark.

Hansen, M. K., A. Aili et al. 2012. PEM steam electrolysis at 130°C using a phosphoric acid doped short side chain PFSA membrane. *International Journal of Hydrogen Energy* 37: 10992–11000.

Haug, A. T., R. E. White. 2000. Oxygen diffusion coefficient and solubility in a new proton exchange membrane. *Journal of the Electrochemical Society* 147: 980–983.

He, R., Q. Li et al. 2003. Proton conductivity of phosphoric acid doped polybenzimidazole and its composite with inorganic proton conductors. *Journal of Membrane Sciences* 226: 169–184.

He, R., Q. Li et al. 2006. Physicochemical properties of phosphoric acid doped polybenzimidazole membranes for fuel cells. *Journal of Membrane Science* 277: 38–45.

Hinatsu, J. T., M. Mizuhata et al. 1994. Water uptake of perfluorosulfonic acid membranes from liquid water and water vapor. *Journal of the Electrochemical Society* 141: 1493–1498.

Ioroi, T., N. Kitazawa et al. 2000. Iridium oxide/platinum electrocatalysts for unitized regenerative polymer electrolyte fuel cells. *Journal of the Electrochemical Society* 147: 2018–2022.

Ito, H., T. Maeda et al. 2010. Effect of flow regime of circulating water on a proton exchange membrane electrolyzer. *International Journal of Hydrogen Energy* 35: 9550–9560.

Ito, H., T. Maeda et al. 2011. Properties of Nafion membranes under PEM water electrolysis. *International Journal of Hydrogen Energy* 36: 10527–10540.

Jang, I. Y., O. H. Kweon et al. 2008a. Covalently cross-linked sulfonated poly (ether ether ketone)/tungstophosphoric acid composite membranes for water electrolysis application. *Journal of Power Sources* 181: 127–134.

Jang, I. Y., O. H. Kweon et al. 2008b. Application of polysulfone (PSf) – and polyether ether ketone (PEEK) – tungstophosphoric acid (TPA) composite membranes for water electrolysis. *Journal of Membrane Sciences* 322: 154–161.

Kerres, J., C.-M. Tang et al. 2004, Improvement of properties of poly (ether ketone) ionomer membranes by blending and cross-linking. *Industrial and Engineering Chemistry Research* 43: 4571–4579.

Kocha, S. S., J. D. Yang et al. 2006. Characterization of gas crossover and its implications in PEM fuel cells. *AIChE Journal* 52: 1916–1925.

Kopitzke, R., C. A. Linkous et al. 2000. Conductivity and water uptake of aromatic-based proton exchange membrane electrolytes. *Journal of the Electrochemical Society* 147: 1677–1681.

Kreuer, K. D. 1997. On the development of proton conducting materials for technological applications. *Solid State Ionics* 97: 1–15.

Kreuer, K. D. 2001. On the development of proton conducting polymer membranes for hydrogen and methanol fuel cell. *Journal of Membrane Sciences* 185: 29–39.

Kreuer, K. D. 2003. Hydrocarbon membranes. In: Vielstich, W., Gasteiger, A, Lamm, A, editors. *Handbook of fuel cells*, vol. 3:420–435, John Wiley & Sons, Cambridge, U.K., Chapter 33.

Kreuer, K. D., S. J. Paddison et al. 2004. Transport in proton conductors for fuel-cell applications: Simulations, elementary reactions, and phenomenology. *Chemical Reviews* 104: 4637–4678.

Kreuer, K. D., M. Schuster et al. 2008. Short-side-chain proton conducting perfluorosulfonic acid ionomers: Why they perform better in PEM fuel cells. *Journal of Power Sources* 178: 499–509.

Lee, K. M., J. Y. Woo et al. 2011. Effect of cross-linking agent and heteropolyacids (HPA) contents on physicochemical characteristics of covalently cross-linked sulfonated poly(ether ether ketone)/HPAs composite for water electrolysis. *Journal of Industrial and Engineering Chemistry* 17: 657–666.

Lehtinen, T., G. Sundholm et al. 1998. Electrochemical characterization of PVDF-based proton conducting membranes for fuel cells. *Electrochimica Acta* 43: 1881–1890.

Leng, Y., G. Chen et al. 2012. Solid-state water electrolysis with an alkaline membrane. *Journal of the American Chemical Society* 134: 9054–9057.

Li, Q., R. He et al. 2003. Approaches and recent development of polymer electrolyte membranes for fuel cells operating above 100°C. *Chemistry of Materials* 15: 4896–4915.

Li, Q., L. H. A. Hjuler et al. 2001. Phosphoric acid doped polybenzimidazole membranes: Physiochemical characterization and fuel cell applications. *Journal of Applied Electrochemistry* 31: 773–779.

Linkous, C. A. 1993. Development of solid polymer electrolytes for water electrolysis at intermediate temperatures. *International Journal of Hydrogen Energy* 18: 641–646.

Linkous, C. A., H. R. Anderson. 1998. Development of new proton exchange membrane electrolysis for water electrolysis at higher temperatures. *International Journal of Hydrogen Energy* 23: 525–529.

Ma, Y. L, J. S. Wainright et al. 2004. Conductivity of PBI membranes for high-temperature polymer electrolyte fuel cells. *Journal of the Electrochemical Society* 151: A8–A16.

Mader, J., L. Xiao et al. 2008. Polybenzimidazole/acid complexes as high-temperature membranes. *Advances in Polymer Science* 216: 63–124.

Masson, J. P., R. Molina. 1982. Obtention and evaluation of polyethylene-based solid polymer electrolyte membranes for hydrogen production. *International Journal of Hydrogen Energy* 7: 167–171.

Merle, G., M. Wessling et al. 2011. Anion exchange membrane for alkaline fuel cells: A review. *Journal of Membrane Science* 377: 1–35.

Millet, P., R., S. A. Grigoriev et al. 2010. PEM water electrolyzers: From electrocatalysis to stack development. *International Journal of Hydrogen Energy* 35: 5043–5052.

Mohamed, H. F. M., K. Ito et al. 2008. Free volume and permeabilities of O_2 and H_2 in Nafion membranes for polymer electrolyte fuel cells. *Polymer* 49: 3091–3097.

Nikiforov, A., E. Christensen et al. 2012. Advanced construction materials for high temperature steam PEM electrolyzers. In: Linkov V., ed. *Electrolysis*, pp. 61–86, InTech, West Sussex, U.K., Chapter 4.

Ogumi, Z., Z. Takehara et al. 1984. Gas permeation in SPE method, I Oxygen permeation through Nafion and NEOSEPTA. *Journal of the Electrochemical Society* 131: 769–773.

Onda, K., T. Murakami et al. 2002. Performance analysis of polymer-electrolyte water electrolysis cell at a small-unit test cell and performance prediction of large stacked cell. *Journal of the Electrochemical Society* 149: A1069–A1078.

Paddison, S. J. 2003. First principle modeling of sulfonic acid based ionomer membranes. In: Vielstich, W., Gasteiger, A., Lamm, A., eds. *Handbook of fuel cells*, vol. 3, pp. 396–411, John Wiley & Sons, West Sussex, U.K., Chapter 31.

Parthasarathy, A., S. Srinivasan et al. 1992. Temperature dependence of the electrode kinetics of oxygen reduction at the platinum/Nafion interface—A microelectrode investigation. *Journal of the Electrochemical Society* 139: 2530–2537.

Pasternak, R. A., M. V. Christensen et al. 1970. Diffusion and permeation of oxygen, nitrogen, carbon dioxide, and nitrogen dioxide through polytetrafluoroethylene. *Macromolecules* 3: 366–371.

Pavel, C. C., F. Cecconi et al. 2014. Highly efficient platinum group metal free based membrane-electrode assembly for anion exchange membrane water electrolysis. *Angewandte Chemie International Edition* 53: 1378–1381.

Piana, M, M. Boccia et al. 2010. H_2/air alkaline membrane fuel cell performance and durability, using novel ionomer and non-platinum group metal cathode catalyst. *Journal of Power Sources* 195: 5875–5881.

Rikukawa, M., K. Sanui. 2000. Proton-conducting polymer membranes based on hydrocarbon polymers. *Progress in Polymer Science* 25: 1463–1502.

Sakai, T., H. Takenaka et al. 1985. Gas permeation properties of solid polymer electrolyte (SPE) membranes. *Journal of the Electrochemical Society* 132: 1328–1332.

Sakai, T., H. Takenaka et al. 1986. Gas diffusion in the dried and hydrated Nafions. *Journal of the Electrochemical Society* 133: 88–92.

Schalenbach, M., M. Carmo et al. 2013. Pressurized PEM water electrolysis: Efficiency and gas crossover. *International Journal of Hydrogen Energy* 38: 14921–14933.

Schroeder, P. 1903. Über erstarrungs und quellungserscheinungen von gelatin. *Zeitschrift für Physikalische Chemie* 45: 75–128 (in German).

Seel, D. C., B. C. Benicewicz et al. 2009. High-temperature polybenzimidazol-based membranes. In: Vielstich, W., Yokoyama, H., Gasteiger, H.A., eds. *Handbook of fuel cells*, vol. 5, pp. 1–13, John Wiley & Sons, Cambridge, U.K., Chapter 19.

Song, S., H. Zhang et al. 2008. Electrochemical investigation of electrocatalysts for the oxygen evolution reaction in PEM water electrolyzers. *International Journal of Hydrogen Energy* 33: 4955–4961.

Springer, T. E., T. A. Zawodzinski et al. 1991. Polymer electrolyte fuel cell model. *Journal of the Electrochemical Society* 138: 2334–2342.

Takenaka, H., E. Torikai et al. 1984. Studies on solid polymer electrolyte in water electrolysis I, Some properties of Nafion ion exchange membranes as a solid polymer electrolyte. *Electrochemistry (Denki Kagaku)* 52: 351–357 (in Japanese).

Valdez, T. I., K. J. Billings et al. 2009. Iridium lead doped ruthenium oxide catalysts for oxygen evolution. *ECS Transactions* 25(1): 1371–1382.

Varoe, J. R., R. C. T. Slade et al. 2006. An alkaline polymer electrochemical interface: A breakthrough in application of alkaline anion-exchange membranes in fuel cells. *Chemical Communications* 2006: 1428–1429.

Wainright, J. S., J. T. Wang et al. 1995. Acid-doped polybenzimidazoles: A new polymer electrolyte. *Journal of the Electrochemical Society* 142: L121–L123.

Watanabe, S., K. Fukuta et al. 2010. Determination of carbonate ion in MEA during the alkaline membrane fuel cell (AMFC) operation. *ECS Transactions* 33(1): 1837–1845.

Weber, A. Z., J. Newman. 2003. Transport in polymer-electrolyte membranes I. Physical model. *Journal of the Electrochemical Society* 150: A1008–A1015.

Wei, G., L. Xu et al. 2010. SPE water electrolysis with SPEEK/PES blend membrane. *International Journal of Hydrogen Energy* 35: 7778–7783.

Wise, D. L., G. Houghton. 1966. The diffusion coefficient of ten slightly soluble gases in water at 10–60°C. *Chemical Engineering Science* 21: 999–1010.

Woo, J. Y., K. M. Lee et al. 2010. Electrocatalytic characteristics of Pt–Co and Pt–Ru–Ni based on covalently cross-linked sulfonated poly (ether ether ketone)/heteropolyacids composite membranes for water electrolysis. *Journal of Industrial and Engineering Chemistry* 16: 688–697.

Wu, X., K. Scott, 2013. A Li-doped Co_3O_4 oxygen evolution catalyst for non-precious metal alkaline anion exchange membrane water electrolysis. *International Journal of Hydrogen Energy* 38: 3123–3129.

Xiao, L., S. Zhang, 2012. First implementation of alkaline polymer electrolyte water electrolysis working only with pure water. *Energy and Environmental Science* 5: 7869–7871.

Xu, W., K. Scott et al. 2011. Performance of a high temperature polymer electrolyte membrane water electrolyzer. *Journal of Power Sources* 196: 8918–8924.

Yoshitake, M., M. Tamura et al. 1996. Studies of perfluorinated ion exchange membranes for polymer electrolyte fuel cells. *Electrochemistry (Denki Kagaku)* 64: 727–736.

Zawodzinski, T. Z., C. Derouin et al. 1993. Water uptake by and transport through Nafion 117 membranes. *Journal of the Electrochemical Society* 140: 1041–1047.

Zhai, Y., H. Zhang et al. 2007. A novel H_3PO_4/Nafion-PBI composite membrane for enhanced durability of high temperature PEM fuel cells. *Journal of Power Sources* 169: 259–264.

7

Bipolar Plates and Plate Materials

Conghua "CH" Wang

CONTENTS

7.1 Introduction

Proton exchange membrane (PEM) electrolyzers were originally developed for oxygen generation and compression for military applications, such as onboard breathing support on aircraft (Harrison 1975) and nuclear submarines (Carlson et al. 2000). In these applications, system operation reliability is critical. The components and system design are focused on reliability, not necessarily on the cost. The commercial application of PEM electrolyzers started as the cylinder replacement of high-purity hydrogen in gas chromatographs and electrical generator cooling gas (Coker et al. 1982) to eliminate the gas-handling cost. Since then, technology developments have continuously reduced PEM electrolyzer costs, which enabled the use of PEM electrolyzers in various commercial applications. The most significant development in recent years is the "power to gas" program supported by the German government to generate hydrogen from the underutilized electricity from renewable sources like photovoltaics and wind turbines (Christopher 2013; Weinmann 2012). The hydrogen generated from an electrolyzer can be stored in natural gas pipeline, salt caverns or used as industrial feeds. However, it has been recognized that the capital investment and the operation cost of the current electrolyzer technology are still too high for "power to gas" application (Colella et al. 2014). Recent technology developments are focused on cost reduction to meet the application requirement.

Detailed cost analysis of the PEM electrolyzer system indicated that the electrolyzer stack accounted for about 53% of the system cost, and the flow field and bipolar separator plates accounted for about 48% of the stack cost, as shown in Figure 7.1 (Ayers et al. 2010). Therefore, the technology development for low-cost, high-performance bipolar plates is important for the commercial success of PEM electrolyzers.

7.2 Functionalities and Requirements of Bipolar Plates

7.2.1 Bipolar Plate Functionalities and Operation Environment

In a PEM electrolyzer, a thin (50–250 µm thick) proton conductive membrane is used as the solid electrolyte. The most widely used membrane is sulfonated tetrafluoroethylene–based fluoropolymer produced by DuPont under the trademark of Nafion® (Mauritz and Moore 2004). Iridium oxide is used as the anode catalyst for oxygen evolution and platinum is used as the cathode catalyst for hydrogen evolution. In order to have the desired gas (hydrogen or oxygen) output capacity, a number of cells are included in an electrolyzer system. They are typically assembled in series as electrolyzer stacks. The component that connects adjacent cells is the

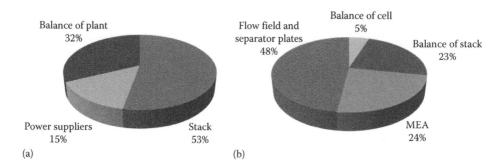

FIGURE 7.1
Capital cost breakdown of Proton Onsite's 13 kg day⁻¹ PEM electrolyzer. (a) System cost breakdown and (b) stack cost breakdown. (Reprinted from Ayers, K.E. et al., *ECS Transactions*, 33, 3, 2010. With permission from the Electrochemical Society.)

bipolar plate. Sometimes, bipolar plate is also called as bipolar separator plate or separator plate.

The operation environment of bipolar plates is determined by the properties of electrolyte membrane and the electrolyzer cell/stack operation conditions. Because Nafion will slowly release sulfate and fluorine ions during operation, the water in electrolyte stack is slightly acidic. Normally, a water-deionizing subsystem is included in modern PEM electrolyzer systems and the deionized water consistently flows through the stack; the pH of the water in stack is between 6 and 7, which is much less acidic than that in PEM fuel cells using the same electrolyte membrane (Abdullah et al. 2008). Unlike catalysts that are in direct contact with electrolyte membrane that has much lower pH value, bipolar plates are only in contact with the water in pH 6–7 at normal operation conditions. PEM electrolyzer typically operates at high pressure, either on cathode (hydrogen electrode for hydrogen evolution reactions) or on anode (oxygen electrode for oxygen evolution reactions). In reversible fuel cell systems, both anode and cathode chambers are at high pressure. Therefore, all components, including bipolar plates, have to have sufficient mechanical strength to withhold the pressure.

The major challenges for electrolyzer stack components come from electrode reactions.

The anode reaction is as follows:

$$H_2O(liq.) \rightarrow \frac{1}{2}O_2(g) + 2H^+ + 2e^- \qquad (7.1)$$

The cathode reaction is as follows:

$$2H^+ + 2e^- \rightarrow H_2(g) \qquad (7.2)$$

And the overall reaction of a PEM electrolyzer is as follows:

$$H_2O(liq.) \rightarrow \frac{1}{2}O_2(g) + H_2(g) \qquad (7.3)$$

As shown in Figure 7.2 of Pourbaix diagram (Pourbaix 1966) of water, operation potentials of anode and cathode reactions are related to the pH of the water. At pH 6–7, the thermodynamic equilibrium potential for electrolyzer anode reaction is between 0.88 and 0.82 V_{NHE} and the thermodynamic equilibrium potential for electrolyzer cathode reaction is between −0.35 and −0.41 V_{NHE}. However, due to the reaction kinetics (requiring overpotentials to drive electrode reactions) and internal ohmic losses, the anode potential could be as high as 1.9 V_{NHE} for oxygen evolution and the cathode potential could be as low as −0.50 V_{NHE} for hydrogen evolution at high current density (high output rate) operation conditions. The potential difference between anode and cathode is the electrolyzer operation voltage that is typically 1.8–2.0 V and possibly goes up to 2.30 V for a short period of time (Carmo et al. 2012). The potential ranges of PEM electrolyzer anode and cathode are schematically illustrated in Figure 7.2.

7.2.2 Requirements on Plate Materials

There are two basic functions for a bipolar plate. One is to electrically connect adjacent cells in a stack. The other is to separate the oxygen in anode chamber of one cell from the hydrogen in cathode chamber of the adjacent cell. Additional functions include the facilitation of mass transport and heat management. These functions have to be maintained in electrolyzer operation environment of high pressure, oxidizing (in anode) and reducing (in cathode) conditions through the life span of electrolyzer systems (typically over 10 years). Due to the high operation pressure, metal bipolar plates are typically required for electrolyzer stack mechanical integrity. The basic requirements of bipolar plates are summarized in Table 7.1.

To electrically connect cells, bipolar plate electrical resistance, including surface contact resistance and bulk resistance, needs to be as low as possible. Using metallic bipolar plates, the electrical resistance predominantly comes from the surface contact resistance of bipolar plate with the component in direct contact,

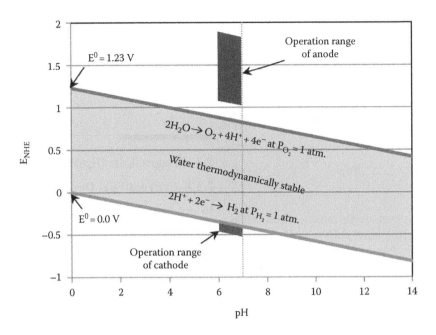

FIGURE 7.2
Pourbaix diagram for water at 25°C and the operation range of PEM electrolyzer anode and cathode.

TABLE 7.1

Summary of Functions and Requirements to Bipolar Plates

Functions	Required Properties
Conduct electrons between cells	Low electrical resistance
Separate H_2 and O_2 between cells	No gas leak through bipolar plate
Facilitate mass transport	High mechanical stability
Facilitate thermal management	High thermal conductance

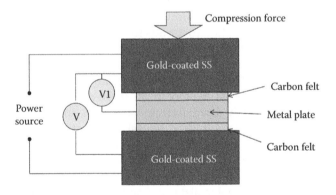

FIGURE 7.3
Schematic drawing of surface contact resistance measurement setup.

such as flow field plate or gas diffusion layer. The actual contact resistance in an electrolyzer is determined by not only bipolar plate material but also the material and the structure of the component that is in direct contact with the bipolar plate. It varies between stacks based on their design and targeted applications. Typically, the electrical contact resistance between conventional platinized metal bipolar plate and gas diffusion layer is below 1 mΩ.cm². For commercial applications, bipolar plates with slight higher electrical contact resistance are still acceptable, if it has the advantage of lower cost.

One challenge for PEM electrolyzer bipolar plate development is that there are no well-recognized evaluation method and target for bipolar plate contact resistance, as that defined by U.S. Department of Energy for PEM fuel cells (US DOE 2012). Another challenge is the measurement error of such low electrical contact resistance. The plate flatness and surface roughness will have significant impacts on the measurement results. The best approach for bipolar plate evaluation is to

compare the performance of electrolyzer cells with the different bipolar plates in real operating cells/stacks (Toops et al. 2014). But this method is complicated and has to have consistent catalyst-coated membrane, gas diffusion layer, cell assembly and testing procedures, etc. Alternatively, surface electrical contact resistance of PEM electrolyzer bipolar plates can be measured in the same way as bipolar plates for PEM fuel cells. As shown in Figure 7.3, the metal plate is sandwiched between two carbon felts, such as Toray carbon paper. The voltage drops between two gold-plated stainless electrodes (V) and between one gold-plated stainless electrode and the metal plate (V_1) are measured under different compression pressures at 1 A/cm² electrical current density (i). The voltage drops with two pieces of carbon felt without metal plate (V_{carbon}) are also measured as the baseline. The surface contact resistance (CR, in $m\Omega.cm^2$) can be calculated as

$$CR = \frac{V_1 - V_{carbon}/2}{i} \qquad (7.4)$$

The contact resistance can also be presented as through plate resistance (TPR, in $m\Omega.cm^2$):

$$TPR = \frac{V - V_{carbon}}{i} \qquad (7.5)$$

The contact resistance measured using carbon felt is higher than that of bipolar plates with metallic components in electrolyzer cells. However, it is very repeatable due to the semi-conformable nature of carbon felt. This method is suitable for ex situ durability tests of metal plate corrosion resistance to detect any degradation of the bipolar plate material.

The challenge to maintain low surface electrical contact resistance mainly comes from the PEM electrolyzer anode side of the bipolar plate that is exposed to highly oxidizing environment. The plate anode side surface may suffer continue oxidization during PEM electrolyzer operation that leads to the electrical contact resistance increase and electrolyzer performance degradation. Because it is difficult to conduct the lifetime (years) durability tests for bipolar plate development, ex situ test is a valuable tool for materials development. The most widely used testing method is to conduct long-term (several hundred hours) polarization test of metal plates at high electrode potential (up to 2.0 V_{NHE}) in pH 2–3 H_2SO_4 solution with trace amount (0.1 ppm) of HF at 80°C, which is similar to the test condition for PEM fuel cell bipolar plates defined by US DOE (2012). Similar test can also be conducted at −0.1 to −0.2 V_{NHE} potential to evaluate the bipolar plate performance in simulated PEM electrolyzer cathode working environment. The surface electrical contact resistance measurement method as shown in Figure 7.3 is a very useful way to evaluate the stability of metal bipolar plates in these long-term ex situ durability tests.

The gas separation requirement of bipolar plate is fairly easy to be fulfilled for metallic plates at the beginning of the life, as long as there are no pinholes and micro-cracks. The challenge comes from possible hydrogen embrittlement of metal bipolar plates during long-term operation. The high-pressure hydrogen and the proton-reducing reactions on cathode promote hydrogen absorption into metal bipolar plates (LaConti and Swette 2003), which may lead to the mechanical failure of bipolar plates under the high-pressure operation conditions. The concentration limit of hydrogen in metal bipolar plates is determined by the type of bipolar plate material, plate thickness, and the stack operation pressure. It was reported that titanium plate will suffer mechanical failure in high-pressure electrolyzer with the hydrogen concentration over 8000 ppm (Hamdan 2012).

7.3 Progress in Bipolar Plate Development

7.3.1 Conventional Design and Material Selections

In order to meet the application requirements in both anode and cathode environments, conventional bipolar plate has a dual-layer metal structure. The anode side of the bipolar plate is a corrosion-resistant valve metal to minimize the oxygen buildup on anode (LaConti et al. 1977). In order to minimize the electrical contact resistance between metallic components, noble metal coating is applied for applications in which high energy efficiency is desired. Among valve metals, titanium is the most common material used in bipolar plates. However, titanium has the risk of vigorous combustion in oxygen-enriched environment (Lutjering and Williams 2007). Therefore, niobium or tantalum is used to avoid the risk in oxygen generation and reversible fuel cell stacks, where high-pressure oxygen is presented. On the cathode side, the hydrogen embrittlement–resistant metal is used to minimize the hydrogen uptake in PEM electrolyzer cathode conditions. Generally, the resistance to hydrogen embrittlement for selected materials appears to be in the order of (LaConti and Swette 2003):

Graphite > Zr = selected stainless steel alloy > Nb > Ti

Therefore, zirconium is the common material used on the cathode side in the conventional dual-layer bipolar plate structure (Hamdan 2010).

In some literatures, the term "bipolar plate" is actually referred to as the bipolar plate assembly. As shown in Figure 7.4 (Narayanan et al. 2011), the bipolar assembly includes the bipolar separator sheet, oxygen–water assembly and hydrogen–water assembly. The gas–water assembly includes frames for cell–stack seals and screens for uniform mass transport through the whole electrode active area, which is critical to avoid hot spot and achieve high current density in cells (Millet et al. 2009). The materials used in oxygen–water screen are typically same as the anode side of the bipolar plates, which have the long-term durability in the oxidizing operation condition. Materials used in hydrogen–water screen could be the same as the cathode side to bipolar plates or graphite-based materials, which have the desired resistance to hydrogen embrittlement.

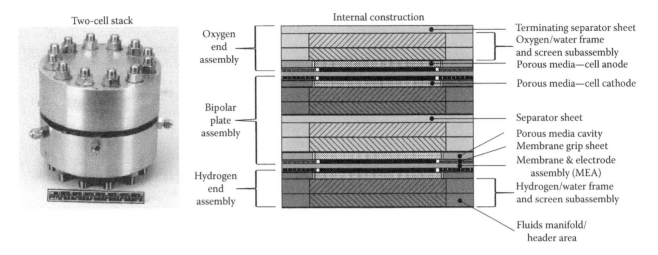

FIGURE 7.4
Picture and schematic drawing of a two-cell stack with a bipolar plate assembly that includes bipolar separator sheet, oxygen–water, hydrogen–water frames and screen assemblies. (Reprinted from Narayanan, S.R. et al., *J. Electrochem. Soc.*, 158, B1348, 2011. With permission from the Electrochemical Society.)

7.3.2 Recent Single-Layer Bipolar Plate Material

The commercial application of PEM electrolyzer requires new technologies that can reduce the cost of all components including bipolar plates. The first action that most electrolyzer developers have been taken is to replace the dual-layer metal bipolar plates with a single-layer plate. The favorite metal for metallic bipolar plate is commercial pure titanium because of its excellent corrosion resistance and reasonable cost. The high surface contact resistance of titanium plate with other components can be reduced by platinizing the plate surface, although further cost reduction is desired. The technical challenge of single-layer titanium bipolar plate is the hydrogen embrittlement of titanium in the hydrogen chamber of PEM electrolyzer anode. It was reported that titanium plate can absorb over 1000 ppm hydrogen in 500 h operation (Hamdan and Norman 2011), which is not acceptable for long-term operation of PEM electrolyzers.

One approach is to have a pinhole-free carbon coating on the cathode side of titanium plate to prevent direct contact of hydrogen with titanium in electrode active area. Experimental data show that it can effectively reduce hydrogen absorption into titanium plate (Hamdan and Norman 2011). The challenge is the processing difficulty and cost for the pinhole-free coating. Another approach is to have a gold coating on titanium plate to prevent hydrogen absorption (Jung et al. 2009). Unfortunately, the cost of the gold coating is obviously an issue for commercial applications.

Stainless steel is an attractive material for bipolar plates. It has lower cost than titanium, which has less concern of hydrogen embrittlement. The challenge is the

corrosion of stainless steel at the high potential in anode environment. It was reported (Papadias et al. 2014) that corrosion of stainless steel in slightly acidic conditions will lead to the continuous growth of surface oxide layer that leads to the surface electrical contact resistance increase and slow ion leaching that may poison membrane and catalysts, resulting in the performance degradation of the electrolyzer. Thermodynamically, stainless steel will suffer transpassivation corrosion at high electrochemical potential (Shih et al. 2004), which is the root cause of its corrosion in PEM electrolyzer anode. However, stainless steel and nickel alloy have been used in alkaline electrolyzer (Gras and Spiteri 1993). Some types of stainless steel and nickel alloy have acceptable corrosion resistance in the high pH alkaline solutions, despite the thermodynamic limitations. Therefore, it is an interesting topic to explore the possibility of using stainless steel bipolar plates in PEM electrolyzer systems that have a high-performance water deionizing subsystem to remove soluble ions from water and maintain the pH of water close to 7.

Stainless steel plate with titanium coating was investigated for PEM electrolyzer applications (Gago et al. 2014). The titanium coating layer was deposited using low-pressure plasma spray (LPPS). Due to the porous nature of thermal-sprayed coating layer, the titanium coating layer was very thick (~30 µm) to protect stainless steel substrate. A thin layer of gold was electroplated on titanium surface to ensure the low electrical contact resistance, as shown in Figure 7.5. Ex situ electrochemical corrosion experiments indicated that the titanium coating could protect stainless steel layer from corrosion in anode environment. However, the coating cost

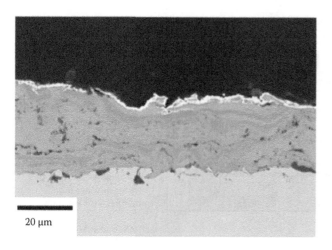

FIGURE 7.5
SEM cross section of stainless steel plate with thermal-sprayed Ti and electroplated Au surface layer. (Reprinted from Gago, A.S. et al., *ECS Trans.*, 64, 1039, 2014. With permission from the Electrochemical Society.)

and the process quality control to minimize the coating defects are challenging issues for the application.

Stainless steel electroplated with a thin layer of tantalum in ionic liquid solution was investigated for its corrosion resistance at high potential in acidic solutions (Ojong et al. 2012) for PEM electrolyzer applications. The corrosion current of tantalum-coated stainless steel 316L was lower than that of bare 316L, but it was still higher than that of pure titanium plates. It is possible that the electroplating process cannot obtain a dense, defect-free tantalum coating.

Nonmetallic bipolar plate materials were also investigated for PEM electrolyzer applications. Graphite, as the most common material in PEM fuel cells, was investigated for PEM electrolyzer applications. The major challenge is the carbon oxidization in the anode environment under high potential with oxygen and water. Another issue is the inherent porous structure of graphite. Thick plates have to be used to prevent gas leak through bipolar plates. Investigations had been carried out to develop a coating on graphite plate to protect graphite from oxidation and prevent gas leak through thin graphite plates (Hamdan and Norman 2011).

Plastic injection molded plates with conductive inserts, such as titanium pins, were also examined in PEM electrolyzer conditions (Ojong et al. 2012). It has the advantage of low cost for flow field forming, but it has challenges in thermal management and the risk of gas leak along the plastic and metal pin interface.

7.3.3 Electrically Conductive Coating for Metal Bipolar Plates

Conventional high-performance bipolar plates are electrically plated with very thin layer of platinum (platinized plate) to reduce the surface electrical resistance.

Electroplating on titanium is a very sensitive process to obtain the desired reliable adhesion of the platinum layer with titanium substrate. Titanium is a very active material and can rapidly form a thin layer of surface oxide as soon as it is in contact with oxygen (air) or water. This oxide surface layer has profound impacts on the adhesion of platinum coating with titanium substrate. Therefore, the production process has to be carefully controlled. The production cost and quality assurance are two major issues for conventional platinized titanium bipolar plates, in addition to the cost of platinum.

In recent years, several researchers have investigated gold plating on titanium (Jung et al. 2009) and titanium-coated stainless steel (Gago et al. 2014) plates. Gold coating can provide a highly conductive surface layer that can also minimize hydrogen absorption into titanium substrate plate (Jung et al. 2009) on cathode side. However, gold suffers oxidization in electrolyzer anode environment, which converts gold surface to high resistive gold oxide surface at the high electrode potential on anode side (Gago et al. 2014).

Conventional platinized titanium bipolar plates have platinum plated on the entire plate surface (or at least on the entire active area). The material (platinum) and processing cost is the barrier for PEM electrolyzer cost reduction. TreadStone Technologies, Inc. (TreadStone) has developed a new approach to reduce the precious metal usage with a low-cost deposition process (Wang 2014). This new technology deposits platinum in the form of small dots that partially cover metal plate surface (dotted plate). Platinum dots are deposited on the metal substrate by thermal spray process (Wang 2009). The process melts platinum particles in the flame and splats platinum droplets onto metal substrate. The velocity and the heat of platinum droplets can reliably bond the platinum in the form as flat dots onto metal plate surface. Figure 7.6 (Wang 2013) shows the SEM picture of platinum dots on titanium substrate surface. The platinum dots thickness is 0.1–0.2 µm. The typical coverage of platinum dots on the plate surface is 5–10%. The platinum loading is 0.02–0.04 mg cm^{-2}, which is significantly lower than that on platinized titanium plate. Same process can also be used to deposit gold dots on the cathode side of the bipolar plates to ensure the low electrical contact resistance on both sides of the plates. In this process, titanium substrate does not need to go through aggressive cleaning, such as hydrofluoric acid etching, which is necessary for platinized titanium bipolar plate production. The simple, reliable fabrication process and low precious metal usage effectively reduce the cost of bipolar plates.

The electrical resistance of titanium bipolar plate with platinum and gold dots was measured in TPR with the setup shown in Figure 7.3 and calculated

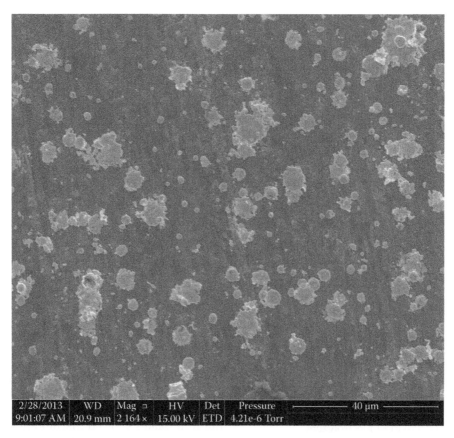

FIGURE 7.6
SEM picture of Pt dots deposited on Ti plate surface to reduce the surface electrical contact resistance.

with Equation 7.5. Carbon felt was used as the contact components for repeatable measurement. As shown in Figure 7.7 (Wang 2013), TreadStone's dotted plate had lower electrical contact resistance than that of fresh titanium plate. The durability of the plate (having 160 cm² active area) was evaluated in a PEM electrolyzer stack with an accelerated testing protocol at Giner Inc. The cell with TreadStone's dotted titanium plates had the same performance as cells with conventional platinized plates in the same stack. Posttest evaluation found that there was no corrosion sign on the plate, and the TPR had little increase (within measurement error) after the accelerated stack test.

The hydrogen embrittlement evaluation of dotted titanium plates was carried out with long-term durability test of the plates in two PEM stacks that operated at hydrogen pressures of 230 and 2400 psi, respectively. There was no noticeable performance degradation during the test. The hydrogen concentration in the plate was analyzed after the stack durability test, as summarized in Table 7.2, in comparison with convention Zr/Ti dual-layer bipolar plates. It shows that the hydrogen concentration in TreadStone's dotted plates is negligible and lower than that of conventional Zr/Ti plate, after long-term,

high hydrogen pressure stack tests. The low electrical contact resistance and the low hydrogen absorption of TreadStone's dotted metal bipolar plate demonstrated in the long term, accelerated stack tests indicate that it is suitable for PEM electrolyzer applications.

The coatings without precious metal are attractive approaches for low-cost bipolar plates. Titanium nitride has gained interests from industry and has demonstrated low resistance and good durability in initial tests (Ayers 2012). The plate did not have surface contact resistance increase after several thousand hours stack test, despite the slight color change. A comprehensive investigation of titanium nitride coating on titanium bipolar plate for PEM electrolyzer applications was conducted by Oak Ridge National Laboratory team with Proton OnSite recently (Toops et al. 2014). Commercial pure titanium was used in the study. The titanium nitride coating was grown on titanium plate surface by thermal nitridation and plasma nitridation processes. These nitridation processes were used because they can treat the entire plate surface with virtually no defects. X-ray diffraction indicated that plasma-nitrided titanium had a TiN-phase surface layer, whereas a mixture of TiN, Ti_2N, and Ti_2O phases was detected on the

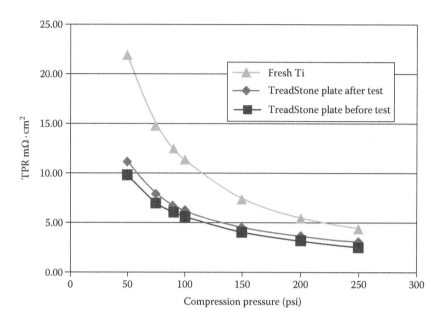

FIGURE 7.7

The comparison of through-plate resistance (TPR) of fresh titanium plate and TreadStone's dotted titanium plates before and after 1000 h accelerated durability test in a PEM electrolyzer.

TABLE 7.2

Comparison of Hydrogen Uptake of TreadStone's Single-Layer Ti Bipolar Plates with Conventional Zr/Ti Dual-Layer Bipolar Plates after Tested in PEM Electrolyzers

Bipolar Separator Plate	H₂ Pressure (psi)	Time (h)	H₂ Uptake (ppm)
TreadStone Ti bipolar plate (120 cm²)	2400	1000	51
TreadStone Ti bipolar plate (250 cm²)	230	5365	55
Zr/Ti (160 cm²)	230	500	140
Ti (as received)	—	0	~60

thermally nitrided titanium plate. XPS analysis indicated that there are a significant amount of carbon and oxygen in the thermally nitrided surface of titanium. The initial performances of PEM electrolyzer cells were similar to that of pure titanium, plasma-nitrided and thermally nitrided titanium plates. After 500 h test, the cell with thermally nitrided titanium plates had the best performance. Both thermal- and plasma-nitrided titanium plates showed discoloration after the cell test, as shown in Figure 7.8. XPS analysis indicated that N- and C-modified TiO₂ were found on the plate surface. It was estimated that the depth of the oxide layer is at least on the order of several hundred nanometers.

Despite the formation of TiO₂ phase on the nitride titanium plate surface, the performance of the PEM electrolyte cell with thermally nitrided titanium plate did not have significant performance degradation after initial 100 h operation at 50°C. The key question for this bipolar plate is the degree to which the nitride surface can contain oxygen and still maintain sufficient low electrical resistance to meet the durability requirement for PEM electrolyzer applications (Toops et al. 2014).

The hydrogen uptake of the nitrided titanium plates was also evaluated in a vessel at 800 psi hydrogen

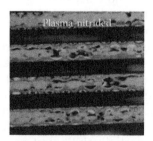

FIGURE 7.8

Optical images of the anode test untreated Ti, thermal-nitrided Ti and plasma-nitrided Ti plates after the 500 h durability measurements showing significant discoloration of the flow field lands. (Reprinted from *Journal of Power Sources*, 272, Toops, T.J., Brady, M.P., Zhang, F.Y., Meyer III, H.M., Ayers, K., Roemer, A., and Dalton, L. Evaluation of nitrided titanium separator plates for proton exchange membrane electrolyzer cells, 954–960, Copyright (2014), with permission from Elsevier.)

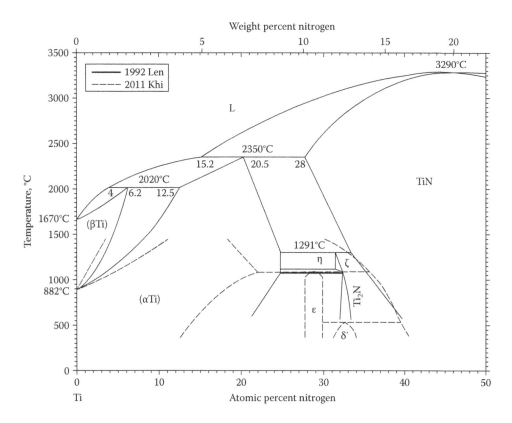

FIGURE 7.9
Titanium–nitrogen phase diagram. (With kind permission from Springer: *Journal of Phase Equilibria and Diffusion*, N–Ti (nitrogen–titanium), 34, 2013, 151–152, Okamoto, H.)

pressures at 150°C up to 1500 h. Posttest analysis indicated that there is no significant hydrogen uptake of nitrided titanium plates. This is a good indication of the resistance to hydrogen embrittlement of the nitrided plate, although further high-pressure PEM electrolyte cell test has to be carried out to verify that there is no hydrogen uptake associated with cathode hydrogen reduction reactions.

The phase diagram of titanium–nitrogen in Figure 7.9 (Okamoto 2013; Wriedt and Murray 1987) shows that titanium will form Ti_2N–TiN or α-Ti–TiN two-phase structures with nitrogen at the nitrogen concentration in titanium between 20 and 40 at.%. At higher nitrogen concentration, it will be in single-phase TiN structure. Ex situ electrochemical corrosion tests indicated that Ti_2N and low nitrogen content TiN phases were not stable, while high nitrogen content TiN has better electrochemical stability (Masami and Kurihara 1992; Taguchi and Kurihara 1991). It was also found that the phase structure and composition of titanium nitride layer was determined by the processing condition. In order to obtain the stable, high nitrogen content single-phase TiN coating, titanium plate has to be processed at very high temperature (>1500 K) in thermal nitridation process. It is expected that the surface layer composition and phase structure of plasma-nitrided titanium

plates will also be determined by processing conditions, such as temperature, nitrogen concentration, oxygen impurity level, and the surface status of the substrate. Therefore, bipolar plates processed at specific conditions have to be evaluated in long-term durability tests to be qualified in commercial application of PEM electrolyzers, which is time consuming and could be cost prohibitive.

Ti–Ag–N coating on pure titanium substrate was also investigated for bipolar plate in a unitized regenerative PEM fuel cell that works as a PEM electrolyzer during charge stage and a PEM fuel cell in discharge stage (Zhang et al. 2012). Silver was added with the intention to enhance the electrical conductance of the coating layer. The Ti–Ag–N coating layer was deposited by reactive co-sputtering of titanium and silver in nitrogen-containing vacuum. XRD analysis detected Ti_2N phase, but did not detect any silver-related structure in the as-coated surface layer, despite the fact that silver was detected by EDX. The hypothesis was that silver or silver-related compounds might be in the form of nanoparticles that cannot be detected by XRD. The plate surface electrical contact resistance was reduced effectively with the Ti–Ag–N coating. The corrosion test was conducted in 0.5M H_2SO_4 solution with 2 ppm HF at 2.0 V_{SEC} and −0.1 V_{SCE} at 70°C. Some pits and micro-cracks were observed on the coating layer after 6.5 h corrosion at 2.0 V_{SEC}. This testing condition was more aggressive than the operation condition of PEM electrolyzer anode in both the electrochemical potential and pH of the solution. But the test time is too short to reach any conclusion of the long-term durability of the Ti–Ag–N coating. It is not clear what the phase structure of silver is in the Ti–Ag–N coating layer. It is possible that the coating is a composite of silver nanoparticle (Zhang et al. 2012) and titanium nitride, because silver does not react with nitrogen to form silver nitride. In this case, the nanoparticle silver will be slowly dissolved under the high potential and the titanium nitride surface layer will have the same challenges as nitride coating formed on pure titanium plates as reported by Toops et al. in 2014.

Other coating materials have also been applied on titanium plate for PEM electrolyzer applications. Boron-doped diamond (BDD) and Ta_2O_5–IrO_2 mixed metal oxide (MMO) were applied on pure titanium substrate to reduce surface electrical contact resistance (Wang et al. 2012). Initial chemical soaking corrosion experiments in 0.5 M H_2SO_4 + 2 ppm HF solutions indicated that the coating had reasonable corrosion resistance and was capable to reduce the surface electrical contact resistance of titanium plates. But more extensive electrochemical evaluations have to be conducted to demonstrate the feasibility of their application in PEM electrolyzers. It is expected that BDD coating is not stable at the high potential of electrolyte anode. MMO coating has the potential if the coating process can ensure the reliable adhesion of MMO coating with titanium substrate.

Despite the promising lower cost, industrial players are very conservative to adopt non-precious metal coating until the demonstration of its lifetime durability. This has been the barrier for the non-precious metal coating technology development. On the other hand, new coating technologies using precious metal at lower cost are easier to be accepted by industrial players because precious metal-coated plates have been used for decades in PEM electrolyzers.

7.3.4 Bipolar Plate Assembly Structure

Conventional bipolar plate assembly includes the separator sheet, oxygen–water screen subassembly and hydrogen–water screen subassembly, as illustrated in Figure 7.4 (Narayanan et al. 2011). The conventional gas–water screen assembly is a multilayer metal screen laminate with gradient structure and pore size for uniform mass transport throughout the cell active area (Roy et al. 1999, 2001). A schematic structure of the multilayer screen laminate is shown in Figure 7.10. It includes several layers of metal screens that have different size

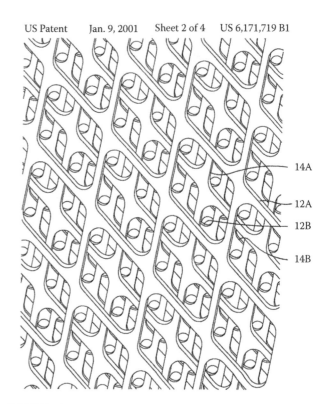

US Patent   Jan. 9, 2001   Sheet 2 of 4   US 6,171,719 B1

14A

12A

12B

14B

FIGURE 7.10
Schematic drawing of multilayer metal screen laminate structure as gas transport layer. 12A, 12B, 14A, and 14B are different layers of metal screens with different diamond patterns and offset to form the proper pore structure for water and gas flow. (Reprinted from Roy, R.J. et al., Electrode plate structure for high-pressure electrochemical cell devices, US Patent 5,942,350, 1999.)

diamond-like pores. The pores of adjacent layers were offset to construct a connected 3D hollow structure for water and gas transport. With a proper design of the 3D hollow structure with the metal screen laminate, the uniform mass transport to the electrodes can be achieved. However, this complicated bipolar plate assembly structure has large part count of electrolyte cells (Hamdan 2012) that need to be simplified to reduce electrolyzer cost.

The design of gas–water assembly is critical to obtain uniform mass transport and electrical current distribution across each cell that is necessary to achieve the high performance of PEM electrolyzers. It also has significant impacts to avoid hot spots in the stack. Although the hydrodynamic in such complex 3D structure is extremely complicated, some simple models can still be used to simulate mass transport in the bipolar plate assembly to provide guidance in the assembly design (Millet et al. 2009). Unfortunately, most developers have their own proprietary designs, and barely disclose it in public media.

In general, there are three types of bipolar plate assembly structures under development. One is to use flat separator sheet with metal meshes or screens laminate for mass transport (Ojong et al. 2012) similar as conventional design. The porous titanium sintering as electrode current collector (Grigoriev et al. 2009) is used to simplify the screen laminate structure. Therefore, much simpler structure than the conventional PEM electrolyzer (Roy et al. 1999, 2001) has been adapted in recent years (Ojong et al. 2012; Schmitt and Norman 2010).

The second structure is to use thick metal separator plate with etched flow field channels. Etched plates can further simplify the bipolar plate assembly structure, but the processing cost is the issue for low-volume production.

The third and the most recent development in bipolar plate assembly structure is to use stamped bipolar separate plates (Ayers 2012; Carmo et al. 2013) to reduce the bipolar plate assembly cost. The flow field design can be formed rapidly by stamping, which has more design flexibility in flow field design and low fabrication cost, especially at large production volume. The challenge is to control the stress distortion of the plate during stamping. The distortion could result in difficulties for stack sealing, especially in high-pressure stacks.

7.4 Summary

Bipolar plate is an important component in PEM electrolyzers. Titanium metal plate is still the most widely used material. Meanwhile, stainless steel–based bipolar plates have attracted significant interests recently. The focus of technology development is the cost reduction of the plate coating with single-layer plate structure. Various new coating technologies have been reported to maintain the low surface electrical contact resistance and prevent hydrogen embrittlement of bipolar plates. The challenge to maintain low surface electrical contact resistance is the plate surface oxidization at the high electrochemical potential in the electrolyzer anode operation environment. The challenge for hydrogen embrittlement is due to not only the high hydrogen gas pressure but also the cathode proton reduction reactions. Coatings with precious metals are the favorite for industrial systems that require 10 years stable operation. The focus is the cost reduction with lower precious metal loading and using cheaper, reliable production process. Some new technologies have made significant progresses and demonstrated excellent durability in PEM electrolyzers. Precious metal-free coating is under development for its potential advantage of cost saving. However, its long-term durability has to be experimentally verified before it can be used commercially.

In addition to the plate coating, plate structure–design development is also important. The purpose is to simplify the bipolar plate assembly structure and to optimize gas–water mass transport in electrolyzer stacks at low cost.

References

Abdullah, A. M., Okajima, T., Kitamura, F., and Ohsaka, T. 2008 Effect of operating condition on the acidity if H_2/air PEM fuel cells' water, *ECS Transactions* 16: 543–550.

Ayers, K. 2012 High performance, low cost hydrogen generation from renewable energy, presentation at *2012 DOE Annual Merit Review*, Arlington, VA, May 16, 2012.

Ayers, K. E., Anderson, E. B., Capuano, C., Carter, B., Dalton, L., and Hanlon, G. 2010 Research advances towards low cost, high efficiency PEM electrolysis, *ECS Transactions* 33: 3–15.

Carlson, H. A., Genovese, J. E., Lunn, M. H. B., and Cassidy, S. 2000 Shelf life experience with solid polymer electrolyte cell stacks Society of Automotive Engineers, Inc. Technical Paper 2000-01-2504, Toulouse, France.

Carmo, M., Fritz, D. L., Mergel, J., and Stolten, D. 2013 A comprehensive review on PEM water electrolysis, *International Journal of Hydrogen Energy* 38: 4901–4934.

Carmo, M., Mergel, J., and Stolten, D. 2012 A review on the recent progress in electrocatalysis of polymer exchange membrane water electrolysis, presentation at *World Hydrogen Energy Conference 2012*, Toronto, Ontario, Canada, June 12, 2012.

Christopher, H. 2013 The role of hydrogen in a renewable energy economy, presentation at *1st International Workshop of Durability and Degradation Issues in PEM Electrolysis Cells and Components*. Freiburg, Germany, March 12–13, 2013.

Coker, T. G., LaConti, A. B., and Nuttall, L. J. 1982 Industrial and government applications of SPE fuel cells and electrolyzers, presentation at *The Case Western Symposium on Membrane and Ionic and Electronic Conducting Polymers*, Cleveland, OH, May 17–19, 1982.

Colella, W. G., James, B.D., Moton, J. M., Saur, G., and Ramsden, D. 2014 Techno-economic analysis of PEM electrolysis for hydrogen production, presentation at *Electrolytic Hydrogen Production Workshop*, Golden, CO, February 27, 2014.

Gago, A. S., Ansar, A. S., Gazdzicki, P., Wagner, N., Arnold, W. J., and Friedrich, K. A. 2014 Low cost bipolar plates for large scale PEM electrolyzers, *ECS Transaction* 64: 1039–1048.

Gras, J. M. and Spiteri, P. 1993 Corrosion of stainless steel and nickel-based alloys for alkaline water electrolysis, *International Journal of Hydrogen Energy* 18: 561–566.

Grigoriev, S. A., Millet, P., Volobuev, S. A., and Fateev, V. N. 2009 Optimization of porous current collectors for PEM water electrolyzers, *International Journal of Hydrogen Energy* 34: 4968–4973.

Hamdan, M. 2010 PEM electrolyzer incorporating an advanced low cost membrane, presentation at *2010 DOE Annual Merit Review*, Washington, DC, June 10, 2010.

Hamdan, M. 2012 PEM electrolyzer incorporating an advanced low cost membrane, presentation at *2012 DOE Annual Merit Review*, Arlington, VA, May 16, 2012.

Hamdan, M. and Norman, T. 2011 PEM electrolyzer incorporating an advanced low-cost membrane, *2011 DOE Hydrogen and Fuel Cells Program Annual Progress Report*, http://www.hydrogen.energy.gov/annual_progress11_production.html#e, accessed on June 6, 2015.

Harrison, J. W. 1975 Aircraft on-board electrochemical breathing oxygen generators, contributed by the Aerospace Division of American Society of Mechanical Engineering for presentation at the *Intersociety Conference on Environmental Systems*, San Francisco, CA, July 21–24, 1975.

Jung, H. Y., Huang, S. Y., Ganesan, P., and Popov, B. N. 2009 Performance of gold-coated titanium bipolar plates in unitized regenerative fuel cell operation, *Journal of Power Sources*, 194: 972–975.

LaConti, A. B., Fragala, A. R., and Boyack, J. R. 1977 Solid polymer electrolyte electrochemical cells: Electrode and other materials considerations, in *Proceedings of the Symposium on Electrode Materials and Processes for Energy Conversion and Storage*, eds. McIntyre, J. D. E., Srinivasan, S., and Will, F. G., The Electrochemical Society, Inc., Pennington, NJ, Vol. 87-12, p. 354.

LaConti, A. B. and Swette, L. 2003 Chapter 55, Special applications using PEM technology, in *Handbook of Fuel Cells – Fundamentals, Technology and Applications*, edited by W. Vielstich, H. A. Gasteiger, and A. Lamm, *Volume 4: Fuel Cell Technology and Applications*, John Wiley & Sons, Ltd. ISBN: 0-471-49926-9, Hoboken, NJ.

Lutjering, G. and Williams, J. C. 2007 *Engineering Materials and Processes: Titanium*, 2nd ed. Series ed, Derby, B., Springer, Berlin, Germany.

Masami, T. and Kurihara, J. 1992 High temperature nitriding and reactive ion plating in sulfuric acid solution, *Materials Transactions, The Japan Institute of Metals* 33: 691–697.

Mauritz, K. A. and Moore, R. B. 2004 State of understanding of Nafion, *Chemical Reviews* 104: 4535–4585.

Millet, P., Dragoe, D., Grigoriev, S., Fateev, V., and Etievant, C. 2009 GenHyPEM: A research program on PEM water electrolysis supported by the European Commission, *International Journal of Hydrogen Energy* 34: 4974–4982.

Narayanan, S. R., Kindler, A., Kisor, A., Valdez, T., Roy, R. J., Eldridge, C., Murach, B. Hoberecht M., and Graf, J. 2011 Dual-feed balanced high-pressure electrolysis of water in a lightweight polymer electrolyte membrane stack, *Journal of the Electrochemical Society* 158: B1348–B1357.

Ojong, E. T., Mayousse, E., Smolinka, T., and Guillet, N. 2012 Advanced bipolar plates without flow channels for PEM electrolyzers operating at high pressure, presentation at *Technoport RERC 2012*, Trondheim, Norway, April 16–18, 2012.

Okamoto, H. 2013 N–Ti, *Journal of Phase Equilibria and Diffusion* 34: 151–152.

Papadias, D. D., Ahluwalia, R. K., Thomson, J. K., Meyer III, H. M., Brady, M. P., Wang, H., Turner, J. A., Mukundan, R., and Borup, R. 2014 Degradation of SS316L bipolar plates in simulated fuel cell environment: Corrosion rate, barrier film formation kinetics and contact resistance, *Journal of Power Sources* 273: 1237–1249.

Pourbaix, M. 1966 Atlas of electrochemical equilibria in aqueous solutions. Nace International, Houston, TX.

Roy, R. J., Critz, K. M., and Leonida, A. 2001 Electrode plate structure for high-pressure electrochemical cell devices, US Patent 6,942,350, January 9, 2001.

Roy, R. J., Leonida, A., Garosshen, T. J., and Molter T. M. 1999 Graded metal hardware component for an electrochemical cell, US Patent 5,942,350, August 24, 1999.

Schmitt, E. W. and Norman, T. J. 2010 Universal cell frame for high-pressure water electrolyzer and electrolyzer include the same, US Patent US2010/0187102 A1, July 29, 2010.

Shih, C. C., Shih, C. M., Su, Y. Y., Chang, M. S., and Lin, S. J. 2004 Effect of surface oxide properties on corrosion resistance of 316l stainless steel for biomedical applications, *Corrosion Science* 46: 427–441.

Taguchi, M. and Kurihara, J. 1991 High temperature nitriding and reactive ion plating in sulfuric acid solution, *Journal of the Japan Institute of Metals* 55: 431–436.

Toops, T. J., Brady, M. P., Zhang, F. Y., Meyer III, H. M., Ayers, K., Roemer, A., and Dalton, L. 2014 Evaluation of nitrided titanium separator plates for proton exchange membrane electrolyzer cells, *Journal of Power Sources* 272: 954–960.

US DOE 2012 Fuel cell technologies office multi-year research, development, and demonstration plan, updated 2012. http://energy.gov/eere/fuelcells/fuel-cell-technologies-office-multi-year-research-development-and-22, accessed on June 6, 2015.

Wang, C. 2009 Highly electrically conductive surface for electrochemical applications, US Patent 20090176120 A1, July 9, 2009.

Wang, C. 2013 Corrosion resistant metallic components for electrochemical devices, presentation at *1st International First International Workshop of Durability and Degradation Issues in PEM Electrolysis Cells and Its Components*, Freiburg, Germany, March 12–13, 2013.

Wang, C. 2014 Corrosion resistant and electrically conductive surface of metallic components for electrolyzers, US Patent 20040224650 A1 August 14, 2014.

Wang, J., Wang, W., Wang, C., and Mao, Z. 2012 Corrosion behavior of three bipolar plate materials in simulated SPE water electrolysis environment, *International Journal of Hydrogen Energy* 37: 12069–12073.

Weinmann, O. 2012 Renewables and energy storage, presentation at *World Hydrogen Energy Conference 2012* Toronto, Ontario, Canada, June 4–7, 2012.

Wriedt, H. A. and Murray, J. L. 1987 The N–Ti (nitrogen–titanium) system, *Bulletin of Alloy Phase Diagrams* 8: 378–379.

Zhang, M., Hu, L., Lin, G., and Shao, Z. 2012 Honeycomb-like nanocomposite Ti–Ag–N films prepared by pulsed bias arc ion plating on titanium as bipolar plates for unitized regenerative fuel cells, *Journal of Power Sources* 198: 196–202.

8

Current Collectors (GDLs) and Materials

Hiroshi Ito

CONTENTS

8.1 Introduction

This chapter provides an overview of current collectors as gas diffusion layers (GDLs) used in proton exchange membrane (PEM) electrolyzers by first listing their functionalities and requirements, then reviewing the various types, and finally, introducing recent research into their pore structure.

8.2 Functionalities and Requirements of Current Collectors (Gas Diffusion Layers)

Figure 8.1 shows a schematic cross section of a PEM water electrolyzer. Its configuration is similar to that of a PEM fuel cell (PEMFC) in that it consists of a membrane electrode assembly (MEA), current collectors, and bipolar plates with flow channels. (Not shown are several other components needed in the cell setup, such as bus plates, manifolds, and end plates.) In a PEM electrolyzer, the PEM serves as the electrolyte as well as the barrier between the hydrogen and oxygen. Because a PEM is acidic, half-cell reactions during electrolysis can be expressed as follows:

$$H_2O \rightarrow 2H^+ + \frac{1}{2}O_2 + 2e^- \text{ at anode} \qquad (8.1)$$

$$2H^+ + 2e^- \rightarrow H_2 \text{ at cathode} \qquad (8.2)$$

At the anode of a PEM electrolyzer, liquid water is transferred through the current collector from the flow channel and thus dissociated into molecular oxygen and proton (Equation 8.1). Produced oxygen gas diffuses back to the flow channel via the anode current collector. Liquid water acts as a reactant in the anode reaction while simultaneously humidifying the membrane to maintain high proton conductivity. If the produced oxygen cannot be removed efficiently, the anode channel will become blocked, thus limiting the mass transport. Therefore, efficient mass transport of liquid (water) and gas (oxygen) through the anode current collector is crucial for the stable operation of a PEM electrolyzer.

At the cathode, hydrogen gas is produced and diffuses through the current collector to the flow channel. In contrast to the anode reaction, the cathode reaction does not require liquid water, although liquid water is transferred (by electro-osmosis) from the anode and is accompanied by protons in the membrane. Thus, hydrogen gas and liquid water are simultaneously transported to the channel through the current collector during electrolysis operation. Because the activation overpotential of the cathode reaction is relatively small, the effect of the properties of the cathode current collector on the cell performance is limited.

Current collectors are porous media placed between the MEA and bipolar plate at both electrode sides. The two major roles of a current collector are similar to those of a GDL of a PEMFC, namely, electric conduction between the electrode (catalyst layer) and the bipolar plate and efficient mass transport of liquid and

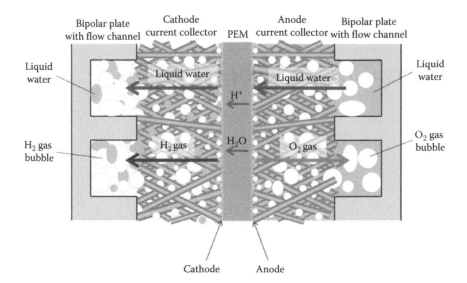

FIGURE 8.1
Schematic cross-section of mass transport in a PEM electrolyzer. (Reprinted from *Electrochimica Acta*, 100, Ito, H., Maeda, T. et al., Influence of pore structural properties of current collectors on the performance of proton exchange membrane electrolyzer, 242–248, Copyright (2013), with permission from Elsevier.)

gas between the electrode and the flow channels. As a technical target of PEMFCs, the U.S. Department of Energy (DOE) has proposed a maximum resistivity of 10 mΩ cm for the bipolar plates including the interfacial contact resistance (DOE 2007; Tawfik et al. 2007). Sufficient electrical contact of the electrode–current collector–bipolar plate is typically achieved by setting the "assembling compression" of the cell at several MPa, even when an electrolyzer is assembled for operation under atmospheric pressure. Care must be taken to ensure that the surface smoothness of the current collector does not damage the catalyst layer while achieving sufficient electrical contact with both the catalyst layers and the bipolar plates. In addition, thickness uniformity is important when the current collector covers a large-scale area. Currently, the differential pressure type PEM electrolyzer is commonly used in which the pressure difference between the cathode (H_2) and the anode (O_2) exceeds 70 bars (Santarelli et al. 2009; Haryu et al. 2011). In such cells, the current collector must be sufficiently rigid so as to support the membrane mechanically against this pressure difference.

In a typical PEMFC, carbon paper or carbon cloth is used as the GDL at both sides of the electrodes. However, in a PEM electrolyzer, carbon material cannot be used for either the electrode (i.e., catalyst layer) or the GDL (i.e., current collector) of the oxygen side (anode), because anodic potential tends to corrode carbon material during electrolysis operation. Titanium is the material for minimal corrosion even under acidic and high anodic potential and relatively easy to process to various types of porous media. Thus, in a PEM electrolyzer, from the practical viewpoint, the material of the anode current collector is limited to titanium (Ti). Contact resistance of Ti is relatively high because the surface of Ti is usually oxidized to TiO_2 whose electric conductance is very low. Therefore, Ti current collectors are often coated (electroplated) with platinum to achieve sufficient conductance (WE-NET Summary of Annual Reports, 1994–2001).

The overpotential of the cathode reaction during electrolysis is much smaller than that of the anode reaction, and thus, the cathode potential during electrolysis is similar to that during fuel cell operation. Therefore, carbon material can be used not only for cathode catalyst layers but also for current collectors, the same as in PEMFCs. In practice, porous carbon material (i.e., carbon paper or carbon cloth) or porous stainless steel has been often used as the cathode current collector, the same as in PEMFCs. In the case of stainless steel, it is often coated with gold to achieve sufficient conductance (WE-NET Summary of Annual Reports, 1994–2001).

8.3 Various Types of Current Collectors

As shown in Figure 8.1, at the anode of a PEM electrolyzer, liquid water is transferred through the current collector from the flow channel while produced oxygen gas diffuses back to the flow channel via the anode current collector. Thus, water transfer and gas removal must be carefully balanced at the anode. Structural properties of the anode current collector are critical for electrolysis operation rather than those of the cathode current collector. In addition, as mentioned in Section 8.2, the material of the anode current collector is actually limited to titanium (Ti). This section provides an overview of three types of anode current collectors made of Ti, namely, screen mesh, sintered powder, and felt types.

8.3.1 Screen Mesh Type

A screen mesh of Ti is a common industrial product and is used in current collectors (Tanaka et al. 2005). The advantage of Ti screen mesh is its low cost and its functionality over a large-scale area (>1 m²). The fiber diameter of the mesh is typically larger than 100 μm and the pore diameter is larger than about 150 μm. In a mesh with lattice structure, the total outer thickness is about double the fiber diameter. However, such large fiber and pore diameters might damage the catalyst layer, because fibers might "sink" in the catalyst layer or membrane during an assembling compression due to the unevenness of the surface being on the same order as the fiber diameter (>100 μm) and thus relatively large.

8.3.2 Sintered Powder Type

Porous Ti media fabricated by the thermal sintering of Ti powder in which the particles are spherical can be applied to the anode current collector. Figure 8.2 shows micrographs of sintered Ti-powder media (Grigoriev et al. 2009). By controlling the sintering conditions (i.e., pressure and temperature) and Ti particle size, both the rigidity and porosity can be adjusted arbitrarily (Thieme et al. 2001; Oh et al. 2003). Rigidity of this sintered powder media can support a membrane against pressure exceeding 100 bars. By adjusting the sintering

conditions, porosity can be set within the range from about 0.30 to 0.40, and pore size can be controlled within the range of 10–25 μm. This type of substrate used for a current collector is typically in the range of 1.0–1.5 mm in total thickness (Grigoriev et al. 2009), and its surface can be smoothed by polishing. This kind of substrate was successfully demonstrated as the current collector in a PEM electrolyzer whose electrode area was 250 cm² (Millet et al. 2010). A highly compact substrate of sintered Ti powder has also been successfully demonstrated (by Honda Motors) for an anode current collector in a high-pressure (350 bar) differential PEM electrolyzer, which exhibited stable performance over the long term of 1 year (1260 h of total operation time and 267 total number of starts and stops) without significant trouble (Haryu et al. 2011; Nagaoka et al. 2012). Despite the advantages of sintered Ti powder, its material cost is high compared to screen mesh.

8.3.3 Felt (Unwoven Fabric)

Figure 8.3 shows micrographs of carbon paper (Toray 090) and Ti felt (Bekinit/Bekaert) (Hwang et al. 2011). Both substrates have a similar structure of unwoven fabric made of fine fibers. A Ti-felt sheet is prepared by evenly distributing fine Ti fibers, then thermally sintering, and finally roll pressing to control the thickness and to smooth the surface. The minimum fiber diameter (ϕ) of Ti is about 20 μm, which is small enough that the effect of surface unevenness on damage to the catalyst layer or membrane is minimal. Porosity (ε) and pore diameter (in the through-plane direction) (d_p) can be adjusted by controlling both the ϕ and the loading

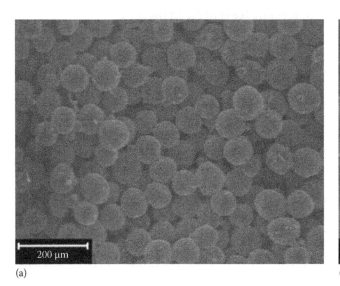

(a)

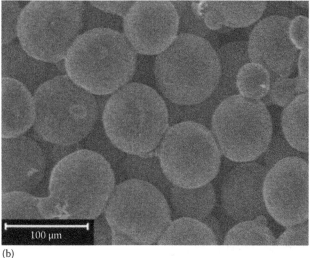

(b)

FIGURE 8.2
Scanning electron micrographs (SEMs) of porous current collectors made from sintered Ti powder (spherical particles). (a) Packing structure of powders and (b) close-up image of packed powder. Mean pore size of 75–100 μm. (Reprinted from Grigoriev, S.A. et al., *Int. J. Hydrogen Energy*, 34, 4968, 2009. With permission from Hydrogen Energy Publications, LLC.)

(a)   Carbon-paper GDL

(b)   Titanium-felt GDL

FIGURE 8.3
SEM images of (a) carbon paper (Toray 090) and (b) Ti felt (Bekinit) used for GDL substrates. Ti felt had a fiber diameter of 20 μm and porosity of 0.75. (Reprinted from Hwang, C.A. et al., *Int. J. Hydrogen Energy*, 36, 1740, 2011. With permission from Hydrogen Energy Publications, LLC.)

amount of fiber. Porosity of Ti felt can be varied in the range of 0.50–0.80, which is a higher range than that for sintered powder media, and pore size depends on ε and φ can be varied in the range of 20–200 μm (Hwang et al. 2011). The minimum outer thickness of the substrate depends on φ. For a fiber diameter of 20 μm, the minimum thickness is about 0.15 mm. Currently, the maximum size of a Ti-felt sheet that can be fabricated is about 400 × 500 mm. The key advantage of Ti felt as the current collector is its capability of mass transport in the in-plane direction due to its fiber orientation. This capability of mass transport in the in-plane direction in addition to that in the through-plane direction is essential to the efficient mass transport of liquid (water) and gas (oxygen) through the anode current collector. In particular, Ti felt is a crucial component as the anode current collector during the electrolysis mode of unitized regenerative fuel cells (Hwang et al. 2011), because it acts as the cathode GDL during fuel cell operation. Ti felt also has exhibited excellent performance as the anode current collector in PEM electrolyzers in Japanese R&D projects of WE-NET (WE-NET Summary of Annual Reports, 1994–2001).

8.4 Degradation Issues

In practical operation of PEM electrolyzers, degradation of the current collector is negligible compared to other components such as the membrane or catalyst.

However, if the surface of the current collector is not sufficiently smooth and if the current is concentrated locally on a small spot, corrosion of the membrane and current collector might occur at that spot. Thus, a smooth surface of the current collector is crucial not only to improve contact resistance but also to prevent degradation.

8.5 Recent Research Efforts

This section introduces recent research on the optimization of the microstructure of current collectors. As described earlier, the role of a current collector is to provide efficient electrical contact between the catalyst layers and bipolar plates and to ensure efficient gas–water transport between them (Figure 8.1).

Grigoriev et al. (2009) tried to optimize the microstructure of current collectors by using Ti porous media substrates (see Section 8.3.2 and Figure 8.2) to fabricate different porous current collectors by changing the particle size and sintering conditions as summarized in Table 8.1. Bulk porosity changed only slightly and was in the range of 0.35–0.40 with one exception (0.28 of No 11), while pore size was relatively widely scattered in the range of 8–25 μm. Figure 8.4 shows the measured current–voltage relationship of PEM electrolysis for these different current collectors under atmospheric pressure. MEAs were fabricated in-house using the catalyst-coated

TABLE 8.1

Properties of Current Collectors Made of Sintered Ti Powder Tested by Grigoriev et al.

No.	Powder Size (μm)	Thickness of Current Collector (mm)	Porosity (%)	Mean Pore Size (μm)	Gas Permeability (m²)	Specific Electric Resistance at I = 2 A cm^{-2} and P = 50 bars (mΩ cm)
1	40–50	1.3	37	10.9	2.4×10^{-12}	6.7
2	75–100	1.2	37	20	—	8.7
3	75–100	1.4	40	21	6.3×10^{-12}	7.2
4	75–100	1.35	35	13.3	4.7×10^{-12}	5.9
5	50–75	1.3	37	10.4	3.2×10^{-12}	—
6	100–125	1.2	40	24.9	1.0×10^{-11}	—
7	75–100	1.3	38	16.2	4.9×10^{-12}	9.5
8	100–200	1.3	37	21.5	9.1×10^{-12}	—
9	75–100	1.4	40	12	6.2×10^{-12}	7.3
10	380–520	1	—	100	—	—
11[a]	40	0.8	28	8	4.1×10^{-13}	9.2

Source: Reprinted from Grigoriev, S.A., Millet, P. et al., *Int. J. Hydrogen Energy*, 34, 4968, 2009. With permission from Hydrogen Energy Publications, LLC.

[a] Powder of irregular shape.

substrate (CCS) method and the catalyst-coated membrane (CCM) method as follows. In the CCS method, ink or slurry that contains catalyst is applied to the current collector (substrate), yielding a basic two-layer structure (CCS). In the CCM method, ink is applied directly to both sides of the membrane, yielding a three-layer structure (CCM). A five-layer complete MEA is formed either by combining two CCSs with a membrane or by combining one CCM with two current collectors (Kocha 2004). The electrolysis data obtained using MEAs with CCM and CCS are shown in Figure 8.4a and b, respectively. Because the difference in outer thickness and porosity among the samples was small, it is difficult to analyze the effect of these two parameters on electrolysis performance. Using the experimental data in Figure 8.4, the present author attempted to evaluate the effect of pore size on the performance by plotting the measured cell voltage at 2.0 A cm^{-2} versus mean pore size of the current collectors. These plots are shown in Figure 8.5, revealing a relationship between cell voltage and pore size when the pore size was larger than 10 μm. Grigoriev et al. also concluded that the optimum pore size for these collectors is 12–13 μm. In these measurements, the difference in electrical resistance among the samples was significantly small, and thus, the effect of resistance difference on cell voltage was negligible (i.e., voltage difference caused by the difference in resistance must be less than 1 mV). In regard to pore size dependency, Grigoriev et al. suggested that pressure at the interface between the catalyst layer and current collector affects the kinetics of the electrode reactions. Results shown in Figures 8.4 and 8.5

also indicate that MEAs prepared with CCM were slightly more efficient than those prepared with CCS. Grigoriev et al. hypothesized that the difference was due to the difference in electrode active area in contact with the membrane.

Ito et al. (2012) also experimentally examined the effect of pore structural properties (i.e., porosity and pore diameter) of the anode current collector on the performance of a PEM electrolyzer. In that study, various Ti-felt substrates with different porosities and pore diameters (measured by capillary flow porometry) were used as the anode current collector (see Section 8.3.3 and Figure 8.3) as summarized in Table 8.2. First, three types of Ti felt of different fiber diameter (ϕ) and porosity (ε) were prepared, in which A1 and A3 had the same ϕ (20 μm) and A1 and A2 had the same ε (0.75). Then to change the pore structure (ϕ and ε), three substrates (B1–B3) were prepared by loading Ti powder (20 μm average particle diameter) onto bare substrates of A1–A3. In the cell setup, the flow field of the bipolar plates at both sides was 26 channels in parallel and each channel had a square cross-sectional area of 0.01 cm^2 (Ito et al. 2010). Figure 8.6 shows an SEM image of a representative Ti-felt current collector with ϕ = 20 μm, ε = 0.75, and Ti-powder loading of 300 mg cm^{-2}. Figure 8.7(a and b) shows the measured *i–V* characteristics during PEM electrolysis under atmospheric pressure with different anode current collectors: bare Ti felt (A1, A2, and A3) and Ti-powder-loaded Ti felt at 0.20 g cm^{-3} (B1-2, B2, and B3). All the MEAs used for these experiments were prepared using the CCM method, and the active electrode area was 27 cm^2. For all the collectors studied here, changes

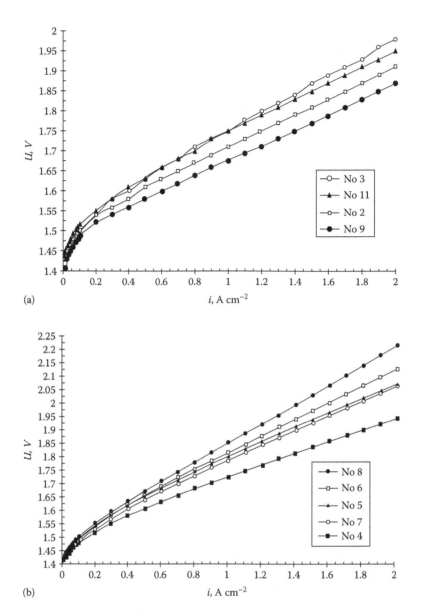

FIGURE 8.4

Current–voltage (*i*–*U*) characteristics of a PEM electrolyzer with different current collectors of sintered Ti powders (parameters listed in Table 8.1). MEA was fabricated by CCM method (a) and CCS method (b). Cell temperature (T_{cell}) was 90°C, $P_{H_2} = P_{O_2} = 1$ bar, after 10 h of continuous operation. (Reprinted from Grigoriev, S.A. et al., *Int. J. Hydrogen Energy*, 34, 4968, 2009. With permission from Hydrogen Energy Publications, LLC.)

in porosity (A1 and A3 in Figure 8.7a, and B1-2 and B3 in Figure 8.7b) had no significant effect on the cell performance. However, cell voltage of A2 (Figure 8.7a) and B2 (Figure 8.7b) was significantly higher than that for the other collectors, about 25–30 mV at around 1.0 A cm⁻², and was beyond the range of experimental reproducibility (±5 mV). In the B1 series (B1-1–B1-3) and in B2, previously reported SEM images of the

Ti-felt substrate revealed that the loaded Ti powder did not form a layer covering the substrate but instead intruded into the bulk of the substrate (Hwang et al. 2012). This suggests that in the earlier study, Ti powders were relatively uniformly distributed not only in the in-plane direction but also in the through-plane direction of the current collectors of the B1 series and of B2. Similar to Figure 8.5, Figure 8.8 shows the plot

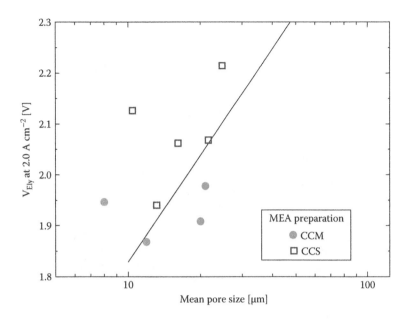

FIGURE 8.5
Cell voltage at 2.0 A cm^{-2} versus mean pore size of Ti-powder-sintered current collectors obtained from *i–U* data (Figure 8.4).

TABLE 8.2

Properties of Anode Current Collectors Made of Ti Felt Tested by Ito et al.[a]

Cell Setup Notation	Current Collector Substrate at Oxygen Side	Fiber Diameter of Ti Felt (φ) [μm]	Porosity of Ti Felt (ε) [—]	Ti Powder Loading[b] [g cm^{-3}]	Bubble Point Diameter (BPD)[c] [μm]	Mean Pore Diameter (MPD)[c] [μm]
A1	Ti felt	20	0.75		90.2	38.6
B1-1	Ti felt + Ti powder		0.73[d]	0.11	72.7	31.9
B1-2	Ti felt + Ti powder		0.71[d]	0.20	95.4	35.5
B1-3	Ti felt + Ti powder		0.68[d]	0.30	70.8	29.5
A2	Ti felt	80	0.75	—	551.5	97.4
B2	Ti felt + Ti powder		0.73[d]	0.20	264.0	106.5
A3	Ti felt	20	0.50	—	43.1	21.2
B3	Ti felt + Ti powder		0.46[d]	0.20	24.6	10.1

Source: Reprinted from Ito, H., Maeda, T. et al., *Int. J. Hydrogen Energy*, 37, 7418, 2012. With permission from Hydrogen Energy Publications, LLC.

[a] Cathode current collector was the commonly used carbon paper.
[b] Calculated based on the amount of loaded Ti powder and outer geometric dimension of Ti-felt substrate.
[c] Measured by capillary flow porometry using the bubble point technique.
[d] Calculation based on the amount of loaded Ti powder.

of cell voltage (at *i* = 1.017 A cm^{-2}) versus mean pore diameter (MPD), revealing a close relationship (despite the scatter in the data) between cell performance and MPD of the anode current collectors regardless of Ti-loading. The cell performance improved with decreasing MPD when MPD > 10 μm. Interestingly, this result agrees with the tendency reported by Grigoriev et al. for Figure 8.5, even though the type of porous media and porosity range were different for the two studies. Ito et al. (2013) also investigated the cell resistance dependency on the current collector structure. They reported that the contact resistance tended to increase as the fiber diameter of the Ti felt increased due to a decrease in the smoothness of the surface. However, the difference in cell voltage caused by the contact resistance was about 4 mV at 1.0 A cm^{-2}, which corresponds to the same order as the experimental reproducibility. Ito et al. hypothesized

FIGURE 8.6

SEM images of Ti-powder-loaded (300 mg cm^{-3}) Ti-felt current collector. Ti felt had a fiber diameter of 20 μm and porosity of 0.75. Average diameter of Ti particles was 20 μm. (Reprinted from *Journal of Power Sources*, 202, Hwang, C.A., Ishida, M. et al., Effect of titanium powder loading in gas diffusion layer of a polymer electrolyte unitized reversible fuel cell, 108–113, Copyright (2012), with permission from Elsevier.)

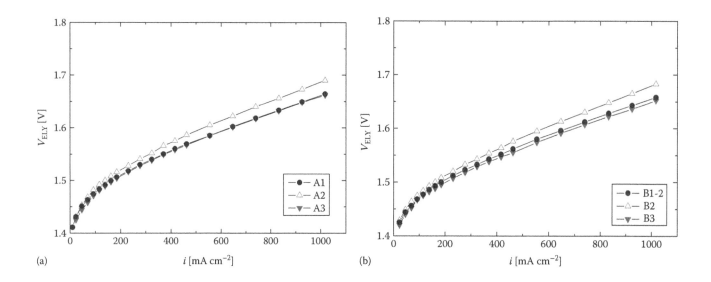

FIGURE 8.7

Current–voltage (*i*–*V*) characteristics of a PEM electrolyzer with different anode current collectors of (a) bare Ti felt (A1, A2, A3) and (b) Ti-powder-loaded Ti felt (B1-2, B2, B3) (parameters listed in Table 8.2). Cell temperature (*T*$_{cell}$) was 80°C. (Reprinted from Ito, H. et al., *Int. J. Hydrogen Energy*, 37, 7418, 2012. With permission from Hydrogen Energy Publications, LLC.)

that larger bubbles generated from larger pores tend to become long slugs and thus hinder the water supply to the membrane.

In summary, the research works of Grigoriev et al. (2009) and Ito et al. (2012, 2013) reveal that when pore diameter of the porous current collector is larger than 10 μm, the electrolysis performance under atmospheric pressure improves with decreasing pore diameter, regardless of the type of porous current collector and porosity. In contrast, changes in porosity have no effect on the cell performance when the porosity exceeds about 0.50. Further research is needed to evaluate and then optimize current collectors under elevated pressure conditions.

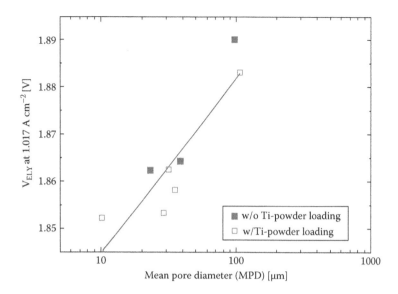

FIGURE 8.8

Cell voltage (V) at 1.017 A cm^{-2} versus mean pore diameter (MPD) for Ti-felt substrates with and without Ti-powder loading as the oxygen-side current collector. (Reprinted from Ito, H. et al., *Int. J. Hydrogen Energy*, 37, 7418, 2012. With permission from Hydrogen Energy Publications, LLC.)

8.6 Summary

The current collector is a critical component of a PEM electrolyzer. From a practical viewpoint, the material of the anode current collector of a PEM electrolyzer is limited to titanium (Ti), whereas various materials such as carbon or stainless steel can be used for the cathode current collector. An efficient current collector of a PEM electrolyzer must be designed by carefully considering the following parameters based on its target role and requirements: (1) bulk electric resistance, (2) contact resistance with the catalyst layer and bipolar plate, (3) corrosion resistance (chemical stability) under acidic and high anodic potentials, (4) surface smoothness, (5) gas permeability and pore structure, and (6) mechanical rigidity when operated at elevated pressure.

References

DOE (U.S. Department of Energy). 2007. Hydrogen, fuel cells and infrastructure technologies program, multi-year research, development and demonstration plan – planned program activities for 2005–2015. Section 3.4. [DOE/GO-102007-102430]. Available at: http://www.nrel.gov/docs/fy08osti/39146.pdf.

Grigoriev, S. A., P. Millet et al. 2009. Optimization of porous current collectors for PEM water electrolysis. *International Journal of Hydrogen Energy* 34: 4968–4973.

Haryu, E., K. Nakazawa et al. 2011. Mechanical structure and performance evaluation of high differential pressure water electrolysis cell. *HONDA R&D Technical Review* 23(2): 90–97 (in Japanese).

Hwang, C. M., M. Ishida et al. 2011. Influence of properties of gas diffusion layers on the performance of polymer electrolyte-based unitized reversible fuel cells. *International Journal of Hydrogen Energy* 36: 1740–1753.

Hwang, C. M., M. Ishida et al. 2012. Effect of titanium powder loading in gas diffusion layer of a polymer electrolyte unitized reversible fuel cell. *Journal of Power Sources* 202: 108–113.

Ito, H., T. Maeda, et al. 2010. Effect of flow regime of circulating water on a proton exchange membrane electrolyzer. *International Journal of Hydrogen Energy* 35: 9550–9560.

Ito, H., T. Maeda et al. 2012. Experimental study on porous current collectors of PEM electrolyzers. *International Journal of Hydrogen Energy* 37: 7418–7428.

Ito, H., T. Maeda et al. 2013. Influence of pore structural properties of current collectors on the performance of proton exchange membrane electrolyzer. *Electrochimica Acta* 100: 242–248.

Kocha, S. S. 2004. Principles of MEA preparation. In: Vielstich, W., Lamm, A., Gasteiger, H. A., editors. *Handbook of Fuel Cells*, vol. 3, Chichester, U.K.: John Wiley & Sons, Chapter 43.

Millet, P., R. Ngameni et al. 2010. PEM water electrolyzers: From electrocatalysis to stack development. *International Journal of Hydrogen Energy* 35: 5043–5052.

Nagaoka, H., N. Yoshida et al. 2012. Verification test of high differential pressure water electrolysis-type solar hydrogen station (SHS2). *HONDA R&D Technical Review* 24(2): 7278 (in Japanese).

Oh, I. K., N. Nomura et al. 2003. Mechanical properties of porous titanium compacts prepared by powder sintering. *Scripta Materialia* 49: 1197–1202.

Santarelli M., P. Medina et al. 2009. Fitting regression model and experimental validation for a high-pressure PEM electrolyzer. *International Journal of Hydrogen Energy* 34: 2519–2530.

Tanaka, Y., K. Kikuchi et al. 2005. Investigation of current feeders for SPE cell. *Electrochimica Acta* 50: 4344–4349.

Tawfik, H., Y. Hung et al. 2007. Metal bipolar plates for PEM fuel cell—A review. *Journal of Power Sources* 163: 755–767.

Thieme, M., K. P. Wieters et al. 2001. Titanium powder sintering for preparation of a porous functionally graded material destined for orthopaedic implants. *Journal of Materials Science: Materials in Medicine* 12: 225–231.

WE-NET Summary of Annual Reports, 1994–2001. Task 8 development of hydrogen production technology, http://www.enaa.or.jp/WE-NET/report/report_e.html. Accessed on July 15, 2015.

9

Proton Exchange Membrane Electrolyzer Stack and System Design

Julie Renner, Kathy Ayers, and Everett Anderson

CONTENTS

9.1 Introduction

9.1.1 Historical Overview of Commercial PEM Electrolyzers

The first water electrolyzer system based on solid polymer electrolyte technology was developed by General Electric (GE) in the 1970s. GE sold their proton exchange membrane (PEM) fuel cell and electrolyzer technologies to Hamilton Sundstrand, a subsidiary of United Technologies. This also led to the creation of new subsidiary of Hamilton Sundstrand, called International Fuel Cells.

In 1996, Proton Energy Systems (currently d.b.a. Proton OnSite) was founded. Hamilton Sundstrand previously employed four of Proton's five founders, where they specialized in designing and demonstrating PEM electrochemical systems for a wide variety of critical military and aerospace life support applications. Proton was founded with the vision of applying ion exchange membrane–based technology and addressing existing commercial markets for hydrogen, as well as emerging applications for energy storage. Two of the original founders are still with the company, bringing substantial continuity to the organization. Proton is now the world leader in hydrogen generation based on polymer membrane technology, with more than 2000 systems installed in over 75 countries around the world. A key aspect of Proton's capability is a reputation for building products that perform consistently over tens of thousands of hours with no safety incidents resulting from their commercial products in their history. Proton also specializes in differential pressure operation including cell stacks operating at up to 2400 psi. Additional commercial level players in the PEM market include Siemens, ITM Power, H-Tec, and Hydrogenics, among others.

9.1.2 Electrolyzer Design Overview

This chapter will focus on PEM electrolyzer stack–system design and manufacturing. The design of both system and stack should consider materials, fluid–gas management, as well as the specific application. System and stack designs affect efficiency, lifetime, capital cost, hydrogen cost, and footprint. Ultimately, successful designs are iterated, and new approaches are developed for superior performance and emerging applications.

General design principles for the stack involve reducing efficiency losses while simultaneously keeping costs low and durability high. For the stack, efficiency is gauged by the total applied voltage that comprises the Nernst potential, anode and cathode overpotentials, and ohmic overpotentials due to membrane ionic resistance, and interfacial resistance (Choi et al. 2004).

$$V = V_0 + \eta_A - \eta_C + \eta_{PEM} + \eta_I + \eta_E \qquad (9.1)$$

where

V is the applied cell voltage

V_0 is the Nernst potential

η_A is the anodic overpotential

η_C is the cathodic overpotential

η_{PEM} is the ohmic overpotential from the membrane ionic resistance

η_I is the ohmic overpotential from the interface

η_E is the ohmic overpotential from electronic conductors (e.g., wires)

The Nernst potential is sometimes called the equilibrium voltage or reversible potential and represents the thermodynamically expected applied voltage. The rest of the terms in the equation account for overpotential, or energy that is required beyond thermodynamic expectations to drive the reaction. Overpotential represents inefficiencies in the cell stack, and a large part of stack design efforts are to limit these overpotential contributions. On the system side, a similar strategy is employed, keeping the hydrogen production efficiency and lifetime of components high, while keeping costs and footprint low. This chapter will discuss these design principles in detail, as well as discuss manufacturing and applications of PEM electrolyzers.

9.2 PEM Electrolyzer Applications

9.2.1 Introduction

Hydrogen is a widely used industrial gas, with a large and expanding global market (MarketsandMarkets 2011). The United States alone manufactures over 9 million tons of hydrogen annually. High-purity hydrogen is used in heat treating, generator cooling, and as a laboratory carrier gas. Power plants use hydrogen-cooled generators that require high-purity hydrogen for effective cooling. With 16,000 hydrogen-cooled generators worldwide generating a two billion dollar market, there is clearly a current and substantial demand for high-purity hydrogen.

The most common way to produce hydrogen is by hydrocarbon steam reforming. Figure 9.1 shows the steam reforming process, which involves reacting hydrocarbons and water in the presence of heat and a catalyst to form carbon monoxide and hydrogen. The carbon monoxide is typically converted to carbon

Hydrocarbon steam reforming:

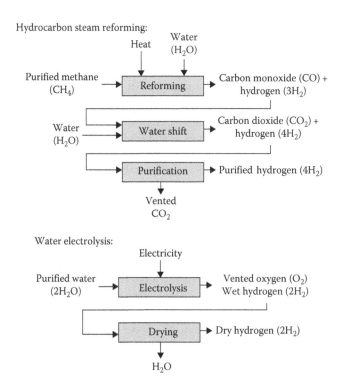

Water electrolysis:

FIGURE 9.1
Simplified block flow diagrams showing the differences between hydrocarbon steam reforming and water electrolysis for hydrogen production.

dioxide in a second step of the process. The carbon dioxide is an unwanted by-product of the reaction, causing this process to have a high carbon footprint. Elemental carbon can also be formed as an unwanted by-product in this process, forming soot. Many hydrogen applications require removal of impurities, such as carbon dioxide from the hydrogen product stream, adding an additional processing step. Because of the high temperatures required, steam reformers require expensive materials of construction and therefore have a high capital cost. As a result, the process is difficult to scale-down.

The hydrogen is often produced at a central location and shipped, which further increases the cost and carbon footprint. Delivered hydrogen can also be cost prohibitive due to handling logistics.

Alternatively, water electrolysis offers a competitive production method for industry because it is on-site, only uses water as the feedstock, and is emission free, as shown in Figure 9.1. Electrolysis also produces very pure hydrogen, with none of the typical impurities that may be found in hydrogen produced from reforming, providing customer reassurance for highly sensitive processes, such as heat treating and fuel cell operation. Electrolyzers allow generation at or near the point of use that enables reduced hydrogen storage inventories and reduction of gas cylinder change outs. On-site generation of hydrogen improves safety and reduces issues with the Occupational Safety and Health Administration requirements, especially for transport of hydrogen cylinders between floors of buildings or storage of hydrogen below the ground floor.

9.2.2 Competition

Within electrolysis technologies, there are two key commercial options: PEM electrolysis and liquid alkaline electrolysis. Figure 9.2 shows schematics of these two technologies for comparison. Alkaline electrolysis is a mature technology, using two electrodes immersed in a liquid alkaline electrolyte of 20%–30% potassium hydroxide (KOH). The electrodes are separated by a diaphragm that keeps the oxygen and hydrogen apart from one another. PEM electrolysis is a relatively new technology and uses a solid polymer as the electrolyte and separator of the product gases. Each technology carries its own advantages and disadvantages as described in the following.

The KOH electrolyte is advantageous because compared to the solid polymer material it is less expensive. In addition, the basic environment that KOH provides

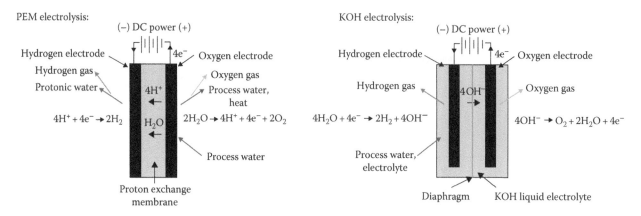

FIGURE 9.2
Comparison between PEM electrolysis operation and liquid KOH electrolysis operation.

requires less costly materials of construction (such as stainless steel and nickel) compared to materials required for the acidic environment in a PEM-based device. Disadvantages associated with the KOH electrolyte are largely associated with its corrosiveness. This increases maintenance, operation costs, and decreases reliability and safety in those systems.

For maintenance, KOH units also typically require complete change out of the electrolyte once a year (~300 L for a 10 N m³ h⁻¹ unit), resulting in several days of down time. Also, the concentration range must be controlled to the correct level for the liquid systems, adding to the operation costs. PEM systems typically only require yearly changes in air filters and guard beds for maintenance and do not require electrolyte change out. Because of the electrolyte corrosiveness and complexity involved with controlling concentration, KOH electrolyzers typically suffer from reliability issues, so that many installations require system redundancy and excess capacity. PEM-based systems have been shown to have high reliability with demonstrated lifetimes in excess of 60,000 h (see Section 9.4).

The KOH electrolyte corrosiveness also affects safety, requiring special personal protective equipment (PPE) and procedures for disposal or in case of leakage. In contrast, PEM units utilize deionized water, eliminating concerns when handling parts that have been exposed to the fluids loop. This makes PEM electrolyzer advantageous where safety is paramount, such as for life support on nuclear submarines.

Finally, there are performance differences between the two technologies as well. Due to the low bubble point between the electrodes, KOH-based systems have a limited turndown range. This condition requires balanced pressure operation and sufficient oxygen generation

to dilute hydrogen crossover from the cathode. High-pressure operation is, therefore, difficult because of the increased hazards associated with high-pressure oxygen gas. Additionally, because of the high ohmic losses across the liquid electrolyte and the diaphragm, alkaline electrolysis is limited to fairly low current densities compared to a typical PEM device. A comparison of rough efficiency curves is shown in Figure 9.3 (Ayers et al. 2010).

A solid polymer electrolyte enables high differential pressure and 100% turndown ratio due to the high bubble point. The lack of requirement for tight pressure balance control also enables fast response times, which allow easy integration with renewable energy sources and grid-buffering capabilities. Current and voltage response of the cell stack in reaction to a command from the power supply have been demonstrated to be less than a 5 ms response (Anderson et al. 2012). Figure 9.4 shows the calculated efficiencies based on known cell stack efficiencies, dryer losses, and other balance of plant loads (Anderson et al. 2012). The efficiency of the PEM-based system is stable over a wide operating range, especially from 30% to 100%.

9.2.3 Current Applications

9.2.3.1 Power Plants

Because of its high heat capacity and low density, hydrogen gas is used to cool large power plant generators. Power plants utilizing hydrogen-cooled generators must maintain optimal hydrogen purity and pressure inside of the generator casing for efficiency, safety, and equipment reliability. There are two ways to supply hydrogen to power plant generators: on-site hydrogen

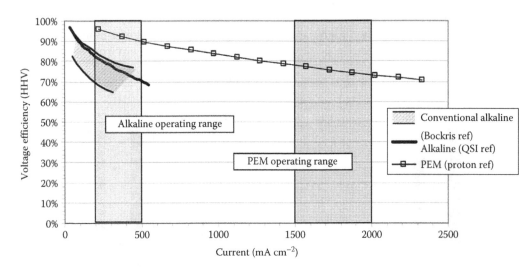

FIGURE 9.3
Liquid alkaline electrolysis efficiency compared to PEM-based electrolysis. (Reprinted from Ayers, K.E. et al., *ECS Trans.*, 33, 3–15, 2010. With permission from the Electrochemical Society.)

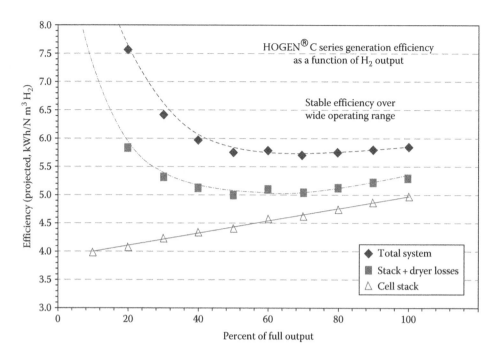

FIGURE 9.4
PEM-based electrolyzers are highly efficient over a wide operating range making them ideal for use in renewable energy capture and grid-buffering applications. (Reprinted from Ayers, K.E. et al., *ECS Trans.*, 41, 75, 2012. With permission from the Electrochemical Society.)

generation, or delivered gas via cylinders or tube trailers. The latter option does not optimize generator capabilities because power plants utilizing hydrogen-cooled generators must maintain optimal hydrogen purity and pressure in the generator casing for efficiency, safety, and equipment reliability. Additionally, moisture in the hydrogen supplied by cylinders can lead to the formation of cracks in the generator retaining rings and possibly generator failure. In contrast, continuous addition of ultra-pure hydrogen from a hydrogen generation system will make up for hydrogen seal losses and can optimize hydrogen purity and pressure for best operating performance and uptime. PEM-based hydrogen generation systems have been shown to reduce inventory maintenance and improve safety. Thus, these electrolyzer units reduce overall hydrogen supply costs while reducing windage loss.

9.2.3.2 Materials Processing

Hydrogen generation systems are an efficient, reliable, and productive means to provide hydrogen for the materials processing industry, especially heat treatment applications. In comparison to dissociated ammonia, exo or endo gas, on-site generated hydrogen gas is a drier and safer alternative. There is no need for a stored inventory of flammable or poisonous gas. Hydrogen generation systems are easy to permit, easy to install, and operate automatically. These systems appeal to a variety of applications including annealing, brazing, flame spray, glass to metal hermetic sealing, metal injection molding, optical fiber processing, and powder metallurgy. By utilizing hydrogen generators, professionals can eliminate the need for delivery and storage of hazardous gases within the industry.

Hydrogen generation systems are also well suited to provide ultrahigh-purity hydrogen for silicon semiconductor and chemical vapor deposition processes. They produce hydrogen at 200 psi or higher without mechanical compression for consistent composition and predictably low levels of oxygen and nitrogen. The hydrogen is available at sufficient pressure to be purified through a palladium or chemical purifier to prevent contaminant reactions.

9.2.3.3 Meteorology

Hydrogen is used as a lift gas for filling weather balloons when helium is unavailable or prohibitively expensive. Hydrogen generation systems are designed to produce the gas at its point of use, using only electricity and water to provide hydrogen at 13.8 barg and 200 psig pressure without the need for mechanical compression. Compared to traditional caustic electrolyzers with mechanical compressors, PEM-based generators are compact, lightweight, one-box automated systems that are small enough to deliver by light plane, and have minimal maintenance requirements.

9.2.3.4 Laboratory

In the laboratory market, professionals seek gas that can be supplied in a safe, cost-effective, and reliable manner. Hydrogen is used as an ultrahigh-purity fuel and reducing agent in analytical labs, as well as a carrier gas for analytical equipment. PEM-based electrolyzers provide consistent hydrogen composition and predictable low levels of oxygen and nitrogen. These features make hydrogen generation units particularly useful for gas chromatography.

9.2.4 Emerging Applications

9.2.4.1 Renewable Energy Storage

Hydrogen is a promising technology for renewable energy capture as it has the capability to store massive amounts of energy in a relatively small volume and is highly flexible (Schiller 2013). Figure 9.5 shows several pathways for the use of electrolysis in energy storage.

Electrolysis can also provide ancillary services to the grid such as frequency regulation and load shifting. In Europe, hydrogen is already being looked upon as a key part of the energy storage solution, providing a link between the electric grid and gas grid infrastructures. Germany has over 70 GW of installed wind capacity, with 20%–40% stranded due to the lack of grid capability to handle excess capacity. Germany is also considered the global leader in biogas energy generation, with 18,244 GWh of generation in 2012 forecasted to grow to 28,265 GWh by 2025.

The stored hydrogen can be used for a variety of energy-related applications. From storage, hydrogen can be used as a transportation fuel, injected into the natural gas pipeline (thus making that energy carrier more green), used in the production of high-value chemicals such as ammonia, or used to upgrade conversion efficiency for methanization-produced biogas.

One of the additional advantages to note for PEM systems is the ability to run at higher current densities and at higher levels of efficiency than liquid KOH systems. This has been documented by the National Renewable Energy Laboratory in the Wind to Hydrogen program at the National Wind Technology Center. Efficiency is a major driver in determining the cost of hydrogen fuel when total life cycle costs are considered. Energy storage applications, particularly renewable energy storage, are also significantly impacted by the efficiency of the electrolyzer. The more efficient the electrolyzer is, the smaller the capacity of the renewable primary power source is required, which can have a major impact on total system cost.

9.2.4.2 Fueling for Fuel Cell Vehicles

Hydrogen can be used as a zero-emission fuel in a variety of vehicles. Car manufacturing companies, such as Hyundai and Toyota, have announced that fuel cell vehicles will be available to consumers in 2015, touting advantages, such as long driving range, short fueling times (similar to gasoline), instantaneous torque, minimal cold-weather effects, quiet operation, durability,

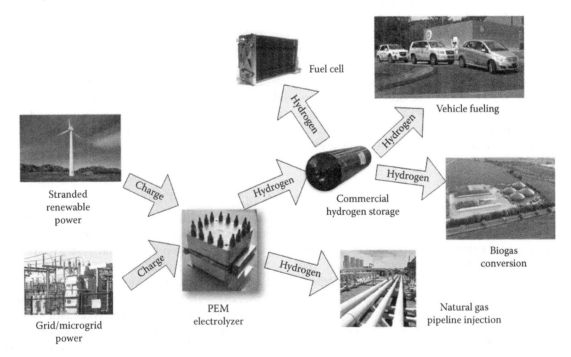

FIGURE 9.5
Hydrogen options for grid stabilization and energy storage.

and reliability. Fuel cell busses and forklifts are already in use by major cities and corporations, and the U.S. Department of Energy has funded the development of a fuel cell delivery truck. These fuel cell vehicles will require a variety of fueling solutions to match vehicle rollout rates (Anderson et al. 2012). Large centralized production facilities will be required to provide sufficient fuel for larger fleets, while small neighborhood and home on-site generators will be needed for early adaptors.

Alkaline liquid electrolysis is emission free and available in relevant size ranges for distributed hydrogen fueling stations overcoming some of the obstacles present in reforming technology. However, as outlined previously in this chapter, liquid systems have a limited turndown range, maintenance issues, safety concerns, and limited capacity to safely produce hydrogen at high pressure. High-pressure electrolyzer capability enables the efficient small-scale production of hydrogen in the neighborhood and "home fueler" concepts. This is because in order to properly fill vehicle tanks, and maintain short fill times, hydrogen has to be dispensed at an adequate pressure.

Hydrogen compression can be done electrochemically or mechanically. Larger stations can tolerate the energy and capital cost of mechanical compression because of the large hydrogen output. However, for smaller stations, the cost of a mechanical compressor is prohibitively expensive as a fraction of the overall unit cost. In addition, compressors have a large footprint, which is typically undesirable in a neighborhood or home fueling setting. Compressors also can decrease the energy efficiency of hydrogen production. Extensive trade studies have been conducted for mechanical and electrochemical compression. Figure 9.6a shows the energy consumption of an electrolyzer as a function of electrochemically generated hydrogen pressure.

From a thermodynamic standpoint, the voltage penalty for electrochemical compression is relatively small. This can be modeled using the following equation (Shen et al. 2011):

$$E_1 = E_0 + \frac{RT}{2F} \ln\left(\frac{Pc}{Pa}\right) \qquad (9.2)$$

where

E_1 is the equilibrium potential
E_0 is the standard potential
R is the gas constant
T is the temperature
F is Faraday's constant
Pc is the cathode pressure
Pa is the anode pressure

Typically, fueling is required at 5,000–10,000 psi (~350–700 bar). This equation can be used to estimate the voltage penalty for differential pressure operation in an electrolyzer. In practice, there is less than a 50 mV difference between an order of magnitude increase in pressure (200 and 2400 psi) (Ayers et al. 2012). Figure 9.6a shows that the efficiency generally remains constant until back diffusion becomes significant around 160 bars. Figure 9.6b shows an energy comparison of pure electrochemical compression versus electrochemical production at ambient pressure followed by mechanical compression. The data show that electrochemical compression is more efficient to over 600 bars (~9000 psi). For mechanical compression, published commercial supplier data were used and are likely optimistic, meaning electrochemical compression may be more efficient to even greater pressures.

These efficiency and monetary trade-offs are important to consider in the design of fueling units. For example, in a home delivery system, 5000 psi pressure may not completely fill a vehicle, but because of cost, footprint, and noise, electrochemical compression to 5000 psi may be more desirable. Through a research program funded by the Department of Energy, a 5000 psi electrolyzer stack at a single cell level was demonstrated. Capital cost projections for the developed system at modest production volumes (1000s per year) are less than $15,000. Ultimately, a product like this will help provide a bridging function in enabling the deployment of vehicles in areas not served by forecourt hydrogen fueling stations. Currently, a configuration with partial electrochemical compression and mechanical compression is often utilized. This is the case for the SunHydro Station located in Wallingford, CT. Figure 9.7 shows a process diagram of the station.

Proton OnSite, as prime contractor to SunHydro LLC, operates the only H70 fast fill and public-accessible hydrogen fueling stations on the U.S. East Coast. The SunHydro stations generate SAE J2719 fuel cell grade hydrogen from water via on-site PEM water electrolysis, powered in part with photovoltaic arrays colocated on site. SunHydro#1 station has dual integrated H70/H35 dispensers for 700 bar fast fills and 350 bar fills. Since its grand opening in 2010, SunHydro#1 has generated and dispensed more than 5800 kg of hydrogen over thousands of H70 and H35 fueling events. Dispenser access is controlled by a personal identification code (PIN) and credit card. The H70 side of the dispenser conforms to SAE J2601/A T-20 (−20°C pre-cooling) and uses an IrDA communications-equipped filling nozzle per SAE J2799 and J2600. The H35 side of the dispenser conforms to SAE J2601/A TA (no pre-cooling) and uses a SAE J2600-compliant nozzle. SunHydro customers experience a

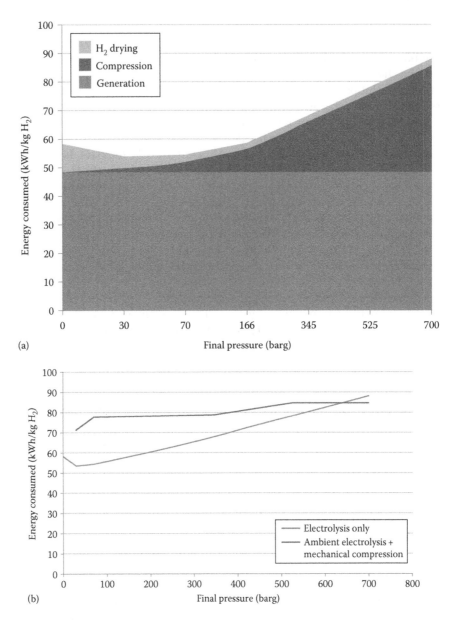

FIGURE 9.6
Electrochemical compression of hydrogen using a PEM electrolyzer. (a) Energy consumption of an electrolysis system as a function of H_2 pressure. (b) Comparison of electrochemical and mechanical compression efficiency. (Reprinted from Ayers, K.E. et al., *ECS Trans.*, 41,27, 2012. With permission from the Electrochemical Society.)

"retail" self-serve fueling experience—they swipe their credit card to initiate an H35 or H70 fast fill, and minutes later they leave with a full tank and a paper receipt showing their charge for the amount of hydrogen they dispensed.

9.2.4.3 Biogas Processing

Biogas is commonly produced through the digestion of organic material with the by-product of carbon dioxide (CO_2), which is typically considered waste. Upgrading of this process can occur by converting the carbon dioxide

to methane via the addition of hydrogen through the following reaction:

$$CO_2 + 4H_2 \rightarrow CH_4 + 2H_2O \qquad (9.3)$$

Hydrogen can be added to the anaerobic digester where the microorganisms that produce methane (hydrogenotropic methanogens) consume hydrogen and excess CO_2. This process has the potential to increase the concentration of methane in the biogas from 60% to 95% (Luo and Angelidaki 2012). Additionally, literature suggests that using the Sabatier process to complete this reaction could increase biogas production by as much as 44% (Mohseni et al. 2012).

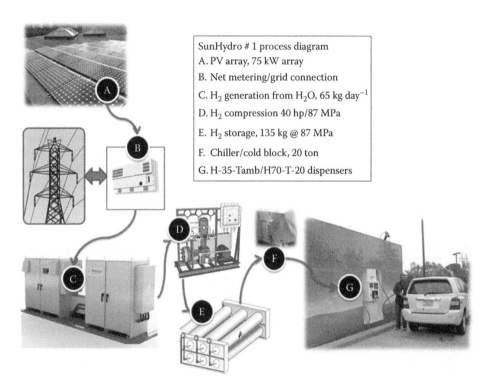

SunHydro # 1 process diagram
A. PV array, 75 kW array
B. Net metering/grid connection
C. H_2 generation from H_2O, 65 kg day^{-1}
D. H_2 compression 40 hp/87 MPa
E. H_2 storage, 135 kg @ 87 MPa
F. Chiller/cold block, 20 ton
G. H-35-Tamb/H70-T-20 dispensers

FIGURE 9.7
Process diagram for the SunHydro station located in Wallingford, CT.

9.3 PEM Electrolyzer Stack and System Components

9.3.1 Introduction

A PEM-based electrolyzer can be split into three main components: the membrane electrode assembly (MEA), the stack, and the system, shown in Figure 9.8. The MEA features two electrodes (the anode and cathode) separated by the PEM. The electrodes catalyze the reactions necessary to produce hydrogen and oxygen products from water and electricity. The MEAs are assembled into a repeating stack structure that increases the reaction area while separating the anode process stream from the cathode product stream. Each repeated unit is commonly referred to as a "cell." The stack is placed in a system (sometimes referred to as the balance of plant) that manages the reactant and product streams entering and exiting the stack, dries the hydrogen, as well as supplies the electricity that drives the reaction. Each component has special considerations in their design and manufacturing, which simultaneously affect cost and performance. These considerations are discussed for each component separately in this section.

9.3.2 Membrane Electrode Assembly (MEA) Material and Design

An MEA consists of a membrane in close contact or bonded with two electrodes on opposite sides. The oxidation and reduction reactions occur at the surface of these electrodes, with the membrane enabling the reactions by selectively conducting protons and separating the product gases. The selection of materials and design of the assembly affects the efficiency, cost, and lifetime of the machine. Often, the MEA is referred to as the "heart" of the electrolyzer.

9.3.2.1 Electrolyzers versus Fuel Cells

While PEM-based electrolyzers and PEM-based fuel cells are distinctly different in many aspects, they also share many similarities. During electrolysis, water enters on the anode side where water is oxidized to protons, electrons, and oxygen gas. The oxygen gas exits the stack, while the protons are selectively conducted by the membrane to the cathode, where they combine with the electrons to form hydrogen gas. During fuel cell operation, an opposite process occurs. Figure 9.9 shows the PEM electrolysis process (operating in hydrogen generation mode) compared to the PEM fuel cell (operating in power generation mode).

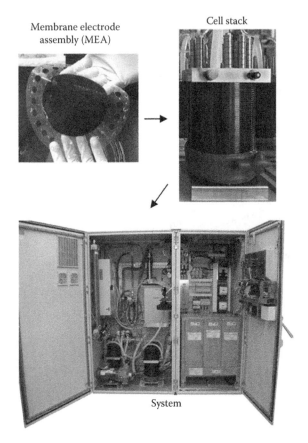

Membrane electrode assembly (MEA)

Cell stack

System

FIGURE 9.8
The main components of a typical PEM electrolyzer consist of the MEA, where the electrochemical reactions occur, assembled into a repeating stack assembly to increase the reaction area, and the system that manages inputs and outputs from the stack.

efficiency is not a strong consideration. PEM-based fuel cells and PEM-based electrolyzers use similar materials of construction, such as perfluorinated sulfonic acid membranes and platinum group metal (PGM)-based catalysts. One important difference is that the electrolysis membrane remains fully hydrated, whereas the fuel cell undergoes relative humidity cycling that causes material wear. Also, in the electrolyzer cell, materials on the oxygen side of the MEA have to withstand greater than 2 V potentials, prohibiting the use of carbon. Electrolysis cells are designed to operate at high differential pressure, with commercial systems available from 200 to 2400 psi hydrogen and ambient oxygen, whereas fuel cells typically do not have a high differential pressure requirement. Finally, lifetime expectations for the electrolyzer stacks are far greater than for fuel cells in vehicles (50,000 h vs. 5,000 h of operation).

These differences help guide the general design principles of the MEA. Generally, materials are selected that limit the ohmic or catalytic efficiency losses within the electrolyzer cell while simultaneously keeping material and manufacturing costs low and having high durability. The activation and ohmic overpotential losses for electrolyzer stacks are depicted in Figure 9.10 (Ayers et al. 2010). Membrane ionic resistance and oxygen evolution overpotential represent the majority of the efficiency losses and therefore will be discussed in detail later in this section.

To help guide membrane and catalyst selection for the MEA, it is important to understand how each contributes to the overall efficiency of the cell stack. The polarization curve (voltage plotted as a function of current density) is a tool often used to assess the efficiency of electrolyzers. Current density represents the amount of hydrogen being produced, and voltage represents energy being applied to the system. Highly efficient systems therefore will have low overpotentials, resulting in lower applied voltage at high current densities.

While electrolysis of water is not a new technology, basic research on electrolysis MEA materials has been limited compared to their PEM fuel cell counterparts. This gap is likely due to the reality that electrolysis has been cost competitive in specific industrial applications for the generation of high-purity hydrogen, where

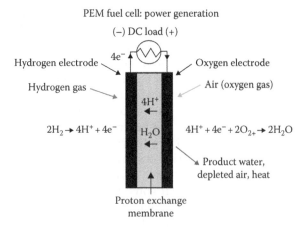

FIGURE 9.9
Comparison of PEM electrolysis and PEM fuel cell operation.

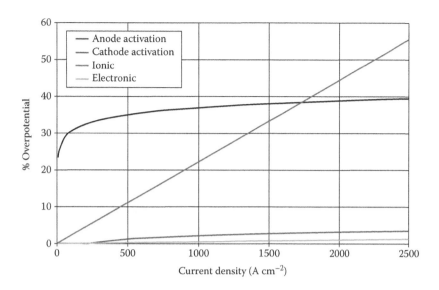

FIGURE 9.10
Overpotential contributions for a typical commercial PEM electrolysis stack. (Reprinted from Ayers, K.E. et al., *ECS Trans.*, 33, 3, 2010. With permission from the Electrochemical Society.)

Contrasting with fuel cells, mass transport losses are generally not seen in electrolyzers, even at very high current densities because of the fully flooded conditions.

9.3.2.2 Membrane Materials

An ideal electrolyzer membrane would have (1) low ionic resistance, (2) high mechanical integrity for differential pressure operation, (3) low hydrogen permeation, (4) no mechanical or chemical degradation with time, (5) high-temperature stability, and (6) low cost. The ability for these materials to maintain all of these qualities while also being thin is attractive. Thin membrane materials will contribute to lower costs and increased efficiency.

The current gold standards in PEM materials are perfluorosulfonic acid (PSFA) polymers, such as Nafion®. While the manufacturing costs can be high for the fluorine-based chemistry, PSFA material has shown superior durability in electrolysis cells, maintaining performance for over 60,000 h with minimal voltage decay. Traditional commercial electrolysis membranes range from 7 to 10 mils (175–250 µm), much thicker than traditional fuel cell membranes, which are about 1–2 mils. The higher operating pressures in the electrolyzer cell require a more robust and often thicker material. This higher thickness directly impacts the materials cost, as well as increases ionic resistance. At the typical operating currents of over 1500 mA cm^{-2}, the ionic resistance can be the largest contributor to the cell overpotential. Commercial 7 mil membranes are known to produce about 350 mV of overpotential at 2 A cm^{-2}. However, thinner membranes commonly

contribute to unacceptably high levels of hydrogen crossover, as well as decreased lifetimes.

Operating at higher temperatures increases ion conduction and decreases anode and cathode activation overpotentials. Therefore, membranes that are robust at high temperatures are ideal. Nafion materials have been shown to degrade at temperatures above 80°C in electrolyzers, which is due to the low glass transition temperature of the material. Many strategies have been employed to increase the temperature stability of proton-conducting membranes, including doping, cross-linking, incorporation of inorganic fillers, and the use of different polymer backbones (Bose et al. 2011).

Temperature stability is often related to mechanical stability. Membranes must be mechanically robust to withstand the differential pressure operation of electrolyzers as well as the seal loads within the cell stack. Superior tensile properties will help resist burst failures and high compressive properties will help withstand seal loading. Tear resistance properties are also important. Local stresses can occur at the edge of sealing gaskets, which can cause failures. Seven to ten mil Nafion has shown adequate mechanical properties to date and often serves as the baseline for new membranes when comparing mechanical properties.

Another membrane property that is also important to consider is ionic conductivity. Ionic conductivity can be measured directly, but another metric that is typically used to gauge ionic conductivity of the membrane is the ion exchange capacity (IEC). The IEC can be measured via a simple titration protocol and is typically expressed in the units of meq/g. Nafion consists of a tetrafluoroethylene backbone and perfluorovinyl ether groups terminated with sulfonate groups. The negatively charged sulfonate groups are what facilitate proton conduction. The IEC is an indicator of how many of these groups are present per mass of material. However, higher IEC does not necessarily mean superior performance as morphology and channels also play a critical role in conductivity. Higher IEC can also result in higher swelling of the material, which can lead to lower mechanical properties.

It should also be noted that ionomers of the membrane material are often incorporated into the electrode layers to aid in ionic conduction. A trade-off exists between superior ionic conduction and hindering electron flow from the catalyst. This trade-off is considered in the design of electrode materials as discussed in the following.

9.3.2.3 Electrode Materials

Electrodes typically consist of a catalyst combined with a binder and/or ionomer in a thin layer directly contacting the ion exchange membrane. Electrodes are very complex structures and are an area of active research in both industry and academic settings. They must achieve a variety of criteria for superior cell performance. They must (1) contain highly active catalysts for anode and cathode reactions, (2) allow reactants to get to the active sites of the catalyst while allowing products to freely leave, (3) remain adhered close to the membrane, and (4) facilitate sufficient electron and ionic transport.

Traditionally, metallic platinum is used for the hydrogen evolution reaction (HER) at the cathode and metallic iridium or iridium oxide is used for the oxygen evolution reaction (OER) at the anode. The highly acidic environment of the PEM requires the use of these expensive noble metals to achieve long-term durability. A common development effort in PEM electrolysis is to reduce the noble metal loading and increase utilization as much as possible. Interestingly, PEM fuel cells typically have much lower loadings than their electrolyzer counterparts. Electrolysis has lagged significantly behind fuel cells in catalyst-loading reduction, for several reasons. First, the capital cost of electrolysis is already competitive in high-purity industrial hydrogen applications for commercial viability, reducing the urgency of immediate change until the energy markets are more mature. Second, there has been resistance to the use of supported catalysts in electrolysis applications, because performance limitations have been noted on the cathode at fuel cell platinum loadings, and carbon supports are not stable on the anode side of the electrolysis cell. Finally, the traditional methods for the manufacture of electrolysis electrodes use techniques that require high loadings in order to achieve uniform catalyst distribution. More recently, however, efforts have demonstrated that significant reductions in catalyst loading are feasible (see Section 9.5).

Selection and screening of catalyst depends on certain physical properties, such as surface area, particle size, purity level, and crystallinity. Ultimately, catalyst materials must be processed and made into electrodes (see Section 9.3.4) for operation testing. Operational testing typically will consist of polarization data, as well as durability tests to ensure the quality of the material. Cyclic voltammetry and global electrochemical impedance spectroscopy have been used in situ to help understand catalyst performance (van der Merwe et al. 2014).

Binder and ionomers are often used in electrodes as well. To achieve highly durable electrolysis stack, catalyst must remain adhered in close contact to the membrane. However, most binders do not conduct electrons efficiently, so the amount must be optimized. Binders will interact differently with catalyst depending on the particle size and surface properties of each material. These interactions can affect adhesion, cell performance, and electrode microstructure. Tuning these interactions often occurs through adjusting electrode formulation ratios, processing temperatures, catalyst specifications, or binder specifications. It is also important to consider the durability of the binder–ionomers in the electrolysis environment. Binders must withstand highly acidic conditions of the PEM and hold the electrode together while bubbles are being formed. On the anode, the binder must not degrade in the high potential oxygen-rich environment.

9.3.3 System Material and Design

A PEM-based water electrolyser system is modular in design and construction. Typically, major subsystems and assemblies are prefabricated and tested prior to final system assembly. The subassemblies are packaged in enclosures according to the component type, functional, and environmental requirements. Major subsystems based on common functional purposes and/or general energy or fluid flow paths are described in the following sections.

9.3.3.1 Electrolysis Cell Stack(s)

As described previously, the electrolysis cell stack uses an electrical current to dissociate water, which creates heat, hydrogen, and oxygen.

9.3.3.2 Water and Oxygen Management System

The water and oxygen management system (WOMS) circulates water through the cell stack. This water is consumed via electrolysis. The WOMS subsystem also removes heat from the cell stack and maintains the stack temperature while regulating system pressure.

Another function of the WOMS is to separate the liquid water phase from the gaseous O_2 phase created during electrolysis. The oxygen gas that is generated can also be cooled, and the resultant condensed water from this gas stream can be returned to the water reservoir for reuse. Material selection is an important consideration in the construction of the WOMS subsystems in order to make sure there is compatibility with PEM electrolysis feed water. Other functions of the WOMS subsystem are to monitor water purity, minimum stack water flow, water quantity/level, stack exit water/oxygen temperature, pressure, and level of combustible gas (CG) in oxygen gas production. Typically, a deionization water conditioner is present in the WOMS subsystem to remove trace ionic contaminants.

9.3.3.3 Hydrogen Gas Management System

The hydrogen gas management system (HGMS) separates the liquid "protonic water" (water that is carried through the cell stack membranes with the gaseous H_2 created during electrolysis) from the hydrogen gas. The HGMS subsystem cools the hydrogen gas and condenses water from this gas stream, returning the condensate to the WOMS subsystem. Hydrogen exiting the HGMS is typically dried to a level of <5 ppmv water. Material compatibility with PEM electrolysis feed water (including dryer manifolds and dryer beds) is a major design consideration. The HGMS subsystem is responsible for creating and regulating system back pressure on the hydrogen side of the cell stack and monitoring system pressures, temperatures, and protonic water levels. Some systems include optional monitoring of hydrogen flow and dew point of the product hydrogen.

A critical element of the HGMS subsystem is the reintroduction of protonic water back to the WOMS subsystem in order to minimize the amount of water consumption in the system. This water reintroduction is typically done in two steps: Step 1 allows for the removal of dissolved hydrogen gas in the pressurized protonic water, which effervesces out of solution at atmospheric pressures when first drained from the system. In Step 2, once the dissolved hydrogen has been allowed to release from solution, the remaining water can be introduced safely to the system water reservoir.

9.3.3.4 Water Input System

The water input subsystem's function is to add water to the system as it is consumed in the electrolysis of water to produce hydrogen and oxygen. This subsystem monitors the water level in the water reservoir.

9.3.3.5 Mounting and Packaging Cabinetry Subsystem

The structure of the hydrogen generator houses and protects system components from the ambient environment and the ambient environment from system components during normal and abnormal conditions. In addition, the subsystem provides structural support for the mounting of system components inside the unit's cabinetry.

9.3.3.6 Cabinet Ventilation System

The ventilation system within the generator cabinet provides a CG detection system to detect any hydrogen leaks in the fluids enclosure. This subsystem ventilates the fluids enclosure to dilute any CG mixture to a safe level in the event of a system upset and minimizes the release of hydrogen to the surrounding atmosphere.

9.3.3.7 Power Electronics

The hydrogen generator's power electronics system filters, controls, transforms, and switches the main AC power input to various components throughout the unit. Depending on the size of the generator, the AC input can be single-phase or three-phase power, low voltage (200–240 V) or high voltage (480–500 V), and 50 or 60 Hz. Typically, the input power is divided to provide low-voltage 24 DC power for control valves, pumps, sensors, etc., and high power for the electrolysis cell stack.

9.3.3.8 Electrolysis Cell Stack Power Supply

The electrolysis cell stack power supply converts incoming AC power to DC power at the proper voltage and current range to match the load characteristics of the electrolysis stack. The power supply subsystem provides monitoring of the stack voltage and amperage and variably controls stack hydrogen output via varying the amperage supply to the stack. The amount of amps applied to the stack directly corresponds to the volume of hydrogen (and oxygen) produced.

9.3.3.9 Controls and Instrumentation

The controls and instrumentation subsystem supplies multiple mission critical features impacting the operation of the hydrogen generator. First and foremost, this subsystem provides high-level, hard-wired safety circuitry for automatic, passive, non-software-controlled, and system shutdowns when safety critical conditions occur. It also warns the user or safely shuts down the system if instrument feedback is out of an acceptable range. Typically, the generator controls allow limited manual control of the system including remote start-up and shutdown and allow manual configuration of control parameters (e.g., number of stacks, nominal pressures, thermostat settings, etc.). The system controls also provide switched and proportional feedback of system parametric measurements or safety shutdowns, automated system control, manual monitoring, automated data acquisition, and software-controlled warnings and system shutdowns. Last, the controls allow chronological recording of instrumentation values and system status for subsequent analysis.

9.3.4 Manufacturing

Manufacturing must be considered in the design of all electrolyzer components. The traditional commercial MEA manufacturing process for electrolysis commonly involves an ink-based decal process, requiring the use of a heated platen press to attach the electrodes to the membrane. Ink solids concentration, viscosity, and binder content all need to be optimized to achieve the desired loading and distribution on the electrode. From there, the MEA is incorporated into the cell stack, under environmentally controlled conditions such that the MEAs form a sufficient seal with the rest of the stack components. Finally, the stacks are inserted into the systems and tested.

9.4 Stack and System Lifetime

Over the last 6 years, evaluation electrolyzer cell stack manufacturers have shown the designs to be robust and reliable, with demonstrated operating lifetimes well in excess of 60,000 h with analysis that predicts

even longer life. Cell stack reliability is estimated using a combination of in-house testing and customer field population data. A summary of the methods typically used to analyze reliability is presented in the following.

Because there are no established accelerated durability tests for electrolysis like there are for PEM fuel cells, long-term cell stack test beds with automated data acquisition are required. Because of the long electrolyzer lifetime, test beds, such as these, have generated over 500,000 stack hours (13 million cell hours) over nearly a decade. Multiple stacks have been operated for more than 60,000 h with 100% duty cycle.

Another component of the cell stack life projection is voltage decay rate. Voltage decay is defined as the rate of increase in the cell stack voltage per unit time. Typically, this decay rate is measured in micro-volts (μV) and averaged over all the cells in the stack. The unit of measure is then μV/cell-h. Voltage decay is important in the determination of cell stack life because in the absence of any other demonstrated failure mechanism, the increase in cell stack voltage to the limit of the power supply compliance voltage determines the life of the stack. By measuring the beginning-of-life cell voltage, knowing the compliance voltage of the power supply, and determining the voltage decay rate, one can calculate a projected cell stack life.

Figure 9.11 shows an example of a stack design that showed a 4 μV/cell-h decay rate, which translated to a projected operational life of greater than 5 years. A new stack design was tested in 2005, showing negligible voltage decay. It is now expected that an industry standard electrolysis cell stack can achieve a 10-year service life if proper feed water quality is maintained for that duration.

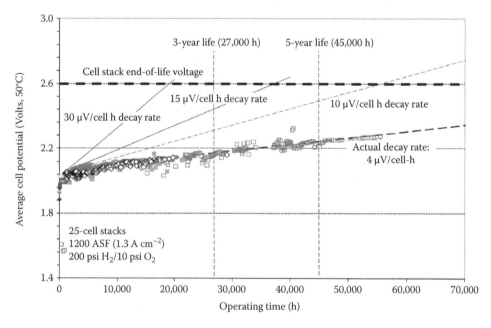

FIGURE 9.11
An example of electrolysis stack durability data at 200 psi differential pressure.

9.5 Recent Research Efforts

9.5.1 Low Noble Metal Content Electrodes for Hydrogen Production

The MEA is currently the most expensive single component of the cell. An important portion of MEA cost is related to the prices of PGMs, which are high and may rise with demand due to limited natural resources. Depending on the cost, the catalyst can typically represent about half of the MEA cost. This high percentage reflects the fact that electrolysis has lagged significantly behind fuel cells in catalyst loading reduction, for several reasons. First, the capital cost of electrolysis is already competitive in high-purity industrial hydrogen applications for commercial viability, reducing the urgency of immediate change until the energy markets are more mature. Second, there has been resistance to the use of supported catalysts in electrolysis applications, because performance limitations have been noted on the cathode at fuel cell platinum loadings, and carbon supports are not stable on the anode side of the electrolysis cell. Finally, the traditional methods for the manufacturing of electrolysis electrodes use techniques that require high loadings in order to achieve uniform catalyst distribution. More recently, however, efforts have demonstrated that significant reductions in catalyst loading are feasible.

Carbon-supported nanocatalysts, such as those developed by Brookhaven National Labs (Hsieh et al. 2013), have been made into gas diffusion electrodes (GDEs) and operated in a PEM electrolyzer. These cathode samples have demonstrated >900 h durability in production quality hardware (Figure 9.12a) while achieving a performance of <2.0 V at 1.8 A cm⁻² at 1/25th the precious metal loading (Figure 9.12b). The data are compared to a baseline made with typical loadings and catalyst materials for a commercial electrolyzer company.

Advanced manufacturing methods are also being explored by PEM electrolysis companies, using more automated coating techniques to deposit the nanocatalysts evenly onto the gas diffusion layer substrate. Using this new technique, cathodes have been manufactured at 1/100th the precious metal loading and tested in bench-scale hardware, showing nearly equivalent performance to the baseline, as seen in Figure 9.13.

Other alternative manufacturing techniques are also being explored, such as reactive spray deposition technology (RSDT) that is being used as a single-step approach to manufacture MEAs or GDEs at lower loadings and lower labor costs. Traditionally, MEAs have represented the most labor and process-intensive components of the stack. As discussed previously, the traditional commercial MEA process for electrolysis involves

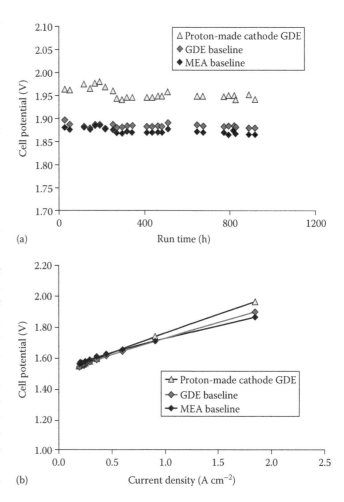

FIGURE 9.12
Ultra-low-loaded cathode shows (a) durability for 917 h with no detected voltage drift and (b) high performance nearly equivalent to baseline at 1/25th the precious metal loading.

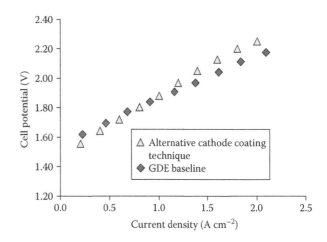

FIGURE 9.13
Performance of an alternatively manufactured cathode with at 1/100th the precious metal loading.

many steps of mixing and grinding, sintering, and pressing at elevated temperature and pressure. Other issues exist with the modern inking process such as the ink shelf life, which can be on the order of hours, and ink waste due to ink loss in the print media. RSDT eliminates most of the traditional processing by forming and subsequently directly depositing nanoparticles onto substrates (membranes or GDLs). This significantly reduces underutilized catalyst material, resulting in lower loadings. The independent control of the components allows for real-time tuning of the support, catalyst, and ionomer ratios in the final electrode. An additional benefit is that the technique can be inserted in a web-processing arrangement for continuous deposition of the electrode layers.

Figure 9.14 shows the RSDT spraying and combustion process (Roller et al. 2014). In summary, nanoparticles are formed from a supersaturated vapor in the flame. The particle formation can be controlled by temperature, oxidant/fuel rate, residence time, and the reactant composition/concentration.

The feasibility of using RSDT for electrolysis components has been evaluated by industry (Roller et al. 2014). Results showed that RSDT deposition onto both membrane and GDEs had impressive durability and performance at >70% precious metal reduction on the cathode compared to the industry standard. Depending on the processing conditions, amorphous or crystalline platinum could be deposited. The study also indicated that binder and a carbon support improved performance, likely due to the increased adhesion and distribution on the electrode.

Based on a comprehensive analysis of the manufacturing steps, measurements also showed that the new process has the potential to reduce the energy usage of the legacy process by up to 90%, based on 0.5 kW required to operate the RSDT bench-scale equipment and measurement of the MEA manufacturing equipment at an electrolyzer manufacturer. The resulting savings in manufacturing cost add to the significant savings in labor and catalyst usage.

Another approach to reducing precious metal in the electrodes is by controlling electrode structure. 3M's (Minnesota Mining and Manufacturing Company) nanostructured thin-film (NSTF) electrode structures effectively increase the available surface area of the catalyst, enabling reduced loadings by an order of magnitude. Briefly, the electrode structure is composed of an organic pigment that forms electrochemically stable, high aspect ratio whiskers. The vertically aligned whiskers are coated by a catalyst. The catalyst is a thin-film coating rather than the dispersed nanoparticles present in a typical carbon-supported catalyst. 3M has demonstrated the ability to scale this process to high volume using high-speed coating technologies.

3M has been developing the NSTF technology for close to two decades, representing tens of millions of dollars of research and development focused on fuel cell electrodes. 3M's investment into NSTF technology has produced high-performance fuel cell cathodes that have met

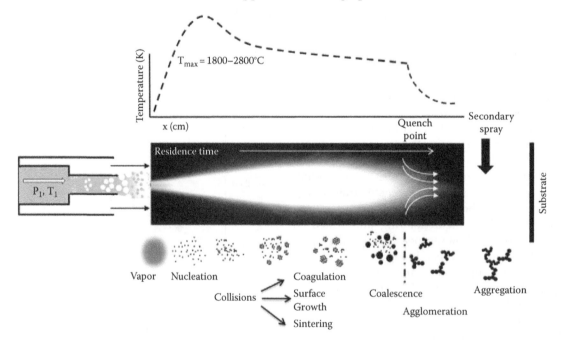

FIGURE 9.14
Temperature distribution for a precursor/solvent solution passing through the RSDT process and transformed into nanoparticles. (Reprinted from *Journal of Power Sources*, 271, Roller, J., Renner, J. et al., Flame-based processing as a practical approach for manufacturing hydrogen evolution electrodes, 366–376, Copyright (2014), with permission from Elsevier.)

or exceeded all of the U.S. Department of Energy fuel cell targets for 2015. 3M has proven 7000 h of operation with this electrode structure based on the challenging fuel cell testing protocols developed for automotive fuel cells. Because of this success, these electrode structures were investigated as electrolysis components (Debe et al. 2012) and have shown a similar durability in electrolysis operation. In fact, these structures may actually be more ideal for electrolysis operation than fuel cell mode, since the electrolysis cell is flooded at all times, and thus, water transport away from the electrode is not a concern.

These NSTF structures enable thinner membranes to be used because the uniform coatings lack large agglomerates that could penetrate the solid polymer under compression. Figure 9.15 shows example data for full NSTF catalyst-coated 3M membrane (50 µm) with 80% reduction of PGM on the cathode and 70%–90% reduction on the anode compared to an industry baseline. The resulting potential was 1.9 V at 4 A cm^{-2} operating at 50°C.

9.5.2 Advanced Membrane Materials for Hydrogen Production

PFSA, the standard membrane material for commercial electrolyzer systems, presents a number of limitations where higher pressure and temperature conditions are desired (Goni-Urtiaga et al. 2012). In electrolysis, operating at higher temperatures increases Goni-Urtiaga reaction kinetics and reduces the electrical energy required. PSFA materials have a low thermal durability limiting the operating temperatures of most units to below 60°C. Additionally, membrane thicknesses for the electrolysis cell are up to five times greater than fuel cell membranes,

due to the need to maintain mechanical stability under higher operating pressures (>200 psi differential pressure on the hydrogen side) and prevent gas crossover. As a result of the large thickness, the ionic resistance represents the largest contribution to overpotential loss at typical commercial operating currents (over 1500 mA cm^{-2}). The need for thicker membranes exacerbates the already high manufacturing costs for PFSA materials.

Alternatives to Nafion are commercially available and manufactured by companies, available such as W.L. Gore and Solvay. Also, a number of alternative polymer systems have been explored for PEMs on an academic level (Hickner et al. 2004). Particularly promising materials are made from wholly aromatic polymers. The rigid aromatic backbone contributes to lower cost and higher mechanical and thermal stability compared to conventional membranes. A variety of chemical compositions can be manufactured, allowing flexibility in design. For example, a study from Pennsylvania State University describes a process for making aromatic ionomers with perfluorinated sulfonic acid side chains. This new material demonstrated proton conductivity and fuel cell performance comparable to Nafion, while showing promising mechanical strength and thermal stability (Xu et al. 2011).

In an effort to reduce cost and increase efficiency, non-perfluorinated hydrocarbon–based materials have also been investigated. One approach includes modifying a commercial polymer backbone (Radel) with ion-conducting groups (Dyck et al. 2002). The backbone chemical structure of the sulfonated Radel polymers is shown in Figure 9.16a. The number of sulfonate groups per monomer unit can be varied to tune the water uptake, ion conductivity, gas permeability, and other transport

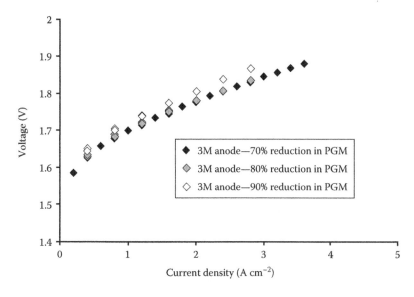

FIGURE 9.15
Performance of MEAs with multiple anode laminations representing a 70%–90% reduction in platinum group metals (PGMs).

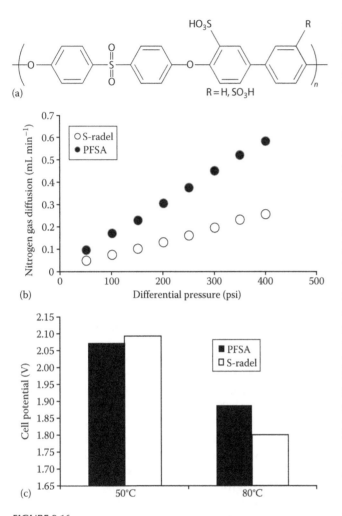

(a)

(b)

(c)

FIGURE 9.16
Hydrocarbon membranes show promise for PEM electrolysis applications. (a) Chemical structure of sulfonated Radel polymers (S-Radel). (b) Hydrocarbon membrane material (S-Radel) has lower gas permeability than PFSA baseline at 70°C despite being ~3 mils thinner. (c) Hydrocarbon membrane material (S-Radel) has greater performance benefit at higher temperatures than PFSA baseline material at 1.8 A cm^{-2}.

properties of the membrane. When these materials have been tested for electrolysis applications, they show superior gas permeability properties (Figure 9.16b) and electrochemical performance (Figure 9.16c) compared to PFSA baseline materials.

Another strategy is to reinforce membranes for better mechanical strength and gas crossover properties. GORE-SELECT® membranes, as an example, use a microporous polytetrafluoroethylene (PTFE) to reinforce standard PEM materials, through W. L. Gore's patented composite technology. Industry has investigated reinforced PFSA in an effort to obtain thinner membranes. Generally, failures will occur under the cell stack loads at thickness of 5 mils and below for PFSA materials. This problem is accentuated at higher temperatures.

The combination of the hydrocarbon-based PEM membrane and the microporous reinforcement is expected to result in stronger, thinner, and less expensive materials compared to their monolithic PFSA counterparts. Figure 9.17 shows polarization data comparing a PFSA baseline to a reinforced hydrocarbon-based sample at 80°C. The approach enabled thinner materials and higher voltage efficiency. These hydrocarbon samples operated in industry standard electrolysis hardware for >500 h without failure or measurable voltage decay.

9.5.3 Alkaline Exchange Membrane (AEM)-Based Systems

Over the past decade, it has been realized that anion exchange membranes (AEMs) can be used as a solid-state electrolyte, enabling AEM fuel cells and other devices (Varcoe and Slade 2005; Merle et al. 2011b). The key difference is a shift to a basic environment instead of the acidic environment provided by the PEM.

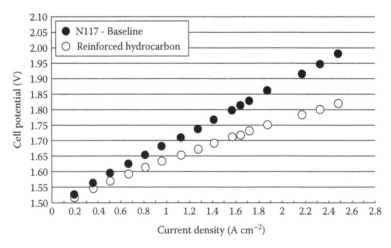

FIGURE 9.17
Comparison of PFSA versus hydrocarbon material at 80°C and 400 psi differential pressure showing increased efficiency.

This combines the advantages of solid polymer electrolysis with the advantages of the liquid KOH systems (see Section 9.2.2). AEMs enable (1) low-cost materials of construction, (2) the utilization of a wider array of low-cost catalysts, and (3) potentially thinner membranes due to increased stiffness and superior gas permeation properties (Ayers et al. 2013). AEM materials often will have lower gas permeation rates than PEM materials of similar thickness. Figure 9.18 shows permeation testing conducted with PEM (Nafion) and AEM materials at 40°C. A reduction of nearly 80% in permeation is seen when using the experimental AEM. Proton OnSite has also conducted permeability studies of commercial AEM material out to 2000 psi without failure. This result provides evidence that AEMs are mechanically robust enough to withstand the differential pressure operation, a key component to qualifying this emerging technology. While the conductivity will be intrinsically lower in AEM materials due to the larger hydroxide ion, it is possible that because of these distinct cost advantages and the capacity for thin materials, efficiency penalties will be mitigated.

Lower capital cost has a distinct advantage when electricity is of low cost, such as in a renewable storage application. There are already wind energy resources in Europe where 20%–40% of the renewable capacity is stranded due to lack of demand, which drives negative electricity pricing. In addition, there is added value from having the electrolyzer system integrated with the grid, such as frequency regulation and grid balancing from more traditional fossil fuel–based sources (see Section 9.2.4). Therefore, many companies and governments are interested in AEM development.

Proton OnSite has demonstrated that the efficiency of the AEM cells falls between the PEM and liquid alkaline systems (Figure 9.19) confirming the advantage of the membrane-based systems in general. However, durability has been an ongoing issue with AEM-based systems compared to PEM, and a variety of strategies are being explored to stabilize the materials (Hickner et al. 2013). Many AEMs, backbone chemistries have been explored; however, polysulfone-based chemistries have provided some insight into the degradation mechanisms possible in the AEM basic environment (Hickner et al. 2004; Arges et al. 2012; Parrondo et al. 2014). Overall, these studies suggest that carbon dioxide can act as a membrane poison in some cases and that the basic environment plays a role in polymer degradation. The mechanisms typically accepted to explain AEM degradation are Hoffman elimination and direct nucleophilic substitutions (Chempath et al. 2010; Merle et al. 2011a). Additionally, it has been proposed that in a basic environment the quaternary ammonium cations may trigger backbone hydrolysis of polysulfone and PPO-based AEMs because the cations are in close proximity to the aromatic rings (Arges and Ramani 2013; Arges et al. 2013). These issues combined with materials and processing challenges have led to the limited commercialization of AEM technology (Arges et al. 2010) except by a few companies, such as FuMA-Tech, and the Tokuyama Corporation.

Currently, there are no viable alternatives to noble metal-based catalysts in PEM-based electrolysis. While still an active area of research, nonnoble metal alternatives are showing promise for AEM-based systems. While more of the focus has been on AEM-based fuel cells, electrolysis-specific studies have also been conducted with promising results, particularly on the cathode. Figure 9.20 shows that the anode is also making progress, with a nonnoble metal anode performing similarly to the precious metal baseline (Renner and Ayers

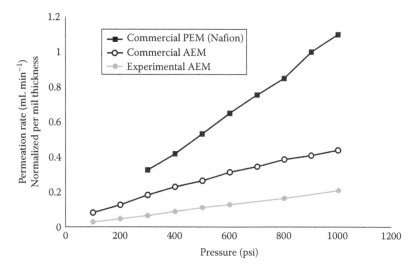

FIGURE 9.18
Comparison of N_2 gas permeation between PEM and AEM materials shows superior AEM performance.

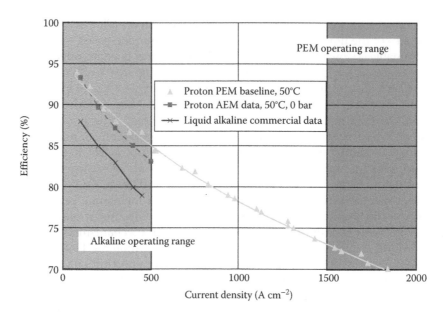

FIGURE 9.19
Comparison of liquid KOH–based electrolysis, PEM electrolysis, and AEM electrolysis.

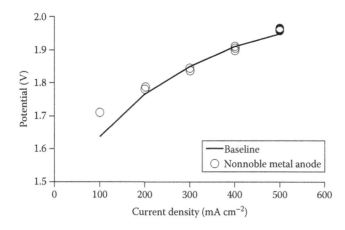

FIGURE 9.20
A nonnoble metal anode has similar performance to the precious metal baseline in an AEM system.

2014). It is anticipated that these catalyst and membrane development efforts will advance AEM technology and eventually lead to a step change in electrolysis cost.

9.6 Summary

Producing oxygen and hydrogen using PEM electrolysis has a long history. Like many technologies, PEM electrolysis can trace its roots back to the U.S. space program. With its reliability proven in space, PEM electrolysis next found utility in other aerospace life support applications, such as aboard nuclear submarines.

Success in these high-cost speciality markets did not necessarily translate to acceptance in broader on-site gas generation applications without a focus on manufacturability and cost reduction. Companies, such as Proton OnSite, transitioned PEM electrolysis into the manufacturing environment where serial production and process optimization reduced material and labor costs to meet an acceptable price point in the marketplace. Over the last decade, PEM electrolysis has established itself as a viable and cost-effective method to produce hydrogen for a diverse number of existing and emerging applications. Major industries adopting PEM electrolysis include electric generator cooling, laboratories, and semiconductor processing.

With the ever-increasing addition of wind and solar renewable energy to the traditional electric grid, the need for energy storage has also grown. Hydrogen from PEM electrolysis is a promising technology for renewable energy capture as it has the capability to store massive amounts of energy in a relatively small volume. In addition, PEM electrolysis can also provide ancillary services to the grid, such as frequency regulation, and load shifting resulting in multiple value streams. The hydrogen produced can alternatively be used as a transportation fuel, injected into the natural gas pipeline (thus making that energy carrier more green), in the production of high-value chemicals, such as ammonia, or in upgrading of methanization-produced biogas.

In order to be competitive in these emerging markets, efficiency improvements, and additional cost reductions in materials and processing are needed. Ongoing research and development efforts in areas, such as reduced precious metal catalyst loadings, thinner, lower-cost polymer exchange membranes, and inexpensive conductive coatings—all show promise to address these needs. Novel cell design approaches that promote the use of composite materials to reduce the amount of semi-precious metals and eliminate labor are equally exciting. Longer-term, new ion exchange membrane materials based on hydroxide exchange instead of proton exchange could provide an order-of-magnitude decrease in cost by totally eliminating the need for precious metals in the electrolyzer cell stack.

In summary, PEM electrolysis has unique characteristics that make it very attractive in the hydrogen markets of today and the energy markets of tomorrow. The need for large-scale renewable energy storage and reduced emission transportation continues to grow and hydrogen can be the enabling linkage that provides the solution to both of these needs. But in order to ensure its success, continued support of research and development is needed to further drive down the capital cost and reduce the operating cost of the PEM technology.

References

Anderson, E. B., L. M. Moulthrop et al. 2012. Hydrogen infrastructure challenges and solutions. *ECS Transactions* 41 (46):75–83.

Arges, C. G., J. Parrondo et al. 2012. Assessing the influence of different cation chemistries on ionic conductivity and alkaline stability of anion exchange membranes. *Journal of Materials Chemistry* 22 (9):3733–3744.

Arges, C. G. and V. Ramani. 2013. Two-dimensional NMR spectroscopy reveals cation-triggered backbone degradation in polysulfone-based anion exchange membranes. *Proceedings of the National Academy of Sciences of the United States of America* 110 (7):2490–2495, S90/1–S90/19.

Arges, C. G., V. Ramani et al. 2010. Anion exchange membrane fuel cells. *The Electrochemical Society Interface* 19:31–35.

Arges, C. G., L. Wang et al. 2013. Best practices for investigating anion exchange membrane suitability for alkaline electrochemical devices: Case study using quaternary ammonium poly(2,6-dimethyl 1,4-phenylene)oxide anion exchange membranes. *Journal of the Electrochemical Society* 160 (11):F1258–F1274.

Ayers, K. E., E. B. Anderson et al. 2010. Research advances towards low cost, high efficiency PEM electrolysis. *ECS Transactions* 33 (1):3–15.

Ayers, K. E., E. B. Anderson et al. 2013. Characterization of anion exchange membrane technology for low cost electrolysis. In *Fuel Cell Membranes, Electrode Binders, and MEA Performance*, edited by P. Pintauro, pp. 121–130. Pennington: Electrochemical Soc Inc.

Ayers, K. E., L. T. Dalton et al. 2012. Efficient generation of high energy density fuel from water. *ECS Transactions* 41 (33):27–38.

Bose, S., T. Kuila et al. 2011. Polymer membranes for high temperature proton exchange membrane fuel cell: Recent advances and challenges. *Progress in Polymer Science* 36 (6):813–843.

Chempath, S., B. R. Einsla et al. 2010. Density functional theory study of degradation of tetraalkylammonium hydroxides. *Journal of Physical Chemistry C* 114:11977.

Choi, P. H., D. G. Bessarabov et al. 2004. A simple model for solid polymer electrolyte (SPE) water electrolysis. *Solid State Ionics* 175 (1–4):535–539.

Debe, M. K., S. M. Hendricks et al. 2012. Initial performance and durability of ultra-low loaded NSTF electrodes for PEM electrolyzers. *Journal of the Electrochemical Society* 159 (6):K165–K176.

Dyck, A., D. Fritsch et al. 2002. Proton-conductive membranes of sulfonated polyphenylsulfone. *Journal of Applied Polymer Science* 86 (11):2820–2827.

Goni-Urtiaga, A., D. Presvytes et al. 2012. Solid acids as electrolyte materials for proton exchange membrane (PEM) electrolysis: Review. *International Journal of Hydrogen Energy* 37 (4):3358–3372.

Hickner, M. A., H. Ghassemi et al. 2004. Alternative polymer systems for proton exchange membranes (PEMs). *Chemical Reviews* 104 (10):4587–4611.

Hickner, M. A., A. M. Herring et al. 2013. Anion exchange membranes: Current status and moving forward. *Journal of Polymer Science Part B—Polymer Physics* 51 (24):1727–1735.

Hsieh, Y. C., Y. Zhang et al. 2013. Ordered bilayer ruthenium–platinum core–shell nanoparticles as carbon monoxide-tolerant fuel cell catalysts. *Nature Communications* 4:9.

Luo, G. and I. Angelidaki. 2012. Integrated biogas upgrading and hydrogen utilization in an anaerobic reactor containing enriched hydrogenotrophic methanogenic culture. *Biotechnology and Bioengineering* 109 (11):2729–2736.

MarketsandMarkets. 2011. Hydrogen generation market – by merchant & captive type, distributed & centralized generation, application & technology – trends & global forecasts (2011–2016). Report Code: EP 1708. Available at: http://www.marketsandmarkets.com.

Merle, G., M. Wessling et al. 2011a. Anion exchange membranes for alkaline fuel cells: A review. *Journal of Membrane Science* 377:1.

Merle, G., M. Wessling et al. 2011b. Anion exchange membranes for alkaline fuel cells: A review. *Journal of Membrane Science* 377 (1–2):1–35.

Mohseni, F., M. Magnusson et al. 2012. Biogas from renewable electricity – increasing a climate neutral fuel supply. *Applied Energy* 90 (1):11–16.

Parrondo, J., C. G. Arges et al. 2014. Degradation of anion exchange membranes used for hydrogen production by ultrapure water electrolysis. *RSC Advances* 4 (19):9875–9879.

Renner, J. N. and K. E. Ayers. 2014. Exploring electrochemical technology: A perspective on the aASEE/NSF Small Business Postdoctoral Research Diversity Fellowship. *Proceedings of 2014 Zone 1 Conference of the American Society for Engineering Education (ASEE Zone 1)*, Bridgport, CT 3-5 April, 2014.

Roller, J., J. Renner et al. 2014. Flame-based processing as a practical approach for manufacturing hydrogen evolution electrodes. *Journal of Power Sources* 271:366–376.

Schiller, M. 2013. Hydrogen energy storage: The holy grail for renewable energy grid integration. *Fuel Cells Bulletin* 2013 (9):12–15.

Shen, M. Z., N. Bennett et al. 2011. A concise model for evaluating water electrolysis. *International Journal of Hydrogen Energy* 36 (22):14335–14341.

van der Merwe, J., K. Uren et al. 2014. Characterisation tools development for PEM electrolysers. *International Journal of Hydrogen Energy* 39 (26):14212–14221.

Varcoe, J. R. and R. C. T. Slade. 2005. Prospects for alkaline anion-exchange membranes in low temperature fuel cells. *Fuel Cells* 5 (2):187–200.

Xu, K., H. Oh et al. 2011. Highly conductive aromatic ionomers with perfluorosulfonic acid side chains for elevated temperature fuel cells. *Macromolecules* 44 (12):4605–4609.

10

Characterization Tools for Polymer Electrolyte
Membrane (PEM) Water Electrolyzers

Pierre Millet

CONTENTS

Nomenclature

a	activity
A	membrane area (m^2)
C	electrical capacitance ($C\ V^{-1}$)
D_i	diffusion coefficient of species i ($m^2\ s^{-1}$)
E	thermodynamic voltage (V)
F	Faraday's constant (96,485 $C\ mol^{-1}$)
f	gas fugacity
G	Gibbs free energy ($J\ mol^{-1}$)
H	enthalpy ($J\ mol^{-1}$)
I	current (A)
j	current density ($A\ cm^{-2}$)
j_0	exchange current density ($A\ cm^{-2}$)
n	number of electrons exchanged in a chemical reaction
P	pressure (Pa)
P^m	membrane permselectivity ($m^2\ Pa^{-1}\ s^{-1}$)
Q	constant-phase element
R	resistance (Ω)
R_{PG}	constant of perfect gas (0.082 $J\ K^{-1}\ mol^{-1}$)
r_f	roughness factor
S	entropy ($J\ mol^{-1}\ K^{-1}$)
t	time (s)
T	absolute temperature (K)
U_{cell}	cell voltage (V)
V_{cell}	thermo-neutral electrolysis voltage (V)
W	electrical work (J)
x,y,z	space coordinates
Z	impedance (Ω)

Greek Symbols

α	charge transfer coefficients
δ	membrane thickness (m)
Δ	difference
ε	cell efficiency (%)
η	overvoltage (V)
λ	thermal conductivity ($W\ m^{-1}\ K^{-1}$)
ω	pulsation ($rad\ s^{-1}$)
φ	electrical potential (V)
ρ	electrical resistivity ($\Omega\ m$)
σ	electrical conductivity ($S\ m^{-1}$)

Subscripts/Superscripts

°	standard conditions (298 K, 1 bar)
cell	electrolysis cell
d	dissociation
el	electrolyte
g	gas
i	species i
l	liquid
r	real
t	theoretical

Abbreviations/Acronyms

Capex	capital expenses or expenditures
CER	contact electrical resistance
CPE	constant-phase element
CV	cyclic voltammetry
EIS	electrochemical impedance spectroscopy
EW	equivalent weight of membrane ($eq\ g^{-1}$)
EXAFS	Extended X-ray absorption fine structure
HER	hydrogen evolution reaction
MEA	membrane electrode assembly

OER   oxygen evolution reaction
Opex   operational expenses or expenditures
PCD   porous current distributors
PEM   polymer electrolyte membrane or proton exchange membrane
PFSA   perfluorosulfonic acid
PFSI   perfluorosulfonic ionomer
PGM   platinum group metal
RDE   rotating disk electrode
rds   rate-determining step
SEM   scanning electron microscopy
SPE   solid polymer electrolyte
TEM   transmission electron microscopy
TPER   through-plate electrical resistance
XPS   X-ray photoelectron spectrometry
XANES   X-ray absorption near-edge structure

10.1 Introduction

Whereas more than 95% of the hydrogen produced in the world comes from natural hydrocarbons, hydrogen of electrolytic grade produced by water electrolysis is receiving increasing attention because it could be used as a universal and carbon-free energy carrier. Energy applications of hydrogen (as a fuel for the automotive industry or as a means for the large-scale storage of renewable energy sources) are emerging markets. The situation is evolving rapidly and business opportunities are appearing. Among a few technologies of appropriate maturity, the so-called PEM water electrolysis is considered as a very promising one because of its high level of performance and because significant cost reductions are possible. Whereas the capacity of existing commercial products is still limited (approximately to <100 Nm^3 H$_2$ h^{-1}), manufacturers are investing to develop production units at the MW scale (~200 Nm^3 H$_2$ h^{-1}). This requires the use of larger cell areas (from ~1,000 cm^2 with current state-of-art technology to 10,000 cm^2). However, their development is not a straightforward task and can potentially be the source of significant performance losses.

The purpose of this chapter is to review the main characterization tools and methodologies that can be used to evaluate the performances of PEM water electrolysis cells (analysis of performance degradation will be provided in Chapter 11). This will be very useful for those working in R&D departments who wish to improve the efficiency and durability of PEM water electrolyzers and also for the manufacturers of PEM water electrolyzers who wish to implement quality control procedures on manufacturing lines (at component, cell, or stack levels) or on-site process monitoring–diagnostics in view of anticipated maintenance operations. Most of the techniques discussed in this chapter are electrochemical ones because charge transport and transfer are the most important properties in these systems. These techniques provide in situ information about the status and performance of the different internal cell components. Some other techniques are also presented and discussed.

10.2 PEM Water Electrolysis Unit Cell

10.2.1 Cell Geometry and the Main Cell Components

Basically, a PEM water electrolysis cell is a device that is used to perform the endergonic (non-spontaneous, electricity-driven) dissociation of liquid water molecules into their gaseous elemental components, molecular hydrogen, and oxygen.

$$H_2O(l) \rightarrow H_2(g) + \tfrac{1}{2}O_2(g) \tag{10.1}$$

A PEM water electrolysis cell uses an acidic polymer electrolyte with protonic mobility. The two half-cell reactions are the following:

$$\text{Anode:} \quad H_2O(l) \rightarrow \tfrac{1}{2}O_2(g) + 2H^+ + 2e^- \tag{10.2}$$

$$\text{Cathode:} \quad 2H^+ + 2e^- \rightarrow H_2(g) \tag{10.3}$$

Equation 10.2 is the hydrogen evolution reaction (HER), and Equation 10.3 is the oxygen evolution reaction (OER). The minimum amount of electricity (nFE) required to split one mole of water is equal to the Gibbs free energy change (ΔG_d).

$$\Delta G_d - nFE = 0 \quad \text{where } \Delta G_d > 0 \tag{10.4}$$

where
   n = 2 (number of electrons exchanged during the electrochemical splitting of water)
   F = ~96 485 C mol^{-1} (Faraday)
   E = thermodynamic voltage (V) associated with the reaction 10.1
   ΔG_d = free energy change in J mol^{-1} associated with the reaction 10.3.

The Gibbs free energy change ΔG_d is a function of two main state variables, operating temperature and pressure, thus

$$\Delta G_d(T,P) = \Delta H_d(T,P) - T\Delta S_d(T,P) > 0 \tag{10.5}$$

In Equation 10.5, ΔH_d (T,P) and ΔS_d (T,P) are the enthalpy change (J mol^{-1}) and entropy change (J mol^{-1} K^{-1})

associated with reaction (1), respectively. To split one mole of water, ΔG_d (J mol^{-1}) of electricity and T ΔS_d (J mol^{-1}) of heat are required from the surroundings.

The thermodynamic electrolysis voltage E in volt is defined as

$$E(T,P) = \frac{\Delta G_d(T,P)}{nF} \qquad (10.6)$$

The thermo-neutral voltage V in volt is defined as

$$V(T,P) = \frac{\Delta H(T,P)}{nF} \qquad (10.7)$$

Under standard conditions of temperature and pressure (T = 298 K, P = 1 atm), water is liquid, and H$_2$ and O$_2$ are gaseous. Standard free energy, enthalpy, and entropy changes for reaction (10.1) are the following:

$$\Delta G_d°(H_2O) = 237.22 \text{ kJ mol}^{-1} \Rightarrow E°$$
$$= \Delta G_d°(H_2O) / 2F = 1.2293 \text{ V} \sim 1.23 \text{ V}$$
$$\Delta H_d°(H_2O) = 285.840 \text{ kJ mol}^{-1} \Rightarrow V°$$
$$= \Delta H_d°(H_2O) / 2F = 1.4813 \text{ V} \sim 1.48 \text{ V}$$
$$\Delta S_d°(H_2O) = 163.15 \text{ J mol}^{-1} \text{ K}^{-1}$$

A total amount of $\Delta H_d°(H_2O)$ kJ of energy is required to electrolyze one mole of water under standard conditions. This energy is provided in the form of electricity ($\Delta G_d°$ kJ) and heat (T $\Delta S_d°$ kJ). In other words, a cell voltage E = $\Delta G_d°/(2F)$ = 1.23 V is required to electrolyze water. An additional voltage term T. $\Delta S_d°/(2F)$ = 0.25 V must be added to the thermodynamic voltage E to provide the heat required by reaction (10.1). In practice, heat is produced when current flows across the electrolysis cell and so the T.DS° kJ of heat is also provided by electricity. The excess heat generated by nonequilibrium processes is transferred to the surroundings due to the temperature gradient. The efficiency ε of the electrolysis cell relates the theoretical amount of energy W_t required to split one mole of reactant to the real amount of energy W_r required by the process. Because of the earlier-mentioned irreversibilities, $W_r > W_t$. The cell efficiency is defined as

$$\varepsilon = \frac{W_t}{W_r} \qquad (10.8)$$

where

$W_r = (U_{cell} \cdot I \cdot t)$, U_{cell} is the actual cell voltage in V, I is current in A, and t is the duration in s

W_t can be defined from the thermodynamic voltage E°: $W_{t,\Delta G} = (E I t)$

W_t can also be defined from the thermo-neutral voltage V°: $W_{t,\Delta H} = (V I t)$

Therefore, two different definitions can be used to express the efficiency of the electrolysis cell. Since E and V are both functions of operating temperature (T) and operating pressure (P), and since U_{cell} is also a function of the operating current density j, the two different cell efficiencies can also be expressed as a function of T, P, j

$$\varepsilon_{\Delta G}(T,P,j) = \frac{E(T,P)}{U_{cell}(T,P,j)} \qquad \varepsilon_{\Delta H}(T,P,j) = \frac{V(T,P)}{U_{cell}(T,P,j)}$$
$$(10.9)$$

At low current densities, cell efficiencies close to 100% are obtained. In conventional water electrolyzers, $\varepsilon_{\Delta H}$ ~70% at 1 A cm^{-2}, T = 90°C, and P = 1 bar. The efficiency of the electrolysis cell is a critical parameter responsible for the energy cost of the process. Operation at high current density is necessary to reduce investment costs but, since efficiency decreases when current density increases, a compromise has to be found between energy and investment costs. The water-splitting reaction (10.1) can be achieved by supplying either thermal energy (thermal or thermo-chemical water splitting) or electricity (water electrolysis). Different technologies are available. The PEM water electrolysis cell is used to electrolyze liquid water at nearly ambient conditions, using an acidic electrolyte (hydrated protons are the ionic charge carriers).

The cross section of a PEM water electrolysis unit cell equipped with conventional cell components (thicknesses not at scale) is pictured in Figure 10.1 (Rozain and Millet 2014).

This unit cell is delimited by two titanium end plates (5 and 5′). Its thickness is usually within the range 4–7 mm, depending on the selection of internal components. The electrochemically active component where water is split into elemental hydrogen and oxygen is the central membrane electrode assembly (MEA). The MEA is made of a proton-conducting membrane at the center (1), surface coated with two porous catalytic layers: the cathodic layer (2) in the cathodic compartment and the anodic layer (2′) in the anodic compartment. The membrane is usually a poly(tetrafluoroethylene)-(PTFE-)reinforced solid polymer electrolyte (SPE). This PTFE reinforcement is used for mechanical purposes, especially during operation at elevated pressure. The cathodic layer is a porous, micrometer-thick 3D structure composed of carbon-supported platinum nanoparticles embedded in perfluorosulfonic acid (PFSA) ionomer chains. The anodic layer is porous 3D structure, a few-micrometers-thick, composed of iridium dioxide nanoparticles also embedded in PFSA ionomer chains. This MEA is clamped between a carbonaceous cathodic current distributor (3) and a millimeter-thick

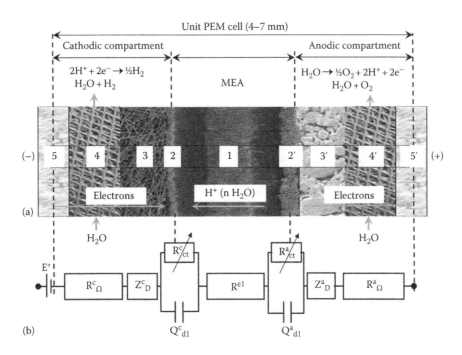

FIGURE 10.1
(a) Cross section of a PEM water electrolysis cell (thicknesses not to scale); 1, PTFE-reinforced SPE; 2, cathodic catalytic layer; 2′, anodic catalytic layer; 3, cathodic current distributor; 3′ anodic current distributor; 4, cathodic spacer; 4′ anodic spacer; 5, cathodic bipolar plate; 5′, anodic bipolar plate. (b) Associated equivalent electrical circuit (see Section 10.2.3 for more details). (Adapted from Rozain, C. and P. Millet., *Electrochim. Acta*, 131, 160, 2014.)

anodic porous titanium current distributor (3′). Two millimeter-thick titanium grid spacers (4 and 4′) through which liquid water is pumped are placed close to the end plates (5 and 5′). Alternatively, bipolar plates with channels can also be used. A cell voltage >E° (the thermodynamic electrolysis voltage) is required. During operation, gas bubbles formed at the catalytic layers diffuse through the porous current distributors (3, 3′) and are collected and transported by liquid water pumped through the grids (4, 4′).

10.2.2 Main Features of PEM Water Electrolysis Cells

A PEM water electrolyzer system has some specific features that are due to the use of a SPE.

- A PEM water electrolysis cell is a compact "zero-gap" cell that uses porous electrodes through which the gas production (hydrogen and oxygen) is evacuated; because gaseous reaction products are released at the rear of the cell, the gas production does not introduce any parasite ohmic resistance and elevated current densities can be achieved.
- The ionic charge carriers (hydrated protons) remain trapped inside the membrane of homogeneous thickness; there is no risk of leakage.

- The SPE used in a PEM water electrolysis cell offers high proton conductivity and low gas permeability, this is why it is possible to use sub-millimeter-thick membranes (typically in the range 100–200 μm). This is a desirable characteristic of a PEM water electrolysis cell. The distance between the anodic and cathodic active layer is small and the ohmic voltage drop across the SPE is minimized.

- The electrolyte is a solid polymer and, as a result, conventional reference electrodes (e.g., saturated calomel reference electrodes) cannot be used to separately investigate the electrochemical features of each interface. The situation is even worse in PEM stacks. The problem can be circumvented by using so-called internal reference electrodes. More details are provided in the following section.

- Contact resistances between internal cell components need to be minimized; this imposes significant constraints in terms of manufacturing tolerances on the different cell components, and then contributes to more expensive cells, compared to alkaline technology. Furthermore, the surface state of cell components is critical; the only way to reduce contact resistances is to apply significant compression forces.

10.2.3 Electrical Analogies

From a thermodynamic viewpoint, the unit PEM water electrolysis cell is a device that is used to convert DC electrical energy into chemical energy. During water electrolysis operation, a stationary DC current is flowing across the two end plates. From each end plate up to each catalytic layer, the electrons are the charge carriers. Within the MEA, the charge carriers are the hydrated protons. Electrochemical processes are conveniently modeled using electrical analogies. Figure 10.1b shows the equivalent electrical circuit of the unit PEM water electrolysis cell of Figure 10.1a.

Individual cell components are modeled by the series or parallel connection of electrical elements. The different circuit components are the following:

R_Ω^c and R_Ω^a (Ω cm^2): electronic resistance of electron-conducting metallic cell components in the cathodic and anodic cell compartments, respectively

R^{el} (Ω cm^2): ionic resistance of the SPE

R_{ct}^c (Ω cm^2): cathodic charge transfer (polarization) resistance associated with the HER; R_{ct}^c decreases as the electric potential of the cathode is lowered during operation

R_{ct}^a (Ω cm^2): anodic charge transfer (polarization) resistance associated with the OER; R_{ct}^a decreases as the electric potential of the anode is raised during operation

Q_{dl}^c (F cm^{-2}): double-layer capacitance associated with the cathode/electrolyte interface

Q_{dl}^a (F cm^{-2}): double-layer capacitance associated with the anode/electrolyte interface

Z_D^c (Ω cm^2): cathodic diffusion impedance due to H$_2$ transport away from the cathode

Z_D^a (Ω cm^2): anodic diffusion impedance due to O$_2$ transport away from the anode and/or to H$_2$O transport to the anode

In this electrical analogy, diffusion impedances (Z_D^c and Z_D^a) are added in series to account for possible mass transport limitations of hydrogen and oxygen gases away from the interfaces across porous current distributors. However, when current distributors of appropriate open porosity are used, such diffusion impedances can be neglected.

An efficient PEM water electrolysis cell is a cell in which internal cell resistances are minimized and maintained at low values for a long period of operation (ideally in the upper range of the 10^4–10^5 h time interval). The role of the electrochemist is, therefore, to minimize these cell resistances in order to maximize cell efficiency (the cell efficiency determines the so-called "operational expenses," or opex, that is, the energy cost of the process expressed in €/kg H$_2$) and durability (the cell durability determines the so-called "capital expenses," [capex], that is, the investment cost for the process expressed in €/kg H$_2$). This can be achieved by selecting appropriate materials and by judicious design of the unit PEM water electrolysis cell.

Electrochemical characterization tools such as those described in this chapter are used to measure, in situ, the different terms that appear in the electrical analogy. Performance degradation of unit PEM water electrolysis cells is always due to a modification of one or several of the resistance–capacitance terms. It is, therefore, necessary to implement methodologies that will facilitate the measurement of electrical resistances of individual cell components (bulk electronic resistances, contact electronic resistances, charge transfer resistances, or bulk ionic resistance of the SPE). The purpose of this chapter is to describe existing characterization tools and methodologies that can be used for the ex situ and in situ measurement of these resistances and to discuss their physical meanings.

10.2.3.1 Electronic Resistance

The objective of the electrochemical engineer is to maximize the energy conversion efficiency of the unit PEM water electrolysis cell or, in other words, to minimize dissipation sources that transform electrical energy into useless heat. This can be achieved by appropriate selection of materials and judicious cell design. Materials used for the manufacturing of non-MEA cell components (bipolar plates, spacers, and current distributor–gas collectors) must be selected by considering two main physical properties: electronic conductivity (bulk and contact resistances must be as low as possible) and corrosion resistance. Contact resistance must be minimized by manufacturing cell components with appropriate shape and surface states (extreme dimensional tolerances, excellent planarity, and appropriate surface roughness). Corrosion resistance properties are also critical. The problem is that these properties tend to deteriorate with time (the aging process), whereas performance durability is the key to the commercial success. Appropriate characterization tools and methodologies are, therefore, required to measure these different properties.

In the electrical analogy of Figure 10.1b, R^c and R^a are the global electronic resistances of metallic cell components, respectively, in the cathodic and anodic cell compartments. A closer look at the situation reveals that each resistance is the sum of individual bulk and inter-face (contact) resistances

$$R_{\Omega}^{c} = \left(R_{bulk}^{bp}\right)_{cath} + \left(R_{contact}^{bp\text{-}gr}\right)_{cath} + \left(R_{bulk}^{gr}\right)_{cath} + \left(R_{contact}^{gr\text{-}cc}\right)_{cath}$$
$$+ \left(R_{bulk}^{cc}\right)_{cath} + \left(R_{contact}^{cc\text{-}CL}\right)_{cath} + \left(R_{bulk}^{CL}\right)_{cath} \quad (10.10)$$

$$R_{\Omega}^{a} = \left(R_{bulk}^{bp}\right)_{anode} + \left(R_{contact}^{bp\text{-}gr}\right)_{anode} + \left(R_{bulk}^{gr}\right)_{anode}$$
$$+ \left(R_{contact}^{gr\text{-}cc}\right)_{anode} + \left(R_{bulk}^{cc}\right)_{anode} + \left(R_{contact}^{cc\text{-}CL}\right)_{anode} + \left(R_{bulk}^{CL}\right)_{anode}$$
$$(10.11)$$

where

R_{bulk}^{bp} (Ω cm^2) is the bulk resistance of the titanium bipolar plate (bp)

$R_{contact}^{bp\text{-}gr}$ (Ω cm^2) is the contact resistance between the bp and grid spacer (gr)

R_{bulk}^{gr} (Ω cm^2) is the bulk resistance of the titanium grid

$R_{contact}^{gr\text{-}cd}$ (Ω cm^2) is the contact resistance between the grid (gr) and the current distributor (cd)

R_{bulk}^{cd} (Ω cm^2) is the bulk resistance of the porous titanium current distributor

$R_{contact}^{cd\text{-}CL}$ (Ω cm^2) is the contact resistance between the cd and the catalytic layer (CL)

R_{bulk}^{CL} (Ω cm^2) is the bulk resistance of the catalyst layer

Usually, these different ohmic resistances are mea-sured ex situ. Within the cell, it is not an easy task to measure them individually and, generally, only their sum (non-MEA electronic cell resistance) can be easily measured, for example, by electrochemical impedance spectroscopy (EIS).

10.2.3.2 Ionic Resistance

In the PEM water electrolysis cell, there are two main ionic resistances: (1) the ionic resistance of the SPE mem-brane and (2) the ionic resistance of the catalytic layers. From the microscopic viewpoint, hydrated protons are the sole charge carriers because sulfonate end groups (covalently bonded to the fluorocarbon backbone) are not mobile species under DC conditions. They differ only by the shape of the diffusion path (which can be more tortuous in the catalytic layers) and by the con-centration (which is less in the catalytic layers). Ionic transport within electrolytes is a purely resistive pro-cess. The nonequilibrium flux–force relationship that relates the voltage force to the ionic current flux fol-lows Ohm's law (Hamann et al. 1998). Although the structure of PFSA materials at the nanoscale is that of a cluster network of charge-conducting hydrophilic

domains embedded in a hydrophobic and nonconduct-ing organic domain (see Section 10.3.1), the SPE can be considered as a homogeneous membrane with isotropic conduction properties.

Therefore, from the electrical viewpoint, the imped-ance of the SPE membrane used in PEM water electrol-ysis cells (cell component 1 in Figure 10.1) is a purely resistive resistance R_{el} (in Ω) that can be normalized to the geometrical surface area (in Ω cm^2). Although the two catalytic layers placed on each side of the SPE mem-brane are thin (in the micrometer range), their ionic resistances are difficult to measure separately. Ex situ accurate measurements show that their ionic resistivity can be much higher than that of the bulk SPE. They both contribute to an increase in the overall ionic resistance of the cell, sometimes by 10%–20%, depending on their structure.

10.2.3.3 Interface Impedance

Electrochemical metal–electrolyte interfaces are usually described by the parallel connection of a charge transfer resistance (R_{ct} in Ω cm^2) and a double-layer capacitance C_{dl} (in F cm^{-2}) (Lvovich 2012).

Roughness factors A key physical characteristic of a metal surface immersed in an electrolyte solution (or in con-tact with a PEM) is its surface state. The surface state is not only characterized by the oxidation level of sur-face atomic layers but also (and importantly) character-ized by its roughness: a geometrical factor r_f expresses the extent to which the surface differs from a perfectly smooth surface. The factor r_f is a dimensionless quan-tity, which is defined as the ratio of the real surface area (A^{real} in cm^2) to the geometrical area (A^{geo} in cm^2)

$$r_f = \frac{A^{real}}{A^{geo}} \quad (10.12)$$

For a smooth interface $r_f = 1$ and rough interfaces have roughness factors $r_f > 1$. The situation is summarized in Figure 10.2. The cell on the left is formed by two smooth ($r_f = 1$) interfaces. The interpolar distance is uniquely defined and has the same value in any place of the cell. The thickness of the electrolytic layer is a constant and the ionic resistance of the cell is constant. It is usually desirable to increase the roughness of the two inter-faces. The reason for this is that when a given voltage (the charge transfer driving force) is set across a given interface, the resulting charge transfer current (electrons flowing across the interface) will be higher across the roughest interface. In other words, rough interfaces are used to reduce charge transfer overvoltages and thus enhance electrochemical efficiency. The price to pay

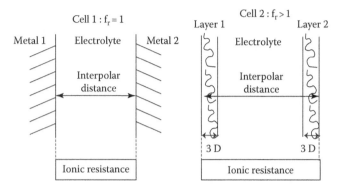

FIGURE 10.2
Schematic drawings: (left) cell with smooth interfaces, (right) cell with rough interfaces.

for this is that rough interfaces are obtained by increasing the thickness of the catalytic layer, and therefore, the ionic resistance of the cell is increased (Figure 10.2, right). But taking advantage of matter structuration at the nanoscale level, very rough interfaces (with roughness factors such as $200 < r_f < 1000$) can be obtained over limited extended distances. The benefit is thus high (especially at low current density when charge transfer resistances are elevated) and disadvantages (a higher cell resistance) remain limited.

It should be noted that the roughness of electrochemical interfaces cannot be extended without limits. The more active catalytic sites within the catalytic layers are those for which the distance (and therefore the ionic resistance of the electrolyte) to the other electrode is the smallest. As the interpolar distance increases, the ionic resistance connected in series to a given catalytic site also increases and the catalytic activity of these sites tends to decrease. Catalytic sites deeply embedded within rough interfaces are penalized in terms of electrochemical activity compared to those that are located close to the interface between the catalytic layer and the electrolyte. Therefore, the challenge is to create as many electrocatalytic sites as possible within a catalytic layer that is as thin as possible. As can be seen from Figure 10.1, this is the situation that prevails in PEM water electrolysis cells (Millet et al. 1993a). Catalytic layers placed on each side of the SPE are porous (to let reaction gases escape), micrometer thick, and have high roughness factors.

Charge transfer resistance The flux–force relationship that prevails at electrochemical interfaces relates the potential difference set across the metal–electrolyte interface to the electron flow (current) that flows across the interface and induces a redox chemical reaction on the electrolyte side. The voltage quantity of interest is not the absolute potential difference (which cannot be determined) but the difference between the actual voltage and the equilibrium voltage. This difference is the overvoltage η (with units of volt). When there is no mass

transfer limitation, there is an exponential relationship between overvoltage and current density j, the so-called Butler–Volmer relationship (Mayneord 1979), for monoelectronic charge transfer reactions

$$j = j_0 \left[\exp\left[\frac{\alpha_c\, nF}{RT}\, \eta \right] - \exp\left[\frac{\beta_c\, nF}{RT}\, \eta \right] \right] \quad (10.13)$$

The exponential flux–force relationship of Equation 10.13 can be approached in stationary operating conditions (η = constant, j = constant) by a pure charge transfer resistance R_{ct}. At constant operating temperature, the value of R_{ct} is a function of the overvoltage η. R_{ct} has units of Ω cm². Surface normalization is usually done by considering the geometrical surface A^{real} of the cell, not the real surface area $A^{real} = r_f \cdot A^{geo}$ (Equation 10.12). In a first approximation (as discussed earlier, catalytic sites away from the catalytic layer–electrolyte interface are penalized by an additional ionic resistance), the higher the roughness factor, the lower the charge transfer resistance and the more efficient the electrochemical cell is. In PEM water electrolysis cells, the measure of the charge transfer resistances associated with the HER (cathode) and OER (anode) provides an indication of the electrochemical activity and integrity of the two catalytic layers.

Interface capacitance metal–electrolyte interfaces are also characterized by the so-called double layer where electrical charges can accumulate. The double layer of an interface of unit roughness factor is a pure capacitance C_{dl} (in F cm⁻²). C_{dl} is placed in parallel with the charge transfer resistance R_{ct} to account for the fact that charge can either remain trapped at the surface of flow across the interface. Metal–electrolyte interfaces found in PEM water electrolysis cells are usually not pure capacitance because of the desirable 3D structure of the catalytic layers, which is used to increase their roughness. Interface capacitances are modeled using so-called constant-phase elements (CPEs) (Brug et al. 1984, Barsoukov and MacDonald 2005). CPEs are used to take into account deviations of double layers from ideal capacitive behaviors due to surface roughness, polycrystallinity, and anion adsorption (sulfonate ions in PEM water electrolysis cells). The impedance Z_Q of a CPE, noted Q, is

$$Z_Q = \frac{1}{Q(j\,\omega)^n} \quad (10.14)$$

where Q is a frequency-independent constant in F cm⁻² s^{n-1} and n is a factor ranging between 0 and 1. Q tends to a pure capacitance C as n tends to 1. In PEM water electrolysis cells, the measure of double-layer capacitances associated with the HER (cathode) and OER (anode) also provides an indication of the electrochemical activity and integrity of the two catalytic layers.

10.2.4 The Need for Reference Electrodes

As discussed in the previous section, it is highly desirable to separately measure charge transfer resistance and double-layer capacitance of each charge transfer process. This requires a reference electrode. In a PEM water electrolysis cell, the anode-to-cathode distance is usually small (<200 μm) and the electrolyte is a sheet of solid polymer. It is, therefore, not possible to introduce a conventional (even miniaturized) reference electrode in the interpolar area, somewhere between the anodic and cathodic catalytic layers. However, there are some alternative options to implement a reference electrode in order to separately measure the different voltage terms that contribute to the polarization curve. By calculating the distribution of electric potential between anode and cathode, it can be shown that the mean electric potential of a membrane strip in the nearby region of the anode and cathode is the same as the mid-distance potential between anode and cathode (Millet 1994). This is done as follows. In the anodic and cathodic zones, the relationship between the potential of the electrode and the potential in the membrane is given by a nonlinear relationship. This relationship is obtained by writing the charge conservation at the interface during the electrochemical process. We obtain the following:

Anode:

$$j = j_0 \exp\left[\frac{\alpha_a\, n\, F}{R\, T}\, \eta\right] = -\lambda \frac{\partial \varphi}{\partial x}\bigg|_{interface} \quad (10.15)$$

(The reverse term in the Butler–Volmer corresponding to the oxygen reduction reaction equation is not considered here because the rate of the reaction is very low.)
Cathode:

$$j = j_0 \left[\exp\left[\frac{\alpha_c\, n\, F}{R\, T}\, \eta\right] - \exp\left[\frac{\beta_c\, n\, F}{R\, T}\, \eta\right]\right] = -\lambda \frac{\partial \varphi}{\partial x}\bigg|_{interface} \quad (10.16)$$

In the membrane, the Laplace equation states that under steady-state conditions, there is no net accumulation of charge in any part of the membrane

$$\Delta\varphi = \frac{\partial^2\varphi}{\partial x^2} + \frac{\partial^2\varphi}{\partial y^2} + \frac{\partial^2\varphi}{\partial z^2} = 0 \quad (10.17)$$

To account for the value of the operating temperature in a simple way, the thermodynamic cell voltage and kinetic parameters at the electrodes were approximated linearly. (Such equations do not have a physical meaning, but allow the influence of temperature to be included in a simple way). These simplifications are correct within a 5% error in the temperature range under consideration (30°C–80°C).

SPE resistivity: $\rho(T)$

$$= -5.75 \times 10^{-4}(\Omega m\, K^{-1})T + 0.268\,(\Omega m) \quad (10.18)$$

Water-splitting voltage: $Ud(T)$

$$= -1.53 \times 10^{-3}\left(V\, K^{-1}\right)T + 1.686\,(V) \quad (10.19)$$

$$j_0^c(Pt) = 3.62 \times 10^{-3}(A\, m^{-2}K^{-1})T - 1.08\, A\, m^{-2}$$
$$j_0^a(IrO_2) = 1.055\left(A\, m^{-2}K^{-1}\right)T - 312.25\, A\, m^{-2} \quad (10.20)$$

Cathodic exchange current density: $j_0^c(Pt)$

$$= -5.46\left(A\, m^{-2}K^{-1}\right)T + 3327\left(A\, m^{-2}\right) \quad (10.21)$$

Figure 10.3a shows the solution of Equations 10.15 through 10.17 for a PEM water electrolysis cell equipped with a lateral membrane strip. The electric potential at the top of the strip is close to the potential value within the SPE at mid-distance between anode and cathode. A reference electrode can be connected to the top of the strip. This can be a simple reversible hydrogen electrode (RHE) obtained by hydrogen bubbling (Figure 10.3b). Better results are obtained when the top of the membrane strip is platinized. Alternatively, it is also possible to perform water electrolysis at low current density on top of the strip (using a Pt/C electrode for the HER and an IrO$_2$ anode for the OER) and to use the cathode as the reference electrode. This is a water electrolysis dynamic reference electrode (Figure 10.3c). The advantage of this configuration over the previous one is that the reference potential is better marked because there is no need for hydrogen gas to reach the interface. A last option is to use the top of the strip to install a hydrogen pump. Gaseous hydrogen is oxidized at the positive and reduced at the negative. This is a H$_2$ pump dynamic reference electrode (Figure 10.3d). Either side of the pump operating at low current density can be used as the reference electrode.

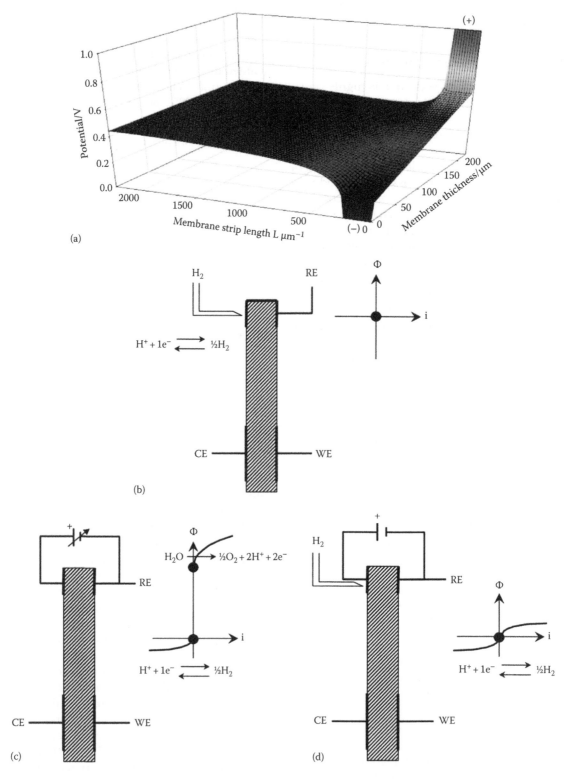

FIGURE 10.3
(a) Calculated 3D potential distribution in a PEM water electrolysis cell equipped with a lateral membrane strip. (b) Principle of a static hydrogen reference electrode. (c) Principle of water electrolysis dynamic hydrogen reference electrode. (d) Principle of a H_2 pump dynamic hydrogen reference electrode.

10.2.5 Global versus Local Information

Although electrochemical processes taking place in PEM water electrolysis cells are somewhat simpler than those encountered in PEM fuel cells (in particular, no mass transport limitation is usually observed), there is also an interest in measuring electrochemical data (ionic and ohmic resistances, charge transfer resistances, and double-layer capacitances) at local scale within the water electrolysis cell. This can help to obtain information about the distribution of current lines, the nonhomogeneous distribution of which can contribute to reduced lifetimes, especially when MEAs with large surface areas are used. As in PEM fuel cell technology, local information can be obtained, in principle, using segmented cells. Mapping surface electrochemical properties is critical to assess that MEAs are operating in homogeneous electrical environments in order to optimize efficiency and durability. Compared to a PEM fuel cell, PEM water electrolysis can potentially be operated at higher current densities (1.0–2.5 A cm^{-2} has been achieved in commercial systems operating in the 1–100 kW power range). Uneven distribution of current lines over the surface of MEAs is potentially a source of inefficiency and accelerated MEA aging. State-of-the-art MEAs of 500–600 cm^2 are commercially available, but recent efforts made to develop PEM water electrolysis stacks on the MW scale require much larger systems (>1000–1200 cm^2) or above. Appropriate cell designs that lead to the even distribution of current lines are required.

10.3 Characterization of Individual Cell Components

10.3.1 Characterization of the Solid Polymer Electrolyte

As in PEM fuel cell technology, most popular SPE materials used in PEM water electrolysis cells are PFSA-based polymer electrolytes. PFSA polymers are made of sulfonated tetrafluoroethylene–based fluoropolymer–copolymer membranes. They were initially developed by E. I. du Pont de Nemours and Company and are commercialized under the Nafion® brand name. The high electronegativity of fluorine and the cumulated inductive effects of the carbon–fluorine bonds facilitate the total dissociation of sulfonic acid end groups. Sulfonyl hydroxide end groups are fully dissociated, which is a desirable property that sets the ionic conductivity of Nafion® products and makes them excellent proton conductors for electrochemical applications. Because of their high acidity, sulfonic acids are very soluble in water, which in turn explains the water-swelling properties of these materials. Carbon–hydrogen bonds found in common polymers cannot sustain the highly oxidizing conditions found at the anode of PEM water electrolysis cells (carbon–hydrogen bonds have a bond length of approximately 1.09 Å and a bond energy of approximately 413 kJ mol^{-1}). The electronegativity difference between carbon (2.55) and hydrogen (2.20) on Pauling's scale is small (0.35), and the C–H bond can be regarded as being non-polar. Fluorine is used to reinforce the carbon backbone and make the material chemically stable during oxygen evolution at the anode at high potential (fluoroalkanes are some of the most unreactive organic compounds; carbon–fluorine bonds are polar–covalent bonds, the strongest single bond found in organic chemistry; this is a relatively short bond due to its partial ionic character). As more fluorine atoms are added to the same carbon, the bonds become increasingly shorter and stronger.

The microstructure of PFSA materials and the relationship between microstructure and physical properties (ionic conductivity, thermal conductivity, and gas permeability) have been extensively studied (Mauritz and Moore 2004). Nafion® products have several physical properties that are necessary for operation in PEM water electrolysis cells: (1) high ionic conductivity (σ > 10 mS cm^{-1}); (2) poor electronic conductivity; (3) good chemical and mechanical stability for long-term operation; (4) high thermal conductivity (σ > 0.1 J s^{-1} m^{-1} K^{-1}); and (5) limited permeability to hydrogen and oxygen. Proton conductivity, relative affinity for water, hydration stability at high temperatures, electro-osmotic drag, and mechanical, thermal, and oxidative stability of these materials are directly determined by their chemical microstructure. Based on the pioneering work of Gierke and others (Gierke et al. 1981), who developed the cluster network model, PFSA materials are known to contain clusters of sulfonate-ended perfluoroalkyl ether groups, diameter ~40 Å that are organized as inverted micelles and arranged on a lattice. These micelles are connected by pores or channels that are ~10 Å in size in the dry state. These –SO$_3$⁻-coated channels were invoked to account for inter-cluster transfer of cations and ion conductivity (Yeager and Steck 1981).

For analyzing the physical properties of these materials, Nafion® can be considered as a homogeneous two-phase medium, a mixture of hydrophobic regions concentrating fluorocarbon backbones and hydrophilic regions containing water, where proton conductivity takes place. Perfluorosulfonic ionomer (PFSI) chains are used for manufacturing SPE membranes. Alcoholic solutions of non-intricated chains are used for the manufacturing of catalytic layers in PEM water electrolysis cells.

10.3.1.1 Hydration and Dimensional Changes

Because during electrolysis the MEAs used in PEM water electrolysis cells are permanently soaked in liquid water, water management issues are less critical than in PEM fuel cells, and risks of partial/total dehydration are less significant. In particular, there is no need to control the hydration level of gases—a simplification in terms of both system complexity and process management.

Native PFSA Films Because of the high water solubility of sulfonic acid end groups, PFSA materials are water-sorbing materials (Vishnyakov and Neimark 2000). As a result, significantly large dimensional changes (anisotropic changes are observed; the commercial materials are less prone to thickness variations than in-plane changes) are observed when the material is hydrated for the first time. The process is reversible and polymer films shrink when dried. Considering the fact that a PEM water electrolysis cell is a compact cell (Figure 10.1a) in which contact resistances (Figure 10.1b) should be minimized, it is clear that dimensional changes must be avoided. The situation is more easily handled during water electrolysis than during fuel cell operation as the cell is used for the electrolysis of liquid water. However, swelling cycling, due to water content changes, can potentially have long-term negative effects and should be given careful attention. Another interesting characteristic of PFSA materials is the electro-osmotic drag of water molecules that solvate protons.

PTFE-reinforced PFSA Films Mechanical reinforcement of PFSA membranes using PTFE fibers makes the membranes less prone to dimensional changes. However, reinforced membranes are thicker and more resistive.

MEAs Once catalytic layers have been deposited on the surface of the membrane, dimensional changes in the SPE are less pronounced. The extent to which dimensional changes take place largely depend on the coating process. Again, results depend on the technique that is used to coat catalyst particles on the surface of the SPE. It is reported that surface chemical precipitation of catalyst particles gives better results than surface spraying of catalytic inks (Millet et al. 1989).

10.3.1.2 Chemical Stability

Although PEM fuel cells operate at relatively low temperature (typically in the range from sub-ambient to 80°C), membrane deterioration due to insufficient chemical stability is one of the most serious contributing factors to MEA degradation (Luan and Zhang 2012). Chemical, mechanical, and thermal factors can lead to accelerated aging, membrane thinning, and pinhole formation. Pinholes open the way to direct H_2/O_2 exothermic combustion and/or cell short circuit, and ultimately irreversible degradation of fuel cell performances and/or a safety hazard. In PEM water electrolysis cells, membranes are usually thicker, but the situation is potentially worse, as more aggressive electrochemical conditions are found at the anodes: elevated electrical potential combined with native oxygen evolution. The strength and inertness of C–F chemical bonds are sufficient for ensuring practical operation over long periods of time, offering compatible economical requirements. However, as in the case of PEM fuel cell technology, membrane thinning is commonly observed and fluoride ions are found in circulating water, indicating that the SPEs are undergoing chemical attack. Details of the chemistry involved in this chemical degradation are provided in Chapter 11.

For the monitoring of these aging processes, different options are available. During water electrolysis operation, it is possible to measure either the ionic conductivity of circulating water or the rate of emission of fluorine ions, in order to evaluate the chemical stability and degradation rate of the SPE. Online monitoring (e.g., using ion conductivity measurements, ion chromatography, or direct chemical analysis) can be easily implemented. EIS provides information on the ionic resistance of the membrane, which may change as the membrane ages. Measurement of contamination levels (H_2 in O_2, and vice versa) also gives information about pinhole formation. Ex situ (postmortem) measurement of the equivalent weight by acid–base titration provides information to those interested in analyzing and modeling degradation mechanisms.

10.3.1.3 Ionic Conductivity

Besides the chemical stability of the SPE, the ionic conductivity is the most important physical property that dictates the level of ohmic losses during operation. The surface resistance of PFSA materials depends mainly on equivalent weight (proton concentration), length of the side chain, thickness, water content, and operating temperature (Tsampas et al. 2006, Brandell et al. 2007). Surface resistances of various commercial products that can be used as SPEs in PEM water electrolysis cells are tabulated in Table 10.1.

The surface resistance of the SPE can be measured in situ using EIS. The Arrhenius plot of the ionic conductivity (σ) of a Nafion® 115 membrane measured at different operating temperatures is shown in Figure 10.4. The change of slope indicates a change in the activation energy of ionic conductance at approximately 50°C (Ferry et al. 1998).

10.3.1.4 Gas Permeability

Gas permeability is another important physical property of PFSA materials used in PEM water electrolysis

TABLE 10.1

Dry Thickness, Equivalent Weight, and Ionic Surface Resistance of Various PFSA Membrane Materials Used as SPE in PEM Water Electrolysis Cells

Trade Name	Dry Thickness (μm)	Equivalent Weight (g eq⁻¹)	Dry Resistance 25°C (Ω cm⁻²)
Nafion® 117	175	1100	0.23
Nafion® 115	115	1100	0.17
Nafion® 1035	90	1100	0.10
Nafion® 112	50	1100	0.036
Flemion®	80	1000	0.10
Aciplex® S	120	1000	0.11
Aquivion®	120	800	0.11

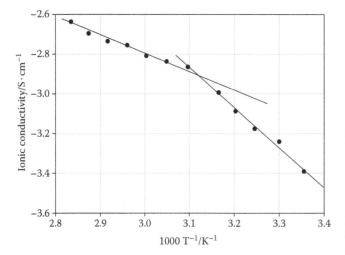

FIGURE 10.4
Arrhenius plots of the ionic conductivity of Nafion® 115.

cells, especially for operation at elevated pressure, as this is the property that determines gas permeability and governs cross-permeation phenomena. The permeability P_i of a gaseous species i across a SPE membrane of thickness L (in m) and surface area A (in m²) is defined as follows:

$$P_i = \frac{f_i}{\Delta P} \frac{L}{A} \text{ in } cm^2\,Pa^{-1}\,s^{-1} \qquad (10.22)$$

where
  f_i is the rate of permeation of species i (expressed in $Nm^3\,s^{-1}$)
  ΔP is the difference in pressure set across the membrane

The hydrogen and oxygen permeability of PFSA materials with different water contents has been extensively studied, especially in view of the practical application in PEM fuel cells. The temperature dependence of hydrogen and oxygen permeability of Nafion® 117 equilibrated

with different water contents is plotted in Figure 10.5 (Ogumi et al. 1984, Sakai et al. 1985, 1986). The permeability varies markedly with both temperature and water content of the polymer membrane. Values bracket those measured on PTFE, which corresponds to the perfluorocarbon backbone of PFSA materials, and liquid water, which corresponds to the hydrophilic sulfonic acid domains. These data show that the transport of dissolved gases across hydrated PFSA materials (Nafion®) is similar to that across liquid water. This is an indication that the transport of dissolved gases takes place mainly through the ionic (hydrated) clusters.

10.3.2 Characterization of Electrocatalysts

Electrodes are used in electrochemical devices for the dual purpose of carrying electric charges (electrons) to (from) the metal–electrolyte interfaces and to facilitate the charge transfer reaction across these interfaces. Because catalysts are usually more expensive than mere conductors, it is preferable to separate these two functions. Higher activity and reduced amounts of precious catalysts can be obtained by using nanoparticles for which the number of bulk (electrochemically inactive) catalyst atoms is minimal. PEM water electrolyzers use acid proton—conducting polymer electrolytes. Due to the acidity of the SPE (pH ~ 0), platinum group metals (PGMs) are used as electrocatalysts. Self-supported platinum nanoparticles or carbon-supported platinum nanoparticles (similar to those employed in PEM fuel cell technology) are used at the cathode for the HER (less expensive palladium is also an interesting alternative material for the HER [Bockris et al. 1982, Grigoriev et al. 2008]) and self-supported iridium dioxide nanoparticles are used at the anode for the OER. There is usually a limited interest to perform electrochemical characterization in liquid electrolytes using rotating disk electrodes (RDEs) because there are no mass transfer imitations.

10.3.2.1 Electrocatalysts for the HER

Physical characterization Electron microscopy is used for the ex situ physical characterization of native or aged catalyst particles (for R&D purposes or for analyzing aging processes). Figures 10.6 and 10.7 show TEM micrographs of carbon-supported platinum nanoparticles used at the cathode of PEM water electrolysis cells. Typical PGM loadings of 0.5 mg cm⁻² are commonly used. Similar to that in PEM fuel cell technology, carbon powder is a convenient and inexpensive platinum carrier but carbon particles tend to agglomerate by surface diffusion and hence appropriate dispersion in catalytic layers is not always easily obtained. Alternatively, more expensive carbon fibers can be used (Grigoriev et al. 2011).

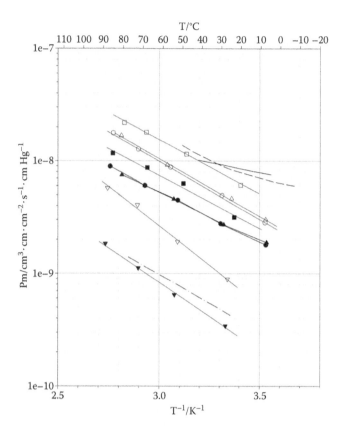

FIGURE 10.5
Temperature dependence of hydrogen and oxygen permeability of Nafion® 117 equilibrated with different water contents.

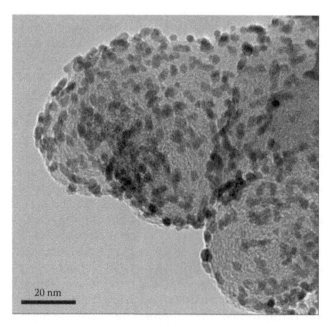

FIGURE 10.6
TEM micrograph of carbon-supported Pt nanoparticles used in PEM water electrolysis cells for the HER.

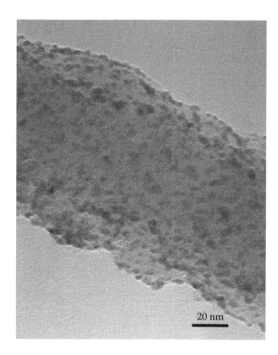

FIGURE 10.7
TEM micrograph of carbon-microfiber-supported Pt nanoparticles used in PEM water electrolysis cells for the HER.

Electrochemical characterization In acidic media, platinum-based catalysts can be characterized by cyclic voltammetry (CV). CV (Figure 10.8) is used to determine the amount of catalytic sites (or the roughness factor) available for the promotion of the HER. Results are usually normalized by unit of mass (to compare mass activity) or by unit of surface (in MEAs). Integration of the cyclic voltammogram in the appropriate potential range provides a coulombic charge that corresponds to the surface formation of hydrogen ad-atoms and hence to the electrochemical active area, a key performance factor that controls the HER overvoltage and hence the efficiency of the cell.

Concerning the HER, only two reaction paths involving two steps each are regarded as likely to occur in acidic media (in the following, H_{ad} denotes a surface ad-atom and M a surface metal site) (Millet and Ngameni 2011).

Mechanism 1 (Volmer–Tafel):

$$H^+ + 1e^- + M \leftrightarrow MH_{ad} \qquad (10.23)$$

$$2MH_{ad} \leftrightarrow 2M + H_2(g) \qquad (10.24)$$

Mechanism 2 (Volmer–Heyrovsky):

$$H^+ + 1e^- + M \leftrightarrow MH_{ad} \qquad (10.25)$$

$$H^+ + 1e^- + MH_{ad} \leftrightarrow M + H_2(g) \qquad (10.26)$$

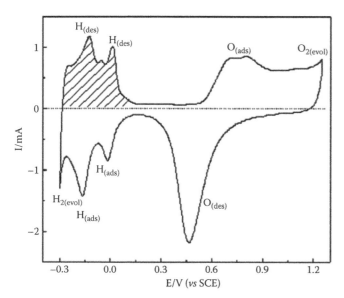

FIGURE 10.8
Cyclic voltammogram measured on a rough Pt electrode.

The first step (Volmer step) is shared by both mechanisms. They only differ by the desorption step (chemical desorption for mechanism 1 via the Tafel desorption step 10.24 that implies surface diffusion of H_{ad} species, or electrochemical desorption for mechanism 2 via the Heyrovsky desorption step 10.26). In the literature, there is a general agreement that on platinum (in acidic media), the HER mechanism is that of a fast proton discharge followed by rate-determining chemical desorption (mechanism 1, step 2 is rds) but there is no experimental evidence that an increased hydrogen pressure can affect the kinetics of the rate-determining step. In some cases, at high current density, mass transport effects are sometimes observed in PEM water electrolysis cells, but this is usually due to mass transport across porous current distributors, the porosity of which is inappropriately open (Millet et al. 2012). Tafel parameters of the HER measured at the cathode of a PEM water electrolysis cell are tabulated in Table 10.2. The exchange current density that is obtained from these measurements is then used to determine the roughness factor of the cathode, which here is close to $r_f = 200$. Both the exchange current density and roughness factor are very sensitive to platinum loadings. They can be used for the online monitoring of performance degradation (see Section 10.4).

TABLE 10.2

Tafel Parameters Measured at the Platinum Cathode of a PEM Water Electrolysis Cell

Temperature (°C)	Tafel Slope (mV dec^{-1})	Measured j_0^c (A cm^{-2})	Roughness Factor (cm^2 cm^{-2})
25	125	0.17	~200
80	105	0.14	~200

10.3.2.2 Electrocatalysts for the OER

Although RuO$_2$ is an excellent and inexpensive catalyst for the OER in acidic media, it lacks sufficient chemical stability. Self-supported iridium dioxide (IrO$_2$) nanoparticles are preferred for operation at the anode of PEM water electrolysis cells. There are few, if any, IrO$_2$ carriers that can sustain the oxidizing conditions found at the oxygen-evolving anodes. The iridium content can be reduced by forming binary or ternary oxide solid solutions with electronic conducting oxides such as SnO$_2$ or Ta$_2$O$_5$ (Tunold et al. 2010). RuO$_2$ is also stabilized in these solid oxide solutions and limited amounts of RuO$_2$ can also be incorporated into the anode composition to achieve higher efficiency.

Physical characterization Figure 10.9 shows the TEM micrograph of self-supported IrO$_2$ nanoparticles. The shape, roughness, and electrochemical activity of these powders depend on the manufacturing process (low-temperature hydrothermal processes, air combustion of precursors at elevated temperatures, etc.).

Electrochemical characterization As for the HER, CV is a convenient tool to determine the electrochemical activity of OER catalysts, to analyze the reaction mechanism, and to determine the roughness factor. Figure 10.10 shows cyclic voltammograms measured at different scan rates during the anodization of an iridium electrode in acidic media. For smooth surfaces (fr ~ 1), the coulombic charge that is measured is independent of the potential scan rate. Figure 10.11 shows the Tafel plot measured at 70°C on an IrO$_2$ electrode in sulfuric acid. Two different Tafel slopes are measured. The first one, obtained at low OER overvoltage, is ~55 mV dec^{-1}. The second one, obtained at higher OER overvoltage, is ~75 mV dec^{-1}.

Different mechanisms have been reported in the literature to account for the kinetics of the OER in acidic media on metal oxides. The two most frequently quoted are as follows:

Mechanism 1 (oxide path on Pt (Bockris 1956))

$$4\,Pt + 4\,H_2O \rightarrow 4\,PtOH + 4\,H^+ + 4e^- \quad (10.27)$$

$$4\,PtOH \rightarrow 2\,PtO + 2\,PtH_2O \quad (10.28)$$

$$2\,PtO \rightarrow O_2 + 2\,Pt \quad (10.29)$$

Mechanism 2 (Krasil'shchikov 1963) path on IrO$_2$; S denotes a reaction site)

$$S + H_2O \rightarrow S - OH + H^+ + 1e^- \quad (10.30)$$

$$S - OH \rightarrow S - O^- + H+ \quad (10.31)$$

FIGURE 10.9
TEM micrograph of self-supported IrO$_2$ nanoparticles used in PEM water electrolysis cells for the OER.

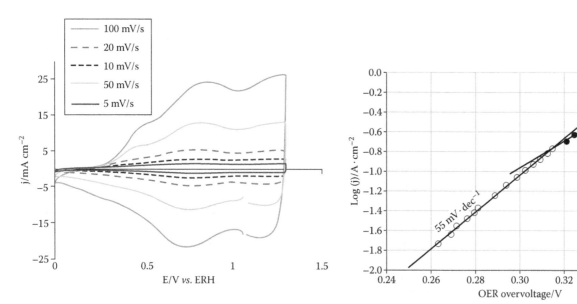

FIGURE 10.10
Cyclic voltammogram measured during the anodization of an Ir electrode.

FIGURE 10.11
Tafel plots for the OER on IrO$_2$.

TABLE 10.3

Krasil'shchikov Mechanism: Theoretical Tafel Slopes at 80°C for the Different rds (Rate-Determining Step)

Reaction Step	Step 10.30 is rds	Step 10.31 is rds	Step 10.32 is rds	Step 10.33 is rds
Tafel slopes				
Low overvoltage	140	70	47	18
High overvoltage	140	70	140	18

$$S - O^- \rightarrow S - O + e^- \quad (10.32)$$

$$S - O \rightarrow S + \tfrac{1}{2}\,O_2 \quad (10.33)$$

The Tafel slope measured experimentally provides an indication on the reaction mechanism and on the rate-determining step (rds) of the OER. Theoretical Tafel slopes for mechanism 2 are compiled in Table 10.3. Tafel slopes at low and high OER overvoltages sometimes differ.

In PEM water electrolyzers where IrO_2 anodes are used for the OER, mechanism 2 usually prevails. Detailed reaction mechanism analysis (not described in detail here) indicates that step 10.32 of mechanism 2 is the rds at low OER overvoltages and that step 10.31 of mechanism 2 is the rds at high OER overvoltages. The effect of oxygen pressure on the kinetics is usually negligible.

10.3.3 Characterization of Electrocatalytic Layers

As in PEM fuel cell technology, catalytic layers found in PEM water electrolysis cells must maximize the number of triple points that gather electronic conductivity (electron transport to cathodic catalytic sites or away from anodic catalytic sites), ionic conductivity (transport of hydrated protons), and mass transport (of gases away from catalytic sites). Each microdomain of each phase (electronic conductor, ionic conductor, mass transport) must be interconnected to others in a non-tortuous way to enhance overall efficiency (percolation theory).

10.3.3.1 Structure of Catalytic Layers

From a macroscopic viewpoint, catalytic layers of PEM water electrolysis cells can be characterized by their thickness, porosity, and adherence to the SPE. From a more microscopic viewpoint, percolation of microdomains and tortuosity should be analyzed to optimize the composition/structure of these layers (although most commercial applications are now using standardized commercial products). The thickness of catalytic layers used in water electrolysis MEAs is in the 0.1–1 µm range. As discussed earlier, a rough interface is desirable to reduce the charge transfer resistance. The anodic catalytic layer is usually thicker because more catalyst is required (the kinetics of the OER is one thousand times slower than the HER). Porosity is not so critical as long as the catalytic layer is not too thick. Usually, no pore promoter is used. The loading in PFSA chains in each catalytic layer is responsible for the adherence and integrity of each layer as well as for the ionic conductivity. The content must be optimized. Adherence of catalytic layers to the SPE is important because they contribute to the long-term durability of electrochemical performances. Adherence depends largely on the process that is used for the manufacturing of the catalytic layers.

Figure 10.12 (left) shows a TEM micrograph of a Pt-SPE interface that was prepared by in situ chemical reduction

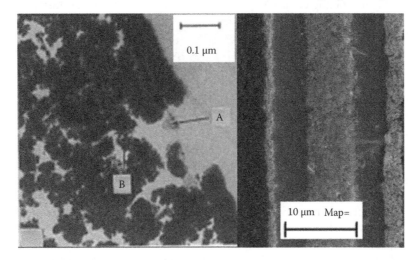

FIGURE 10.12
(left) TEM micrograph showing surface (A) and bulk (B) catalyst particles; (right) SEM micrograph showing a cross-sectional view of a membrane electrode assembly with central PTFE reinforcement.

of cationic platinum species (Millet et al. 1990). In this case, there are platinum particles located in the sub surface region of the SPE. These particles (when they are in electric contact with the catalytic layer) can contribute to the electrochemical process. They also contribute to reinforcing the adherence of the cathode and reducing dimensional changes when the water content of the SPE changes.

Figure 10.12 (right) shows the cross section of a conventional PEM water electrolysis MEA that was obtained by the spraying of catalytic ink onto the surface of the SPE. The layer of the anodic catalytic layer on the right is larger, as discussed earlier. With such MEAs, the adherence of the surface catalytic layers is not as good as that of those prepared by in situ precipitation of cationic salts but there are few indications in the literature on performance durability obtained with different types of MEAs.

10.3.3.2 Conductivity of Catalytic Layers

Catalytic layers play a critical role in PEM water electrolysis cells. They are the main cell components that determine cell performance and durability. Besides electrochemical activity, a catalytic layer can be characterized by its electronic and ionic conductivity, which are also important parameters. Measuring the ionic conductivity of thin catalytic layers is, however, not a simple task. It can be measured ex situ by AC impedance spectroscopy, for example, after deposition of a catalytic layer onto appropriate substrates, or in situ, directly on the MEA, but there are few results available in the literature (those available are mainly for PEM fuel cells) (Hongsirikarn et al. 2010). Electronic conductivity can be measured using the four-point probe method, a method commonly used for measuring the conductivity of thin metal or semi-conducting layers (Zhang et al. 2006).

Some estimated surface resistances are plotted in Figure 10.13 as a function of the thickness of the catalytic layer. Because powders dispersed in a polymeric matrix were used instead of bulk electrodes, the surface resistance values are much larger than those of bulk materials. Besides bulk conductivity, film structure (porosity, tortuosity, chemical composition, shape of particles) also plays a critical role. By comparison, surface resistance of bulk platinum and carbon is very low (ρ_{Pt} = 105 nΩ m at 20°C; ρ_C = 7.8 $\mu\Omega$ m at 20°C), hence, at the cathode, bulk electronic conductivity is not a significant matter of concern. At the anode, the situation is significantly different. The electronic resistivity of IrO$_2$ nanoparticles is much higher and depends on the oxidation level of iridium (ρ_{IrO_2} ~ 10^4 Ω m at 20°C for Ir^{III} and ρ_{IrO_2} ~ 1 Ω m at 20°C Ir^V) (Chow et al. 2014). A maximum surface resistance of 10 mΩ cm^2 is desirable for good operation.

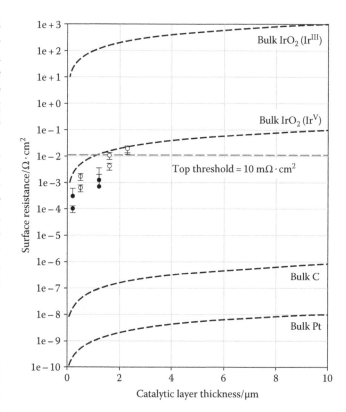

FIGURE 10.13
Electronic resistance of catalytic layers as a function of thickness. (•) Cathode, (○) anode.

10.3.4 Characterization of Other Cell Components

Because nickel and stainless steel are not chemically stable in deionized water, titanium is most frequently used for manufacturing non-MEA internal cell components in PEM water electrolysis cells (bipolar plates, spacers, and current distributors; see Figure 10.1). Although titanium has fairly low electrical and thermal conductivities compared to other metals, it has an appropriate corrosion resistance in hydrogen- and oxygen-saturated deionized water media, due to the dense TiO$_2$ surface layer that forms spontaneously at the surface (surface passivation). Some of the most important physical properties of bulk metallic titanium are tabulated in Table 10.4.

Values of elastic modulus (Young's, shear, and bulk) show that titanium is a rigid material, the mechanical properties of which can be used to exert significantly high internal compression forces within the PEM water electrolysis cell and thus contribute to reduce contact resistances between cell components. The desired open porosity of the current distributors does not significantly change the situation. The Poisson ratio (a measure in % of material expansion divided by the fraction in % of compression) and the different hardness values show that titanium is well suited for

TABLE 10.4

Some Important Physical Properties of Titanium

Crystal Structure	Hexagonal Close Packed
Thermal expansion	8.6 µm m^{-1} K^{-1} (25°C)
Thermal conductivity	21.9 W m^{-1} K^{-1}
Electrical resistivity	420 nΩ m (20°C)
Young's modulus	116 GPa
Shear modulus	44 GPa
Poisson's ratio	0.32
Vickers hardness	970 MPa

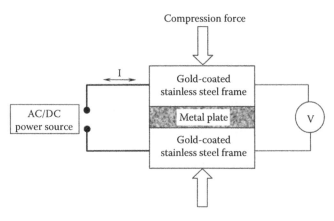

FIGURE 10.14

Schematic diagram of a compression tool used for the accurate measurement of bulk and contact electronic resistances.

application as bipolar plates, grid spacers, and current distributors.

Electrical conductivity measurements: Besides chemical and mechanical stability, electronic conductivity is the key parameter of non-MEA cell components. The "through-plate electrical resistance" (TPER) and "contact electrical resistance" (CER) of these cell components can be measured accurately using a specific Ohm's law setup that is used to set compression forces accurately (see Figure 10.14).

A challenge associated with the use of titanium non-MEA cell components arises from the semi-conducting surface oxide layer of high resistivity that spontaneously forms in contact with air or water. The electrical resistivity of bulk TiO$_2$ at ambient temperature is reported to be very high ($\rho_{TiO_2} = 10^{12}$ Ω cm). A nanometer-thick surface layer should introduce a much higher electric resistance than what is observed experimentally. Current carriers in titanium dioxide are electrons (Earle 1942). The variation of electronic conductivity within the temperature range of interest for PEM water electrolysis applications is limited and can be neglected. XPS analysis of aged titanium bipolar plates confirms that the surface oxide layer remains very thin (see Figure 10.15).

Operation at high electric potential, in contact with native oxygen, does not increase its thickness. However, the behavior of these surface oxide layers during operation of the PEM water electrolysis cell is not fully

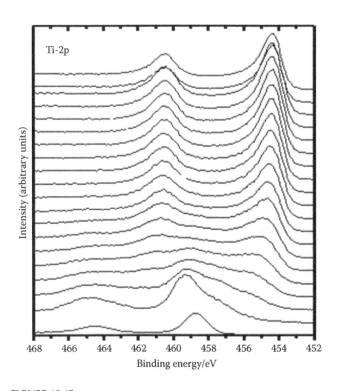

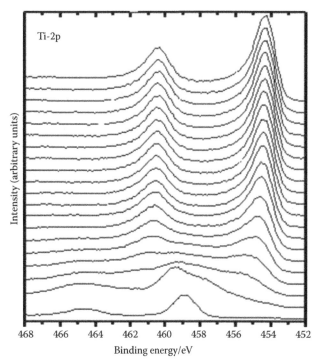

FIGURE 10.15

XPS analysis of titanium bipolar plates after 5000 h of operation: (left) H$_2$ side; (right) O$_2$ side.

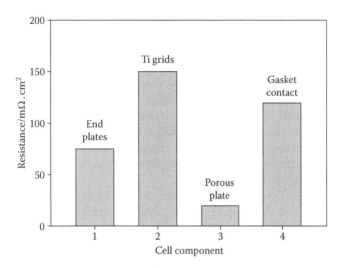

FIGURE 10.16
Indication of surface resistance of non-MEA cell components used in a PEM water electrolysis cell.

understood. What is unclear is the charge transfer mechanism across the oxide layer when the PEM cell is operating, the stability of the film, and what is its resistivity compared to bulk values. Surface resistances of aged cell components are plotted in Figure 10.16. At the cathode, a risk of hydrogen embrittlement (for H_2 concentrations >8000 ppm) should be noted.

10.3.4.1 Current Distributors

Optimization of porosity: Porous current distributors (PCDs) used in PEM water electrolysis cells are usually millimeter-thick titanium plates made of sintered titanium particles. These PCDs play a key role in the PEM cell (Figure 10.1); they transfer electricity from the bipolar plates to the electrochemical interfaces. Simultaneously,

they allow the transport of liquid water to the anode and the release of oxygen gas in the opposite direction. At the cathode, they allow the transport of a biphasic mixture (liquid water resulting from the electro-osmotic drag across the SPE and gaseous hydrogen) away from the cathode. Inadequate thickness and/or porosity results in excessive ohmic losses. Depending on the size (~10 – 100 μm) and shape of these particles (Figure 10.17), it is possible to adjust the open porosity and determine the optimum in terms of gas transport and electronic conductivity (Grigoriev et al. 2009). Optimal open porosity varies depending on operating conditions (nominal operating current density and pressure).

During electrolysis, these porous current distributors are firmly pressed against the catalytic layers to obtain good electrical contacts. As can be seen from Figure 10.18, gas evolution does not take place homogeneously over the entire surface of the catalytic layers but, instead, gas is collected through cracks located in the neighborhood of sintered titanium particles, at contact points with the current distributors. When current distributors are pressed too firmly against the MEA, the catalytic layer can be locally crushed (or even destroyed), and this can reduce the lifetime of the MEA. A homogeneous and not too high compression pressure is, therefore, required.

10.3.4.2 Cell Spacers

In the PEM water electrolysis cell of Figure 10.1, the role of cell spacers is to let liquid water flowing parallel to the membrane surface, to let current flowing perpendicular to the membrane surface, and to facilitate gas collection at the rear of cathodic and anodic cell compartments. Water is pumped through (to feed the water-splitting reaction and also for thermal management and gas collection). Different designs of spacers

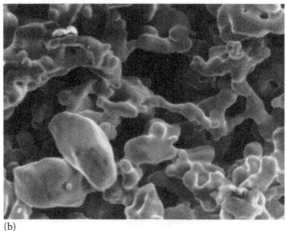

FIGURE 10.17
SEM micrographs of porous titanium current distributors: (a) low open porosity, (b) large open porosity.

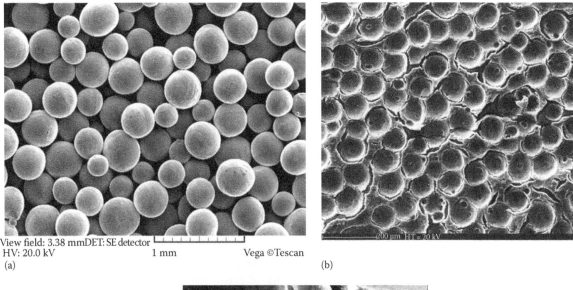

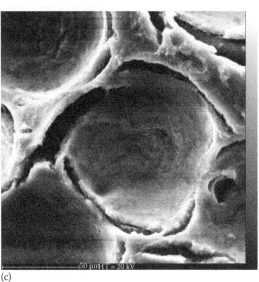

FIGURE 10.18
(a) SEM micrograph of a porous titanium current distributor made of spherical titanium particles. (b) SEM micrograph of the surface of the anodic catalytic layer after 1000 h of operation. (c) SEM micrograph of the catalytic layer in the neighborhood of a titanium particle.

can be used. Thicker bipolar plates with an appropriate network of machine-made channels like in PEM fuel cell technology can be used for that purpose. Titanium grids can also be used (Figure 10.1). In spite of a strong open porosity, the bulk electronic conductivity is sufficiently high to introduce a limited surface resistance but contact resistances are more elevated than those measured on type-1 bipolar plates. There is no specific need to characterize these cell components except from a hydrodynamic viewpoint, because they can contribute to the pressure drop across the cell and thus contribute to a homogeneous distribution of water inside the cells. However, most of the pressure drop is usually by design across the water outlets, so surface conductivity remains the main physical parameter that needs to be monitored.

10.3.4.3 End Plates

The TPER and CER of non-MEA cell components can be measured using the dedicated compression device of Figure 10.14 that is used to set accurately compression forces. Some data are provided in Figure 10.19. Results show that resistances differ significantly with compression.

10.4 Characterization of Individual PEM Cells

The individual cell components described in the previous section are then stacked together to form a compact unit PEM water electrolysis cell (Figure 10.1a). The main

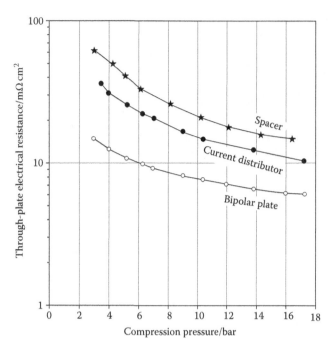

FIGURE 10.19
Through-plate electrical resistance measurements on Ti end plates, current distributors and grids under pressure at ambient temperature.

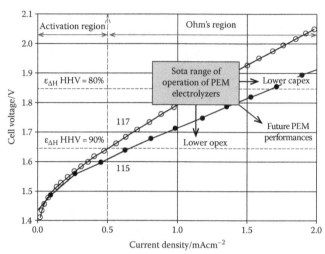

FIGURE 10.20
Typical polarization curves measured on a PEM water electrolysis cell equipped with MEAs of two different thicknesses.

challenge here is to measure some key physical properties (mainly charge conductivity) of individual cell components in this compact cell environment, taking into account the different interfaces. The purpose of this section is to describe tools and methodologies that can be used for that purpose.

10.4.1 Measurement of Polarization Curves and Analysis

10.4.1.1 Measurement of Polarization Curves

Similar to PEM fuel cell technology, the polarization curve (the relationship between cell voltage and current density across the cell) is the main characteristic of a PEM water electrolysis cell. It should be measured at constant operating temperature and pressure. Typical polarization curves measured with two SPEs of different thicknesses (Nafion® 117 and Nafion® 115) are plotted in Figure 10.20. Average performances of PEM water electrolysis cells are those indicated by the gray rectangle in the middle of the figure. This is usually a compromise between opex (opex increases as the operating current density is raised) and capex (capex decreases as the operating current density is raised). By decreasing cell voltage at constant current density, it is possible to reduce opex at constant capex. This requires more efficient catalysts. By increasing current density at constant cell voltage, it is possible to reduce capex at constant opex. This requires more efficient catalysts, thinner

SPEs, less resistive cells, and appropriate cell design. Today, R&D is required to reduce both opex and capex.

From the experimental viewpoint, there are two main experimental conditions that should be satisfied for the proper measurement of polarization curves: (1) the condition of isothermicity—although it might sometimes be difficult to perform isothermal measurements at high current density and low operating temperature (because the heat released by the reaction cannot be evacuated rapidly enough) and (2) the condition of stationarity—a difference should be made between polarization curves measured using sweep voltammetry (even at low sweep rates) and real stationary curves (recording of individual data point takes at least minutes).

10.4.1.2 Energy Levels and Energy Requirements

Figure 10.21 shows a description of the electronic structure of a PEM water electrolysis cell during operation. The band structure of the two metallic electrodes (Fermi levels, work functions) and the energy level of the two redox couples are provided. By applying an appropriate DC cell voltage between the two electrodes, the anodic and cathodic charge transfer overvoltages are adjusted so that a constant current flows across the cell.

Figure 10.22 shows the plot of a model polarization curve. $\Delta H_d°(H_2O) = 285$ kJ of energy is required to dissociate one mole of water under standard conditions. Part of that energy must be supplied to the cell as electricity ($\Delta G_d° = 238$ kJ mol⁻¹) and the remaining part as heat ($T\Delta S_d° = 47$ kJ mol⁻¹) to satisfy the entropy change. This is the reason why the electrolysis voltage $V_d(T,P) = \Delta G_d/2F + T \Delta S_d/2F = \Delta H_d/2F$ is called the "thermoneutral voltage" of the cell.

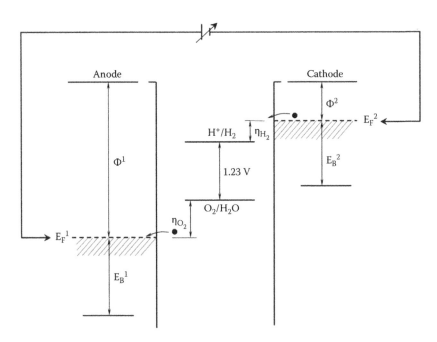

FIGURE 10.21
Electronic structure of a PEM water electrolysis cell. Φ = work function, E_F = Fermi level, and η = overvoltage.

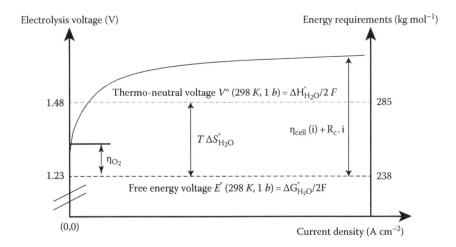

FIGURE 10.22
Model PEM electrolysis polarization curve showing key thermodynamic voltages and energy requirements.

However, a PEM water electrolysis cell is exothermic and the heat required for the entropy change is provided by the cell itself. Should both the anodic and the cathodic half-cell reactions be fully reversible, then current would start flowing across the cell as soon as the external applied voltage (U_{cell}) reaches the Gibbs free energy electrolysis voltage ($E = \Delta G/2F = 1.23$ V under standard conditions). However, mainly due to the low reversibility of the water/oxygen redox couple at the temperature of operation of a PEM water electrolysis cell, a significant overvoltage is required to let a current flow across the interface and a minimum of 100–200 mV overvoltage is required before the current starts flowing. As U_{cell} is further raised, increasingly more current flows across the cell. The excess voltage U_{cell}–E is used to overcome the different cell resistances: ohmic, ionic, and charge transfer resistances (Figure 10.1b), and the excess energy $2F(U_{cell}-E)$ J mol^{-1} is dissipated as heat to the surroundings. In practice, the thermo-neutral voltage is reached once a current density that satisfies Equation (10.34) is reached.

$$2F(\eta_{H_2} + \eta_{O_2} + R_{cell} \cdot i) = T\Delta S_d \qquad (10.34)$$

10.4.1.3 Analysis of Polarization Curves

A typical polarization curve measured on a 250 cm² laboratory PEM water electrolysis cell is plotted in Figure 10.23. On the same figure, a typical polarization curve measured on a laboratory hydrogen/oxygen PEM fuel cell is also plotted, for comparison. In both cases, three distinct regions are observed on the current–voltage (U–I) plot. The first one is observed at low current density. This is a domain where the ohmic voltage drop due to the internal resistance of the PEM cell is small compared to the charge transfer resistances associated with metal–electrolyte processes. In this region, the U–I relationship is logarithmic in shape and corresponds to interface activation. The second region is observed

in the intermediate current density range. This is a domain where charge transfer resistances are negligible because each interface (anode and cathode) has reached a potential value far from the equilibrium potential, and the linear shape of the U–I relationship is dictated by the internal resistance of the cell. The slope of the polarization curve is equal to the cell resistance measured at low frequencies by EIS. In the third region, at elevated current densities, mass transfer limitations are observed. Whereas in the fuel cell these limitations are due to the rate at which gaseous reactants are transferred from external reservoirs to the reactions sites, the situation is different during water electrolysis. Mass transport limitations are due to the formation of resistive gaseous films at the interface between at least one of the two catalytic layers and the surface of the adjacent porous current distributors.

A polarization curve measured on a laboratory cell up to several A cm⁻² is plotted in Figure 10.24 (curve a). Its first-order derivative versus current density is also plotted (curve b) as a function of operating current density. In the activation region, the cell resistance is high due to the charge transfer resistances. As the current density increases, the cell resistance becomes constant. This is the sum of the ionic resistance of the SPE and the electronic resistance of non-MEA components.

10.4.1.4 Different Terms of the Polarization Curves

Using an internal reference electrode (as described in Section 2.4), it is possible to measure separately the different terms of the polarization curve. Typical results are plotted in Figure 10.25. At 80°C, the Gibbs

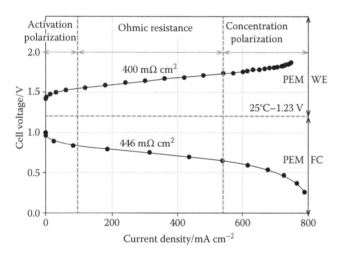

FIGURE 10.23
Polarization curves: (top) a PEM water electrolysis cell, (bottom) a PEM fuel cell.

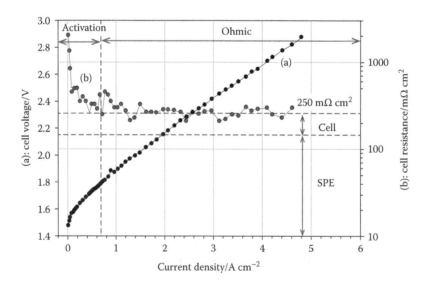

FIGURE 10.24
(a) PEM water electrolysis polarization curve; (b) cell resistance versus current density.

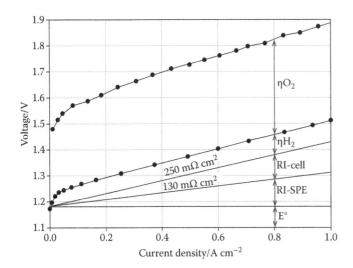

FIGURE 10.25
Polarization curve measured on a PEM water electrolysis cell at 80°C showing the main contributing voltage terms.

free energy electrolysis voltage is 1.18 V. Two constant ohmic terms are added: the SPE surface resistance (130 mΩ cm²) and the non-MEA surface resistance (120 mΩ cm²) for a total cell resistance of 250 mΩ cm². The two last contributions are the HER overvoltage η_{H_2} (the smallest charge transfer contribution) and finally the OER overvoltage η_{O_2} (the most important term after the thermodynamic voltage). Each term of the cell voltage tends to decrease when the operating temperature is increased (Figure 10.26) (Millet et al. 1993b).

10.4.1.5 Model Polarization Curves

Experimental current–voltage polarization curves of PEM water electrolysis cells can be approached using the following simplified equation:

$$U_{cell}(T,P) = E_{T,P} + \frac{RT}{\alpha^a F} Ln\left(\frac{j}{j_0^a}\right) + \frac{RT}{\alpha^c F} Ln\left(\frac{j}{j_0^c}\right) + j\sum_{k=1}^{n} R_k$$

(10.35)

where
  U_{cell} (T,P) is the water electrolysis voltage (V) at temperature T and pressure P of operation
  $E_{T,P} = \Delta G(T,P)/nF$ is the thermodynamic voltage (V), at T,P; $\alpha^a = \alpha^c \sim 0.5$ are the charge transfer coefficients of the anodic OER and cathodic HER, respectively
  j is the current density in A cm⁻²
  j_0^a and j_0^c are the exchange current densities of the OER and HER, respectively, in A cm⁻²
  F is the Faraday constant (~96,500 C mol⁻¹)
  R_k values are the different ionic and electronic cell resistances in Ω cm⁻²

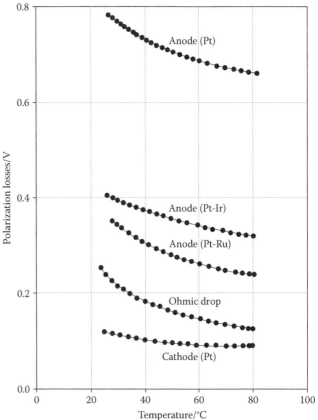

FIGURE 10.26
Temperature dependence of main cell voltage terms.

Equation 10.35 applies for sufficiently large current densities because it is assumed there that the reverse terms of the Butler–Volmer equations can be neglected.

Figure 10.27 shows a plot of the potential versus current density relationship for the anode and the cathode.

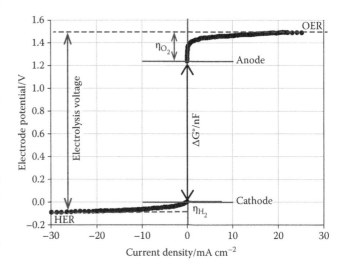

FIGURE 10.27
Plot of electrode potentials versus current density during PEM water electrolysis.

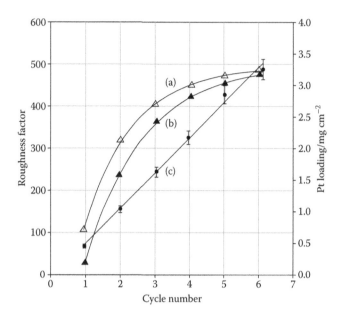

FIGURE 10.28

PEM water electrolysis cathode: (a and b) roughness factor, (c) Pt loading as a function of the number of deposition cycles.

During operation, the current flowing across each interface is the same (otherwise, a net charge accumulation would take place). To let the same current flowing, across each interface, the OER overvoltage (η_{O_2}) must be higher than the cathodic overvoltage (η_{H_2}). For a given catalyst, the best way to reduce the overvoltage required for a given current is to increase the roughness of the catalyst–electrolyte interface. The value of the roughness factor depends on catalyst particle size and the process used for manufacturing the MEA.

Figure 10.28 shows how the roughness of a PEM water electrolysis cathode increases by successive deposition

of platinum (Millet et al. 1995). Changing the roughness factor of an interface is equivalent to changing the current density of the redox process that takes place at this interface.

10.4.2 Characterization by Cyclic Voltammetry

Two key parameters that have a significant effect on the efficiency of the water-splitting reaction are the roughness factors of the anode and the cathode.

10.4.2.1 Roughness Factors

Figure 10.29 shows cyclic voltammograms of platinum electrodes in acidic media. Curve (a) is that of carbon-supported platinum nanoparticles at the cathode of a PEM water electrolysis cell and, for comparison, curve (b) is that of a platinum foil in sulfuric acid. In the PEM cell, the shape of the cyclic voltammogram is somewhat distorted but the main features of the curve in the hydrogen adsorption region are sufficiently clear to determine (at least approximately) the roughness factor r_f. The reference charge is 210 μC cm^{-2} (Trasatti and Petri 1992). The roughness factor of PEM water electrolysis cathodes is usually within the 100–600 range, depending on the morphology of the catalyst particles, thickness of the catalytic layer, and plating parameters.

Half-cell measurements can also be used to investigate PEM anodes. A typical cyclic voltammogram measured on a PEM water electrolysis anode containing unsupported IrO_2 is plotted in Figure 10.30. Integration of the cyclic voltammogram in the 100–1200 mV potential range yields a coulombic charge that can be used to evaluate the IrO_2 SPE roughness factor. The reference

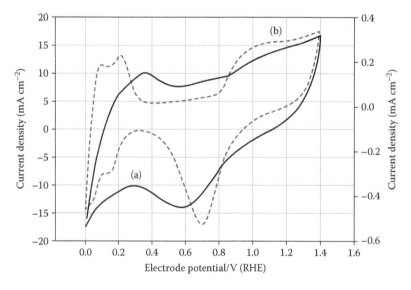

FIGURE 10.29

(a) Cyclic voltammogram of Pt in H_2SO_4 1 M and (b) cyclic voltammogram of Pt/C in a PEM water electrolysis cell.

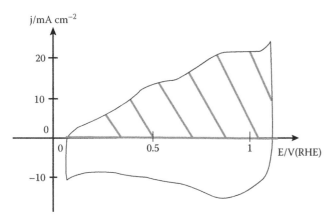

FIGURE 10.30
Cyclic voltammogram of IrO_2 in a PEM water electrolysis cell.

charge for IrO_2 is 310 µC cm^{-2} (Savinell 1990). Roughness factors of unsupported IrO_2 particles are usually ~700–800. Here again, values can differ significantly, depending on the morphology of the powder, catalyst loading, layer thickness, etc.

It should be noted here that the charge density that is measured is a function of the potential scan rate (see Figure 10.31). This is an indication that the different catalytic sites are located at different positions within the catalytic layers. At low scan rate, electrolyte resistances are minimized and more charges are scanned. As the scan rate increases, catalytic sites located deep inside the catalytic layer are negatively affected compared to those located at the front, close to the SPE membrane. The conclusion is that the roughness factor that is measured by CV is creating an inventory of all catalytic sites that are connected to the electronic and ionic circuits of the cell but, during operation, only a fraction of these

sites is active. It is, therefore, recommended that roughness factors be measured at different scan rates to gain more reliable information of active and nonactive catalytic sites.

10.4.2.2 Tafel Parameters

Some Tafel parameters measured on PEM cells are compiled in Table 10.5 (Millet et al. 1993b). They are in agreement with values made in aqueous solutions.

10.4.3 Characterization by Electrochemical Impedance Spectroscopy

Figure 10.32a shows the cross section of a symmetrical PEM water electrolysis cell and its equivalent electric circuit. Experiments have shown that the electrical circuit of Figure 10.32a can be simplified. The charge transfer resistance at the cathode is small compared to that of the anode where the OER takes place. Also, when current distributors of appropriate porosity are used, the diffusion impedances can also be neglected. A simplified model is shown in Figure 10.32b.

Experimental EIS impedance spectra measured on a PEM water electrolysis cell for different cell voltages are plotted in Figure 10.33. At low voltages (less than the thermodynamic voltage), spectra are mostly capacitive in shape (deviations from perfectly capacitive shape are due to the rough catalyst/electrolyte interfaces). Gradually, as the cell voltage increases, the low frequency limit bends down to the real axis. Whereas the high-frequency impedance on the real axis remains unchanged (this is the internal ohmic resistance of the cell), the diameter of the impedance semi-circle

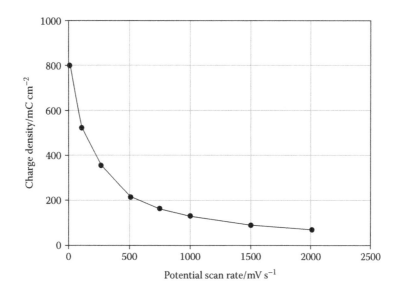

FIGURE 10.31
Charge density measure on an MEA as a function of potential scan rate.

TABLE 10.5

Tafel Parameters for the OER in PEM Water Electrolysis Cell Measured on Platinum and Iridium at Two Different Operating Temperatures (Geo = Geometrical Area; Real = Real Surface Area; r_f = Roughness Factor)

Electrode/Temp.	$\delta\eta^a/\delta\log j$ (mV dec^{-1})	α	Measured j_0^{geo} (A cm^{-2})	Measured j_0^{real} (A cm^{-2})	j_0^{real} (Damjanovic et al., 1966) (A cm^{-2})	r_f
Pt						
25°C	110	0.54	10^{-7}	6×10^{-10}	10^{-9}	166
80°C	130	0.45	2×10^{-5}	10^{-7}	—	200
Ir						
25°C	110	0.54	2×10^{-4}	10^{-6}	10^{-6}	200
80°C	130	0.45	6×10^{-3}	4×10^{-5}	—	150

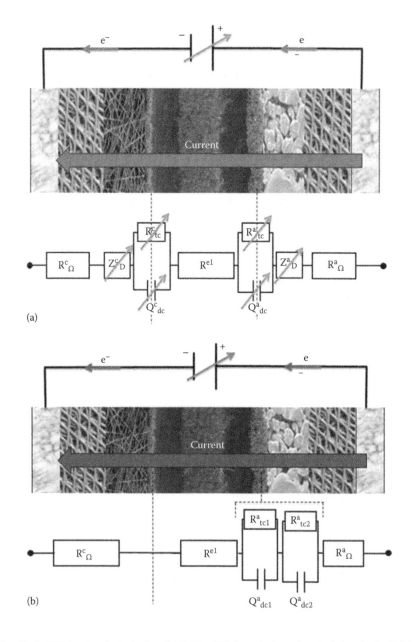

FIGURE 10.32
(a) Equivalent electrical circuit of a PEM water electrolysis cell. (b) Simplified equivalent electrical circuit of a PEM water electrolysis cell.

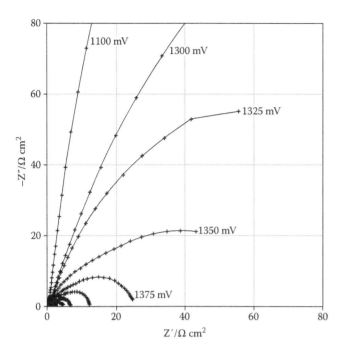

FIGURE 10.33
Experimental EIS impedance diagrams measured at 80°C on a 23 cm² PEM water electrolysis cell at different cell voltages.

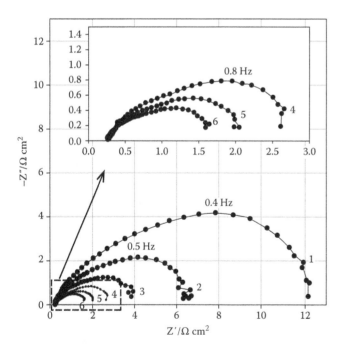

FIGURE 10.34
Experimental EIS impedance diagrams measured on a PEM water electrolysis cell at high cell voltage (>1400 mV).

decreases as the cell voltage increases (see Figure 10.34). At sufficiently elevated cell voltages (typically >1800 mV), charge transfer resistances R_{ct} become negligible and the impedance of the cell is purely ohmic and equal to the cell resistance.

10.4.4 Measurement of Gas Cross-Permeation Flows

Gas cross-permeation through the SPE during water electrolysis is due to hydrogen and oxygen solubility and diffusivity in PFSA materials (Mann et al. 2006). This is a diffusion-controlled (Fickian) process. As a result, the hydrogen concentration in oxygen is not zero, and vice versa, the oxygen content in hydrogen is not zero. For safety reasons and proper management of explosion hazards, both contamination levels should be continuously monitored. The main physical parameters that determine contamination levels are the equivalent weight, thickness, and water content of the SPE. Main operating parameters are current density, temperature, and pressure. Pinhole formation in the SPE can be detected by progressive or sudden changes in contamination levels. Contamination levels tend to increase at low operating current densities (Fateev et al. 2011). Contamination levels can be measured by gas chromatography. More recently, Bensmann et al. (2014) have proposed an electric method for measuring the hydrogen crossover. The measurement principle is based on the electrochemical compensation of the crossover flux. The mass flux is proportional to an electric current that is more simply and accurately measured.

10.4.5 Additional Nonelectrochemical Characterization Tools

Most of the characterization tools that are used in PEM fuel cell technology can also provide some interesting information on PEM water electrolysis cells (PEM Fuel Cell Diagnostic Tools 2012.). Some of these are briefly reviewed in this section.

10.4.5.1 EXAFS and XANES

Among a set of physical characterization tools, extended x-ray absorption fine structure (EXAFS) and x-ray absorption near-edge structure (XANES) can both be used for ex situ and in situ characterization of electrocatalysts. Figure 10.35 illustrates the different types of information that can be gained on Pt nanoparticles used at the cathode of PEM water electrolysis cells for the HER with these nonelectrochemical characterization tools. Whereas transmission electronic microscopy (TEM) provides structural information (size, surface, distribution, etc.) about catalyst nanoparticles, EXAFS provides mainly geometric information: nature of nearest neighbors, number of nearest neighbors, and distances between nearest neighbors. XANES provides information on the electronic structure of catalytic sites: oxidation state and adsorbate coverage. Crystallinity is not required for EXAFS measurements, and this technique provides one of the few structural probes available for the characterization of noncrystalline and

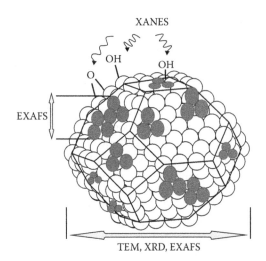

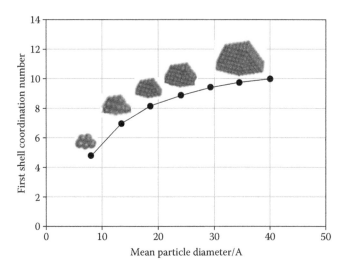

FIGURE 10.35
Combination of several nonelectrochemical techniques for the characterization of catalyst nanoparticles.

FIGURE 10.36
Aging Pt/C cathodes used in PEM water electrolysis cells.

highly disordered materials. EXAFS experiments can be performed either in transmission mode or in fluorescence mode. In the latter case, the sensitivity can be very high and sufficient to detect surface adsorption at the monolayer scale. It is not possible here to provide a thorough review of the benefits that can be gained from EXAFS measurements for the characterization of catalysts used in PEM water electrolysis cell. Briefly, besides deeper basic understanding of electrocatalysis, EXAFS measurements can also be used to optimize the manufacture of MEAs (Millet et al.1993b), or to investigate aging and degradation processes. Similar to that in PEM fuel cell technology, carbon-supported Pt nanoparticles used at the cathode of PEM water electrolysis cells are prone to surface migration and coalescence (aging phenomena). As a result, there is a reduction in the electrochemical active area and associated performance losses. The problem is illustrated in Figure 10.36, where a plot of particles mean diameter is provided as a function of the average first shell coordination number. Before aging, the mean particle diameter is close to 15 Å (coordination number of 7–8 and mean Pt–Pt distance of 2.74 Å). The same measurement performed on an aged material yields a mean particle diameter close to 35 Å (coordination number of ~10, with unchanged mean Pt–Pt distance).

10.4.5.2 Temperature Profiles Using IR Thermography

Infrared (IR) thermography can be used to analyze thermal dissipation in PEM water electrolyzers. Considering an energy efficiency of 70% at 1 A cm⁻², heat dissipation amounts to approximately 500–600 mW cm⁻². At the MEA level, the thermal conductivity of PFSA materials (Price and Jarratt 2000, Khandelwal and Mench 2006) is

sufficient to maintain temperature peaks within the subdegree Celsius range, even during operation at several A cm⁻² (Millet 1990). At the unit PEM cell level, IR thermography can be used to verify that current lines are homogeneously distributed over the active area of the cell and that no local hot spot forms. It can also be used to gain information on heat dissipation paths. At the stack level, it can be used to identify low-performance cells where more heat is dissipated.

10.4.5.3 Distribution of Current Lines

As for any electrochemical device operating at elevated current density, it is important to ensure in PEM water electrolysis cells the homogeneous distributions of current lines over the active surface area. If this condition is not satisfied, then both opex and capex are immediately penalized for two reasons: the efficiency of the cell will decrease and degradation rates will be increased. Performance durability will then not be as good as it should. The problem can be understood by looking at the cross section of a PEM water electrolysis cell (see Figure 10.1). The current is flowing across the cell from one end plate to the other, perpendicular to the plane of the cell. Electrons are the charge carriers in metallic cell components (from the end plate up to the catalytic layers) and hydrated protons are the mobile species within the SPE. Under stationary conditions of operation, the mean current density across the cell is defined as the ratio (in A cm⁻²) of the total current to the geometrical surface area of the cell. The question is to determine how the local current density (in any place of the cell) deviates from the mean value. As discussed in Section 10.4.5.2, IR thermography can be used for that purpose, but only on unit PEM cells equipped with appropriate transparent end plates and special current distributors. Local current density distributions can also be considered using segmented cells (Natarajan and Van Nguyen 2012).

Uneven distribution of current lines in PEM water electrolysis cells comes mainly from inappropriate cell designs or uneven cell stacking. They can be detected in principle using segmented cells (a segmented cell is a cell equipped with discontinuous current distributors that allow a local mapping of current lines). Some other specific characterization tools (such as magnetic resonance imaging, neutron imaging) may also be used to gain a better understanding of PEM water electrolysis processes during operation, but few (if any) results have been published so far on the subject in the literature.

10.4.6 Durability Tests

Durability (the ability of a PEM water electrolysis cell or stack to maintain its electrochemical performances over a significant period) is a critical property for practical applications. There are some reports in the literature (Arkilander and Molter 1984) that PEM water electrolyzers can operate in the upper range of 10^4–10^5 h. Similar to that in PEM fuel cells, degradation rates are a function of the aging test protocol used for the experiments. A convenient way to evaluate durability is to measure the time dependency of the cell voltage required for operation at a given current density. As in PEM fuel cell technology, degradation can be quantified by measuring the cell voltage drift (variation of cell voltage with time at constant current density), expressed in $\mu V\ h^{-1}$. Whereas the sole polarization curve provides a global measure of PEM performance

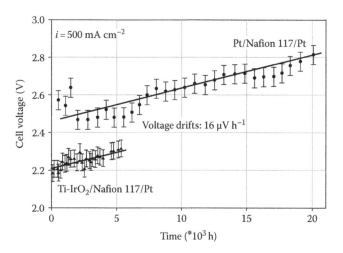

FIGURE 10.37
Durability tests performed: (top) on a Pt/Nafion® 117/Pr MEA, (bottom) on an IrO₂/Nafion® 117/Pr MEA.

(see Figure 10.37), it is also possible to use internal reference electrodes to gain a more detailed vision of the situation (Andolfatto et al. 1994). Results usually show that anodes are prone to faster degradation rates than cathodes (see Figure 10.38). Degradation mechanisms will be discussed in Chapter 11.

10.5 Characterization of Stacked PEM Cells

10.5.1 PEM Water Electrolysis Stacks

Figure 10.39 shows a cross-sectional view of a PEM water electrolysis stack bolted to a stainless steel front flange. The cells are pressed together under pressure to ensure

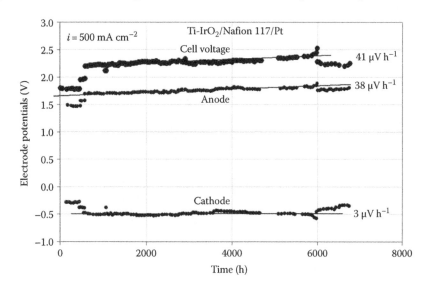

FIGURE 10.38
Durability test performed on an IrO₂/Nafion® 117/Pr MEA at 500 mA cm⁻² showing cathode and anode degradation rates.

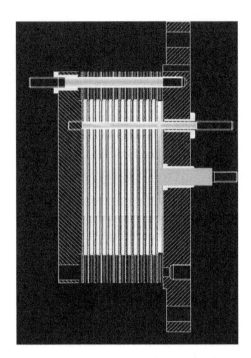

FIGURE 10.39
Cross-sectional view of a PEM water electrolysis stack of 2.5 Nm³ H₂ h⁻¹.

peripheral liquid water and gas tightness and to minimize contact resistances between the different cell components. Figure 10.40 shows the assembly of a PEM water electrolysis stack using a mechanical press. The main challenge when individual PEM water electrolysis cells are series connected (stacked) to increase production capacity is to ensure homogeneous distribution of electrochemical performances throughout the entire stack. Characterization tools are used for two purposes: (1) for control quality to verify that the cells are appropriately mounted in the stack and (2) for maintenance to measure performance and monitor degradation rates and to

anticipate possible maintenance operations. The challenge is that internal reference electrodes are usually not implemented in PEM stacks; it is, therefore, difficult to measure separately the behavior of anodes and cathodes.

10.5.2 Electrochemical Tools for the Characterization of PEM Water Electrolysis Stacks

Electrochemical tools (e.g., CV, EIS, etc.) presented in the previous sections can also be interesting tools for the characterization and optimization of PEM water electrolysis stacks. Figures 10.41 and 10.42 show prototype PEM water electrolysis stacks equipped with 600 cm² MEAs, and power potentiostats used for their characterization.

FIGURE 10.41
Prototype PEM water electrolysis stack of 10 Nm³ H₂ h⁻¹ with 250 cm² MEAs. With permission from CETH2 Co.

FIGURE 10.40
Assembling a PEM water electrolysis stack using a mechanical press.

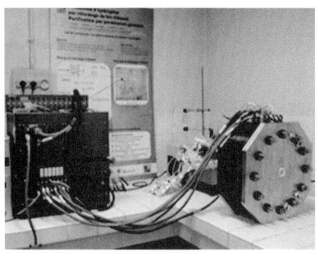

FIGURE 10.42
Prototype PEM water electrolysis stack with 600 cm² MEAs. With permission from CETH2 Co.

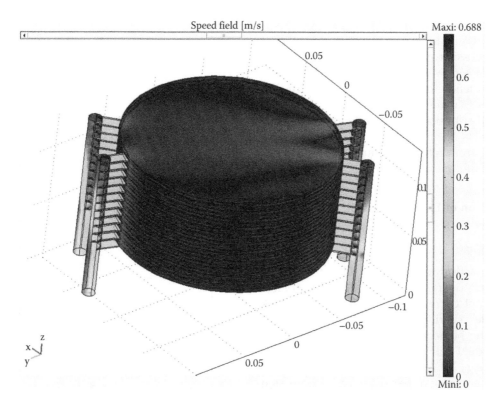

FIGURE 10.43
Numerical representation of a 12-cell PEM water electrolysis stack.

There are two options for experimental measurements: (1) characterization without external polarization, using a conventional potentiostat equipped with an amperometric booster (commercial system can deliver several hundred Amp—this is sufficient to record cyclic voltammograms at elevated scan rates in spite of large capacitive currents—this is also sufficient to measure impedance spectra); and (2) characterization during operation (e.g., by sur-imposing an AC voltage or current modulation to measure impedance diagrams).

10.5.3 Hydrodynamics

A first condition for a correct operation of the stack is to determine whether water pumped through the cells is homogeneously distributed. Figure 10.43 shows a digital representation of a PEM water electrolysis stack equipped with 12 cells. Suitable cell designs are those that allow a homogeneous distribution of water over the entire operating surface area in each individual cell. Numerical modeling is used by electrochemical designers as a supporting tool to satisfy that condition. Homogeneous water distribution is a problem dominated by the geometrical disposition of water inlets and outlets. Although individual measurements at cell level are difficult to implement, monitoring of pressure drops across the stack on the cathodic and anodic loops provides information on hydrodynamics inside the stack.

Although the modeling of biphasic liquid–gas mixtures is not a trivial task, due to the complex geometry of the cell, it is possible to develop models that are useful for determining reference conditions.

A detailed analysis of the problem is not provided here, but some experimental and model data are plotted in Figure 10.44, showing some pressure versus

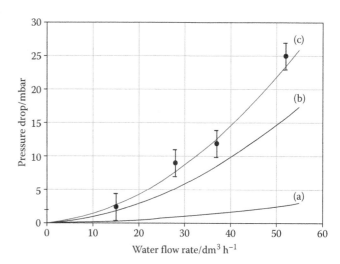

FIGURE 10.44
Pressure drops due to friction losses within a unit PEM cell as a function of the water flow rate: (a) empty cell, (b) with outlets, (c) with grid spacer. Experimental (●) and model (–).

water flow relationships in unit PEM cells. By monitoring the hydrodynamic resistance of a PEM cell or a PEM stack, it is possible to assess that nominal hydrodynamic conditions prevail in the long term. Dilatation of gaskets' edges tends to introduce parasite hydrodynamic resistances that are not easily identified and may induce uneven distribution of current lines and accelerated aging. There is no general rule, except that values should not differ from reference values recorded at the beginning of the lifetime of the stack.

10.5.4 Distribution of Electrochemical Characteristics

10.5.4.1 Cell Voltage

For safety and practical reasons, most PEM water electrolysis stacks are operated at constant stack voltage instead of constant current density. At constant stack voltage, a simple criterion that indicates whether all cells are operating homogeneously or not is the distribution of individual cell voltages among the stack. If for any reason (e.g., MEAs are not the same, titanium cell components are more or less oxidized, compression forces and cell voltages are not homogeneously distributed along the stack, etc.) adjacent cells have different impedances, then automatically cell voltages will start to differ. Let us now consider a PEM stack that contains 56 individual cells. If the cells are not correctly stacked, then there is a large distribution of cell voltages (Figure 10.45a) and the situation does not improve with time. When the cells are better connected, then the distribution of cell voltages is narrower and tends to improve with time of operation (Figure 10.45b).

10.5.4.2 Charge Density

In the previous section, it was shown that the charge density (and the roughness factor of each interface) can be individually measured. Implementation of reference electrodes on PEM cells of large commercial water electrolyzers is not a trivial task and, usually, no reference electrode is available to scan individual electrodes. In order to measure the activity of the electrodes and follow degradation rates, it is nevertheless possible to measure non-faradaic (capacitive) electric charges that are related to the roughness of the electrodes. When no reference is available, there are two options: (1) the cathode (the potential of which, under zero-current conditions, is that of RHE) can be used as the reference electrode to characterize the anode by CV; (2) a cyclic voltammogram of the entire PEM cell (within a voltage range where no faradaic process takes place) can be measured (Millet et al. 2011). For example, Figure 10.46 shows cyclic voltammograms measured between ±1 V on the

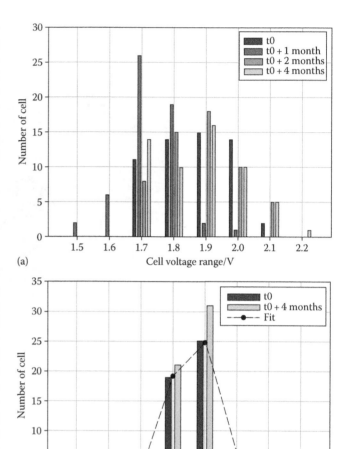

(a)

(b)

FIGURE 10.45
(a) Example of uneven cell voltage distribution at different time intervals. (b) Example of even cell voltage distribution at two time intervals.

12 cells of a PEM water electrolysis stack. Without entering into details, the integral of each cycle provides a charge which is a function of the PEM cell, the potential range of the scan, the scan rate, operating temperature and pressure. The charge can then be normalized to the surface area of the cell to make comparisons.

Figure 10.47 shows plots of the anodic charge densities of each individual cell (obtained by integration of cyclic voltammograms measured on the anode side) and the high-frequency cell resistances (ohmic internal cell resistance) measured at open potential voltage and in the early stages of the activation process on a prototype PEM water electrolysis stack that contains 56 cells. There is a significant discrepancy of data points from one cell to the other because the SPEs are not yet fully hydrated. At the end of the activation process, which may take several days, cell resistances are more homogeneous throughout the stack. This illustrates that routine EIS

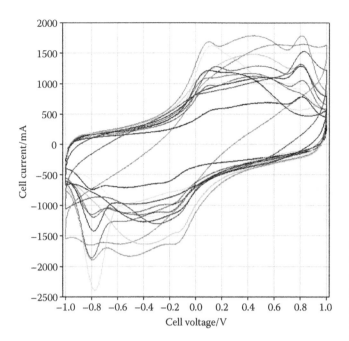

FIGURE 10.46
Cyclic voltammograms measured at 50 mV s^{-1} on a 12-cell PEM water electrolysis stack (active area = 250 cm^2) without reference electrodes.

measurements can be used to evaluate the homogeneity of the electrical environment in PEM water electrolysis stacks. They can also be used to quantify and follow degradation processes during long-term operation.

10.5.4.3 Cell Impedance

At stack level, it is important to assess that each individual cell is operating at the same level of performances as the others. EIS can be used for that purpose. Commercial EIS analyzers are usually not equipped with internal current boosters of sufficient current

capacity. However, it is possible to use such analyzers as external auxiliaries to measure impedances of stacks powered by power DC supplies, but care must be taken with potential levels and floating grounds to avoid parasite high voltage or high currents that could damage the equipment. A first option is to measure the impedance of the entire stack. For N cells connected in series, the total impedance is simply

$$Z_{stack} = \sum_{j=1}^{N} Z_j \qquad (10.36)$$

However, it becomes increasingly difficult to extract useful information as N increases. When possible (individual bipolar plates must be equipped with appropriate connectors), the measurement of individual cell impedance may provide an indication about possible electrical heterogeneous environments. In order to highlight the benefits that can be gained from EIS measurements for the determination of electrochemical environments experienced by the different cells in a stack, Figure 10.48 shows the impedance diagrams measured (between 300 mHz and 300 Hz) on the 12-cell prototype stack of Figure 10.40 (equipped with 250 cm^2 MEAs) at three different cell voltages.

Although the high-frequency limit on the real axis (cell resistance ~250 mΩ cm^2) is approximately the same for all cells (this is a good indication of electrical homogeneity), a significant discrepancy of charge transfer resistances (proportional to the diameter of the semi-circle along the real axis) is observed for the three cell voltages under consideration. This is an indication that the different catalytic layers of the different MEAs exhibit different performances. At the beginning of the lifetime (this is the case for the data of Figure 10.48),

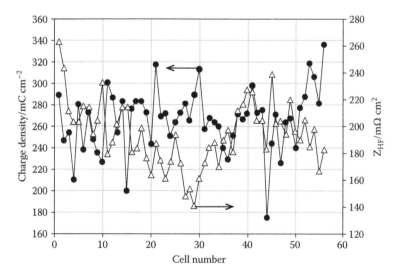

FIGURE 10.47
Plots of anodic charge density and high-frequency cell impedance as a function of cell number in the stack of Figure 10.41.

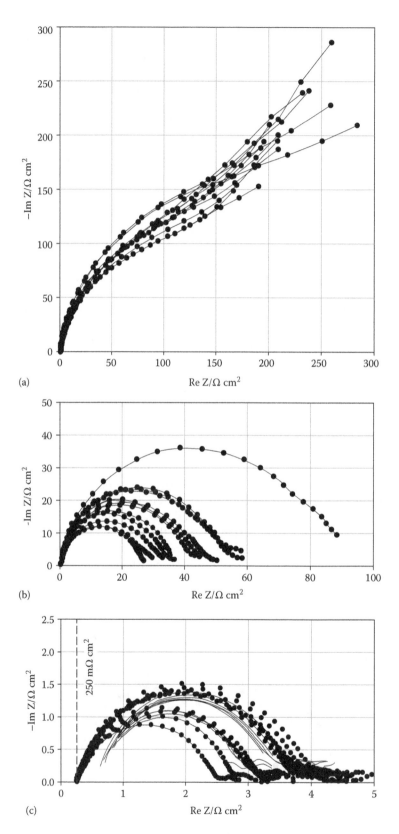

FIGURE 10.48
Individual cell impedances measured on a 12-cell stack at (a) $U^{cell} = 1.40$ V, (b) $U^{cell} = 1.45$ V, and (c) $U^{cell} = 1.50$ V.

this is usually due to the fact that MEAs are activated differently (water swelling). At the end of lifetime, this is an indication that some aged MEAs are not performing as they should. Impedance spectra of Figure 10.48a were measured at U^{cell} = 1.40 V. The discrepancy is already large although the current density is small. Charge transfer resistances of approximately 250–300 Ω cm^2 and the low-frequency behavior suggest diffusion-controlled mass transport of reaction products away from the interfaces. Impedance spectra of Figure 10.48b were measured at U^{cell} = 1.45 V. The discrepancy is larger than in the case of the 1.40 V. Charge transfer resistances are significantly smaller (~30–60 Ω cm^2) and low-frequency limits are close to the real axis where some mass transport limitations of limited magnitude can be seen. Impedance spectra of Figure 10.48c were measured at U^{cell} = 1.50 V. Charge transfer resistances are further reduced (~2.5–4 Ω cm^2) and low-frequency mass transport limitations through the porous current distributors are clearly seen. As the activation process proceeds (this may require several hundred hours of operation), impedance spectra usually tend to homogenize.

10.5.5 Performance Measurements and Durability Tests

There are some reports in the literature of PEM water electrolysis technology operating in the upper range of 10^4–10^5 h. However, the durability of performance depends strongly on operating conditions (stationary operating conditions favor durability). An example is provided in Figure 10.49, where the performances are recorded during the first 800 h of operation of a PEM stack (1 Nm^3 H$_2$ h^{-1} capacity) at constant stack voltage of 22.5 V (mean cell voltage = 1.88 V) and 40°C (Millet et al. 2010). Periodically, fresh water at room temperature is injected into the anodic loop. As a result, the mean temperature decreases (b in Figure 10.40) and the current density decreases (a in Figure 10.49) before returning to its stationary value within a few hours. In the long term, current density changes are correlated with internal temperature changes that may arise from changes in external room temperature. Over the short period of this test, however, performances remain unchanged (see Figures 10.37 and 10.38 for long-term results). Degradation processes will be analyzed in Chapter 11.

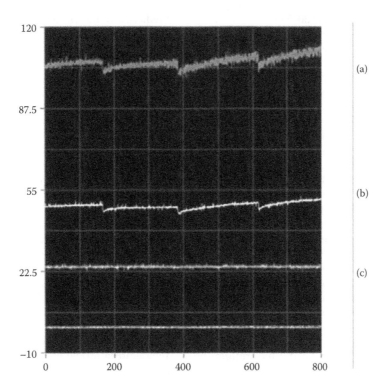

FIGURE 10.49

(a) Stack current versus time measured during the first 800 h of operation of the GenHy1000 electrolyzer, operating at (b) a temperature T ~ 40°C and (c) a constant stack voltage (~23 V). P = 3 bars. (Reprinted from *International Journal of Hydrogen Energy*, 35, Millet, P., Ngameni, R., Grigoriev, S.A., Mbemba, N., Brisset, F., Ranjbari, A., and Etiévant, C., PEM water electrolyzers: From electrocatalysis to stack development, 5043–5052, Copyright (2010), with permission from Elsevier.)

10.6 Conclusions

In conclusion, a PEM water electrolysis cell is an electrochemical device that is used to perform the endergonic splitting of water molecules into elemental hydrogen and oxygen. The cell contains different components. Each cell component is characterized by a resistance that needs to be minimized in view of the practical applications that require high dissociation efficiencies and high durability. In this chapter, several characterization tools and methodologies that can be used to measure some key physical properties (mainly the charge conduction) of the different cell components have been described.

References

Andolfatto, F., R. Durand, A. Michas, P. Millet and P. Stevens. 1994. Solid polymer electrolyte water electrolysis: Electrocatalysis and long-term stability. *International Journal of Hydrogen Energy* 19(5):421–427.

Arkilander, W. N. and T. M. Molter. 1984. Oxygen generator cell design for future submarines. The World's Knowledge, London: The British Library 961440, 530–535.

Barsoukov, E. and J. R. Macdonald. 2005. *Impedance Spectroscopy: Theory, Experiment, and Applications.* Hoboken, NJ: John Wiley & Sons.

Bensmann, B., R. Hanke-Rauschenbach, and K. Sundmacher. 2014. In-situ measurement of hydrogen crossover in polymer electrolyte membrane water electrolysis. *International Journal of Hydrogen Energy* 39:49–53.

Bockris, J. O'M. 1956. Kinetics theory of adsorption intermediates in electrochemical catalysis. *Journal of Chemical Physics* 24:817–822.

Bockris, J. O'M. and A. K. N. Reddy. 1982. *Comprehensive treatise of electrochemistry.* New York: Plenum Press Ed.

Brandell, D., J. Karo, A. Livat, and J. O. Thomas. 2007. Molecular dynamics studies of the Nafion, Dow and Aciplex fuel cell polymer membrane systems. *Journal of Molecular Modeling* 13(10):1039–1046.

Brug, G. J., A. L. G. Van Den Eeden, M. Sluyters-Rehbach, and J. H. Sluyters. 1984. The analysis of electrode impedances complicated by the presence of a constant phase element. *Journal of Electroanalytical Chemistry and Interfacial Electrochemistry* 176:275–295.

Chow, F.-F., M. Carducci, and R. W. Murray. 2014. Electronic conductivity of films of electrofloculated 2 nm iridium oxide nano-particles. *Journal of the American Chemical Society* 136(9):3385–3387.

Damjanovic, A., A. Dey, and J.O'M. Bockris. 1966. Electrode kinetics of oxygen evolution and dissolution on Rh, Ir and Pt-Rh alloy electrodes. *Journal of the Electrochemical Society* 113:739–746.

Earle, M. D. 1942. The electrical conductivity of titanium dioxide. *Physical Review* 61:56–63.

Fateev, V., S. A. Grigoriev, P. Millet, S. Korobtsev, V. Porembskiy, and F. Auprêtre. 2011. High pressure PEM water electrolysis and corresponding safety issues. *International Journal of Hydrogen Energy* 36:2721–2728.

Ferry, A., M. M. Doeff, and L. C. DeJonghe. 1998. Transport property measurements of polymer electrolytes. *Electrochimica Acta* 43:1387–1393.

Gierke, T. D., G. E. Munn, and F. C. Wilson. 1981. The morphology in Nafion perfluorinated membrane products, as determined by wide- and small-angle X-ray studies. *Journal of Polymer Science B: Polymers Physics* 19(11):1687–1704.

Grigoriev, S., L. I. Ilyukhina, P. H. Middleton, P. Millet, T. O. Saetre, and V. N. Fateev. 2008. A comparative evaluation of palladium and platinum nanoparticles as catalysts in PEM electrochemical cells. *International Journal of Nuclear Hydrogen Production and Application* 1–4:343–354.

Grigoriev, S. A., M. S. Mamat, K. A. Dzhus, G. S. Walker, and P. Millet. 2011. Platinum and palladium nanoparticles supported by graphitic nano-fibers as catalysts for PEM water electrolysis. *International Journal of Hydrogen Energy* 36:4143–4147.

Grigoriev, S. A., P. Millet, S. A. Volobuev, and V. N. Fateev. 2009. Optimization of porous current collectors for PEM water electrolysers. *International Journal of Hydrogen Energy* 34:4968–4973.

Hamann, C. H., A. Hamnett, and W. Vielstich. 1998. *Electrochemistry*, 2nd edn. New York: Wiley VCH.

Hongsirikarn, K., X. Mo, Z. Liu, and J. G. Goodwin Jr. 2010. Prediction of the effective conductivity of Nafion in the catalyst layer of a proton exchange membrane fuel cell. *Journal of Power Sources* 195:5493–5500.

Khandelwal, M. and M. M. Mench. 2006. Direct measurement of through-plane thermal conductivity and contact resistance in fuel cell materials. *Journal of Power Sources* 161:1106–1115.

Krasil'shchikov, I. 1963. Intermediate stages in the anodic evolution of oxygen. *Russian Journal of physical Chemistry* 37:273–276.

Luan, Y. and Y. Zhang. 2012. *Membrane Degradation, PEM Fuel Cell Mode Degradation Analysis.* London, UK: CRC Press.

Lvovich, V. F. 2012. *Impedance Spectroscopy: Applications to Electrochemical and Dielectric Phenomena.* Hoboken, NJ: Wiley Online Library.

Mann, R. E., J. C. Amphlett, B. A. Peppley, and C. P. Thurgood. 2006. Henry's law and the solubilities of reactant gases in the modeling of PEM fuel cells. *Journal of Power Sources* 161:768–774.

Mauritz, K. A. and R. B. Moore. 2004. State of understanding of Nafion. *Chemical Reviews* 104:4535–4585.

Mayneord, W. V. 1979. John Alfred Valentine Butler. *Biographical Memoirs of Fellows of the Royal Society* 25:126–144.

Millet, P. 1990. Water electrolysis using EME technology: Temperature profile inside a Nafion membrane during electrolysis. *Electrochimica Acta* 36:263–267.

Millet, P. 1994. Water electrolysis using EME technology: Electric potential distribution inside a Nafion membrane during electrolysis. *Electrochimica Acta* 39:2501–2506.

Millet, P., T. Alleau, and R. Durand. 1993a. Characterization of membrane-electrodes assemblies for solid polymer electrolyte water electrolysis. *Journal of Applied Electrochemistry* 23:322–331.

Millet, P., F. Andolfatto, and R. Durand. 1995. Preparation of slid polymer electrolyte composites. Investigation of the precipitation process. *Journal of Applied Electrochemistry* 25:233–239.

Millet, P., F. de Guglielmo, S. A. Grigoriev, and V. I. Porembskiy. 2012. Cell failure mechanisms in PEM water electrolyzers. *International Journal of Hydrogen Energy* 37:17478–17487.

Millet, P., R. Durand, E. Dartyge, G. Tourillon, and A. Fontaine. 1993b. Precipitation of metallic platinum into Nafion ionomer membranes: Experimental results. *Journal of the Electrochemical Society* 140:1373–1380.

Millet, P., R. Durand, and M. Pinéri. 1989. New solid polymer electrolyte composites for water electrolysis. *Journal of Applied Electrochemistry* 19:162–166.

Millet, P., R. Durand, and M. Pinéri. 1990. Preparation of new solid polymer electrolyte composites for water electrolysis. *International Journal of Hydrogen Energy* 15:245–253.

Millet, P., N. Mbemba, S. A. Grigoriev, V. N. Fateev, A. Aukauloo, and C. Etievant. 2011. Electrochemical performances of PEM water electrolysis cells and perspectives. *International Journal of Hydrogen Energy* 36:4134–4142.

Millet, P. and R. Ngameni. 2011. Non-harmonic electro-chemical and pneumato-chemical impedance spectroscopies for analyzing the hydriding kinetics of palladium. *Electrochimica Acta* 56(23):7907–7915.

Millet, P., R. Ngameni, S. A. Grigoriev, N. Mbemba, F. Brisset, A. Ranjbari, and C. Etiévant. 2010. PEM water electrolyzers: From electrocatalysis to stack development. *International Journal of Hydrogen Energy* 35:5043–5052.

Natarajan, D. and T. Van Nguyen. 2012. *PEM Fuel Cells Diagnostic Tools, Current Mapping.* London, UK: CRC Press.

Ogumi, Z., Z. Takehara, and S. Yoshizawa. 1984. Gas permeation in SPE method. *Journal of the Electrochemical Society* 131:769–773.

Price, D. C. and M. Jarratt. 2000. Thermal conductivity of PTFE and PTFE composites. *Proceedings of the Twenty-Eight Conference of the North-American Thermal Analysis Society,* October 4–6, Orlando, FL.

Rozain, C. and P. Millet. 2014. Electrochemical characterization of polymer electrolyte membrane water electrolysis cells. *Electrochimica Acta* 131:160–167.

Sakai, T., H. Takenaka, and E. Torikai. 1986. Gas diffusion in the dry and hydrated Nafion. *Journal of the Electrochemical Society* 133:88–92.

Sakai, T., H. Takenaka, N. Wakabayashi, Y. Kawami, and E. Torikai. 1985. Gas permeation properties of solid polymer electrolyte (SPE) membranes. *Journal of the Electrochemical Society* 132:1328–1332.

Savinell, R. F., R. L. Z. Iii, and J. A. Adams. 1990. Electrochemically active surface area voltammetric charge correlations for ruthenium and iridium dioxide electrodes. *Journal of the Electrochemical Society* 137:1–6.

Trasatti, S. and O. Petri. 1992. Real surface area measurement in electrochemistry. *Journal of Electroanalytical Chemistry* 327:353–376.

Tsampas, M. N., A. Pikos, S. Brosda, A. Katsaounis, and C. G. Vayenas. 2006. The effect of membrane thickness on the conductivity of Nafion. *Electrochimica Acta* 51(13):2743–2755.

Tunold, R., A. T. Marshall, E. Rasten M. Tsypkin, L.-E. Owe, and S. Sunde. 2010. Materials for electrocatalysis of oxygen evolution process in PEM water electrolysis cells. *Journal of the Electrochemical Society* 25:103–117.

Vishnyakov, A. and A. V. Neimark. 2000. Molecular simulation study of Nafion membrane solvation in water and methanol. *Journal of Physical Chemistry B* 104:4471–4476.

Wang, H., X.-Z. Yuan, and H. Li (eds.). 2012. *PEM Fuel Cell Diagnostic Tools.* Boca Raton, FL: CRC Press.

Yeager, H. L. and A. Steck. 1981. Cation and water diffusion in Nafion ion-exchange membranes: Influence of polymer structure. *Journal of the Electrochemical Society* 128:1880–1884.

Zhang, L. M., Y. S. Gong, C. B. Wang, Q. Shen, and M. X. Xia. 2006. Substrate temperature dependent morphology and resistivity of pulsed laser deposited iridium oxide thin films. *Thin Solid Films* 496(2):371–375.

11

Degradation Processes and Failure Mechanisms in PEM Water Electrolyzers

Pierre Millet

CONTENTS

Nomenclature

E	thermodynamic voltage (V)
F	Faraday's constant (96,485 $C \cdot mol^{-1}$)
G	Gibbs free energy ($J \cdot mol^{-1}$)
H	enthalpy ($J \cdot mol^{-1}$)
I	current (A)
i	current density ($A \cdot cm^{-2}$)
i_0	exchange current density ($A \cdot cm^{-2}$)
j	imaginary number

n	number of electron exchanged in a chemical reaction
P	pressure (Pa)
Q	pseudo-capacitance ($F \cdot cm^{-2} \cdot s^{n-1}$)
R	resistance (Ω)
R_{PG}	constant of perfect gas (0.082 $J \cdot K^{-1} \cdot mol^{-1}$)
S	entropy ($J \cdot mol^{-1} \cdot K^{-1}$)
t	time (s)
T	absolute temperature (K)
U_{cell}	cell voltage (V)

V	thermo-neutral electrolysis voltage (V)
W	electrical work (J)

Greek Symbols

$\overleftarrow{\alpha_a}$	oxidation charge transfer coefficient
$\overrightarrow{\alpha_a}$	reduction charge transfer coefficient
Δ	difference
ε	cell efficiency (%)
η	overvoltage (V)
ω	pulsation (rad$\cdot$s^{-1})

Subscripts or Superscripts

$^\circ$	standard conditions (298 K, 1 bar)
a	anode
c	cathode
cell	electrolysis cell

Acronyms

AST	accelerated stress test
BoL	beginning-of-life
Capex	capital expenses or expenditures
CV	cyclic voltamogram
EIS	electrochemical impedance spectroscopy
EoL	end-of-life
EW	equivalent weight of membrane (eq$\cdot$g^{-1})
HER	hydrogen evolution reaction
HHV	high heating value
KPI	key performance indicator
LHV	low heating value
OER	oxygen evolution reaction
Opex	operational expenses or expenditures
PEM	proton exchange membrane
PEMFC	polymer electrolyte membrane fuel cell
PFSA	perfluorosulfonic acid
PFSI	perfluorosulfonic ionomer
PGM	platinum group metals
PTFE	Polytetrafluorethylene
rds	rate-determining step
RES	renewable energy sources
RHE	reversible hydrogen electrode
SEM	scanning electronic microscopy
SPE	solid polymer electrolyte

11.1 Introduction

The purpose of this chapter is to review the state-of-the-art knowledge regarding the degradation of PEM water electrolysis cells and stacks. Whereas significant efforts have been made over the last years to better understand degradation mechanisms in PEM fuel cells (in particular for applications in the automotive industry) (Wang et al. 2012), the degradation of PEM water electrolysis cells has been less considered. Although both technologies share common features, they differ significantly. A better understanding of microscopic phenomena that tend to degrade both performance and durability of PEM water electrolysis cells is critical not only for those working in R&D departments who wish to improve the technology for new market applications, but also for the manufacturers of PEM water electrolyzers who wish to implement quality control procedures on production lines (at component, cell, or stack level) or on-site process monitoring–diagnostic in view of anticipated maintenance operations.

The KPI of a PEM water electrolysis cell is the current–voltage polarization curve measured at reference operating temperature and pressure. The term "degradation" used in this chapter is making reference to the effect of quite diverse microscopic processes that individually or collectively tend to induce a loss of performance with time. A difference is made between degradation (progressive loss of performances that tend to degrade cell efficiency) and failure (sudden and irreversible loss of performances that can eventually lead to the entire destruction of the cell). Different techniques used to evaluate the level of performance of PEM water electrolysis cells (see Chapter 10) can also be used to investigate degradation processes. They can provide either in situ information about the status and level of performance for the different internal cell components or postmortem information.

The rate at which degradation proceeds is strongly related to operating conditions. Whereas up to now most PEM water electrolyzers in operation in the industry sector were operated under (quasi-)stationary conditions of temperature, pressure, and current density, emerging markets related to the implementation of renewable energy sources (RES) are calling for systems that can sustain more aggressive operating conditions, that is, the coupling of temperature, pressure, and power load cycles of different magnitude and frequency. The impact of such operation cycles on cell components can be extremely deleterious and can contribute to reduce the life span of the electrolysis unit, in terms of both performances and durability. A distinction has to be made between operation in (mostly) stationary conditions (according to the literature, U.S. Navy test cells have exceeded 120,000 h, following General Electric Program pioneering experiments in the 1960's Arkilander and Molter 1984) and operation in transient conditions for which durability results are not available. To some extent, accelerated stress test (AST) protocols that can be implemented to accelerate cell aging through degradation processes in order to gain a better understanding of underlying microscopic phenomena in a limited amount of time, tend to become a real mode of operation. There is a common agreement among the water electrolysis community that new end-uses are requiring adjustment of performances. It is beyond the scope of this chapter to discuss strategies that can be implemented to facilitate the service. The focus is on the description of degradation mechanisms and the identification of underlying microscopic processes.

11.2 Overview of Degradation Processes and Failure Mechanisms

11.2.1 Water-Splitting Cell Designs

First, let us briefly review water-splitting cell designs to better outline the specific features of PEM electrolysis compared to incumbent technologies. The water splitting reaction is a nonspontaneous endergonic transformation that leads to the decomposition of water molecules into elemental hydrogen and oxygen. PEM is a low-temperature technology that is used to electrolyze liquid water according to

$$H_2O(l) \rightarrow H_2(g) + \frac{1}{2}O_2(g) \qquad (11.1)$$

During electrolysis, the energy required for the process is provided in the form of electrical energy. From the engineering viewpoint, three main cell designs can be used for that purpose (Figure 11.1). All three share similar features: two electrodes of planar geometry are placed face-to-face and separated by an ion-conducting electrolyte. The larger the distance between anode and cathode, the larger the ohmic resistance of the cell. To avoid the spontaneous recombination of gaseous hydrogen and oxygen, a separator is introduced in the interpolar gap. The first design (Figure 11.1a) is called a "gap cell" (Millet 2011). The term "gap" is making reference to the distance between the electrodes and the cell separator. This gap is filled with liquid electrolyte. The cell separator is usually a microporous inert medium impregnated with the liquid electrolyte. During water electrolysis, gaseous oxygen evolves at the anode/electrolyte interface on the left and gaseous hydrogen

evolves at the cathode/electrolyte interface on the right. As current density increases, gas bubbles tend to form a screening and nonconducting film over electrode surfaces. This cell design is, therefore, not appropriate for water electrolysis and when used, limited to low current density values.

The situation can be improved by using the second cell design of Figure 11.1b called a "zero-gap cell." A liquid electrolyte is still used but two porous electrodes are pressed onto the separator to reduce as much as possible the distance between anode and cathode (and corresponding ohmic losses). Gaseous reaction products are released through the pores at the rear of the electrodes. Such cells are more compact and higher current densities can be reached without significant extra losses and gas screening effects are avoided.

Both "gap cell" and "zero-gap cell" designs require a liquid electrolyte. For water electrolysis, this is usually a concentrated aqueous solution of potassium hydroxide. Whereas there is an interest at using strongly acid electrolyte solutions (e.g., proton transport by hoping mechanism is more effective than the transport of hydroxyl ions), acidic electrolyte is strongly corrosive and tends to attack metallic components of the circuitry through which the electrolyte is pumped. A first solution that has been used to circumvent the problem was to use acid-impregnated porous cell separators. However, the problem was not fully solved since electrolyte leaks were difficult to manage. A decisive step was made when appropriate proton-conducting polymers became available. The third design is called a "solid polymer electrolyte (SPE cell)" (Figure 11.1c). In such cells, an ion-conducting membrane material is used to combine the role of electrolyte and cell separator. Ions that convey electric charges from the electrode to the other are immobilized inside the membrane, which acts as a solid electrolyte. In PEM water electrolysis, a proton-conducting membrane is used as the cell separator. The acidity of the electrolyte remains confined within the thin membrane sheet. There is no liquid electrolyte in circulation in the electrolyzer. Only deionized water is circulated in the anodic chamber to feed the electrochemical reaction. Similar concepts using hydroxyl ion–conducting polymers are also under development.

11.2.2 Strengths and Weaknesses of PEM Water Electrolysis Cells

Figure 11.2 shows the cross section of a conventional PEM water electrolysis cell (a refers to anode and c to cathode). More details are provided in Chapter 10. Briefly, the unit cell is delimited by two titanium end plates (4a and 4c). The cell is compact, having a typical thickness of 4–7 mm, depending on the selection of internal

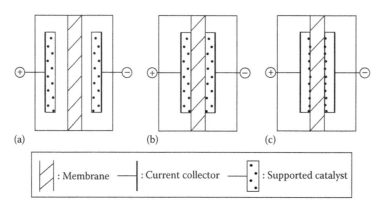

FIGURE 11.1
2D Schematic diagrams of (a) a gap cell; (b) a zero-gap cell; (c) a SPE cell.

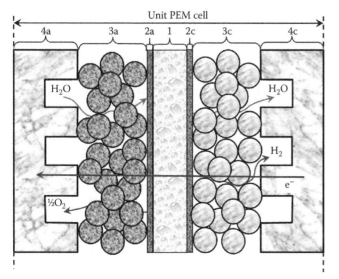

FIGURE 11.2
Schematic diagram showing the cross-section of a PEM water electrolysis cell.

components. The ohmic resistance of a PEM cell is usually in the range of 150–250 mΩ·cm², yielding a voltage loss of 150–250 mV at 1 A·cm⁻², and a heat source of 150–250 mW·cm⁻² (Millet 1990). The central component (the electrochemical cell itself where water dissociation takes place) is the membrane electrode assembly (MEA). The MEA is made of a thin (100–250 μm thick) proton-conducting membrane of perfluorosulfonic acid (PFSA) at the center (1) (Vishnyakov and Neimark 2000; Mauritz and Moore 2004). The membrane is surface coated with two porous catalytic layers where half-cell reactions take place, the cathodic layer (2c) in the cathodic compartment and the anodic layer (2a) in the anodic compartment. These layers are a mixture of catalysts particles and perfluorosulfonated ionomer (PFSI) chains. The membrane or SPE is usually reinforced using a PTFE web. This Polytetrafluorethylene (PTFE) reinforcement

is used for mechanical purpose, especially during operation at elevated pressure. The cathodic layer is a porous and micrometer thick 3D structure composed of carbon-supported platinum nanoparticles embedded in PFSA ionomer chains. The anodic layer is a few micrometer thick porous 3D structure composed of iridium dioxide nanoparticles also embedded in PFSA ionomer chains. This MEA is clamped between a cathodic current distributor (3c) and a millimeter thick anodic porous titanium current distributor (3a) of appropriate porosity (Grigoriev et al. 2009). The cell is delimited by two titanium end plates (4a and 4c). A void through which liquid water is pumped is provided. This can be either directly machine made within thick end plates of alternatively millimeter thick titanium grid spacers. A DC voltage > E° (the thermodynamic electrolysis voltage) is applied to the cell. During operation, gas bubbles formed in the catalytic layers diffuse through the porous current distributors (3a, 3c) and are collected and transported by liquid water pumped through the grids (4a, 4c).

A PEM water electrolysis cell presents some specific characteristics that are due to the use of a thin film of acidic SPE and to the use of highly deionized water.

Strengths:

- This is a compact cell that uses porous electrodes through which the gas production (hydrogen and oxygen) is evacuated; because gaseous reaction products are released at the rear of the cell, the gas production does not introduce any parasite ohmic resistance.

- The SPE used in PEM water electrolysis cell has a high proton conductivity and a low gas permeability, this is why it is possible to use sub-millimeter thick membranes (typically in the 100–200 μm range); this is a desirable characteristic of a PEM water electrolysis cell; the distance between anodic and cathodic active layers is small and the ohmic voltage drop across the SPE is minimized.

- The ionic charge carriers (hydrated protons) remain trapped inside the membrane; there is no risk of electrolyte leak that could induce corrosion in the circuitry.
- The cell can be operated at elevated current density; whereas commercial electrolyzers are already operating between 0.8 and 2.5 $A \cdot cm^{-2}$, laboratory experiments have been performed up to 10 $A \cdot cm^{-2}$, opening the way to significant capex reduction.

Weaknesses:

- Due to the strongly acidic environment of the proton-conducting membranes, only platinum group metals (PGM) can be used as electrocatalysts; Pt that is used at the cathode is scarce and expensive; in addition, Pt is sensitive to surface contamination by traces of amounts of metallic cations; this is also an active catalyst for the recombination of H_2 and O_2 into either hydrogen peroxide (a chemical responsible for membrane degradation) or water; Pt particles can potentially ignite direct combustion of H_2 in O_2.
- The membrane is prone to incorporate cationic species found in feed water; as a result, there is a need to use strongly deaerated/deionized water, a situation that is energy consuming, expensive, and requires the use of self-passivating metals such as titanium for cell components; however, top input water quality is a good investment.
- Charge carriers are confined within the thin SPE sheet of very accurate thickness; the electric contact between the MEA and the current collectors is obtained by compression; the reduction of parasitic ohmic contact resistances imposes significant constraints in terms of manufacturing tolerances on the different cell components, and this makes the cells more expensive compared to those that use a liquid electrolyte.
- The electrolyte is a solid polymer and the distance between anode and cathode is sub-millimetric; as a result, conventional reference electrodes (e.g., saturated calomel reference electrodes) cannot be used to separately investigate the electrochemical features of each interface (the problem can be circumvented by using so-called internal reference electrodes; more details are provided in Chapter 10); the situation is even worth in PEM stacks.

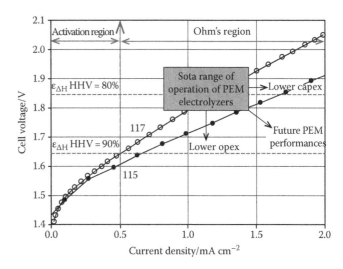

FIGURE 11.3
Two typical polarization curves measured on PEM water electrolysis cells. (o) MA with Nafion 117; (•) MEA with Nafion 115.

Current–voltage polarization curves are the main KPIs used to assess the performance level of PEM water electrolysis cells. Figure 11.3 shows two typical polarization curves measured on two MEAs with the same electrocatalysts (Pt-C at the cathode for the hydrogen evolution reaction (HER) and IrO_2 at the anode for the oxygen evolution reaction (OER)) but different membranes. The efficiency of the cell is a function of operating current density. At low current density (close to equilibrium), the efficiency of the cell is high but the amount of water that is decomposed per unit of time and surface is small. At high current density, the gas production increases but the efficiency is lower. An efficient PEM water electrolysis cell is a cell in which internal cell resistances are minimized and maintained at low values for a long (ideally in the upper range of the 10^4 to 10^5 h time interval) period of operation. Degradation means that the efficiency of the cells is decreasing with time. In other words, performances measured at beginning-of-life (BoL) are better than those measured at end-of-life (EoL). The role of the electrochemist engineer is therefore to minimize loss sources in order to maximize cell efficiency (the cell efficiency determines the so-called "operational expenses" (opex), that is, the energy cost of the process expressed in €/kg H_2) and durability (the durability of a PEM water electrolysis cell determines the so-called "capital expenses" (capex), that is, the investment cost for the process expressed in €/kg H_2). This can be achieved by selecting appropriate materials and by a judicious design of the unit PEM cell.

11.2.3 Performance of PEM Water Electrolysis Cells

A simple but nevertheless accurate expression of the current–voltage relationship can be derived as follows. For a given redox couple, the Butler–Volmer relationship relates the current density i to the overvoltage η set across an interface (i_0 is the exchange current density, β the symmetry factor, R_{PG} the constant of perfect gas and T the absolute temperature) for a mono-electronic reaction:

$$i = i_0 \left\{ \exp\left[\frac{(1-\beta)F}{R_{PG}T}\eta\right] - \exp\left[-\frac{\beta F}{R_{PG}T}\eta\right] \right\} \quad (11.2)$$

According to Bockris (Bockris and Reddy 1982), Equation 11.2 can be generalized to multistep reactions in which there may be electron transfers in steps other than the rate-determining step (rds) and in which the rds may have to occur ν times per occurrence of the overall reaction:

$$i = i_0 \left\{ \exp\left[\frac{\overleftarrow{\alpha} F}{R_{PG}T}\eta\right] - \exp\left[-\frac{\overrightarrow{\alpha} F}{R_{PG}T}\eta\right] \right\} \quad (11.3)$$

In Equation 11.3, oxidation and reduction transfer coefficients $\overleftarrow{\alpha}$ and $\overrightarrow{\alpha}$ are a function of n, which is the number of single-electron transfer steps in the overall reaction. During water electrolysis, η_{O_2} at the anode (OER) is positive and η_{H_2} at the cathode (HER) is negative:

$$i^a = i_0^a \left\{ \exp\left[\frac{\overleftarrow{\alpha_a} F}{R_{PG}T}\eta_{O_2}\right] - \exp\left[-\frac{\overrightarrow{\alpha_a} F}{R_{PG}T}\eta_{O_2}\right] \right\} \quad (11.4)$$

$$i^c = i_0^c \left\{ \exp\left[\frac{\overleftarrow{\alpha_c} F}{R_{PG}T}\eta_{H_2}\right] - \exp\left[-\frac{\overrightarrow{\alpha_c} F}{R_{PG}T}\eta_{H_2}\right] \right\} \quad (11.5)$$

Figure 11.4 shows that although water splitting is a reversible process at elevated (>800°C) temperatures, the overall reaction is strongly irreversible at

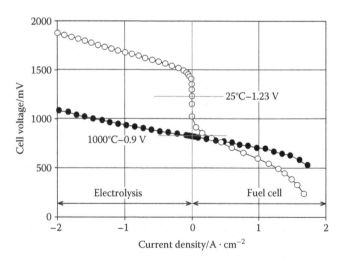

FIGURE 11.4
Reversibility of H_2O/O_2 and H_2O/H_2 redox couples at two different temperatures.

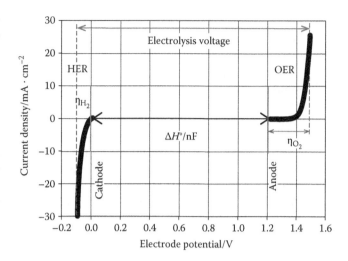

FIGURE 11.5
Plot of electrode potentials versus current density during PEM water electrolysis (IR is neglected).

close-to-ambient temperature. This is due to the low kinetics of the H_2O/O_2 redox couple at the anode that requires significant overvoltage η_{O_2} to get a detectable current of oxidation (Figure 11.5). In other words, the exchange current density of the H_2O/O_2 redox couple is very small. Therefore, at the anode, the kinetics of the OER at any current density can be approached by neglecting the reverse term of Equation 11.2 (the reduction of oxygen into water):

$$i^a \approx i_0^a \exp\left[\frac{\overleftarrow{\alpha_a} F}{R_{PG}T}\eta_{O_2}\right]$$

which is equivalent to

$$\eta_{O_2} \approx \frac{R_{PG}T}{\overleftarrow{\alpha_a} F} Ln\left(\frac{i^a}{i_0^a}\right) \quad (11.6)$$

At the cathode, the HER is more reversible than the OER and its kinetics is significantly larger (in other words, the exchange current density of the H_2O/H_2 redox couple is high). Equation 11.2 can be simplified only at sufficiently high current densities:

$$i^c \approx i_0^c \exp\left[-\frac{\overrightarrow{\alpha_c} F}{R_{PG}T}\eta_{H_2}\right]$$

which is equivalent to

$$\eta_{H_2} \approx \frac{R_{PG}T}{\overrightarrow{\alpha_c} F} Ln\left(\frac{-i^c}{i_0^c}\right) \quad (11.7)$$

Experimental current–voltage polarization curves of PEM water electrolysis cells can be modeled using Equation 11.8 in which the different cell voltage contributions are summed up:

$$U_{cell}(T,P) = V_{T,P} + i\sum_{k=1}^{n} R_k + \eta_{H_2}i + \eta_{O_2}i \qquad (11.8)$$

Equation 11.8 is similar to the one used in PEM fuel cell technology (Barbir 2005). U_{cell} (T,P) is the water electrolysis voltage (V) at temperature T and pressure P of operation; $V_{T,P} = \Delta H(T,P)/nF$ is the thermo-neutral voltage (V), at T,P; i is the current density in A·cm^{-2}; R_k are the different ionic and electronic cell resistances in Ω·cm^{-2}. Equations 11.6 and 11.7 can be used in Equation 11.8 only for sufficiently large current density values (approximately 10 times the exchange current density of OER and HER). At low current densities, it is necessary to use Equations 11.4 and 11.5.

Figure 11.6 shows the two polarization curves of Figure 11.3 and their best fits using Equation 11.8. Fit parameters are compiled in Table 11.1. Reference exchange current density for the OER on iridium dioxide was taken from Damjanovic and Bockris (1966), and reference exchange current density for the HER on platinum was taken from Bockris and Reddy (1982). r_f^a and r_f^c are the roughness factors of anode and cathode, respectively. The two fits were obtained by changing only the cell resistance R_{cell}. According to the literature, Nafion 117 has a dry thickness of 178 μm and a surface resistance of 230 mΩ·cm². Nafion 115 has a dry thickness of 127 μm and a surface resistance of 165 mΩ·cm². Best-fit values are consistent with these data.

TABLE 11.1

Parameters of Equation 11.5 Used to Fit Experimental Polarization Curves of Figure 11.6

$R_{cell} = \sum_{k=1}^{n} R_k$ (mΩ·cm²)	i_0^a (A·cm^{-2})	r_f^a (adim)	i_0^c (A·cm^{-2})	r_f^c (adim)
Curve 1	230	5×10^{-6}	800	1×10^{-3}	800
Curve 2	155	5×10^{-6}	800	1×10^{-3}	800

Note: $T = 80°C$; $\beta^a = \beta^c = 0.5$.

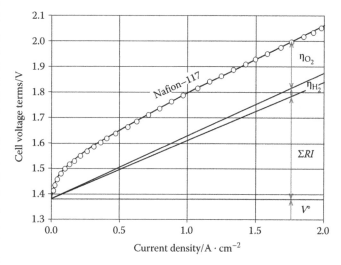

FIGURE 11.7
Different voltage terms contributing to the cell voltage of a PEM water electrolysis cell.

Figure 11.7 shows the different terms of the polarization curve obtained with Nafion 117. In conventional PEM water electrolysis cells, besides the thermodynamic voltage V°, the oxygen evolution overvoltage η_{O_2} is the predominant term. The cathodic overvoltage η_{H_2} is much smaller. The ohmic drop ΣRI across the cell (where ΣR is making reference to the different internal cell resistances) is a linear function of operating current density.

Performance degradation of PEM water electrolysis cells can be analyzed by considering Equation 11.8. For different reasons that will be discussed in detail in the next sections, the internal cell resistance and charge transfer overvoltages tend to increase with time.

11.2.4 Impedance of PEM Water Electrolysis Cells

Electrochemical impedance spectroscopy (EIS) analysis provides valuable information on the status of PEM water electrolysis cells. A cross section of a conventional PEM cell is provided in Figure 11.8, together with an electrical analogy (adapted from Rozain and Millet 2014). Electrochemical metal-electrolyte interfaces are usually described by the parallel connection

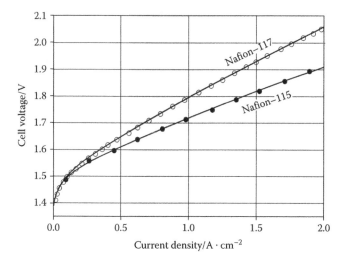

FIGURE 11.6
Experimental (o,•) and model (–) polarization curves. Fit parameters of Table 11.1.

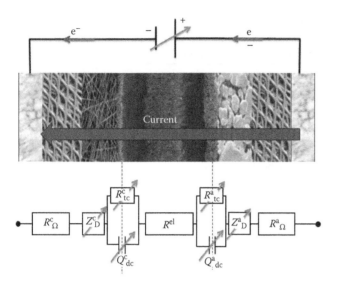

FIGURE 11.8

(top) Cross-section of a PEM water electrolysis cell (thicknesses not to scale); (bottom) equivalent electrical circuit (see Section 11.2.4 for the definition of the different terms of the circuit).

of a charge transfer resistance (R_{ct} in $\Omega \cdot cm^2$) and a double-layer capacitance C_{dl} (in $F \cdot cm^{-2}$) (Lvovich 2012). As already discussed in Chapter 10, the different terms used in the analogy are as follows:

- R^c_Ω and R^a_Ω ($\Omega \cdot cm^2$) are current nondependant electronic resistances of electron-conducting metallic cell components, respectively, in the cathodic and anodic cell compartments.

- R^{el} ($\Omega \cdot cm^2$) is the current nondependant ionic resistance of the SPE.

- R^c_{ct} ($\Omega \cdot cm^2$) is the current-dependant cathodic charge transfer (polarization) resistance associated with the HER; R^c_{ct} decreases as the electric potential of the cathode is lowered during operation.

- R^a_{ct} ($\Omega \cdot cm^2$) is the current-dependant anodic charge transfer (polarization) resistance associated with the OER; R^a_{ct} decreases as the electric potential of the anode is raised during operation.

- Q^c_{dl} ($F \cdot cm^{-2}$) is the potential-dependant double-layer capacitance associated with the cathode/electrolyte interface.

- Q^a_{dl} ($F \cdot cm^{-2}$) is the potential-dependant double-layer capacitance associated with the anode/electrolyte interface.

- Z^c_D ($\Omega \cdot cm^2$) is the current-dependant cathodic diffusion impedance due to H_2 transport away from the cathode.

- Z^a_D ($\Omega \cdot cm^2$) is the current-dependant anodic diffusion impedance due to O_2 transport away from the anode and/or to H_2O transport to the anode.

Metal-electrolyte interfaces are characterized by a charge transfer resistance and the so-called double layer where electric charges can accumulate. The double layer of an interface of unit roughness factor is a pure capacitance C_{dl} (in $F \cdot cm^{-2}$). C_{dl} is placed in parallel with the charge transfer resistance R_{ct} to account for the fact that charge can either remain trapped at the surface of flow across the interface. Metal-electrolyte interfaces found in PEM water electrolysis cells are usually not pure capacitance because of the desirable 3D structure of the catalytic layers, which is used to increase their roughness. Interface capacitances are modeled using so-called constant-phase elements (CPEs) (Brug et al. 1984). CPEs are used to take into account deviations of double layers from ideal capacitive behaviours due to surface roughness, polycrystallinity, and anion adsorption (sulfonate ions in PEM water electrolysis cells). The impedance Z_Q of a CPE noted Q is

$$Z_Q = \frac{1}{Q(j\omega)^n} \qquad (11.9)$$

where

Q is a frequency independent constant in $F \cdot cm^{-2} \cdot s^{n-1}$

n is a factor ranging between 0 and 1

Q tends to a pure capacitance C as n tends to 1. In PEM water electrolysis cells, the measure of double-layer capacitances associated with the HER (cathode) and OER (anode) also provides an indication on the electrochemical activity and integrity of the two catalytic layers.

In this electrical analogy, diffusion impedances (Z^c_D and Z^a_D) are added in series to account for possible mass transport limitations of hydrogen and oxygen gases away from the interfaces across porous current distributors. However, when current distributors of appropriate open porosity are used, such diffusion impedances can be neglected. Individual terms of the electrical analogy can, in principle, be measured using EIS. During operation, both resistances and capacitances are prone to changes with time. Resistances (ohmic or charge transfer) tend to increase. As a result, ohmic losses increase and the cell becomes less efficient. Capacitances (related to the surface of electrode/electrolyte interfaces) tend to decrease because less reaction sites remain available. Analysis of cell resistances and capacitances evolution with time is a good option to investigate degradation mechanisms.

It should be noted that the information that can be gained from the analysis of polarization curves is the same as the information gained by EIS. The advantage of EIS is that in favorable cases, individual cell resistances can be measured. Charge transfer resistances measured by EIS are related to the polarization curve as follows. At low overpotential, Equation 11.2 can be linearized into

$$\lim_{\eta \to 0} \eta(i) \equiv \eta = \left(\frac{R_{PG}T}{nFi_0} \right) i \qquad (11.10)$$

The slope of Equation 11.6 is the charge transfer resistance at $\eta = 0$:

$$R_{ct}^{o} = \frac{R_{PG}T}{nFi_0} \qquad (11.11)$$

Equation 11.7 indicates that the charge transfer resistance at equilibrium is not zero. The higher the exchange current density, the lower the charge transfer resistance. At any η value, the charge transfer resistance R_{ct}^{η} is

$$R_{ct}^{\eta} = \left(\frac{\partial \eta}{\partial i} \right)_{T} \qquad (11.12)$$

Therefore,

$$\eta = \frac{2.3 R_{PG}T}{nF} \log \left(\frac{R_{ct}^{o}}{R_{ct}^{\eta}} \right) \qquad (11.13)$$

The slope of the polarization curve in the activation area (when the voltage drop due to the cell resistance is small) is the sum of the charge transfer resistances of both anode and cathode processes.

11.2.5 Cell Life Factors

Performance degradation means that the efficiency of the PEM cell (or the PEM stack) is decreasing more or less rapidly during operation. In other words, the polarization curve is gradually shifted to higher potential values. Degradation processes are either reversible or irreversible. The reason for such loss of efficiency is that the resistances of Figure 11.8 tend to increase while capacitances tend to decrease during operation. Ohmic resistances tend to increase because the membrane is either contaminated by chemical pollutants or degraded, or because contact resistances are increasing. In Figure 11.8, charge transfer resistances and catalyst capacitances are directly related to the number of active sites that are available for the electrochemical reactions. If for any reason the number of these sites decreases, then charge transfer resistances will increase and interfacial capacitances will decrease.

The main factors that impact cell life are as follows: (i) water quality (dissolved metal ions in the feed water are incorporated by the membrane and the result is a greater membrane resistance, measured as voltage degradation); (ii) heat (over-heated conditions will cause the membrane material to degrade); (iii) hydration (cell–stack hydration level should be maintained to avoid damage from shrinkage); and (iv) freezing (freezing can cause physical damage due to expansion of ice). According to PEM water electrolysis suppliers (Anderson et al. 2013), some factors are less critical: (i) continuous duty at 100% duty cycle; (ii) starts and stops (there is no corrosion associated with shutting down a system; there are no life limits to the number of starts and stops possible); (iii) long-term storage without operation (cells can be stored indefinitely as long as they are hydrated).

Regarding operating conditions, cells operating in stationary conditions of temperature, pressure, and current density have demonstrated their ability to operate over significantly long periods of time (>20,000 h). However, intermittent and flexible operation of PEM water electrolyzers tends to significantly increase the rate of degradation. Some reports indicate a 30% performance loss due to nonstationary conditions (DOE annual progress report on hydrogen and fuel cells program 2013). Investigation and analysis of underlying degradation processes is a somewhat recent matter of interest. Whereas it can be assumed that coupling effects tend to accelerate degradation processes, there is still a need for additional studies in the field to identify the impact of individual parameters on degradation rates.

11.2.6 Techniques and Methodologies for Investigating Degradation Mechanisms

A list of main experimental techniques that can be used to evaluate the performance of PEM water electrolysis cells is provided in Chapter 10. They can also be used to measure degradation rates. Since degradation processes are strongly related to the conditions of operation of the PEM water electrolysis cells or stack, it is not possible to have a unique picture of degradation phenomena. Some steps may prevail in some specific cases of operation. ASTs can be defined according to target applications and used to investigate specific aging phenomena.

11.2.7 From Cell Degradation to Failure

Degradation is a term that usually makes reference to a gradual loss of electrochemical performances with time. Ultimately, a degradation process can end up with a cell–stack failure. Figure 11.9 shows plots of cell voltage, anodic, and cathodic overvoltages with time for a PEM cell operating at 500 mA·cm^{-2} (Andolfatto et al. 1994; Millet et al. 1996). The rate of performance degradation is measured with units of $\mu V \cdot h^{-1}$. The mean voltage drift of $41\ \mu V \cdot h^{-1}$ measured on the cell voltage is mainly due to the degradation of anode.

Failure is a term that makes reference to irreversible cell–stack degradation that can in turn leads to the destruction of the electrolyzer (explosion or combustion). This is usually the ultimate step of degradation

processes. Figure 11.10 shows an example of a PEM water electrolysis cell after combustion provoked by the spontaneous combustion of hydrogen and oxygen (Millet et al. 2012).

Risks of cell–stack failure can be predicted and prevented to a certain extent by monitoring individual cell voltages (e.g., a cell electrical short circuit or a high contamination levels of impurities in the gaseous production are clear indications of cell failure with the risk of formation of hazardous hydrogen–oxygen mixtures). More information is provided in Section 11.6.

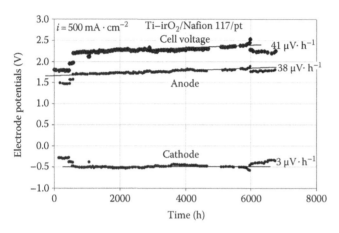

FIGURE 11.9
Endurance testing performed on a PEM water electrolysis cell at 500 mA·cm⁻². (Reprinted from *Int. J. Hydrogen Energ.*, 19, Andolfatto, F., Durand, R., Michas, A., Millet, P., and Stevens, P., Solid polymer electrolyte water electrolysis: Electrocatalysis and long term stability, 421–427. Copyright 1994, with permission from Elsevier.)

11.3 Catalyst Degradation, Consequences, and Risk Mitigation

11.3.1 General Characteristics of PEM Water Electrolysis Catalysts

In PEM water electrolyzers, the half-cell reactions are

$$\text{Anode:}\quad H_2O(l) \rightarrow \frac{1}{2}O_2(g) + 2H^+ + 2e^- \quad (11.14)$$

$$\text{Cathode:}\quad 2H^+ + 2e^- \rightarrow H_2(g) \quad (11.15)$$

(a)

(b)

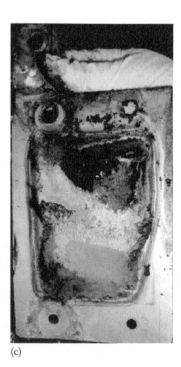

(c)

FIGURE 11.10
Photographs of a PEM water electrolysis cell after combustion: (a) the MEA; (b) the cathodic cell compartment with Ti grid; (c) bipolar plate. (Reprinted from *Int. J. Hydrogen Energ.*, 37, Millet, P., de Guglielmo, F., Grigoriev, S.A., and Porembskiy, V.I., Cell failure mechanisms in PEM water electrolyzers, 17478–17487. Copyright 2012, with permission from Elsevier.)

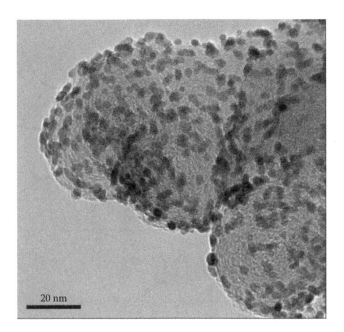

FIGURE 11.11
SEM micrograph of Pt nanoparticles on carbon used for the HER. (Reprinted from *Int. J. Hydrogen Energ.*, 35, Grigorieva, S.A., Milletb, P., Dzhusc, K.A., Middletond, H., Saetred, T.O., and Fateeva, V.N., Design and characterization of bi-functional electrocatalytic layers for application in PEM unitized regenerative fuel cells, 5070–5076. Copyright 2010, with permission from Elsevier.)

Unsupported platinum particles are used at the cathode for the HER. Alternatively, carbon-supported platinum nanoparticles (Figure 11.11) can also be used. At the anode, unsupported iridium dioxide nanoparticles are usually used as oxygen-evolving centers (Figure 11.12).

The key factor that determines the efficiency of the cell is the number of active sites at the cathode and anode. They can be determined by cyclic voltammetry. At the cathode, the electric charge measured by integration of the cyclic voltammogram (CV) in the region of

hydrogen adsorption ($0 < E < +400$ mV/RHE) provides an information on the number of cathodic sites. At the anode, the electric charge measured by the integration of the CV in the oxide formation–reduction region provides information on the number of anodic sites. CVs plotted in Figures 11.13 and 11.14 show that the number of catalytic sites available at BoL is much larger than the number available at EoL.

Instead of measuring the absolute number of catalytic sites in each catalytic layer, it is usually more convenient to determine the roughness factor r_f of an electrode. This is a dimensionless quantity defined as the ratio of the charge density Q (in $C \cdot cm^{-2}$) measured on a rough (real) interface to the charge density Q_{ref} (in $C \cdot cm^{-2}$) measured on a smooth reference interface:

$$r_f = \frac{Q}{Q_{ref}} \tag{11.16}$$

Roughness factors are, therefore, an indirect measure of the number of catalytic sites available for the electrochemical reactions in each catalytic layer. For a given interface, a reduction of the roughness factor is equivalent to a reduction of the interface capacitance and to an increase in the charge transfer resistance. All three quantities provide an indication on the electrochemical activity and integrity of the two catalytic layers. There are different microscopic phenomena that tend to reduce the number of catalytic sites available on each side of the PEM cell between BoL and EoL. Surface contamination is a first example. The sintering of catalyst particles is a second example. Catalyst particles can also be gradually washed out during operation.

Using Equation 11.8, it is possible to visualize the effect of different microscopic rate parameters on the shape of the polarization curve. The different plots of Figure 11.15a and b show what happens when the roughness factors of the cathode (r_f^c) and anode (r_f^a) are decreased. The effect appears mainly at low current densities in the activation region of the polarization curve.

11.3.2 Example of Reversible Degradation Process

In PEM water electrolyzers, it is necessary to pump deionized water because the SPE can incorporate alien cationic species by simple contact with the solution. However, deionized water is a somewhat aggressive and corrosive medium. When contacting stainless steel, a steady dissolution of metallic ions is observed. Different factors such as temperature, the presence of oxygen or hydrogen make the process more severe. Analysis of the content of MEA in contact with water circulating in stainless steel piping reveals that at equilibrium, ($\approx$24% of SPE sites were occupied by Fe^{3+} species, $\approx$3% by Ni^{2+} species and $\approx$1% by Cr^{3+} species). The impact

FIGURE 11.12
SEM micrograph of unsupported IrO_2 nanoparticles for the OER.

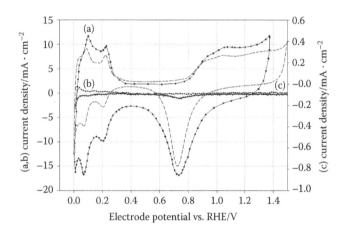

FIGURE 11.13
(a,b) CVs measured on C/Pt PEM cathodes at BoL and EoL and (c) Reference cyclic voltammogram measured on bulk Pt in H_2SO_4.

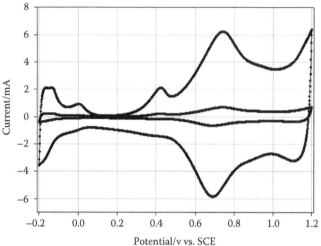

FIGURE 11.14
Cyclic voltammograms measured on IrO_2 in a PEM MEA at BoL and EoL.

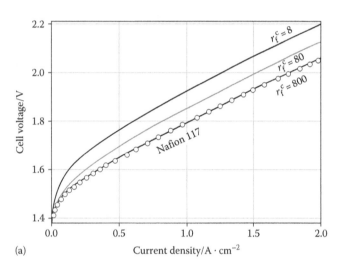

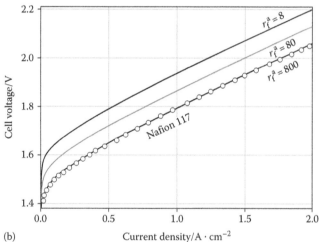

FIGURE 11.15
(a) Impact of the cathodic roughness factor r_f^c on cell performances; (o) reference curve of Figure 11.3 with Nafion 117. (b) Impact of the anodic roughness factor r_f^a on cell performances; (o) reference curve of Figure 11.3 with Nafion 117.

on the performance of the PEM can be significant. For example, Figure 11.16 shows what happens when an MEA is equilibrated with an aqueous solution of nickel chloride of increasing concentration (Millet et al. 1993). The polarization curve is shifted toward higher potential values. Incorporation of alien cationic species has several negative effects. First, the ionic conductivity of the SPE is reduced. Second, metallic cations tend to reduce to the metallic state at the cathode, covering Pt nanoparticles, and hence increase the HER overvoltage. Also, metallic cations are known to facilitate the formation of peroxide, which is a threat for PFSA materials.

Although possible, in situ (or even ex situ) regeneration of MEAs is not a simple task. The circulation of adequately concentrated acidic solutions within the PEM

cells is not sufficient to remove these cationic species, and cell disassembly to treat the MEAs in acidic solutions should be avoided.

11.3.3 Example of Irreversible Degradation Process

The number of catalytic sites available in both anodic and cathodic catalytic layers directly determines the level of performance of a PEM water electrolysis cell (see Figure 11.15). As in PEM fuel cell technology, a catalytic site is a triple point located at the interface of three different phases, namely, electronic, ionic, and gas transport phases. A triple point is basically a catalyst surface site connected to the underlying network of electronic conduction paths, and also in contact with the ion-conducting electrolyte

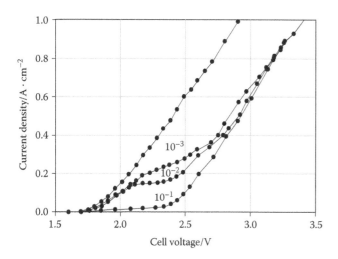

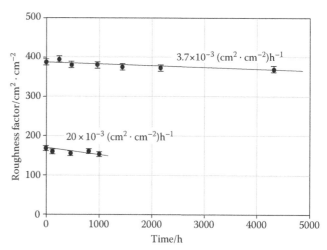

FIGURE 11.16
Polarization curve measured on an MEA equilibrated with an aqueous solution of nickel chloride of increasing concentration. (Reproduced from Millet, P. et al., *J. Appl. Electrochem.*, 23, 322, 1993.)

FIGURE 11.18
Plots of roughness factor of a PEM cathode as a function of time measured at two different current densities.

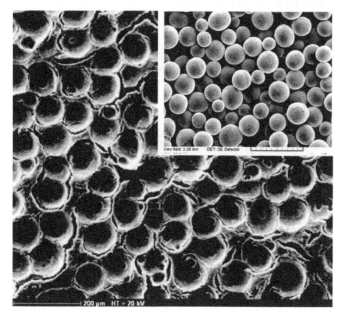

FIGURE 11.17
SEM micrograph of the surface of an IrO$_2$ catalytic layer after $\approx$ 100 h of operation; insert: SEM micrograph of the titanium current collectors made of sintered spherical titanium particles. (Reprinted from *Int. J. Hydrogen Energ.*, 36, Millet, P., Mbemba, N., Grigoriev, S.A., Fateev, V.N., Aukauloo, A., and Etiévant, C., Electrochemical performances of PEM water electrolysis cells and perspectives, 4134–4142. Copyright 2011, with permission from Elsevier.)

surface titanium sphere. As discussed earlier, the roughness factor of the catalytic layer can be used as a measure of the number of active sites available for the reaction. Using an internal reference electrode (see Chapter 10 for details), it is possible to measure separately the roughness factors of anode and cathode in a PEM water electrolysis cell. These factors tend to decrease with time during operation. This is exemplified in Figure 11.18, for two PEM cathodes with different initial roughness. There are different factors such as temperature, the presence of oxygen or hydrogen tend to accelerate the process, that can contribute, individually or collectively, to this loss of activity (e.g., the sintering of Pt nanoparticles at the surface of the carbon carrier or the loss of particles that are washed out during operation and pulled along the pores of the current collectors by the gas production). It is not always an easy task to establish a clear relationship between one of these processes and the observed loss of performance, in order to find a solution to the problem and improve durability. An empirical optimization of the chemical composition and microstructure of the catalytic layers, although requiring time and efforts, is probably the best approach.

and the pore structure that collects gaseous hydrogen or oxygen end-product molecules. Figure 11.17 shows the top view of an IrO$_2$ catalytic layer after operation (Millet et al. 2011). The crater-like surface is due to the shape of the porous current collector (Figure 11.17, inset), which is made of sintered spherical titanium particles. During operation, gaseous oxygen produced within the catalytic layer is collected through the cracks located around each

11.4 Membrane Degradation, Consequences, and Risk Mitigation

11.4.1 General Characteristics of PEM Water Electrolysis Membranes

Membranes used in PEM water electrolysis cells are invariably based on PFSA such as those originally developed by DuPont and known as Nafion®. Short-side-chain

PFSA materials commercially available from Solvay under the Aquivion® trade name are also used. These short-side-chain ionomers are known to exhibit superior ionic conductivity and higher thermal resistance compared to Nafion-like structures. Due to the higher glass transition temperature, higher operating temperatures (up to at least 120°C) can be reached (Arico et al. 2010). The nature of PFSA ionomers is a key to achieving levels of performance and durability required by applications. Several key physical properties have been tuned by lowering the equivalent weight EW (which is equivalent to increasing the concentration of sulfonic acid groups) from the initial 1100–1200 $g \cdot eq^{-1}$ of Nafion down to 800 $g \cdot eq^{-1}$ in today state-of-the-art PFSA materials for PEM fuel cells. The ionic resistance is a function of equivalent weight, thickness, temperature, and percentage of relative humidity. Some data are compiled in Table 11.2. For a given EW, the ionic cell resistance is directly related to the thickness of the membrane (Tsampas et al. 2006).

Whereas significant progress has been made over the recent years in the fuel cell industry, leading to the implementation of 15–20 μm thick membranes in fuel cell vehicles, the situation is somewhat different in PEM water electrolyzers, and much thicker membranes (150–200 μm) are commonly used (especially for operation under pressure to avoid deleterious gas cross-permeation phenomena [Sakai et al. 1985, 1986]). Looking closer at the situation in PEM water electrolysis cells, there are two main ionic resistances that are difficult to measure separately: (i) the ionic resistance of the SPE membrane; (ii) the ionic resistance of the catalytic layers. From the microscopic viewpoint, hydrated protons are the sole charge carriers since ending sulfonate groups (covalently bonded to the fluoro-carbon backbone) are not mobile species under DC conditions. They differ only by the shape of the diffusion path (which can be more tortuous in the catalytic layers) and by the concentration (which is less in the catalytic layers). Ionic transport within electrolytes is a purely resistive process. The nonequilibrium

flux–force relationship that relates the voltage force to the ionic current flux follows Ohm's law (Hamann et al. 1998). Although the structure of PFSA materials at the nanoscale is that of a cluster network of charge conducting hydrophilic domains embedded in a hydrophobic and nonconducting organic domain (Figure 11.19), the SPE can be considered as a homogeneous membrane with isotropic conduction properties (Gierke et al. 1981). Therefore, from the electric viewpoint, the impedance of the SPE membrane used in PEM water electrolysis cells (cell component n°1 in Figure 11.8) is a purely resistive resistance R_{el} (in Ω) that can be normalized to the geometrical surface area (in $\Omega \cdot cm^2$). Although the two catalytic layers placed on each side of the SPE membrane are thin (in the micrometer range), their ionic resistance is difficult to measure separately. Ex situ accurate measurements show that their ionic resistivity can be much higher than that of the bulk SPE. They both contribute to increase the overall ionic resistance of the cell, sometimes by 10%–20%, depending on their structure.

In PEM fuel cell technology, membranes must have high mechanical strength to withstand premature failure due to dimensional changes that tend to occur on repeated hydration/dehydration cycles (Subianto et al. 2013). Although it can be considered that dimensional changes of PFSA membranes in PEM water electrolysis cells are limited because MEAs are constantly kept hydrated by liquid water, there are dimensional changes due to temperature changes. Dimensional changes can be limited by incorporating an inert, nonconducting reinforcement material (e.g., an expended PTFE film or web into which the PFSA ionomer is impregnated to form the membrane). An appropriate impregnation of the PTFE matrix is critical to avoid the formation of voids or pinholes that form weak points and make the membrane more susceptible to hydration-induced stresses and PFSA–PTFE exfoliation. PFSA membranes are also prone to chemical degradation due to the attack of PFSA chains by reactive species that form either during water

TABLE 11.2

Thickness, Equivalent Weight (EW), Ionic Resistance, and Voltage Drop at 1 $A \cdot cm^{-2}$ of Different Commercial PFSA Membranes

Supplier	Trade name	Thickness (μm)	EW ($g \cdot mol^{-1}$)	Resistance ($m\Omega \cdot cm^2$)	Voltage @ 1 $A \cdot cm^{-2}$ (mV)
Ion Power	Nafion® N1110	254	1100	328	328
Ion Power	Nafion® 117	178	1100	230	230
Ion Power	Nafion® 115	127	1100	168	168
Ion Power	Nafion® NR212	50.8	1100	66	66
Ion Power	Nafion® NR211	25.4	1100	33	33
Fumatech	F10180[a]	150–180	1000	<500	<500
Solvay	Aquivion® E87-12S	120	870	117	117
Solvay	Aquivion® E98-18S	180	980	207	207

[a] With PTFE reinforcement.

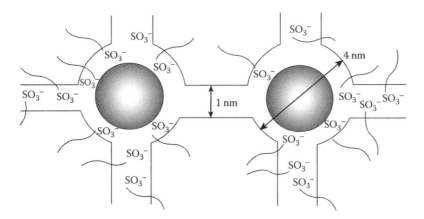

FIGURE 11.19
Schematic diagram showing the network microstructure of PFSA materials.

electrolysis or at rest potential (e.g., peroxy radicals). The process can be monitored by measuring the concentration of fluoride ions released in the main water stream. In conclusion, all these microscopic phenomena contribute to overall degradation processes that tend to spoil membrane properties and make PEM water electrolysis cells less efficient.

11.4.2 Example of Reversible Degradation Process

One example of reversible degradation process that takes place in PEM water electrolysis membranes is that of an increasing ionic resistivity that comes from contamination by foreign metallic cations. These cations are found in the stream water. They are released by the corrosion of stainless steel piping components used in the circuitry and ancillary equipment in contact with highly deionized water. PFSA materials used in PEM water electrolysis cells are proton conductors. Hydrated protons are the only mobile species. When the PFSA membrane is equilibrated with an aqueous solution of diluted cationic species, Donnan exclusion conditions prevail (in other words, cation–anion pairs are excluded from the membrane). However, an ion exchange process takes place, the driving force of which is the difference of chemical potential (concentration) of the different species in the two (water and membrane) phases. The process can be characterized by an equilibrium constant that sets, for a given PFSA material and a given temperature, the cation-to-proton ratio within the membrane. Examples are provided in References section in the later part of the article (Steck and Yeager 1980; Millet et al. 1995). Metallic cations tend to incorporate the membrane, and protons are released in the stream water. The resulting acidification of water tends to accelerate corrosion. Once within the membrane, metallic cations contribute to reduce ionic conductivity. This can be detected by conductivity measurement, using, for example, EIS spectroscopy. They also tend to migrate down to the cathode in response to the electrical field set across the PEM cell where they form metallic monolayers at the surface of platinum catalyst particles. Since the HER on non-Pt surface requires significantly larger overvoltages, the efficiency of the PEM cells tends to decrease. The process is somewhat reversible in the sense that metallic cations can be removed by soaking the MEA in an acidic aqueous solution. But of course this is a difficult, time-consuming, and expensive process that requires cell–stack disassembly and it is by far more convenient to maintain the ionic conductivity of the stream water below adequate threshold to avoid degradation.

11.4.3 Example of Irreversible Degradation Process

There are different degradation modes (chemical, thermal, mechanical) that contribute to membrane degradation, either alone or in combination. Membrane deterioration due to a lack of chemical stability is one of the most serious contributing factors to MEA degradation (Luan and Zhang 2012). Chemical, mechanical, and thermal factors can lead to accelerated aging, membrane thinning and pinhole formation. Pinholes open the way to direct H_2/O_2 exothermic combustion and/or cell short-circuit, and ultimately irreversible degradation of cell performances and/or safety hazard. Compared to PEM fuel cell technology, membranes used in PEM water electrolysis cells are thicker but the situation is potentially even worse since very aggressive electrochemical conditions are found at the anodes: elevated electrical potential combined with native oxygen evolution. The strength and inertness of C–F chemical bonds are sufficient for practical operation over long periods of time, compatible with economical requirements, but like in PEM fuel cell technology, membrane thinning is commonly observed and quite different degradation products are found in circulating water, indicating that the SPEs are undergoing chemical attack. Figure 11.20 shows the chemical composition of PFSA materials.

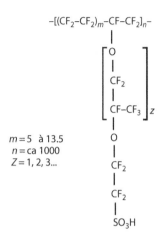

FIGURE 11.20
Chemical composition of PFSA materials.

$m = 5$ à 13.5
$n = $ ca 1000
$Z = 1, 2, 3...$

They mainly contain C–C ($\approx$350–400 kJ·mol^{-1}), C–O ($\approx$360–380 kJ·mol^{-1}), and C–F ($\approx$480 kJ·mol^{-1}) bonds. In spite of these stable bonds, PFSA materials are prone to chemical degradation. For example, in PEM fuel cells where operating conditions are somewhat less aggressive than in PEM water electrolysis cells (the maximum potential of the positive at rest conditions is 1.24 V whereas the potential of oxygen-evolving anodes is in the 1.8–2.0 V range), a quite large number of degradation products have been identified, including formic acid, carbon dioxide, sulfur dioxide, sulfate ions, and even carboxylic acids (Chen and Fuller 2009). According to the literature, degradation rates strongly depend on operating conditions but reaction mechanism involves radicals and species such as HO• or HOO• formed within the membrane are considered to be the main cause of chemical degradation. Hydrogen peroxide H_2O_2 can lead to the formation of radicals in the presence of metallic (e.g., ferrous) ions. Since H_2O_2 can be formed by reduction of oxygen through a two-electron process at platinum surfaces (Equation 11.17), it is usually considered as a key contributor to membrane degradation in both fuel cells and water electrolysis cells (Hommura et al. 2008):

$$O_2 + 2H^+ + 2e^- <=> H_2O_2 (E° = 0.695 \text{ V/ENH}) \quad (11.17)$$

In PEM fuel cells, the reaction can take place at the cathode side during operation. In PEM water electrolysis cells, due to hydrogen and oxygen crossover, especially during pressurized water electrolysis, it can take place either at the cathode or somewhere within the membrane where isolated platinum particles are sometimes found. Membrane degradation can be followed by EIS spectroscopy, postmortem SEM analysis, or by measuring the equivalent weight of the membrane by acid–base titration.

11.5 Other Cell Components Degradation, Consequences, and Risk Mitigation

11.5.1 General Characteristics of Other Cell Components in PEM Water Electrolyzers

The objective of the electrochemical engineer is to maximize the energy conversion efficiency of the unit PEM water electrolysis cell or, in other words, to minimize dissipation sources that transform electrical energy into useless heat. This can be achieved by an appropriate selection of materials and a judicious cell design. Materials used for the manufacturing of non-MEA cell components (bipolar plates, spacers, and current distributor/gas collectors) must be selected by considering mainly electronic conductivity (bulk and contact resistances must be as low as possible) and corrosion-resistance. Due to its appropriate electronic conductivity (the resistivity of bulk titanium is 20 nΩ·m at 20°C; a 3 mm thick titanium bipolar plate has a surface resistance of only 0.4 $\mu\Omega$·cm^{-2}), mechanical and corrosion resistance properties, and cost (4.5–5.0 €/kg in 2014), titanium is widely used as the raw material for the manufacturing of non-MEA cell components in PEM water electrolyzers. Looking closer at the situation, there is in fact a large variety of commercially available "pure" titanium, with different grades (see American Society for Testing and Materials (ASTM) for details) that correspond to different chemical compositions (Table 11.3) and different mechanical–corrosion properties. According to titanium suppliers, grade 2 is typically used in applications that require superior corrosion resistance in various aggressive media. Corrosion resistance is similar between these four grades but mechanical properties vary along with varying oxygen and iron contents. An even larger number of so-called alpha and beta-titanium alloys (solid solutions of aluminum, vanadium, niobium, molybdenum, etc.) with quite diverse mechanical properties are also commercially available but not specifically used in PEM water electrolyzers.

At nominal current densities of 1–2 A·cm^{-2}, the anodes and cathodes of conventional PEM water

TABLE 11.3

Chemical Composition of Commercial Titanium

Grade	1	2	3	4
C (max. %)	0.08	0.08	0.08	0.08
Fe (max. %)	0.20	0.30	0.30	0.50
N (max. %)	0.03	0.03	0.05	0.05
H (max. %)	0.015	0.015	0.015	0.015
O (max. %)	0.18	0.25	0.35	0.40
Ti	Remainder	Remainder	Remainder	Remainder

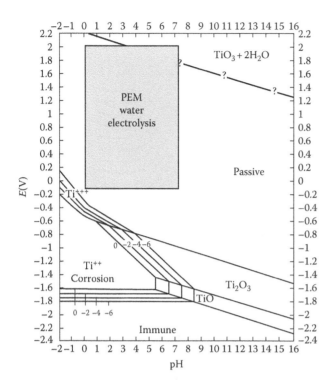

FIGURE 11.21
Potential pH Pourbaix diagram of the titanium–water system.

electrolysis cells are operating at +1.8 to 2.0 V/RHE and −0.2 V/RHE, respectively. Only neutral water is pumped through PEM water electrolysis cells, so most cell components are immerged in neutral water. However, current collectors are in contact with catalytic layers (Figure 11.8), and the pH of PSA materials in acid form is ≈0. According to the Pourbaix diagram for the titanium–water system (Figure 11.21), Ti is passivated at pH and potential conditions found in PEM water electrolysis cell (Pourbaix 1974).

Acidic solution would corrode titanium in acidic environments at much lower potential values only. Several chemicals tend to increase the corrosion sensitivity of titanium. For example, fluoride ions (that may be released in feed water following PSA degradation) are known to attack TiO_2 passive film, a process that leads to the formation of nonprotective titanium fluoride (Boere 1995). Even low concentrations at ppm level are sufficient to destroy the passive oxide film (Nakagawa et al. 1999). Hydrogen peroxide that is responsible for PFSA degradation is also known to attack the passive oxide TiO_2 film (Fonseca and Barbosa 2001).

In a PEM water electrolysis cells, the cathodic compartment is saturated with hydrogen and the anodic compartment is saturated with oxygen. The phase diagram of the titanium–hydrogen system is of the eutectoid type (San-Martin and Manchester 1987). The phase

diagram consists of different phases, including two interstitial solid solutions (based on the allotropic α and β forms of pure titanium), and different hydride phases. Titanium is prone to hydrogen embrittlement (becomes brittle and fracture following exposure to hydrogen). Hydrogen embrittlement of commercial titanium of various grades has been reported in the literature (Briant et al. 2002). It was found that a layer of titanium hydride precipitated at grain boundaries and the elongation to failure decreased with increasing hydrogen content. In the temperature range of operation of PEM water electrolyzers, the hydrogen content in titanium increases with pressure and time. Above approximately 8000 ppm of hydrogen in titanium, risks of hydrogen embrittlement failure are considered as severe. However, according to the literature, the passive TiO_2 film also protects titanium from hydrogen corrosion over a large range of operating temperature and pressure. Hydrogen embrittlement of titanium is, therefore, not considered as a critical issue for operation in PEM water electrolyzers. On the anodic side, the presence of oxygen favors titanium passivation (Nakagawa et al. 1999). The surface passivation mechanism of titanium in aqueous environments involves several steps and leads to the formation of nanometer thick surface TiO_2 layer (Briant et al. 2002, Franz and Göhr 1963). Figure 11.22 shows the results of XPS measurements made on bulk titanium samples immersed for 30 days in an autoclave at 100°C in deionized and neutral water, under 64 bars of oxygen and hydrogen. At the end of the experiments, no trace of titanium species was detected in water. Results show that the surface of the samples is protected by a nanometer thick film of titanium oxide, which acts as a protective layer against bulk dissolution.

11.5.2 Degradation Process

In the electrical analogy of Figure 11.8, R^c and R^a denote the global electronic resistances of metallic cell components, respectively, in cathodic and anodic cell compartments. As discussed in Chapter 10, a closer look at the situation shows that each of these two resistances is the sum of individual bulk and interface (contact) resistances:

$$R_\Omega^c = \sum_{i=1}^{3} R_{bulk}^{c,i} + \sum_{j=1}^{3} R_{contact}^{c,j}$$

$$= \left(R_{bulk}^{bp}\right)_{cath} + \left(R_{contact}^{bp-gr}\right)_{cath} + \left(R_{bulk}^{gr}\right)_{cath}$$

$$+ \left(R_{contact}^{gr-cc}\right)_{cath} + \left(R_{bulk}^{cc}\right)_{cath} + \left(R_{contact}^{cc-CL}\right)_{cath} + \left(R_{bulk}^{CL}\right)_{cath}$$

$$(11.18)$$

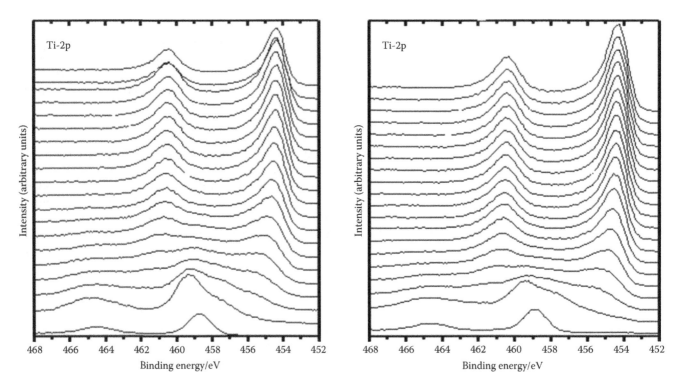

FIGURE 11.22
2p titanium XPS spectra measured on (left): sample in O_2; (right) sample in H_2. Each spectrum was obtained after 100 s of surface erosion.

$$R_\Omega^a = \sum_{i=1}^{3} R_{bulk}^{a,i} + \sum_{j=1}^{3} R_{contact}^{a,j}$$

$$= \left(R_{bulk}^{bp}\right)_{anode} + \left(R_{contact}^{bp-gr}\right)_{anode} + \left(R_{bulk}^{gr}\right)_{anode} + \left(R_{contact}^{gr-cc}\right)_{anode}$$

$$+ \left(R_{bulk}^{cc}\right)_{anode} + \left(R_{contact}^{cc-CL}\right)_{anode} + \left(R_{bulk}^{CL}\right)_{anode}$$

$$(11.19)$$

where

R_{bulk}^{bp} ($\Omega \cdot cm^2$) is the bulk resistance of titanium bipolar plate (bp)

$R_{contact}^{bp-gr}$ ($\Omega \cdot cm^2$) is the contact resistance between bipolar plate (bp) and grid spacer (gr)

R_{bulk}^{gr} ($\Omega \cdot cm^2$) is the bulk resistance of the titanium grid

$R_{contact}^{gr-cd}$ ($\Omega \cdot cm^2$) is the contact resistance between the grid (gr) and the current distributor (cd)

R_{bulk}^{cd} ($\Omega \cdot cm^2$) is the bulk resistance of the porous titanium current distributor

$R_{contact}^{cd-CL}$ ($\Omega \cdot cm^2$) is the contact resistance between the current distributor (cd) and the catalytic layer (CL)

R_{bulk}^{CL} ($\Omega \cdot cm^2$) is the bulk resistance of the catalytic layer

Regarding the measurement of these different cell resistances, some characterization tools and methodologies are available (see Chapter 10). The ohmic resistance of individual cell components can be measured ex situ. Within the cell, it is not an easy task to measure them individually and most of the time, only their sum (non-MEA electronic cell resistance) can be measured, for example, using EIS. Bulk conductivity of non-MEA titanium cell components is sufficiently large for application in PEM water electrolyzers but contact resistances are much larger and need to be minimized. The corrosion-protecting TiO_2 film is strongly resistive (5 M$\Omega \cdot cm^2$ according to Pan et al. 1996). This is a critical issue although DC current can flow across such interfaces (the charge transfer process across the film under DC polarization is not fully understood yet). Figure 11.23 shows the aspect of a Ti mesh used in a PEM water electrolyzer at BoL (titanium is a lustrous transition metal with a silver color) and EoL. At EoL, the yellowish color of the Ti mesh indicates the presence of thick layers of different resistive titanium oxides. The electrical resistance of titanium cell components tends to increase with time (aging process) and this is the main source of degradation of non-MEA cell components.

Bulk resistances can be minimized by reducing the thickness of cell components but an optimum has to be found by taking into account mechanical properties (in addition to electricity transport, non-MEA cell components contribute to the homogeneous distribution of current lines, especially in cells of large surface areas). Contact resistances can be minimized partly by

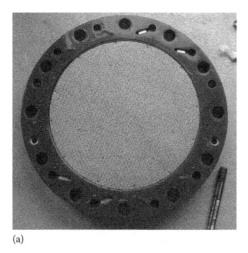
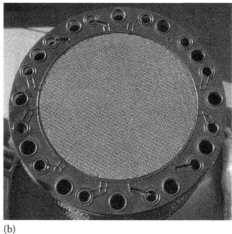

(a)   (b)

FIGURE 11.23
Photographs of Ti mesh at BoL (a) and at EoL (b).

manufacturing cell components with appropriate shape and surface states (severe dimensional tolerances, excellent planarity, and appropriate surface roughness are required), partly using appropriate surface treatments such as protective coatings. Different surface treatments have been tested over the last decades. In the early days of technology development, PEM water electrolysis was used for oxygen generation in anaerobic environments and homogeneous noble metal (e.g., Pt) surface deposits were used to prevent surface oxidation. The large-scale deployment of PEM water electrolysis for civil applications needs to satisfy severe cost constraints and alternative techniques have been tested (e.g., the coating of dispersed platinum or gold dots that bridge bulk titanium to surface environment, titanium carbide or nitride, sub-stoichiometric titanium dioxide phases that are acceptable electronic conductors, etc.). Also, although titanium is an appropriate material for the manufacturing of non-MEA cell components in PEM

water electrolyzers, cheaper alternative materials (compressed carbon powder covered by a surface tantalum film, zirconium, or stainless steel surface coated with titanium, etc.) have also been tested. So far, no unique solution prevails and the use of more or less expensive coatings depends on the cost constraints imposed by each application.

11.6 Stack Failure Mechanisms, Consequences, and Risk Mitigation

11.6.1 PEM Water Electrolysis Stacks

In a PEM water electrolyzer, individual PEM water electrolysis cells are stacked together to increase the production capacity. The filter-press configuration with lateral main anode and cathode (Figure 11.24) is mostly

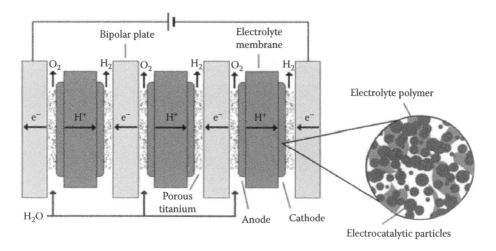

FIGURE 11.24
Schematic diagram of a PEM water electrolysis stack.

FIGURE 11.25
Photographs of PEM water electrolysis stacks. (Reproduced from Millet, P. et al., Int. *J. Energ. Res.*, 37, 449, 2013.)

used. Systems that can deliver several tens of N m³ H₂/h are commercially available (Figures 11.25 and 11.26). Efforts are currently made by PEM water electrolysis manufacturers to put on the market systems that can deliver 100–200 N m³ H₂/h (>1 MW) for application in hydrogen refueling stations.

The management of hydrogen in PEM water electrolyzers is challenging due to its ease of leaking, its low energy ignition, the wide range of combustible fuel–air mixtures, and the presence of pure oxygen. Regarding leaks, hydrogen is odorless, colorless, and tasteless, and is therefore difficult to be detected by humans. Since known odorants contaminate fuel cells, new methods are required for hydrogen detection. Hydrogen leaks

can support combustion at very low flow rates, as low as 4 µg·s⁻¹ (Butler et al. 2009). A water electrolyzer is not a closed system. Water is introduced in the circuit while hydrogen and oxygen are produced (Figure 11.27). However, hydrogen leaks from inside to outside can be easily monitored and vented. The main operational risk is the formation of hazardous H₂/O₂ gaseous mixtures having a composition within the flammability range, within the circuitry of the electrolyzer. As the size of the electrolyzer increases, the amount of gas that remains within the entire system also increases and risks become more serious.

Regarding ignition, hydrogen–air mixtures can ignite with very low energy input (as low as ≈0.02 mJ for spark ignition at atmospheric pressure). This makes the ignition of hydrogen–air mixtures more likely than the ignition of most hydrocarbons, in spite of a higher temperature of autoignition.

Regarding the range of flammability, the limits at 1 atm in air are approximately 4 and 75 vol.%. In oxygen, limits at 1 atm are approximately 4% and 94 vol.%. A deflagration (subsonic combustion propagating through heat transfer) is prone to evolve into a detonation (accelerating supersonic exothermic front), even in the absence of confinement.

11.6.2 Failure of PEM Water Electrolysis Stacks

There are few reports available in the literature on cell–stack failure mechanisms and consequences. The formation of hydrogen–oxygen gaseous mixtures within the flammability range inside the electrolysis unit can

FIGURE 11.26
Photograph of a 24 N m³ H₂/h PEM water electrolyzer. (Courtesy CETH2 Co, France.)

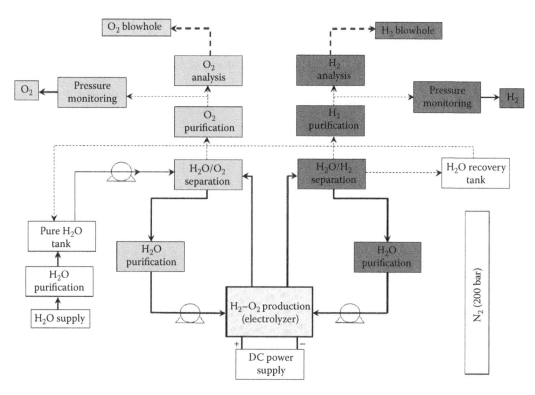

FIGURE 11.27
Schematic PEM water electrolysis process flow sheet.

lead to combustion. As discussed by Millet et al. (2012), although the operating temperature of the PEM stack is low (<100°C) and although hydrogen and oxygen are saturated with water vapor, the presence of finely divided platinum catalytic particles at the cathode of the polymer electrolyte can promote the H_2–O_2 recombination. There is enough energy stored in a PEM cell operating at atmospheric pressure to destroy the entire stack. If combustion takes place, the temperature reached inside the stack is such that most cell components, including metallic parts, can melt (Figure 11.10).

The weakest part of a PEM water electrolysis unit is the solid polymer membrane. Under conventional conditions of operation, because of gas cross-permeation through the membrane, hydrogen produced at the cathode is contaminated by oxygen and oxygen produced at the anode is contaminated by hydrogen. This is a risk that can be monitored by using thicker membranes. However, membrane degradation can lead to membrane perforation and in turn to the formation of larger amounts of H_2/O_2 atmospheres within the cells, stacks, and liquid–gas separation units. Several factors such as membrane aging (see discussion in Section 11.4), hot points due to the uneven distribution of current lines, local drying, etc., can potentially lead to membrane perforation. To prevent the formation of flammable hydrogen–oxygen gas mixtures, there is a need to monitor either the stack

performance (to detect cells at risk) or directly the gaseous atmospheres within the circuit to detect the possible formation of hazardous mixtures.

11.7 Conclusions

Water electrolysis can be used as a source of hydrogen and oxygen of electrolytic grade in different sectors. Hydrogen can be used directly as the raw material or as an energy carrier. It can also be used as a chemical reducer for the treatment of carbon dioxide emissions and for the synthesis of synthetic fuels. Low-temperature water electrolyzers can produce hydrogen using conventional power sources or directly from RES. Therefore, this technology is called to play a central role in future energy infrastructures. There are three major markets of quite different maturity for which PEM water electrolysis can potentially compete with other water technologies and with steam methane reforming: (i) on-site hydrogen production for miscellaneous industrial applications, requiring 1–30 N m³ H_2/h (150 kW) capacity; (ii) hydrogen refueling stations, requiring 60–200 N m³ H_2/h (100 kW–1 MW); (iii) services to power grids (flexibility, frequency regulation, etc.); (iv) large-scale energy storage, requiring multi MW systems. There is, therefore, a need to increase the production capacity of PEM electrolyzers from ≈ 60 N m$^3 \cdot$ h^{-1} (state-of-the-art) toward 200 N m$^3 \cdot$ h^{-1} and more. The key driver for the implementation of water electrolyzers is the cost (expressed in €/kg H$_2$). According to the US DOE and European FCH-JU roadmaps, capital expenses (capex) of PEM electrolysis technology should be reduced on the short term from ≈ 2000 €/kW down to ≈ 1400 €/kW for systems at the 1 MW/200 N m$^3 \cdot$ h^{-1} scale, with an ultimate goal of ≈ 700 €/kW in 2020 (DOE Hydrogen and Fuel Cells Program 2014). On large systems, the operational expenses (opex) are cost predominant and the cost of the kWh of electricity alone dictates the cost of hydrogen. A target cost for on-site delivery is 4–5 €/kg. In some favorable cases (e.g., grid regulation service for which electricity cost is zero or even negative), water electrolysis can compete directly with steam reforming.

There are still some remaining limitations in view of large-scale deployment of PEM water electrolyzers. To meet capex cost targets, PEM water electrolysis must increase durability toward the upper range of the 10^4 to 10^5 h time duration of operation. To reach these goals, there is a need to better understand microscopic processes that tend to degrade performances on the long term and to find appropriate solutions to these problems. Main degradation sources are reviewed in this chapter, and some contingency solutions are proposed. Keeping in mind that much larger water electrolysis production capacities will be required in the future to take the full advantage of hydrogen as an energy carrier, it is clear that substantial R&D efforts will be needed in order to favor the large-scale implementation of clean RES. Regarding future PEM water electrolyzers, progress is still needed in material science (mechanically stable, thinner, and less gas permeable polymer electrolyte membranes and active, robust, and less expensive electrocatalysts are still needed), in the development of automated manufacturing tools and assembly procedures of larger cell components with more reproducible levels of performance, and in overall cost reduction.

References

Anderson, E., K. Ayers, and C. Capuano. 2013. R&D focus areas based on 60,000 hr life PEM water electrolysis stack experience. *Proceedings of the First International Workshop on Durability and Degradation Issues in PEM Electrolysis Cells and its Components*, 11–12 March, Freiburg, Germany.

Andolfatto, F., R. Durand, A. Michas, P. Millet, and P. Stevens. 1994. Solid polymer electrolyte water electrolysis: Electrocatalysis and long term stability. *International Journal of Hydrogen Energy* 19:421–427.

Arico, A. S., A. Di Blasi, G. Brunaccini, F. Sergi, G. Dispenza, L. Andaloro, M. Ferraro, V. Antonucci, P. Asher, S. Buche, D. Fongalland, G. A. Hards, J. D. B. Sharman, A. Bayer, G. Heinz, N. Zandona, R. Zuber, M. Gebert, M. Corasaniti, A. Ghielmi, and D. J. Jones. 2010. High temperature operation of a solid polymer electrolyte fuel cell stack based on a new ionomer membrane. *Fuel Cells* 10:1013–1023.

Arkilander, W. N. and T. M. Molter. 1984. *Oxygen generator cell design for future submarines*. The world's knowledge, n°961440. The British Library, London.

Brug, G. J., A. L. G. Van Den Eeden, M. Sluyters-Rehbach, and J. H. Sluyters. 1984. The analysis of electrode impedances complicated by the presence of a constant phase element. *Journal of Electroanalytical Chemistry and Interfacial Electrochemistry* 176:275–295.

Chen, C. and T. F. Fuller. 2009. The effect of humidity on the degradation of Nafion membrane. *Polymer Degradation and Stability* 94:1436–1447.

Barbir, F. 2005. *PEM Fuel Cells, Theory and Practice*. Elsevier Academic Press, London.

Bockris, J. O'M. and A. K. N. Reddy. 1982. *Comprehensive Treatise of Electrochemistry*. London, U.K.: Plenum Press.

Boere, G. 1995. Influence of fluoride on titanium in an acidic environment measured by polarization resistance technique. *Journal of Applied Biomaterials* 6:283–288.

Briant, C., Z. Wang, and N. Chollocoop. 2002. Hydrogen embrittlement of commercial purity titanium. *Corrosion Science* 44:1875–1888.

Butler, M. S., C. W. Moran, P. B. Sunderland, and R. L. Axelbaum. 2009. Limits for hydrogen leaks that can support stable flames. *International Journal of Hydrogen Energy* 34:5174–5182.

Damjanovic, A. and J. O'M. Bockris. 1966. Electrode kinetics of oxygen evolution and dissolution on Rh, Ir, and Pt–Rh alloy electrodes. *Journal of the Electrochemical Society* 113:739–746.

Fonseca, C. and M. Barbosa. 2001. Corrosion behavior of titanium in biofluids containing H$_2$O$_2$ studied by electrochemical impedance spectroscopy. *Corrosion Science* 43:547–599.

Franz, D. and H. Göhr. 1963. Stromstoffumsätze am Titan in wäßrigen Lösungen. *Berichte der Bunsengesellschaft für Physikalische Chemie* 67:680–690.

Gierke, T. D., G. E. Munn, and F. C. Wilson. 1981. The morphology in Nafion perfluorinated membrane products, as determined by wide and small-angle X-ray studies. *Journal of Polymer Science B: Polymers Physics* 19(11):1687–1704.

Grigoriev, S. A., P. Millet, S. A. Volobuev, and V. N. Fateev. 2009. Optimization of porous current collectors for PEM water electrolysers. *International Journal of Hydrogen Energy* 34:4968–4973.

Hamann, C. H., A. Hamnett, and W. Vielstich. 1998. *Electrochemistry*. Wiley-VCH Ed., Weinheim, Germany.

Harrison, K. and M. Peters, 2014. DOE annual progress report on hydrogen and fuel cells program, II.A.2 renewable electrolysis integrated systems development and testing, FY 2014 Annual Progress Report, DOE, Washington.

Hommura, S., K. Kawahara, and Y. Teraoka. 2008. Development of a method for clarifying the perfluorosulfonated membrane degradation mechanism in a fuel cell environment. *Journal of the Electrochemical Society* 155:29–33.

Hydrogen pathways analysis for polymer electrolyte membrane (PEM) electrolysis. 2014. *DOE Hydrogen and Fuel Cells Program*, June 2014, Washington.

Luan, Y. and Y. Zhang. 2012. *Membrane Degradation, PEM Fuel Cell Mode Degradation Analysis*. CRC Press, Boca Raton, FL.

Lvovich, V. F. 2012. *Impedance Spectroscopy, Application to Electrochemical and Dielectric Phenomena*. Wiley, Hoboken, NJ.

Mauritz, K. A. and R. B. Moore. 2004. State of understanding of Nafion. *Chemical Reviews* 104:4535–4585.

Millet, P. 1990. Water electrolysis using EME technology: Temperature profile inside a Nafion membrane during electrolysis. *Electrochimica Acta* 36:263–267.

Millet, P. 2011. Electrochemical technologies for energy storage and conversion, Chapter 9. *Water Electrolysis for Hydrogen Generation*, R-S. Liu, X. Sun, H. Liu, L. Zhang, and J. Zhang, editors. New York: John Wiley & Sons.

Millet, P., T. Alleau, and R. Durand. 1993. Characterization of membrane-electrodes assemblies for solid polymer electrolyte water electrolysis. *Journal of Applied Electrochemistry* 23:322–331.

Millet, P., F. Andolfatto, and R. Durand. 1995. Preparation of solid polymer electrolyte composites. Investigation of the ion-exchange process. *Journal of Applied Electrochemistry* 25:227–232.

Millet, P., F. Andolfato, and R. Durand. 1996. Design and performances of a solid polymer electrolyte water electrolyzer. *International Journal of Hydrogen Energy* 21:87–93.

Millet, P., F. de Guglielmo, S. A. Grigoriev, and V. I. Porembskiy. 2012. Cell failure mechanisms in PEM water electrolyzers. *International Journal of Hydrogen Energy* 37:17478–17487.

Millet, P., N. Mbemba, S. A. Grigoriev, V. N. Fateev, A. Aukauloo, and C. Etiévant. 2011. Electrochemical performances of PEM water electrolysis cells and perspectives. *International Journal of Hydrogen Energy* 36:4134–4142.

Nakagawa, M., S. Matsuya, T. Shiraishi, and M. Ohta. 1999. Effect of fluoride concentration and pH on corrosion behavior of titanium for dental use. *Journal of Dental Research* 78(9):1568–1572.

Pan, J., D. Thierry, and C. Leygraf. 1996. Electrochemical impedance spectroscopy study of the passive oxide film on titanium for implant application. *Electrochimica Acta* 41:1143–1153.

Pourbaix, M. 1974. *Atlas of Electrochemical Equilibria in Aqueous Solutions*, second edition. National Association of Corrosion Engineers, New York.

Rozain, C. and P. Millet. 2014. Electrochemical characterization of polymer electrolyte membrane water electrolysis cells. *Electrochimica Acta* 131:160–167.

Sakai, T., H. Takenaka, and E. Torikai. 1986. Gas diffusion in the dry and hydrated Nafion. *Journal of the Electrochemical Society* 133:88–92.

Sakai, T., H. Takenaka, N. Wakabayashi, Y. Kawami, and E. Torikai. 1985. Gas permeation properties of solid polymer electrolyte (SPE) membranes. *Journal of the Electrochemical Society* 132:1328–1332.

San-Martin, A. and F. D. Manchester. 1987. The H–Ti (hydrogen–titanium) system. *Bulletin of Alloy Phase Diagrams* 8(1):30–47.

Steck, A. and H. L. Yeager. 1980. Water sorption and cation-exchange selectivity of a perfluorosulfonate ion-exchange polymer. *Analytical Chemistry* 52:1215–1218.

Subianto, S., M. Pica, M. Casciola, P. Cojocaru, L. Merlo, D. J. Jones, and G. Hards. 2013. Physical and chemical modification routes leading to improved mechanical properties of perfluorosulfonic acid membranes for PEM fuel cells. *Journal of Power Sources* 233:216–230.

Tsampas, M. N., A. Pikos, S. Brosda, A. Katsaounis, and C. G. Vayenas. 2006. The effect of membrane thickness on the conductivity of Nafion. *Electrochimica Acta* 51(13):2743–2755.

Vishnyakov, A. and A. V. Neimark. 2000. Molecular simulation study of Nafion membrane solvation in water and methanol. *Journal of Physical Chemistry B*, 104:4471–4478.

Wang, H., H. Li, and X.-Z. Yuean, (Eds.). 2012. *PEM Fuel Cell Failure Mode Analysis*. CRC Press, Boca Raton, FL.

12

Modeling of PEM Water Electrolyzer

Ravindra Datta, Drew J. Martino, Yan Dong, and Pyoungho Choi

CONTENTS

Nomenclature

a_i — activity of species i

A_{Cell} — MEA area in a single cell of a stack, cm^2 total MEA area in the stack, cm^2

A_{MEA} — total MEA area in the stack, cm^2

B_0 — PEM d'Arcy permeability, cm^2

$c_{i,0}$ — concentration of species i at equilibrium, $mol\,cm^{-3}$

$c_{i,\alpha}$ — concentration of species i in layer α, $mol\,cm^{-3}$

$c_{i,b}$ — concentration of species i in bulk phase, $mol\,cm^{-3}$

$c_{i,ref}$ — reference concentration of species i, $mol\,cm^{-3}$

$d_{M,A}$ — anode metal catalyst nanoparticle diameter, cm

$d_{M,C}$ — anode metal catalyst nanoparticle diameter, cm

$D^e_{i,\alpha}$ — effective diffusion coefficient of species i in layer α, $cm^2\,s^{-1}$

E_μ — activation energy for viscosity of water, $J\,mol^{-1}$

E_{ρ,Φ_0} — activation energy for electrode reaction ρ at equilibrium (Nernst) potential, $J\,mol^{-1}$

E_{C,Φ_0} — cathode effective activation energy for oxygen reduction reaction, $J\,mol^{-1}$

F — Faraday's constant, 96,487 $C\,eq^{-1}$

i — current density, $A\,cm^{-2}$ of geometric MEA area

i_ρ — electrode ρ current density, $A\,cm^{-2}$ geometric MEA area

$i_{\rho,X}$ — electrode ρ crossover current density, $A\,cm^{-2}$

$i_{\rho,0}$ — exchange current density for electrode reaction ρ, $A\,cm^{-2}$

$i_{\rho,L}$ — electrode ρ limiting current density, $A\,cm^{-2}$

i^*_ρ — current density, $A\,cm^{-2}$ ECSA

$i^*_{\rho,0}$ — exchange current density, $A\,cm^{-2}$ ECSA

$i^*_{A,0,ref}$ — anode reference exchange current density, $A\,cm^{-2}$ ECSA

I_{Stack} — total current in stack, A

k_i — permeability of gaseous species i in PEM

$\overrightarrow{k}_\rho$ — forward rate constant for electrode reaction ρ

$\overleftarrow{k}_\rho$ — reverse rate constant for electrode reaction ρ

K_ρ — thermodynamic equilibrium constant for electrode reaction ρ

L_B — electrolyte layer thickness, cm

L_D — GDL layer thickness, cm

$m_{M,A}$ — anode catalyst loading, $g\,cm^{-2}$

$m_{M,C}$ — cathode catalyst loading, $g\,cm^{-2}$

N_{Av} — Avogadro's number

N_{Cell} — number of cells in a stack

$N_{i,z}$ — flux of species i in the membrane along the z direction, $mol\,s^{-1}\,cm^{-2}$ MEA

N_{H_2} — molar flux of H_2 generated in the PEM-WE, $mol\,s^{-1}\,cm^{-2}$ MEA

$\dot{n}_i$ — molar flow rate of species i, $mol\,s^{-1}$

243

P	power density, W cm^{-2}
p_i	partial pressure of species i, atm
$P_{i,\alpha}$	permeance of species i in layer α, mol cm^{-1}
Q	heat exchange between PEM-WE and surroundings, J mol^{-1}
$\langle Q \rangle$	vehicle, or carrier, molecule for diffusing ion in electrolyte layer
r	ratio of PEM to water partial molar volume
R	gas constant
R_I	interfacial resistance, Ω
T	temperature, K
T_ref	reference temperature, 298 K
V	cell voltage, V
V_0	thermodynamic (Nernst) cell voltage, V
v	water convective velocity in PEM, cm s^{-1}
V_max	maximum theoretical cell voltage determined from ΔH, V
V_Stack	stack voltage, V
W_t	electric work per mol of H$_2$ generation, or specific energy consumption, J mol^{-1}
$\dot{W}_\mathrm{t}$	total power consumption in water electrolyzer, W
x_i	mole fraction of species i

Greek Symbols

β	degree of acid dissociation in Nafion
β_ρ	symmetry factor for electrode reaction/step ρ, typically 1/2
$\beta_\rho^{\cdot}$	symmetry factor of the RDS in electrode step ρ, typically 1/2
γ_M	roughness factor, ratio of active ECSA area to MEA area
δ	ratio of mutual to matrix effective diffusion coefficients
ΔG	Gibbs free energy change KJ mol^{-1}
ΔH	enthalpy change KJ mol^{-1}
ε	fuel cell efficiency
ε_i	Faradaic efficiency
ε_V	voltage efficiency
$\varepsilon_{\Delta G}$	Second law efficiency
$\varepsilon_{\Delta H}$	First law efficiency
η_A	anode overpotential, V
η_ρ	overpotential of electrode reaction $\rho = \Phi_\rho - \Phi_{\rho,0}$, V
η_B	Ohmic overpotential in electrolyte layer, V
η_C	overpotential, cathode, V
$\kappa_{i,\alpha}$	partition coefficient of species i in layer α
$\overrightarrow{\Lambda}_\rho$	forward pre-exponential factor for reaction ρ
$\overleftarrow{\Lambda}_\rho$	reverse pre-exponential factor for reaction ρ
λ	number of water molecules per sulfonic acid group in Nafion

$\nu_{\rho i}$	stoichiometric coefficient of species i in reaction ρ
$\nu_{\rho e^-}$	stoichiometric coefficient of electrons in reaction ρ
$\nu_{\rho,e^-}^{\cdot}$	stoichiometric coefficient of electrons in the RDS in electrode reaction ρ
ξ	electro-osmotic drag coefficient of water
$\rho_\mathrm{M,A}$	anode catalyst metal density, g cm^{-3}
$\rho_\mathrm{M,C}$	cathode catalyst metal density, g cm^{-3}
σ_B	effective electrolyte conductivity, S cm^{-1}
φ_I	fraction of metal surface in contact with ionomer
Φ	electrode potential, V
$\Phi_{\rho,0}$	half-cell thermodynamic (Nernst) potential of electrode reaction ρ, V
$\Phi_{\rho,0}^{\circ}$	standard half-cell thermodynamic (Nernst) potential of electrode reaction ρ, V
ω_I	mass fraction of ionomer in catalyst layer

Subscripts/Superscripts

$\bullet$	rate-determining step
0	equilibrium
A	anode layer
B	electrolyte layer
C	cathode layer
D	diffusion/gas diffusion layer
e$^-$	electron
i	species
K	kinetic
L	limiting
o	standard conditions
ref	reference
X	crossover
z	charge number
ρ	reaction step, or electrode reaction (anode or cathode)

Abbreviations

AEM	anion-exchange membrane
DGM	dusty-gas model
DMFC	direct methanol fuel cell
GDL	gas-diffusion layer
ECSA	electrochemical surface area
HER	hydrogen evolution reaction
HOR	hydrogen oxidation reaction
LAE	liquid alkaline electrolyte
LAEWE	liquid alkaline electrolyte water electrolyzer

MEA   membrane electrode assembly
OCV   open circuit voltage
OR   overall reaction
OER   oxygen evolution reaction
ORR   oxygen reduction (or electrode) reaction
PEM   polymer electrolyte membrane
PEM-FC polymer electrolyte membrane fuel cell
PEM-WE polymer electrolyte membrane water electrolyzer
QE   quasi-equilibrium
RDS   rate-determining step
SHE   standard hydrogen electrode
SPE   solid polymer electrolyte
TPI   three-phase interface

12.1 Introduction

The realization of the so-called hydrogen economy (Bockris, 1975; Turner, 2004; Bockris and Veziroglu, 2007), a potentially environmentally benign and sustainable energy scenario, is predicated on economical and widespread availability of hydrogen as the energy vector. While hydrogen is produced industrially on a very large scale, principally from fossil resources, that is, via natural gas and coal gasification at the moment (Kothari et al., 2008; Holladay et al., 2009), its storage and transportation from such centralized hydrogen generation plants present significant technological and cost challenges. Consequently, distributed hydrogen generation is attractive, for which polymer electrolyte membrane (PEM), also called solid polymer electrolyte (SPE), water electrolyzers have significant potential, especially when coupled with renewable but intermittent sources of electricity such as solar cells and wind turbines (Turner, 2004; Barbir, 2005; Bockris and Veziroglu, 2007; Levene et al., 2007; Abbasi and Abbasi, 2011; Andrews and Shabani, 2012) (Figure 12.1) to provide carbon-free hydrogen. Another advantage of PEM water electrolyzer (PEM-WE) is that the same stack can serve the dual purpose of hydrogen generation in the electrolysis mode and power generation in the PEM fuel cell (PEM-FC) mode (Choi et al., 2006), that is, as a unitized regenerative fuel cell, or URFC (Mitlitsky et al., 1998; Ioroi et al., 2002). This chapter describes the basic modeling of PEM-WE individual cells or stacks (Selamet et al., 2011), so that appropriate design and analysis may be facilitated. Such quantitative models can also provide insights that can lead to improved performance and design.

In fact, water electrolyzers are already quite common, being utilized for roughly 1% of the worldwide hydrogen generation. However, the conventional electrolyzers are based on free liquid alkaline electrolyte (FLAE) solutions (~30% KOH), operating at roughly 80°C, and with a current density of around 300 mA cm^{-2} at a voltage of about 2 V (Ivy, 2004; Marini et al., 2012). The cell electrodes and overall reaction (OR) in a FLAE water electrolyzer (FLAE-WE) may be described by the following:

Electrode	Reaction	Potential (V)	ΔG°_{298} (kJ/mol)
Anode:	$2OH^- \leftrightarrows \frac{1}{2}O_2 + \langle H_2O \rangle + 2e^-$	$\Phi^\circ_{A,0} = +0.401$ V	+77.4
Cathode:	$H_2O + \langle H_2O \rangle + 2e^- \leftrightarrows H_2 + 2OH^-$	$\Phi^\circ_{C,0} = -0.828$ V	+159.8
Overall:	$H_2O(l) \leftrightarrows H_2 + \frac{1}{2}O_2$	$V^\circ_0 = 1.229$ V	+237.2

(12.1)

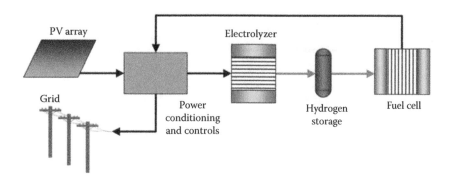

FIGURE 12.1
Photovoltaic hydrogen generation and utilization. (Reprinted from *Solar Energy*, 78, Barbir, F., PEM electrolysis for production of hydrogen from renewable energy sources, 661–669, Copyright 2005, with permission from Elsevier.)

where the reactant water molecule H_2O is distinguished from the ion vehicle molecule $\langle H_2O \rangle$, the latter identified by the enclosing brackets, $\langle \ \rangle$, the charge carrying ionic species here being OH^-.

The conventional FLAE-WE comprises two electrode chambers with Ni-based catalysts (Li et al., 2011), separated by a porous separator, but immersed in the same aqueous electrolyte solution. This causes two issues: (1) long ion diffusion path lengths and (2) additional transport resistance caused by the H_2 and O_2 gas bubbles present in the ion diffusion path in the free liquid electrolyte. In the newer "zero gap" alkaline electrolyzer cell, or supported liquid alkaline electrolyte water electrolyzer (SLAE-WE), on the other hand, the electrodes are in direct contact with the separator, much like in the PEM-WE, so that the ion diffusion paths are minimal

and further there are no gas bubbles present between the two electrodes. An anion exchange membrane (AEM) may alternately be used in place of the porous separator soaked with the liquid alkaline electrolyte (Leng et al., 2012), that is, in an AEM-WE.

The PEM water electrolyzer (PEM-WE) (Grigoriev et al., 2006; Carmo et al., 2013), shown schematically in Figure 12.2 (Ito et al., 2013), is typically based on the Nafion® 115 or 117 membrane, flanked on either side by porous electrodes comprising a gas diffusion electrode (sintered porous titanium at the anode and/or porous graphite at the cathode), and with Pt-B or Pt/C cathode and Ir-B-based anode as-state-of-the art catalysts (Carmo et al., 2013). The electrode and ORs are then described by the following:

Electrode	Reaction	Potential (V)	ΔG^o_{298} (kJ/mol)
Anode:	$H_2O + 2\langle H_2O \rangle \leftrightarrows \frac{1}{2}O_2 + 2H_3O^+ + 2e^-$	$\Phi^o_{A,0} = +1.229$ V	+237.2
Cathode:	$2H_3O^+ + 2e^- \leftrightarrows H_2 + 2\langle H_2O \rangle$	$\Phi^o_{C,0} = 0$ V	0.0
Overall:	$H_2O(l) \leftrightarrows H_2 + \frac{1}{2}O_2$	$V^o_0 = 1.229$ V	+237.2

$$(12.2)$$

where again the reactant water molecule H_2O is distinguished from the ion vehicle molecule $\langle H_2O \rangle$, which carries the proton H^+ in the form of the hydronium ion, H_3O^+. In reality, of course, more than one water molecule is involved in ferrying a proton (Figure 12.2), resulting in electroosmosis of water that can also carry with it some dissolved H_2 and O_2.

The key advantages of PEM-WE, as compared to the traditional FLAE-WE, include higher current densities at

lower cell voltages and production of a high pressure, up to 100 bar (Millet et al., 2009; Bensmann et al., 2013), high purity (>99.99% pure H_2) gas, along with environmentally better and safer operation because of the absence of any corrosive free liquid electrolyte. This makes the PEM-WE technology particularly attractive for small-scale and distributed hydrogen generation systems.

The scheme in Equation 12.2, in fact, simply represents the reverse of the reactions in a PEM-FC, and the

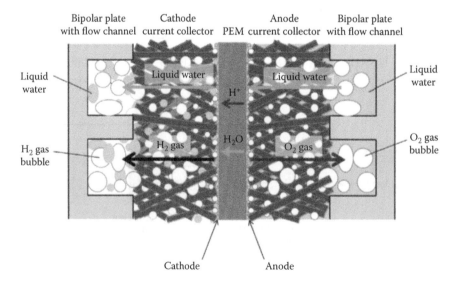

FIGURE 12.2
Schematic of the membrane electrode assembly (MEA) in a PEM water electrolyzer (PEM-WE). (Reprinted from *Electrochim. Acta*, 100, Ito, H., Maeda, T., Nakano, A., Kato, A., and Yoshida, T., Influence of pore structural properties of current collectors on the performance of proton exchange membrane electrolyzer, 242–248, Copyright 2013, with permission from Elsevier.)

membrane electrode assembly (MEA) structure (Figure 12.2) is, thus, identical to that of the PEM-FC (Thampan et al., 2001), with the exception that the oxygen electrode is not Pt/C as in a PEM fuel cell, but rather typically IrO$_2$ black, sometimes with Ta$_2$O$_5$ and/or SnO$_2$, for example, Ir$_x$Ru$_{0.5-x}$Sn$_{0.5}$O$_2$ (Marshall et al., 2006), and Ir$_x$Ru$_{0.7-x}$Ta$_{0.3}$O$_2$ (Marshall et al., 2007), along with porous titanium rather than porous graphite as the electrode at the anode. This is because at the high operating potential (and low pH) of the anode in a PEM-WE, the conventional Pt used in PEM-FC, the carbon support, and the porous graphite gas diffusion layer (GDL) are not adequately stable. Thus, carbon supports for catalysts cannot be used at the anode, but oxides with electronic conductivity that are more stable may be used. Nonetheless, the same cell may be utilized for both PEM-WE and PEM-FC, that is, as a URFC (Mitlitsky et al., 1998; Ioroi et al., 2002). Typical cycle performance of a URFC operating as a PEM-WE as well as a PEM-FC is shown in Figure 12.3 (Ioroi et al., 2002).

A key issue in water electrolysis for hydrogen generation, of course, is its high standard thermodynamic cell potential, $V_0^\circ = 1.229$ V, the minimum cell voltage needed for water electrolysis, which translates into a minimum electricity consumption of around 3 kWh Nm^{-3} H$_2$, while the actual energy consumption is more of the order of around 5 kWh Nm^{-3} H$_2$. In fact, the actual open circuit voltage (OCV) in the PEM-WE mode may be even higher (~1.4 V) (Figure 12.3), while that in PEM-FC mode is lower (~1.0 V) (Figure 12.3) than the standard thermodynamic cell potential of 1.229 V. In the case of PEM-FC, this is a result of the gas crossover across the PEM to the opposite electrode (Vilekar and Datta, 2010), where it causes a significant overpotential of ~0.2 V, despite the relatively small hydrogen permeability of Nafion. The issue of the OCV for a PEM-WE has not yet been carefully investigated.

When all the other kinetic and transport losses in the PEM-WE are accounted for, the actual cell voltage is even higher as is the actual energy consumption. We will presently discuss these losses in quantitative detail, but it is useful to first discuss these qualitatively. Thus, the various losses are graphically represented as internal resistances in Figure 12.4 showing the electrical analog of the PEM-WE, while Figure 12.5 shows the corresponding overpotentials, or potential drops, over the different MEA components, as a function of current density, along with the resulting typical polarization plot for the PEM-WE cell. As discussed earlier, the OCV may possibly be larger than cell thermodynamic voltage, $V_0 = \Phi_{A,0} - \Phi_{C,0}$. For instance, Selamet et al. (2013) note that the electrolysis process starts at 1.48 V at 25°C and at around 1.4 V at 80°C. The kinetic overpotential, or activation overpotential, at an electrode $\eta_\rho \equiv \Phi - \Phi_{\rho,0}$, where Φ is the electrode potential and $\Phi_{\rho,0}$ is the equilibrium potential for the electrode reaction ρ, is caused by the finite rate of the electrode reactions, namely,

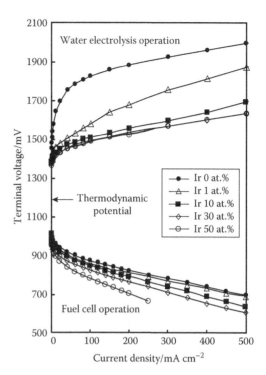

FIGURE 12.3
Cell voltage versus current density of a URFC operating at 80°C as a PEM-WE as well as a PEM-FC, with various iridium contents in the anode catalyst. The PEM-WE is operated with purified water at atmospheric pressure, while the PEM-FC operated with pure H$_2$ and O$_2$ at 100 mL min^{-1} and at 0.30 MPa. (Reprinted from *J. Power Sources*, 112, Ioroi, T., Yasuda, K., Siroma, Z., Fujiwara, N., and Miyazaki, Y., Thin film electrocatalyst layer for unitized regenerative polymer electrolyte fuel cells, 583–587, Copyright 2002, with permission from Elsevier.)

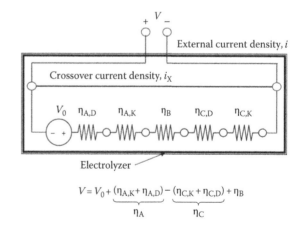

FIGURE 12.4
An electrical analog of the PEM water electrolyzer including a thermodynamic voltage and internal kinetic and diffusion resistances.

a finite current density, and represents the extra potential energy needed to force the electrode reaction to proceed in a desired direction at the desired rate.

Of course, the two electrode reactions, at the anode and the cathode, must proceed at the same rate under steady-state conditions, the relative overpotentials at either electrode being determined by their intrinsic

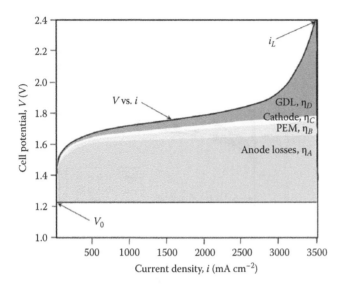

FIGURE 12.5
A schematic polarization plot and the various overpotentials within the PEM water electrolyzer.

kinetics. For PEM-WE and PEM-FC, the kinetic overpotential associated with the hydrogen electrode is usually small (Grigoriev et al., 2008), while that associated with the oxygen electrode is very significant (Figure 12.5). The Ohmic overpotential is due primarily to the resistance to the transport of protons in the PEM, although electrolyte transport resistance in the catalyst layer can also contribute (Makharia et al., 2005).

In general, the diffusion overpotential is characterized by the limiting current density i_L. In principle, this could be due to the mass transport limitations of any of the reactant or product species at either electrode, for example, H_2O, H_3O^+, H_2, or O_2, for the case of PEM-WE. In PEM fuel cells, the dominant mass transfer resistance that determines the limiting current, i_L, is the resistance to diffusion of O_2 through the GDL that is loaded with water produced at the cathode. However, the PEM-WE would not suffer from this diffusion limitation, although at high current densities, bubbling of exiting oxygen could limit the permeation of water to the anode.

In principle, the limiting current could also be due to proton conductivity. For instance, for PEM-FC under dry conditions, the limiting current could be determined by the transport of protons both through the membrane and in the electrolyte in the catalyst layer. However, in the PEM-WE, due to the presence of liquid water, the PEM is fully hydrated and the membrane conductivity is high (Thampan et al., 2000). Consequently, unlike in PEM fuel cell, the diffusion limitations in PEM water electrolyzer may be expected to be relatively small except perhaps at very high current densities, as shown schematically in Figure 12.5.

Modeling of PEM-WE (Choi et al., 2004), and indeed for PEM-FC (Thampan et al., 2001; Vilekar and Datta, 2010;

Rosenthal et al., 2012), follows two alternate approaches: (1) spatially lumped models providing simple analytical solutions (Choi et al., 2004), often with the help of an electric analog with lumped resistors representing the various layers (Figure 12.4) and (2) spatially distributed models involving numerical analysis or the use of computational fluid dynamics (CFD) software (Oliveira et al., 2013).

Among the earliest models, Choi et al. (2004) developed a lumped model based on an electrical analog (Figure 12.4) and involving Butler–Volmer electrode kinetics, nanoscale electrocatalytically active area, along with Ohmic resistance in the PEM. They were able to fit the model to experimental data with literature parameters. Also around that time, Busquet et al. (2004) proposed an empirical model with fitted parameters. Onda et al. (2002) had earlier developed a spatially distributed 2D nonisothermal model for the simulation of a large unit cell, but it was based on empirical overpotential relations. The more recent spatially distributed model of Marangio et al. (2009), on the other hand, involves theoretical correlations. Further, they provide a comparison of numerical predictions with experimental polarization data. More recently, multiphysics approach has been utilized in CFD analysis including multiphase flow and heat effects. Thus, Agbli et al. (2011) used a graphical approach to visualize the interconnection among the various phenomena, namely, electrochemical, thermodynamic, thermal, and fluidic. Oliveira et al. (2013) have provided a multiscale model that includes a detailed molecular mechanism for the anode and cathode reactions, nanoscale catalyst electrolyte interface, and a continuum description of transport phenomena. They utilized it to investigate the effect of various design and operating parameters on PEM-WE performance.

In the following, we provide a detailed lumped analytical model that includes detailed consideration of thermodynamics, electrode kinetics, proton transport, as well as hydrogen and oxygen crossover. We also provide validation of the model by comparing it to recent experimental results and utilizing literature model parameters without any fitting.

12.2 Thermodynamic Analysis

For the thermodynamic analysis, we start by considering a generic electrode reaction ρ

$$\sum_{\substack{i=1 \\ i \neq e^-}}^{n} \underbrace{\nu_{\rho i} B_i^{z_i}}_{\text{In solution}} + \underbrace{\nu_{\rho e} e^-}_{\text{In metal}} = 0 \qquad (12.3)$$

among n species, $B_i^{z_i}$ ($i = 1, 2, ..., n$), with a charge number z_i and a stoichiometric coefficient $\nu_{\rho i}$ ($\nu_{\rho i} > 0$ if B_i is a product, and $\nu_{\rho i} < 0$, if B_i is a reactant). Further, $\nu_{\rho e^-}$ is the stoichiometric coefficient of electrons in the electrode reaction. Thus, $\nu_{\rho e^-} > 0$ for an anodic reaction, that is, when electrons are produced, while $\nu_{\rho e^-} < 0$ for a cathodic reaction, where electrons are consumed. The electrode potential for this reaction is given by the Nernst equation

$$\Phi_{\rho,0} = \Phi_{\rho,0}^\circ + \frac{RT}{\nu_{\rho e^-}F} \ln \prod_{\substack{i=1 \\ i \neq e^-}}^{n} a_i^{\nu_{\rho i}} \tag{12.4}$$

where the subscript "0" denotes equilibrium, while the superscript "o" denotes standard conditions, that is, unit activities, a_i. The standard electrode potential for the electrode reaction ρ, $\Phi_{\rho,0}^\circ$, is given by

$$\Phi_{\rho,0}^\circ = \frac{\Delta G_{\rho,\Phi=0}^\circ}{\nu_{\rho e^-}F} \tag{12.5}$$

where the standard Gibbs free energy change (in the absence of potential), $\Delta G_{\rho,\Phi=0}^\circ \equiv \sum_{\substack{i=1 \\ i \neq e^-}}^{n} \nu_{\rho i} G_{f,i}^\circ (T,p)$, where $G_{f,i}^\circ$ is the Gibbs free energy of formation of i.

Thus, the standard electrode potential for the anode for liquid water (Equation 12.2), $\Phi_{A,0}^\circ = (+2.372 \times 10^5) / (+2)(96,487) = +1.229$ V, while that for the cathode, $\Phi_{C,0}^\circ = 0$ by definition, in fact, being the standard hydrogen electrode (SHE). Consequently, at 298 K the standard cell potential $V_{0,298}^\circ = \Phi_{A,0}^\circ - \Phi_{C,0}^\circ = 1.229$ V.

Defining the cell voltage $V_0 \equiv \Phi_{A,0} - \Phi_{C,0}$, thus, and further using Equation 12.4, $\nu_{Ae^-} = -\nu_{Ce^-}$, along with $\Delta G_{OR}^\circ = \Delta H_{OR}^\circ - T\Delta S_{OR}^\circ$ and assuming the OR enthalpy change ΔH_{OR}° and the entropy change ΔS_{OR}° to be independent of temperature,

$$V_0 = V_{0,298}^\circ - \frac{\Delta S_{OR}^\circ}{\nu_{Ae^-}F}(T - 298) + \frac{RT}{\nu_{Ae^-}F} \ln \prod_{\substack{i=1 \\ i \neq e^-}}^{n} \left(a_i^{\nu_{Ai}}\right)\left(a_i^{\nu_{Ci}}\right) \tag{12.6}$$

where $\nu_{Ae^-} = +2$, for the OR as written (Equation 12.2). As discussed earlier, the standard thermodynamic cell voltage at 298 K

$$V_{0,298}^\circ = \frac{\Delta G_{OR,298}^\circ}{\nu_{Ae^-}F} \tag{12.7}$$

More correctly, of course, both ΔH_{OR}° and ΔS_{OR}° vary with temperature. Accounting for this variation provides a more accurate calculation of the effect of temperature on cell equilibrium potential. However, for our purposes, it is adequate to assume that ΔH_{OR}° and ΔS_{OR}° are constant.

Thus, for the OR for the low-temperature PEM-WE using liquid water, $H_2O(l) \leftrightarrows H_2 + 1/2O_2$, with $\Delta G_{OR}^\circ = +237.2$ kJ mol^{-1} and $\Delta S_{OR}^\circ = +163.3$ J mol^{-1} K^{-1}, this provides

$$V_0 = 1.229 - 8.46 \times 10^{-4}(T - 298) + \frac{RT}{2F} \ln \left(\frac{p_{H_2}}{p_C} \right) \sqrt{\frac{p_{O_2}}{p_A}} \tag{12.8}$$

where we have assumed the activity of liquid water to be unity, while the activities of permanent gases are approximated by their mole fraction in the gas phase, $x_i = p_i/p$, p_i is the partial pressure of i and p_A and p_C are the total pressure in anode and cathode chamber, respectively.

It is noteworthy that the partial pressures of hydrogen and oxygen decline quickly with operating temperature in the case of a low-pressure electrolyzer with liquid water, as the vapor pressure of the water vapor present increases rapidly, as given (in atm.), for example, by the Antoine equation

$$\ln p_{H_2O}^\circ = 11.676 - \frac{3816.44}{T - 46.13} \tag{12.9}$$

On the other hand, when water *vapor*, or steam, is used for electrolysis (Hansen et al., 2012), true for higher-temperature WE, $H_2O(g) \leftrightarrows H_2 + 1/2O_2$, for which $\Delta G_{OR}^\circ = +228.6$ kJ mol^{-1} and $\Delta S_{OR}^\circ = +44.4$ J mol^{-1} K^{-1}, the corresponding expression is

$$V_0 = 1.185 - 2.302 \times 10^{-4}(T - 298)$$
$$+ \frac{RT}{2F} \ln \left(\frac{p_{H_2}}{p_C} \right) \sqrt{\frac{p_{O_2}}{p_A}} \left(\frac{p_{H_2O}^\circ}{p_{H_2O}} \right) \tag{12.10}$$

where we have assumed that the water vapor activity is equal to the ratio of partial to vapor pressure, $a_i = p_i/p_i^\circ$, where p_i° is the vapor pressure of i (superscript o refers to unit activity).

The temperature variation of the standard equilibrium potential for PEM-WE (or, for that matter, for PEM-FC), V_0°, as calculated from the earlier two expressions (Equations 12.8 and 12.10) for liquid water and for water vapor feeds, respectively, is shown in Figure 12.6. It is seen that there is a considerable drop in thermodynamic potential at higher operating temperatures. Additionally, although it has a higher initial potential at 300 K, the voltage decline with temperature

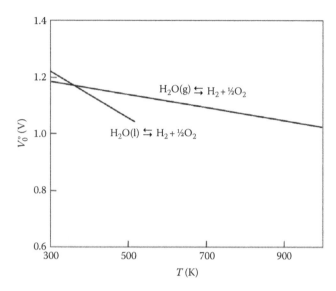

FIGURE 12.6
Standard (for unit activities) equilibrium cell voltage versus temperature for liquid-phase and vapor-phase water electrolysis, as calculated from Equations 12.8 and 12.10.

for liquid water is steeper than that when the feed is in the vapor form. Clearly, higher-temperature operation is indicated for electrolysis with both liquid and vapor feeds, especially, when account is made of the dramatic reduction in kinetic overpotentials with temperature (discussed in Section 12.3), so that higher performance and overall efficiencies are obtained at higher temperatures.

12.3 Electrode Reaction Mechanism and Kinetics

The electrode ORs, namely, the oxygen electrode reaction (OER) and the hydrogen electrode reaction (HER), are themselves comprised of several elementary steps in a given sequence, and often with more than one parallel pathway. In general, this molecular level detail cannot be ignored, because the electrode reactions do not always faithfully follow the single Butler–Volmer equation, strictly applicable to an elementary step in an overall scheme, not to the OR (Vilekar et al., 2010). If a single rate-determining step (RDS) exists in a sequence of steps, however, we can still often use the Butler–Volmer equation as being applicable for the OR, at least over a limited range, since the OR rate is then controlled by that of the elementary RDS. However, the OER and HER rates vary over such a large range by virtue of the strong effect of potential that a single Butler–Volmer equation cannot in general describe the electrode OR kinetics over the complete range of potentials of interest.

The Butler–Volmer expression valid for an elementary electrode reaction step ρ (Equation 12.3) is (Thampan et al., 2001)

$$i_\rho^* = i_{\rho,0}^* \left[\exp\left\{ \frac{(\beta_\rho)\nu_{\rho e^-}F\eta_\rho}{RT} \right\} - \exp\left\{ \frac{(\beta_\rho - 1)\nu_{\rho e^-}F\eta_\rho}{RT} \right\} \right]$$

$$\frac{A}{cm^2\ ESCA} \tag{12.11}$$

where

β_ρ is the symmetry factor
F is Faraday's constant

This stems from the use of the relation with $i_\rho^* = F\nu_{\rho e^-}r_\rho^*$, where r_ρ^* is the rate of the electrode reaction per unit area, and the exchange current density $i_{\rho,0}^* = F\nu_{\rho e^-}r_{\rho,0}^*$. These relations are written for the actual electrocatalytic surface area, or ECSA, as denoted by the superscript asterisk in the current density and in the reaction rate.

Often, for elementary reactions, we can assume $\beta_\rho = 1/2$, so that in that case, the Butler–Volmer expression is rewritten in the convenient form

$$i_\rho^* = i_{\rho,0}^* \left\{ 2\sin h\left(\frac{\beta_\rho \nu_{\rho e^-}F\eta_\rho}{RT} \right) \right\} \tag{12.12}$$

In this equation, $\beta_\rho = 1/2$, and the exchange current density is a function of species composition as given by

$$i_{\rho,0}^* = i_{\rho,0}^{o*} \left\{ \prod_{\substack{i=1 \\ i\neq e^-}}^{n} a_i^{-\vec{v}_{\rho i}(1-\beta_\rho)} \prod_{\substack{i=1 \\ i\neq e^-}}^{n} a_i^{(\vec{v}_{\rho i})\beta_\rho} \right\} \frac{A}{cm^2\ ESCA} \tag{12.13}$$

where $\vec{v}_{\rho i}$ is the stoichiometric coefficient of the species i in reaction step s_ρ (Equation 12.3) as a *reactant* and $\overleftarrow{v}_{\rho i}$ is that as a *product*. This accounts for the possibility that a species, for example, water, may be present both as a reactant and as a product in a reaction. The net stoichiometric coefficient of the species i in step s_ρ (Equation 12.3) is then $v_{\rho i} = \vec{v}_{\rho i} + \overleftarrow{v}_{\rho i}$. In Equation 12.13, the *standard* exchange current density, that is, with unit species activities

$$i_{\rho,0}^{o*} = F\nu_{\rho e^-}\left(\frac{c_t^*}{N_{Av}} \right)\vec{\Lambda}_\rho \exp\left(-\frac{\vec{E}_{\rho,\Phi_0^o}}{RT} \right) \frac{A}{cm^2\ cat} \tag{12.14}$$

where

$\vec{E}_{\rho,\Phi_0^o}$ is the thermal activation energy
$\vec{\Lambda}_\rho$ is the pre-exponential factor
c_t^* is the number of reaction sites/cm² of catalyst surface
N_{Av} is Avogadro's number

In other words, the standard exchange current density is akin to the electrochemical equivalent of the rate constant at standard electrode potential, and consequently, varies strongly as a function of temperature as given by its Arrhenius form.

These relations show the powerful effect of electrode overpotential η_ρ and that of temperature on reaction thermodynamics and kinetics. In other words, temperature and potential are complementary tools for accomplishing favorable kinetics and thermodynamics. Thus, performance can be substantially enhanced by operating the electrolyzer at a higher temperature, the latter being usually limited by the performance and stability of the electrolyte used. Thus, the PEM-WE is usually limited to an operating temperature less than 100°C.

For a sequence of steps with a single RDS, since the OR rate is then determined by that of the RDS, we can use the Butler–Volmer equation for the OR, which is written in the following inverted form:

$$\eta_\rho = \left(\frac{RT}{\beta_\rho^{\cdot} v_{\rho e^-}^{\cdot} F} \right) \sinh^{-1} \left(\frac{i_\rho^*}{2 i_{\rho,0}^*} \right) \quad (12.15)$$

where $\beta_\rho^{\cdot} = 1/2$ and effective $v_{\rho e^-}^{\cdot}$ correspond to the RDS and are so identified by the superscript "dot." Often, these two parameters are combined into effective transfer coefficients, $\alpha_\rho^{\cdot} = \beta_\rho^{\cdot} v_{\rho e^-}^{\cdot}$.

Since the HER mechanism and kinetics have been considered by us in considerable detail recently (Vilekar et al., 2010), we will very briefly discuss in the following the example of the OER. This is, of course, a very complex OR, the mechanism of which is far from settled yet, although it is currently a very active area of investigation (Hansen et al., 2014; Katsounaros et al., 2014).

While it can involve many steps, as an illustration, let us consider a particularly simple example of an OER mechanism comprising a single pathway of only four sequential steps (Oliveira et al., 2013; Hansen et al., 2014):

$$
\begin{aligned}
s_1: & \quad H_2O + S \leftrightarrows OH \cdot S + H^+ + e^- \\
s_2: & \quad OH \cdot S \leftrightarrows O \cdot S + H^+ + e^- \\
s_3: & \quad H_2O + O \cdot S \leftrightarrows OOH \cdot S + H^+ + e^- \quad (12.16) \\
s_4: & \quad OOH \cdot S \leftrightarrows O_2 + H^+ + e^- + S \\
\hline
OR: & \quad 2H_2O \leftrightarrows O_2 + 4H^+ + 4e^-
\end{aligned}
$$

In general, each elementary step in this would follow the Butler–Volmer equation, but not the OR (Vilekar et al., 2010). If we were to assume, for example, that step s_1 is the RDS, so that others are at quasi-equilibrium (QE), then we could effectively combine the QE steps

that follow the RDS in a sequence into an intermediate reaction (IR), since thermodynamic equilibrium, which characterizes these steps, will not change, as per Hess's law. As a result, the mechanism may be written as a two-step sequence as follows:

$$
\begin{aligned}
s_1 & \quad H_2O + S \leftrightarrows OH \cdot S + H^+ + e^- \quad (RDS) \\
IR_2: & \quad H_2O + OH \cdot S \leftrightarrows O_2 + 3H^+ + 3e^- + S \quad (QE) \\
\hline
OR: & \quad 2H_2O \leftrightarrows O_2 + 4H^+ + 4e^-
\end{aligned}
$$

$$(12.17)$$

the first of which is the RDS, so that Equation 12.15 applies with $\beta_\rho = 1/2$ and $v_{\rho e^-}^{\cdot} = +1$, while the second step is QE.

On the other hand, let us consider the case when step s_3 is the RDS, so that the remaining steps in the sequence are at QE. We could consequently combine the QE steps preceding the RDS with the RDS into an IR, so that the mechanism may now be written as the two-step sequence

$$
\begin{aligned}
IR_1: & \quad 2H_2O + S \leftrightarrows OOH \cdot S + 3H^+ + 3e^- \quad (RDS) \\
s_4: & \quad OOH \cdot S \leftrightarrows O_2 + H^+ + e^- + S \quad (QE) \\
\hline
OR: & \quad 2H_2O \leftrightarrows O_2 + 4H^+ + 4e^-
\end{aligned}
$$

$$(12.18)$$

so that Equation 12.15 applies with $\beta_\rho = 1/2$ and $v_{\rho e^-}^{\cdot} = +3$. In other words, the location of the RDS in a sequence determines the effective stoichiometric coefficient of electrons in the Butler–Volmer equation.

In fact, as discussed earlier, the RDS in a sequence can change with temperature or potential. This would be indicated by a change in the Tafel plot. An example of this is shown in Figure 12.7 for the case of OER on an Ir anode, where the data of Damjanovic et al. (1966) are described by two-piecewise Butler–Volmer equations with two different RDSs.

Also provided in Figure 12.7 are the corresponding Tafel approximations of the Butler–Volmer equation, that is,

$$\eta_\rho \approx \left(\frac{RT}{\beta_\rho^{\cdot} v_{\rho e^-}^{\cdot} F} \right) \ln \left(\frac{i_\rho^*}{i_{\rho,0}^*} \right) \quad (12.19)$$

which is strictly applicable only when $i_\rho^* > 2 i_{\rho,0}^*$ (Figure 12.7).

At any rate, the low current densities are characterized by a Tafel slope, $b \equiv \partial\Phi/\partial\log i^*$ of around $b = 2.303(2RT/3F)$, or around 40 mV/decade, while the higher current density $i^* > 5 \times 10^{-4}$ A cm^{-2}, the Tafel slope changes to ~120 mV/decade. Although used commonly in modeling,

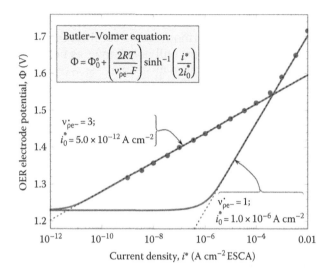

Butler–Volmer equation:

$$\Phi = \Phi_0^\circ + \left(\frac{2RT}{\nu_{pe^-}'F}\right)\sinh^{-1}\left(\frac{i^*}{2i_0^*}\right)$$

$\nu_{pe^-}' = 3;$
$i_0^* = 5.0 \times 10^{-12}\ \mathrm{A\ cm^{-2}}$

$\nu_{pe^-}' = 1;$
$i_0^* = 1.0 \times 10^{-6}\ \mathrm{A\ cm^{-2}}$

FIGURE 12.7
Electrode potential versus current density for oxygen evolution reaction (OER) on Ir in 1 N HClO$_4$, comparison of experimental data of Damjanovic et al. (1966) with the Butler–Volmer equation for two different rate-determining steps and exchange current densities, along with their Tafel approximations shown via dotted lines.

a limitation of the Tafel equation is that it is, of course, erroneous at low current densities and does not converge to the thermodynamic potential, as evident from Figure 12.7. Further, the change of the RDS also changes the corresponding exchange current density (Figure 12.7).

For the purpose of PEM-WE modeling, we will simply assume that mechanism in Equation 12.18 is applicable for the OER, so that the Butler–Volmer equation, Equation 12.15, is applicable, along with $i_{O_2,0}^* = 5.0 \times 10^{-12}$ A cm^{-2} ESCA, $\beta_p' = 1/2$, and $\nu_{pe^-}' = +3$. As seen from Figure 12.7, this is clearly not strictly applicable at current densities in excess $i^* > 5 \times 10^{-4}$ A cm^{-2} ESCA. However, as shown in the comparison to experimental results here, this seems adequate for current densities in the range considered here.

For the case of HER too, although the reality is more complex (Vilekar and Datta, 2010), we will simply assume the Butler–Volmer equation with $\beta_p' = 1/2$ and $\nu_{pe^-}' = -1$, and the standard exchange current density of 1×10^{-3} A cm^{-2} (Thampan et al., 2001).

This current density and the exchange current density may be written on the basis of the MEA area (A cm^{-2} MEA) by using the catalyst roughness factor (γ_M), that is, $i_p = \gamma_M i_p^*$. Thus (Thampan et al., 2001),

$$i_{p,0} = \gamma_M \left(\frac{a_i}{a_{i,\text{ref}}}\right)^{\bar{\nu}_{pi}} \exp\left\{-\frac{E_{p,\Phi_0}}{R}\left(\frac{1}{T} - \frac{1}{T_{\text{ref}}}\right)\right\} i_{p,0,\text{ref}}^{\circ*} \quad (12.20)$$

where a_i is the activity of the limiting reactant, for example, water, and $i_{p,0,\text{ref}}^{\circ*}$ is the standard exchange current density at the reference temperature.

Further, the roughness factor is the ratio between the active electrocatalyst area (ECSA) and the geometric area of the electrode/MEA. It is given in terms of catalyst particle diameter by (Thampan et al., 2001)

$$\gamma_M = \varphi_I m_M \frac{6}{\rho_M d_M} \quad (12.21)$$

where
ρ_M is the catalyst density
m_M is the catalyst leading
d_M is the supported or unsupported catalyst crystallite diameter
φ_I is the fraction of metal catalyst surface in contact with the ionomer, forming the three-phase interface, or TPI

12.4 Transport Limitations

The electrode overpotentials discussed are further affected by the following diffusional considerations: (1) due to any diffusional limitations of the reactants/products in the GDL (Thampan et al., 2001; Ito et al., 2013); (2) the crossover of reactants to the opposite electrode across the PEM and any resulting effect on the electrode potential there (Vilekar and Datta, 2010); and (3) the effect of electrolyte resistance in the PEM (Thampan et al., 2000; Choi et al., 2005) as well as in the catalyst layer (Makharia et al., 2005). These are discussed in turn in the following. Additionally, of course, there is often a finite interfacial contact resistance (Makharia et al., 2005), although here we assume it as negligible, along with the electrolyte resistance in the catalyst layers.

The flux of the reactant species i across the GDL (denoted as layer D) is given by the integrated (for constant flux) form of Fick's law

$$N_{i,z} = \frac{D_{i,D}^e}{L_D}\left\{c_{i,D}(0) - c_{i,D}(a)\right\}$$

$$= P_{i,D}\left\{c_{i,b} - \frac{c_{i,D}(a)}{\kappa_{i,D}}\right\}\frac{\text{mol}}{\text{cm}^2\ \text{MEA s}} \quad (12.22)$$

where
D_i^e is the effective diffusion coefficient of i in the GDL layer of thickness L_D
$c_{i,D}(0)$ and $c_{i,D}(a)$ are its concentrations at either end within the GDL
$c_{i,b}$ is the bulk-phase concentration in the electrode chamber

The partition coefficient $\kappa_{i,D}$ and the permeance $P_{i,D}$ of species i in the GDL layer are

$$\kappa_{i,D} \equiv \left(\frac{c_{i,D}}{c_{i,b}}\right)_0 ; \quad P_{i,D} \equiv \frac{\kappa_{i,D} D_{i,D}^e}{L_D} \quad (12.23)$$

The flux is next written in terms of the current density, $i_\rho = (\nu_{\rho,e^-}/-\nu_{\rho,i})FN_{i,z}$, and equated to the kinetic current density (electrocatalytic rate) in the catalyst layer. This is solved for the unknown interfacial concentration $c_i(a)$, followed by its use in kinetics, resulting in the Butler–Volmer equation (Equation 12.15) in the inverted form (Thampan et al., 2001)

$$\eta_\rho = \left(\frac{RT}{\beta_\rho \nu_{\rho e^-} F}\right) \sinh^{-1}\left[\frac{1}{2}\left\{\frac{i_\rho/i_{\rho,0}}{1 - i_\rho/i_{\rho,L}}\right\}\right] \quad (12.24)$$
$$= \eta_{\rho,K} + \eta_{\rho,D}$$

which is the combined overpotential associated with kinetic and diffusional processes at the electrode (Figure 12.4). Here, the GDL diffusion limiting current density (when $c_{i,D}(a) \to 0$) for the electrode reaction ρ

$$i_{\rho,L} \equiv \left(\frac{\nu_{\rho,e^-}}{-\nu_{\rho,i}}\right) FP_{i,D}c_{i,b} \quad (12.25)$$

In the PEM-FC, this limiting current is usually determined by the diffusion across the GDL of O_2 from air to the catalyst layer, of the order of around 2 A cm^{-2}, made possible by keeping the GDL from becoming waterlogged via PTFE treatment. On the other hand, for the case of PEM-WE with liquid water feed, the current densities limited by the permeation of liquid water across the GDL to the anode catalyst layer, with counter flow bubbling of the pure O_2 produced at the anode CL, are likely to be much higher, perhaps as much as 5 A cm^{-2} (Rozain and Millet, 2014), depending on bipolar plate design and water flow rate. For this reason, Choi et al. (2004) neglected the diffusion limiting current considerations in their model. However, there are cases when it could be significantly lower and, consequently, affect the polarization, for example, if the flow rate of water in the anode chamber is low (Ito et al., 2010), leading to a buildup of a blanket of oxygen at the catalyst layer and, thus, impeding the permeation of water to it, as well as for the case of higher-temperature steam electrolysis (Hansen et al., 2012; Wang et al., 2014). In fact, mass transfer limitation in water electrolysis is an issue that involves two-phase co- or counter-diffusion and flow of water and oxygen in a porous layer (Ito et al., 2013), as well as in the bipolar plate channels (Ito et al., 2010), a topic that has not yet been adequately dealt with in the literature.

12.5 Gas Crossover

The hydrogen and oxygen produced at the cathode and anode, respectively, of course, have a finite, albeit small, solubility in water as well as in the PEM (Kocha et al., 2006; Weber and Newman, 2007; Ito et al., 2011), so that, driven by their large partial pressure difference, they crossover to the opposite electrode (Bensmann et al., 2014), a phenomenon that has serious implications from the viewpoint of PEM-WE efficiency and, more importantly, safety (Schalenbach et al., 2013), as H_2–O_2 mixtures have broad flammability limits. This is especially true for higher-pressure PEM-WEs (Grigoriev et al., 2009, 2011) that seek to directly produce compressed gases via electrochemical co-compression, and thus avoid or reduce compression equipment and costs (Onda et al., 2007; Bensmann et al., 2013). The crossover and the resulting safety concerns hence limit the pressurization obtainable in a PEM-WE and can call for appropriate mitigation measures, for example, use of catalytic oxidizers following the electrolyzers (Grigoriev et al., 2009, 2011), in order to reduce the mixture composition so as to lie outside the flammability limits.

The transport of gases across a well-hydrated PEM is primarily via the solution–diffusion mechanism through both the liquid water channels and the polymer phases within the electrolyte membrane layer (Weber and Newman, 2007), as driven mainly by activity or partial pressure difference between the two chambers. In addition, there could be some convective transport as well of the dissolved gases via permeation of water through the water channels as a result of any pressure difference plus electroosmosis of water, that is, via water molecules dragged along with the proton, which would carry along dissolved hydrogen and oxygen. Thus, we may write, for the flux of a dissolved gas i across the PEM (layer B) (Weber and Newman, 2007),

$$N_{i,z} = -D_{i,B}^e \nabla c_{i,B} - c_{i,w}\left(\frac{B_0}{\mu}\nabla p + \xi\frac{i}{F\bar{V}_w}\right) \quad (12.26)$$

where

$c_{i,B}$ is the concentration in the membranes layer
$D_{i,B}^e$ is the effective diffusivity of the dissolved gas in the membrane layer considered as a pseudo-homogeneous medium that accounts for diffusion both via the water-filled channel and through the polymer matrix

Further, $c_{i,w}$ is the concentration of the gas dissolved in water, μ is the water viscosity, and $B_0 \approx \varepsilon^{1.5}(a^2/8)$, where a is the water channel/cluster radius (~2 nm), is the d'Arcy permeability of the PEM. The term in the parentheses in the earlier expression is the total convective velocity

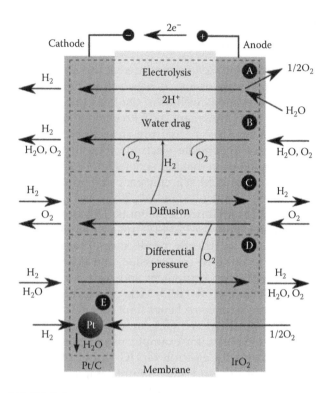

FIGURE 12.8
Schematic of the crossover and fate of hydrogen and oxygen across the PEM in a PEM-WE via diffusion plus convective transport involving permeation and electroosmosis. It is shown that while the oxygen transported to the cathode is electrocatalytically oxidized by hydrogen to water, the hydrogen permeating to the anode is not electrocatalytically or catalytically oxidized despite a large excess of oxygen and the presence of an effective catalyst. (Reprinted from *Int. J. Hydrogen Energy*, 38, Schalenbach, M., Carmo, M., Fritz, D. L., Mergel, J., and Stolten, D., Pressurized PEM water electrolysis: Efficiency and gas crossover, 14921–14933, Copyright 2013, with permission from Elsevier.)

of water v_z as a result of pressure (d'Arcy flow) and potential (electroosmosis) gradients (i.e., Schlögl equation, e.g., see Thampan et al., 2000), the latter written in terms of proton flux (current density), with ξ being the electro-osmotic drag coefficient, that is, the number of water molecules dragged along per proton crossing the PEM. These various transport mechanisms are shown schematically in Figure 12.8 (Schalenbach et al., 2013).

It is, thus, clear that the permeability and transport of gases are, to at least some extent, dependent on operating conditions, namely, current density that determines proton and hence electroosmotic flux, RH, and any differential pressure between the anode and the cathode chambers. If both the anode and the cathode compartments were at the same pressure, the d'Arcy flow is zero, although convective transport via electroosmosis is still present. On the other hand, if the cathode were at a higher pressure than, say, the anode, then d'Arcy flow would oppose electroosmosis (Figure 12.8).

Although some progress has recently been made in modeling these various gas transport modes (Schalenbach et al., 2013), for the purpose of our model here, we will neglect convective transport, and simply assume the solution and diffusion model (Ito et al., 2011) as the dominant mode of dissolved gas transport across the PEM. This is because of the small water permeability of Nafion ($B_0 \approx 4 \times 10^{-16}$ cm^2 and $\xi \approx 2.5$ at $\lambda = 20$, as measured by Meier and Eigenberger, 2004) coupled with small hydrogen and oxygen solubility in water (Ito et al., 2011). We shall presently show that the hence reduced model adequately represents the case of equal electrode chamber pressures, that is, with symmetric pressure operation. In this case, following the development earlier for diffusive transport in the GDL, the crossover current due to diffusion of H_2 and O_2 across the PEM layer B is given by (Vilekar and Datta, 2010)

$$i_{\rho,X} = \left(\frac{\nu_{\rho,e^-}}{-\nu_{\rho,i}}\right)\left(\frac{\kappa_{i,B}D_{i,B}^e}{L_B}\right)Fc_i^G \qquad (12.27)$$

where

$\kappa_{i,B}$ is the partition coefficient (solubility) of the species i in the electrolyte membrane
L_B is the membrane thickness
c_i^G is its concentration in the gas phase in the electrode chamber where it is the dominant species

This assumes that the concentration in the opposing chamber to where it is diffusing is vanishingly small.

The partition coefficient $\kappa_{i,B} = (c_{i,B}/c_i^G)_0$ represents the solubility of the gaseous species i within the hydrated membrane. Further, Equation 12.27 may be written instead in terms of the species partial pressure, $p_i = c_i^G(RT)$, and with the membrane permeability defined as $k_i = D_{i,B}^e \kappa_{i,B}/RT$ (mol bar^{-1} cm^{-1} s^{-1}), so that Equation 12.27 may be written in the alternate form

$$i_{\rho,X} = \left(\frac{\nu_{\rho,e^-}}{-\nu_{\rho,i}}\right)\left(\frac{k_i}{L_B}\right)Fp_i \qquad (12.28)$$

For example, the hydrogen permeability in Nafion provided by Kocha et al. (2006) under water vapor–saturated conditions

$$k_{H_2} = 6.6 \times 10^{-8} \exp\left(-\frac{21{,}030 \text{ J mol}^{-1}}{RT}\right) \text{ mol bar}^{-1}\text{cm}^{-1}\text{s}^{-1}$$

$$(12.29)$$

Further, they suggest that the permeability of oxygen in Nafion is simply roughly one-half that of hydrogen in Nafion, that is, $k_{O_2} = k_{H_2}/2$. Additional discussion of

H_2 and O_2 permeation in Nafion is provided by Ito et al. (2011) and Baik et al. (2013).

Now that we are in a position to make reasonable estimates of gas crossover for both H_2 and O_2, the key question to ponder next is the fate of these species upon crossover. Keeping in mind that H_2 and O_2 mixtures are mutually explosively reactive, particularly in the presence of a finely divided excellent oxidation catalyst such as Pt and IrO_2, the evident assumption would be that they would react immediately catalytically or electrocatalytically upon crossover. However, the evidence appears to be to the contrary, that is, H_2 is found in the O_2 product at the anode, and although to a lesser extent, some O_2 is found in the H_2 produced at the cathode. Clearly, a more careful consideration of the various possibilities is called for.

Let us first consider the case of H_2 arriving from the cathode to the anode. Once at the anode, it has several possible fates: (1) it could simply dilute the O_2 product there; (2) it could *chemically* combine with the abundant O_2 there in the presence of catalyst; (3) it could *electrochemically* combine with the O_2, that is, via electrocatalytic hydrogen oxidation reaction (HOR) along with oxygen reduction reaction (ORR) proceeding at the anode concomitantly, as is the case in the PEM-FC (Vilekar and Datta, 2010); or (4) it could simply undergo HOR, with the resulting protons and electrons joining others produced via WE to be pumped back to the cathode (Kocha et al., 2006). The actual fate of a H_2 molecule crossing over to the anode among these competing options would depend on the relative ease with which these parallel pathways could proceed.

Let us first consider the possibility of occurrence of the electrocatalytic oxidation of H_2 at the anode, a discussion of which is facilitated by Figure 12.9, which provides a schematic of the individual electrode potentials along with kinetic overpotentials for the H_2 and O_2 electrode reactions, hence indicating regions of electrocatalytic reaction feasibility.

As evident from the Butler–Volmer equation in the conventional form for reaction ρ (HER or OER) at a given electrode α (for anode α = A, while for cathode, α = C), along with the common assumption that $\beta_\rho' = 1/2$

$$
\begin{aligned}
i_{\rho,\alpha} &= \vec{i}_{\rho,\alpha} - \overleftarrow{i}_{\rho,\alpha} \\
&= i_{\rho,0} \exp\left\{ \left(\frac{F}{2RT}\right) v_{\rho e^-}(\Phi_\alpha - \Phi_{\rho,0}) \right\} \\
&\quad - i_{\rho,0} \exp\left\{ -\left(\frac{F}{2RT}\right) v_{\rho e^-}(\Phi_\alpha - \Phi_{\rho,0}) \right\}
\end{aligned}
\tag{12.30}
$$

A finite rate in the forward direction thus requires the product in the term $v_{\rho e^-}\eta_{\rho,\alpha} > 0$, coupled with

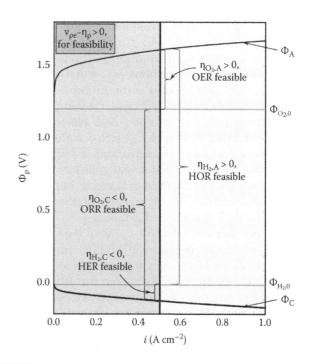

FIGURE 12.9
Schematic of electrode potentials and kinetic overpotentials, indicating regions of electrocatalytic reaction feasibility.

adequate overpotential for reaction ρ (HER or OER) at the electrode α (A or C), $\eta_{\rho,\alpha} = \Phi_\alpha - \Phi_{\rho,0}$, so that $\vec{i}_{\rho,\alpha} \gg \overleftarrow{i}_{\rho,\alpha}$. If $v_{\rho e^-}\eta_{\rho,\alpha} < 0$, on the other hand, $\vec{i}_{\rho,\alpha} \ll \overleftarrow{i}_{\rho,\alpha}$, and the reaction would proceed in the reverse direction, that is, as HOR or ORR, since $i_{\rho,\alpha} = \vec{i}_{\rho,\alpha} - \overleftarrow{i}_{\rho,\alpha} << 0$.

For the crossing over H_2 to react with O_2 electrocatalytically at the anode or vice versa at the cathode, both HOR and ORR must occur sequentially on the *same* electrode

$$
\begin{aligned}
\text{HOR:} \quad & H_2 \leftrightarrows 2H^+ + 2e^- \\
\text{ORR:} \quad & \frac{1}{2}O_2 + 2H^+ + 2e^- \leftrightarrows H_2O \\
\hline
\text{OR:} \quad & H_2 + \frac{1}{2}O_2 \leftrightarrows H_2O
\end{aligned}
\tag{12.31}
$$

The rate of this electrocatalytic OR would be limited by that of the slower of the two steps in the sequence. Both these reactions readily occur under the operating electrode potentials at either electrode in a PEM-FC and, in fact, account for the OCV being 200 mV less than the equilibrium thermodynamic potential (Vilekar and Datta, 2010). However, such is not the case in the PEM-WE, as the following analysis reveals.

At the anode (α = A), thus, where under WE conditions the electrode potential Φ_A the equilibrium thermodynamic potentials for both the OER, $\Phi_{O_2,0}$, and the HER, $\Phi_{H_2,0}$. Thus, the overpotential for OER, $\eta_{O_2,A} = \Phi_A - \Phi_{O_2,0} > 0$, as well as that for the HER,

$\eta_{H_2,A} = \Phi_A - \Phi_{H_2,0} \gg 0$. Consequently, only electron-producing reactions ($v_{pe^-} > 0$) can occur at the anode, namely, HOR, and oxygen evolution reaction (OER), that is, electron-consuming reactions ($v_{pe^-} < 0$), HER as well as ORR, are unlikely to occur at the anode.

In other words, while hydrogen crossing over can readily dissociate into protons and electrons, it is unlikely to further react with O_2 electrocatalytically via ORR at the anode, since as calculations readily indicate, the ORR can proceed only at a negligible rate under the typical conditions at the PEM-WE anode, both because of the overpotential $\eta_{O_2,A} > 0$ coupled with the very small $i_{O_2,0}$. For the ORR to occur at the anode, $\eta_{O_2,A} < 0$, that is, the electrode potential $\Phi_A < \Phi_{O_2,0}$, as in the case of the PEM-FC, where this occurs readily (Vilekar and Datta, 2010). In short, the H_2 crossing over to the anode will not react with O_2 electrocatalytically, but the other options mentioned earlier may still be possible. We show in the following that the H_2 crossing over simply dilutes the O_2 produced via electrolysis at the anode.

On the other hand, at the cathode ($\alpha = C$) under WE conditions (Figure 12.9), the electrode potential $\Phi_C <$ both $\Phi_{O_2,0}$ and $\Phi_{H_2,0}$. Thus, $\eta_{O_2,C} = \Phi_C - \Phi_{O_2,0} \ll 0$ and $\eta_{H_2,C} = \Phi_C - \Phi_{H_2,0} < 0$. Consequently, electron-consuming reactions ($v_{pe^-} < 0$) are mainly indicated, namely, HER and ORR. However, a quick calculation shows that in fact HOR can proceed at the cathode at a finite rate despite $\eta_{H_2,C} < 0$. This is because of the large exchange current density for the HER. Thus, O_2 crossing over to the cathode does undergo electrocatalytic reduction, especially at low current densities. However, the overpotential caused by this crossover at the cathode is negligible because of the huge overpotential for ORR at the cathode, and the large exchange current density for HOR. This pathway may be especially significant at low current densities when $\eta_{H_2,C} \to 0$, but would reduce in effectiveness at higher current densities when the cathode potential becomes more negative.

In summary, calculations with the Butler–Volmer equation (Equation 12.30) under typical conditions at the two electrodes indicate that at the cathode, ORR is highly favorable and HOR is possible, while at the anode, although the HOR is highly favorable, ORR is highly unlikely. Since for electrocatalytic reaction, both the HOR and the ORR must proceed concomitantly (Equation 12.31) at either electrode for electrocatalytic reaction to proceed, it is clear that while electrocatalytic reduction of O_2 is possible at the cathode, the H_2 oxidation does not occur at the anode. Further, the overpotential due to this crossover at the cathode is small, although it can measurably impact the Faradaic efficiency (Grigoriev et al., 2011).

Let us also briefly consider the alternate fates mentioned earlier for the H_2 crossing over to the anode. In principle, catalytic combustion could occur because of

the presence of the catalyst IrO_2 at the anode. However, this would require oxygen that comes from water electrolysis (although a part could come from any oxygen dissolved in the water feed), with an equivalent amount of hydrogen coproduced as protons and electrons. Finally, because of high overpotential for the HOR at the anode, it is conceivable that the H_2 crossing over dissociates into protons and electrons and gets pumped back to the cathode. However, as shown in the following equations, the H_2 crossing over simply exits the anode with the O_2 gas produced there. Since the flammability limit of H_2 in O_2 is 3.9–95.8% (Grigoriev et al., 2011), a consequence of this is that it limits the extent to which cathode gas can be pressurized (Schalenbach et al., 2013).

Assuming, thus, that the hydrogen crossing over simply dilutes the anode gas, the mole fraction of H_2 in the anode gas (O_2) may be obtained from

$$x_{H_2} = \frac{i_{H_2,X}}{i_{H_2,X} + i/2} \tag{12.32}$$

where from Equation 12.28, the equivalent current for hydrogen crossover

$$i_{H_2,X} = \frac{2Fk_{H_2}}{L_B} p_{H_2,C} \tag{12.33}$$

which is based on assuming that the partial pressure of H_2 in the anode chamber, $p_{H_2,A} \to 0$. Here, the H_2 permeability in Nafion is given by Equation 12.29.

The resulting prediction of mole fraction is shown in Figure 12.10 versus the experimental measurements of

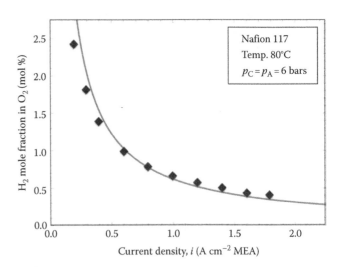

FIGURE 12.10

Mole fraction of H_2 in O_2 at the anode versus operating current density in a PEM-WE. A comparison of theoretical model (Equation 12.32) with experimental data of Schalenbach et al. (2013) under conditions of balanced pressure (6 bars) in anode and cathode chambers.

the H_2 mole fraction in the oxygen gas produced at the anode (Schalenbach et al., 2013), and convincingly demonstrates that the crossover flux of H_2 is accounted for in its entirety in the O_2 product gas. This is at odds with what happens in a PEM-FC, where the H_2 crossing over is completely consumed *electrochemically* at the oxygen electrode via ORR that accounts for an OCV in PEM-FC that is lower than the thermodynamic cell potential. As Grigoriev et al. (2009) observed, based on their experiments, "Even hydrogen, which is so easily oxidized at the platinum-iridium anodes, can reach the output oxygen stream on anode circuits."

This crossover of H_2 and O_2, thus, apparently does not significantly affect the overpotential and only marginally impacts the Faradaic efficiency (Grigoriev et al., 2011). The magnitude of the crossover current in PEM-WE was, in fact, directly measured in situ by Bensmann et al. (2014) as provided in Figure 12.11, which shows hydrogen crossover flux as a function of the pressure difference between the cathode and anode chambers, and the corresponding current density, measured via a unique technique devised by Bensmann et al. (2014), in which they applied the small currents shown in Figure 12.11 in a PEM-WE with the cathode chamber closed off, so that H_2 generated at the cathode accumulates, causing a rise in cathode pressure until reaching a plateau where the rate of electrochemical generation of hydrogen at the cathode is equal to its flux across the MEA.

Bensmann et al. (2014) used a commercial 240 μm thick fumea® F-10180 rf catalyst-coated-reinforced PFSA membrane in their PEM-WE experiments, for which they back-calculated the H_2 permeability.

Their values for the permeability for the fumea F-10180 rf *reinforced* membrane used by them are roughly 50% higher than the value indicated by Equation 12.29 for Nafion membranes. However, with this higher permeability, the agreement of Equation 12.33 with experiments of Bensmann et al. (2014) is good, as seen in Figure 12.11.

Clearly, nonetheless, even with higher permeability and significant compression, the loss of Faradaic efficiency is relatively small

$$\varepsilon_i = 1 - \left(\frac{i_{H_2,X} + i_{O_2,X}}{i} \right) \qquad (12.34)$$

since $i_{H_2,X}$, $i_{O_2,X} \ll i$ (Figure 12.11).

However, the main concern of crossover is safety, that is, to keep the H_2 content in O_2 below ignition composition, which limits the extent of co-compression that can be accomplished in the PEM-WE. Combining the earlier results thus allows us to determine the conditions of maximum cathode pressure to stay below the explosion limit ($x_{H_2,Crit} \approx 0.04$)

$$p_{H_2,C} \leq L_B \left\{ \frac{1}{4Fk_{H_2}} \left(\frac{x_{H_2,Crit}}{1 - x_{H_2,Crit}} \right) \right\} i \qquad (12.35)$$

Since the term in the curly brackets in this is a constant, this shows that higher-pressure H_2 may be realized only when operating at a high current density or by using a thicker membrane, although the latter increases the operating voltage and reduces efficiency. A final possibility is to design PEMs with lower hydrogen permeability (Grigoriev et al., 2009).

An alternate safety measure is to catalytically oxidize the H_2 outside of the electrolyzer. Thus, Grigoriev et al. (2009) experimented with both platinized anode GDL as well as use of external oxidizers (Figure 12.12) based on Pt, both of which oxidize this hydrogen present in the oxygen very effectively. It is indeed curious that this chemical catalysis does not occur at the anode catalyst layer, but rather external to the MEA, because of the significant overpotential for the needed oxygen produced via WE.

Compared to the issue of H_2 crossover that of O_2 crossover from the anode to the cathode is less critical due to two reasons: (1) the permeability of O_2 is lower than that of H_2 and (2) the O_2 reaching the cathode is electrocatalytically oxidized via HOR at least to some extent, as discussed earlier. Thus, Grigoriev et al. (2009) state that, "During electrolysis from atmosphere pressure up to 6 bar, oxygen was not detected in hydrogen (less than 5 ppm was measured…). But as the operating

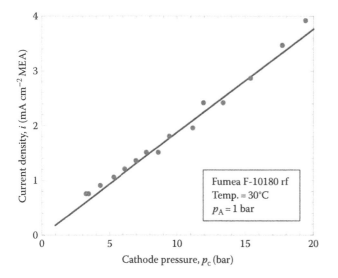

FIGURE 12.11
Crossover hydrogen flux versus cathode (pure hydrogen) pressure and the corresponding crossover current density in a MEA with 240 μm thick fumea F-10180 rf catalyst-coated-reinforced PFSA membrane. Data from Bensmann, B. et al., *Int. J. Hydrogen Energy*, 39, 49, 2014.

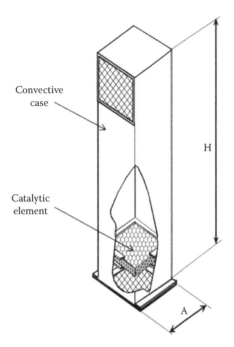

Convective case

Catalytic element

H

A

FIGURE 12.12
Schematic of a H_2–O_2 catalytic combustor for potential use to remove trace gases via catalytic oxidation from compressed gases produced by a PEM-WE. (Reprinted from *Int. J. Hydrogen Energy*, 34, Grigoriev, S.A. et al., Hydrogen safety aspects related to high-pressure polymer electrolyte membrane water electrolysis, 5986–5991, Copyright 2009, with permission from Elsevier.)

pressure was increased (up to 50 bar), the oxygen concentration in hydrogen rose to 2–4 vol.%, especially at low (<0.4 A cm^{-2}) current densities."

12.6 Ohmic Resistance

The Ohmic resistance through the PEM is, of course, proportional to its thickness L_B and inversely proportional to the membrane conductivity σ_B. Although more complex forms are available in the literature (e.g., Choi et al., 2005), here we assume the simpler expression of Thampan et al. (2000)

$$\sigma_B = (\varepsilon_B - \varepsilon_{B,0})^{1.5}\left(\frac{349.8}{1+\delta}\right)\exp\left\{-\frac{E_\mu}{R}\left(\frac{1}{T}-\frac{1}{298}\right)\right\}\left(\frac{1}{18\lambda}\right)\beta$$

(12.36)

where E_μ is the activation energy for the viscosity of water, and the parameter $\delta =1.65$ is based on fitting the conductivity data of Ren et al. (2000). In this expression, the volume fraction of water in the PEM ε_B depends upon λ, the number of water molecules sorbed by the electrolyte layer per acid site,

$$\varepsilon_B = \frac{\lambda}{\lambda + r}$$

(12.37)

where $r = \overline{V}_B/\overline{V}_W = 537/18 \approx 30$ is the ratio of membrane partial molar volume of polymer electrolyte to that of water. Further, $\varepsilon_{B,0}$ is the percolation threshold, roughly corresponding to $\lambda_0 = 1.8$.

The degree of dissociation of acid sites, or β, is

$$\beta = \frac{(\lambda+1)-\sqrt{(\lambda+1)^2-4\lambda(1-1/K_a)}}{2(1-1/K_a)}$$

(12.38)

where λ is the number of water molecules sorbed by the electrolyte layer per acid site, and the dissociation equilibrium constant for acid sites is

$$K_a = \exp\left[\frac{-52,300}{R}\left(\frac{1}{T}-\frac{1}{298}\right)\right]$$

(12.39)

In the PEM-WE, assuming a liquid water feed, we assumed the membrane to be saturated, so that $\lambda = 20$ (Ren et al., 2000).

In addition to the Ohmic resistance in the PEM, there is some ionomer resistance in the catalyst layer, along with some interfacial contact resistance, as well as some minor electronic resistance in the MEA (Makharia et al., 2005). The appropriate manner to describe the ionomer resistance in the catalyst layer is via a spatially distributed transport and reaction model in the catalyst layer, although Makharia et al. (2005) provide a simpler estimate of this for use in a lumped model such as the one employed here. However, it was found to affect the model predictions in a negligible manner due to the small catalyst layer thickness and is so neglected here. In other words, the lumped model assumed here for the catalyst layer is adequate.

12.7 Cell Polarization

The minimum standard thermodynamic cell voltage needed for water electrolysis is 1.229 V, although the real voltage and power requirement is significantly higher as a result of the various kinetic and transport losses. We provide next a model for voltage and power requirement for a given current density or hydrogen production rate. The current density and the molar hydrogen production rate $\dot{n}_{H_2}$ are interrelated via

$$i(A_{MEA}) = \varepsilon_i(2F\dot{n}_{H_2})$$

(12.40)

Here as discussed earlier, the Faradaic efficiency, or current efficiency, $\varepsilon_i \to 1$, because the permeability of H_2 in PEM is relatively small. In this equation, A_{MEA} is the total MEA area in the electrolyzer, whether arranged in a single cell or in a stack.

The current density, in turn, determines the cell voltage and power requirement. It is most common to discuss the performance characteristics of a PEM-WE in terms of a polarization plot depicted in Figure 12.5 as resulting from the various overpotentials within the different layers of the MEA. This analysis follows our earlier approach (Thampan et al., 2001; Choi et al., 2004; Vilekar and Datta, 2010), in which we consider the MEA as consisting of the five layers, as shown in Figure 12.2, an electrical analog for which is shown in Figure 12.4.

In the absence of current (internal and external), $V = V_0$, the thermodynamic equilibrium cell potential. As the external current I is applied, the increase in potential registered V is equal to V_0 plus the sum of the potential drops across all the branches in series (Figure 12.4). Thus, from $V = \Phi_A - \Phi_C$, and using the definition of overpotentials, along with that for the equilibrium cell voltage, $V_0 = \Phi_{A,0} - \Phi_{C,0}$

$$V = V_0 + \underbrace{(\eta_{A,K} + \eta_{A,D} + \eta_{A,X})}_{\eta_A} + \eta_B - \underbrace{(\eta_{C,K} + \eta_{C,D} + \eta_{C,X})}_{\eta_C} + \eta_I$$

$$= V_0 + \eta_A + \eta_B - \eta_C + \eta_I$$

(12.41)

which accounts for kinetic (subscript, K) and diffusion (D) losses (for reacting species, including protons) at each electrode. Further, the sign of η_C is negative, since $v_{Ce^-} < 0$. Here, the subscripts A, B, and C refer, respectively, to the anode, electrolyte, and the cathode layers. Finally, η_I represents potential losses due to interfacial contact resistance among the various layers of the MEA, and any electronic resistance (Makharia et al., 2005), all small.

The overpotential at an electrode is given by Equation 12.24 in the absence of crossover current $i_{p,X}$. As discussed at length earlier, even though there is gas crossover, there is no corresponding electrochemical parasitic current at the anode. While there is electrochemical crossover current at the cathode, its effect on the overpotential is small. Consequently, unlike in a PEM-FC (Vilekar and Datta, 2010), we neglect any crossover current effect on electrode overpotential in PEM-WE, so that Equation 12.24 applies to both the anode and the cathode.

Further, as discussed earlier, for cathode $v_{Ce^-}^\cdot = -1$ for the RDS in the cathode mechanism for PEM-WE, while the symmetry factor is as usual, $\beta_C = 1/2$. For anode as discussed earlier, the evidence (Figure 12.7) is that $v_{Ae^-}^\cdot = +3$ for the RDS in the mechanism for the OER for IrO_x catalyst up to moderately high current densities, while $\beta_A^\cdot = 1/2$ (the superscript dots in these quantities refer to those for the RDS in the mechanistic sequence). We will assume here for simplicity that these parameters for OER are adequate even for higher current densities, even though experimental evidence indicates otherwise (Figure 12.7).

For the electrolyte layer B of conductivity σ_B and thickness L_B, assuming Ohm's law applies

$$\eta_B = i_B \left(\frac{L_B}{\sigma_B} \right) \tag{12.42}$$

where the conductivity is given by Equation 12.36.

Combining these overpotentials, there finally results (Thampan et al., 2001; Vilekar and Datta, 2010)

$$V = V_0 + \frac{RT}{\beta_A^\cdot v_{Ae^-}^\cdot F} \sinh^{-1} \left\{ \frac{1}{2} \left(\frac{i/i_{A,0}}{1 - i/i_{A,L}} \right) \right\}$$
$$- \frac{RT}{\beta_C^\cdot v_{Ce^-}^\cdot F} \sinh^{-1} \left\{ \frac{1}{2} \left(\frac{i/i_{C,0}}{1 - i/i_{C,L}} \right) \right\} + i \left(\frac{L_B}{\sigma_B} \right) + iR_I$$

(12.43)

This is an expression for the cell voltage V versus the current density i for a PEM-WE in terms V_0 and the characteristic parameters, namely, the exchange current densities for the anode and the cathode reactions, and the limiting current densities for the two electrodes, the reactant permeability in the membrane, and the electrolyte conductivity.

This model, with a priori parameters listed in Table 12.1, correctly predicts the performance of a PEM-WE as shown in Figures 12.13 and 12.14 against the data of van der Merwe et al. (2014). Thus, Figure 12.13 shows the effect of temperature on the cell polarization for an MEA based on Nafion 117. Figure 12.14 depicts experimental results versus model predictions for different membrane thicknesses. It is, thus, seen that the model predictions, without any fitted parameters, are in good agreement with experiments.

12.8 Cell Power Requirement and Efficiency

For a thermodynamic analysis, we consider the steady-state system shown in Figure 12.15, comprising a PEM-WE unit operating at a temperature T and a pressure p, to which electric work at the rate of $\dot{W}_t$ is supplied,

TABLE 12.1

PEM-WE Model Parameters Employed

Parameter	Value	Units	Reference/Comment
ΔG°	+237.2	kJ K^{-1} mol^{-1}	Standard Gibbs free energy change
ΔH°	+285.8	kJ K^{-1} mol^{-1}	Standard enthalpy change
$V_{0,298}^\circ$	$\Delta G^\circ/2F = 1.229$	V	Standard thermodynamic potential
V_{max}°	$\Delta H^\circ/2F = 1.48$	V	Standard maximum potential
$m_{M,A}$	1.0×10^{-3}	g cm^{-2}	van der Merwe et al. (2014)
$m_{M,C}$	0.3×10^{-3}	g cm^{-2}	Rasten et al. (2003)
ρ_{Pt}	21.45	g cm^{-3}	Platinum density
ρ_{Ir}	22.56	g cm^{-3}	Iridium density
$d_{M,C}$	2.7	nm	Typical for C supported Pt (Rosenthal et al., 2012)
$d_{M,A}$	2.9	nm	Siracusano et al. (2010)
φ_I	0.75	—	Thampan et al. (2001)
ω_I	0.10	—	Ioroi et al. (2002)
$i_{A,0,ref}^*$	5.0×10^{-12}	A cm^{-2}	Fitted to data of Damjanovic et al. (1966) in Figure 12.7
$i_{C,0,ref}^*$	1.0×10^{-3}	A cm^{-2}	Thampan et al. (2001)
T_{ref}	298	K	Standard
$p_{H_2,ref}$	1	atm.	Standard
$p_{O_2,ref}$	1	atm.	Standard
β_A	1/2	—	Typical for elementary RDS
v_{A,e^-}	+3	—	Fitted to data of Damjanovic et al. (1966) in Figure 12.7
β_C	1/2	—	Typical for elementary RDS
v_{C,e^-}	−1	K	Thampan et al. (2001)
E_{A,Φ_0}	76	kJ mol^{-1}	Thampan et al. (2001)
E_{C,Φ_0}	4.3	kJ mol^{-1}	Thampan et al. (2001)
L_B	125×10^{-4}	cm	Nafion 115 thickness
L_B	175×10^{-4}	cm	Nafion 117 thickness
L_B	250×10^{-4}	cm	Nafion 1110 thickness
L_D	200×10^{-4}	cm	Standard GDL thickness
L_E	200×10^{-4}	cm	Standard GDL thickness
$\varepsilon_D, \varepsilon_E$	0.65	—	Typical ADL/CDL porosity upon compression
B_0	4×10^{-16}	cm^2	Meier and Eigenberger (2004)
k_{H_2}	$6.6 \times 10^{-8} \exp\left(-\dfrac{21,030 \text{ J mol}^{-1}}{RT}\right)$	mol bar^{-1} cm^{-1} s^{-1}	Kocha et al. (2006)
k_{O_2}	$k_{H_2}/2$		Kocha et al. (2006)
ξ	2.5	—	Meier and Eigenberger (2004)
E_μ	14	kJ mol^{-1}	Water viscosity activation energy
λ	20	—	Ren et al. (2000)
δ	1.65	—	Fitted to conductivity data of Ren et al. (2000)
$D_{O_2,E}$	$0.357 \times \left(\dfrac{T}{352}\right)^{1.823}$	cm^2 s^{-1}	Gas-phase oxygen diffusion coefficient
p_W	$\exp\left(11.676 - \dfrac{3816.44}{T-46.13}\right)$	atm.	Water vapor pressure in cathode
r	$\bar{V}_B/\bar{V}_W = 537/18$	—	Thampan et al. (2000)
R_I	0	Ω cm^{-2}	MEA interfacial resistance, assumed

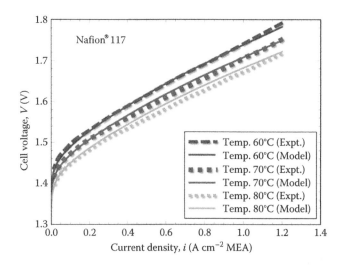

FIGURE 12.13
Polarization plot of a PEM-WE based on Nafion 117 at different temperatures. Comparison of the data of van der Merwe et al. (2014) with theoretical model involving a priori parameters listed in Table 12.1.

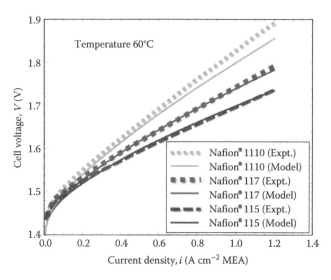

FIGURE 12.14
Polarization plot of a PEM-WE operating at 60°C with different polymer electrolyte membranes. Comparison of the data of van der Merwe et al. (2014) with theoretical model involving a priori parameters listed in Table 12.1.

along with the heat at the rate of $\dot{Q}$. According to convention, both heat and electric work are shown as *inputs* to the thermodynamic system in Figure 12.15, although in reality a PEM-WE system dissipates heat. This simply means that the sign of the heat term would, in general, be negative.

Assuming, further, the flow rates to the system to be in the stoichiometric proportion indicated by the OR shown in Figure 12.15, steady-state energy balance, or first law analysis for the system, neglecting the potential and kinetic energy changes between outlet and inlet

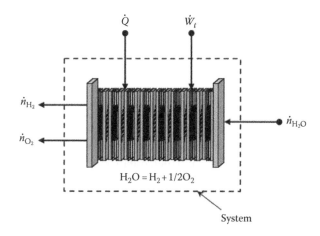

FIGURE 12.15
Thermodynamic analysis of a PEM-WE system operating at a temperature T and a pressure p.

streams, assuming no mechanical work and no moisture in the gases produced, provides

$$\dot{n}_{H_2}\underbrace{\left(H_{f,H_2}+\frac{1}{2}H_{f,O_2}-H_{f,H_2O}\right)}_{\Delta H_{OR}}=\dot{Q}+\dot{W}_t\ (W) \quad (12.44)$$

This may further be divided through by the molar flow rate of H_2, $\dot{n}_{H_2}$, to obtain the first law on a per mol H_2 basis

$$\Delta H_{OR}=Q+W_t\ \text{J mol}^{-1} \quad (12.45)$$

Here, Q and W_t are the heat and electric work per mol of the H_2, that is, specific energy consumption.

The first law efficiency then

$$\varepsilon_{\Delta H}\equiv\frac{\Delta H_{OR}}{W_t} \quad (12.46)$$

while the heat rejected

$$-Q=W_t(1-\varepsilon_{\Delta H}) \quad (12.47)$$

On the other hand, for a reversible PEM-WE, $Q_0 = T\Delta S_{OR}$, so that reversible electrical work, $W_{t,0} = \Delta G_{OR} = \Delta H^o_{OR} - T\Delta S^o_{OR}$, where we have further assumed that the enthalpy change and entropy change are independent of temperature so that their standard state values may be adopted, indicated by the superscript "o." The corresponding second law efficiency for the PEM-WE then

$$\varepsilon_{\Delta G}\equiv\frac{\Delta G_{OR}}{W_t}=\varepsilon_{\Delta H}-\left(\frac{T\Delta S^o_{OR}}{W_t}\right) \quad (12.48)$$

The specific energy consumption per mol H_2 generation, in turn, may be determined from

$$W_t = \frac{P}{N_{H_2}} \text{ J mol}^{-1} \qquad (12.49)$$

where the power density P (W cm^{-2}), $P = iV$, that is, it simply results from multiplying cell voltage V predicted by Equation 12.43 by i. Further, the production rate of H_2 per cm^2 of MEA, or H_2 flux

$$N_{H_2} = \left(\frac{\varepsilon_i}{2F}\right)i \quad \text{mol s}^{-1} \text{ cm}^{-2} \text{ MEA} \qquad (12.50)$$

In this, the Faradaic efficiency is given by Equation 12.34, typically of the order of unity except at very low current densities or at rather high pressures, due to the relatively small H_2 and O_2 crossover current densities (Figure 12.11). Combining the last two relations provides

$$W_t = 2F\left(\frac{V}{\varepsilon_i}\right) \quad \text{J mol}^{-1} \qquad (12.51)$$

which, with the use of Equation 12.43 for V and of 12.34 for ε_i, predicts the work requirements per mol H_2 produced, or specific energy consumption, versus the current density, which in turn may be related to hydrogen generation flux N_{H_2} through Equation 12.50.

Thus, Figure 12.16 provides a prediction of W_t as a function of N_{H_2} for an operating temperature of 90°C

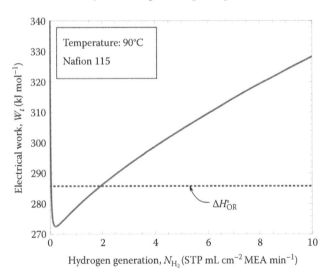

FIGURE 12.16
A prediction of the specific energy consumption W_t as a function of the hydrogen generation rate N_{H_2} in a PEM-WE with Nafion 115 and operating at a temperature of 90°C, for the list of parameters provided in Table 12.1.

and with Nafion 115 as the PEM, for the list of parameters provided in Table 12.1. Clearly, except at very low current densities, when the crossover dominates, the energy requirement/mol H_2 increases monotonically with hydrogen generation flux. The operating point for a PEM-WE stack would then be determined by a compromise between a high initial cost for a large PEM-WE stack versus the high operating cost for operating at high current densities. It is noteworthy that at low H_2 production rates (or current densities), the specific electrical energy consumption is less than the reaction enthalpy change. This is discussed further in the following.

Finally, by using $V_0 = \Delta G_{OR}^\circ / 2F$ and $V_{max} = \Delta H_{OR}^\circ / 2F$ in Equations 12.46 and 12.48, the first law and second law efficiencies of the PEM-WE unit may be written, respectively, as

$$\varepsilon_{\Delta H} = \varepsilon_i \varepsilon_V \varepsilon_0 \qquad (12.52)$$

and

$$\varepsilon_{\Delta G} = \varepsilon_i \varepsilon_V \qquad (12.53)$$

where the voltage efficiency

$$\varepsilon_V = \frac{V_0}{V} \qquad (12.54)$$

where V is given by Equation 12.43, while the thermodynamic efficiency

$$\varepsilon_0 = \frac{V_{max}}{V_0} \qquad (12.55)$$

Thus, the first law and second law efficiencies, $\varepsilon_{\Delta H}$ and $\varepsilon_{\Delta G}$, respectively, of the PEM-WE are plotted versus the current density in Figure 12.17, along with comparison with the data of Millet et al. (2009) for second law efficiency for a PEM-WE involving a Nafion 115 membrane and operating at a temperature, $T = 90$°C, based on parameters listed in Table 12.1 except for $m_{M,C} = 0.2 \times 10^{-3}$ g Pt cm^{-2}, and $m_{M,A} = 2.4 \times 10^{-3}$ g Ir cm^{-2}. The agreement, without any fitted parameters, is good.

Further, it is noteworthy in Figure 12.17 that at low current densities the first law efficiency can exceed 100%. This may be explained via Equation 12.48, since $\Delta S_{OR}^\circ > 0$, which implies that heat is transferred to the PEM-WE system from the surroundings (Figure 12.15). This may be further gleaned from a plot of heat

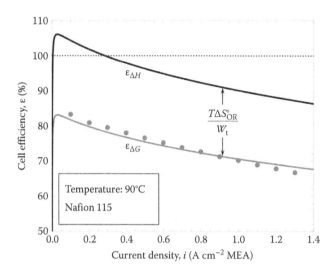

FIGURE 12.17
Predicted first law ($\varepsilon_{\Delta H}$) and second law ($\varepsilon_{\Delta G}$) efficiencies for a PEM-WE, along with a comparison with the data of Millet et al. (2009) for second law efficiency for a PEM-WE with a Nafion 115 membrane and operating at $T = 90°C$, based on parameters in Table 12.1 except for catalyst loading, $m_{M,C} = 0.2 \times 10^{-3}$ g Pt cm^{-2}, and $m_{M,A} = 2.4 \times 10^{-3}$ g Ir cm^{-2}.

dissipated ($-Q$) versus current density in Figure 12.18 based on combining Equation 12.47 with Equation 12.51, which provides

$$-Q = 2F\left(\frac{V}{\varepsilon_i}\right)(1 - \varepsilon_{\Delta H}) \quad \text{J mol}^{-1} \quad (12.56)$$

Thus, Figure 12.18 shows that at low current densities, heat loss ($-Q$) is negative, explaining a first law efficiency > 100%. This also explains why at low H_2

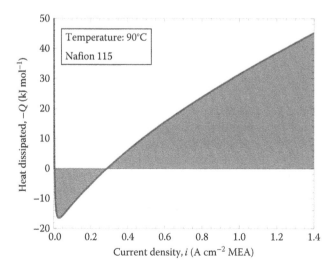

FIGURE 12.18
A prediction of heat loss ($-Q$) versus the current density in a PEM-WE with Nafion 115 as the PEM and operating at a temperature of 90°C, for the list of parameters provided in Table 12.1.

production rates or current densities, the electrical energy consumption is predicted as being less than the reaction enthalpy change (Figure 12.16).

12.9 Compressed Hydrogen Generation

High-pressure H_2 is desirable for ease of storage and transportation and is also needed in many processes. Thus, production of compressed H_2 from PEM-WE is desired. Bensmann et al. (2013) considered the theoretical analysis and comparison of energy requirements and efficiency for three alternate methods (Figure 12.19) for compressed H_2 generation: (a) ambient pressure PEM-WE followed by conventional mechanical multistage compressions (Path I); (b) use of pressurized liquid water feed to the anode in a pressurized PEM-WE (Path II), that is, with symmetric pressure between the cathode and the anode; and (c) use of ambient pressure liquid water feed to the anode in a PEM-WE with H_2 generation and concomitant electrochemical co-compression within the same unit (Path III), that is, with differential pressure between the cathode and the anode. They considered detailed first law and second law analysis, including the effect of humidification of the gas stream produced (neglected in the thermodynamic analysis here), PEM-WE irreversibilities, as well as H_2 crossover.

Their results showed that for the practical range of up to 40 bar, Pathway III outperforms the other two (Figure 12.20). Pathway I involves the inefficiencies of mechanical compression, while Pathway II suffers from the generation of compressed oxygen. On the other hand, at higher delivery pressures, Pathway III is affected by hydrogen crossover resulting in a decrease of the Faradaic efficiency.

12.10 Stack Design and Analysis

The total MEA ara for a given H_2 molar production rate, $\dot{n}_{H_2}$, is obtained directly from rearranging Equation 12.40 (or from Equation 12.50)

$$A_{MEA} = 2F\left(\frac{\varepsilon_i}{i}\right)\dot{n}_{H_2} \quad (12.57)$$

In other words, it is inversely proportional to the operating current density, which determines the operating cell voltage V (Equation 12.43). The capital cost of the unit is, of course, proportional to this MEA area, which

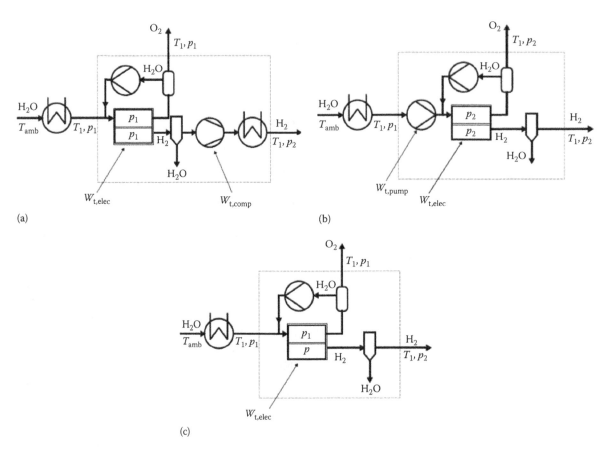

(a)   (b)

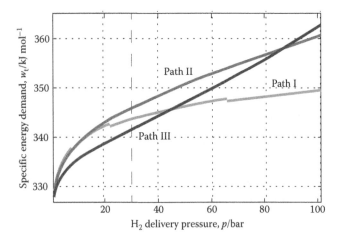

(c)

FIGURE 12.19
Three alternate pathways for the generation of compressed H_2 via water electrolysis. (a) Low-p electrolysis + H_2 compression (path I). (b) H_2O pressurization + high-p electrolysis (path II). (c) Asymmetric-p electrolysis (path III). (Reprinted from *Electrochim. Acta*, 110, Bensmann, B., Hanke-Rauschenbach, R., Peña Arias, I.K., and Sundmacher, K., Energetic evaluation of high pressure PEM electrolyzer systems for intermediate storage of renewable energies, 570–580, Copyright 2013, with permission from Elsevier.)

FIGURE 12.20
A theoretical comparison of the work requirement at 333 K on the three alternate pathways shown in Figure 12.19. (Reprinted from *Electrochim. Acta*, 110, Bensmann, B., Hanke-Rauschenbach, R., Peña Arias, I.K., and Sundmacher, K., Energetic evaluation of high pressure PEM electrolyzer systems for intermediate storage of renewable energies, 570–580, Copyright 2013, with permission from Elsevier.)

thus reduces as the cell voltage V increases. On the other hand, the total energy consumption is proportional to the specific energy consumption

$$\dot{W}_t = (W_t)\dot{n}_{H_2} = (\Delta H_{OR} - Q)\dot{n}_{H_2} \qquad (12.58)$$

where the specific energy consumption, which determines the operating cost, is in turn directly proportional to the cell voltage V (Equation 12.51). Thus, the total cost, the sum of capital, and operating costs may be minimized by carefully selecting the cell operating voltage or current density. In other words, the most important parameter determined by the electrolyzer designer is the operating voltage V of the individual cell.

Once the cell V and the corresponding i have been determined based on the polarization plot, either calculated or experimental, the total MEA area may then be calculated from Equation 12.57. This MEA area may be arranged into a stack of N_{Cell} cells in parallel to directly obtain a desired voltage V_{Stack}, that is,

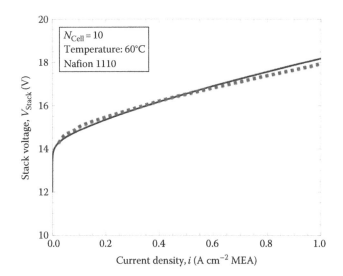

FIGURE 12.21
A comparison of the predicted stack voltage versus current density versus the data of Selamet et al. (2013) for a PEM-WE stack containing 10 cells based on Nafion 1110 and operating at a temperature of 60°C, for the list of parameters provided in Table 12.1.

$$A_{MEA} = N_{Cell}A_{Cell} \ (cm^2) \tag{12.59}$$

Thus, the stack voltage

$$V_{Stack} = N_{Cell}(V) \tag{12.60}$$

where the individual cell voltage V is given by Equation 12.43, while the total stack current

$$I_{Stack} = A_{Cell}i \tag{12.61}$$

Of course, the total power

$$\dot{W}_t = V_{Stack}I_{Stack} = (N_{Cell}V)(A_{Cell}i) = A_{MEA}Vi \tag{12.62}$$

remains unchanged, regardless of how the stack, or the total A_{MEA}, is arranged in a number of cells.

Simply for illustrative purposes, Figure 12.21 provides a comparison of the predicted stack voltage versus current density based on Equation 12.60 along with Equation 12.43 in a PEM-WE stack with 10 cells and with Nafion 1110 as the PEM, while operating at a temperature of 60°C and using the list of parameters provided in Table 12.1, versus the data of Selamet et al. (2013). It is seen that the prediction is reasonable. If the stack is not well assembled, however, there can be additional Ohmic voltage drop between cells in a stack.

12.11 Conclusion

PEM-WE is a promising method for distributed generation of compressed hydrogen for use in energy storage, in the laboratory, or as a feedstock in chemical synthesis. This chapter provides a lumped, isothermal, analytical model for the performance evaluation or design of PEM-WE single cells or stacks. The model is simple yet robust and is able to make reasonable predictions of single cell and stack performance under a variety of operating conditions and with different membranes as well as catalyst loading. The kinetics, transport, and thermodynamic parameters used in these predictions are taken from the literature with no further fitting. While the approach we have taken here can no doubt be further refined, it provides a good starting point for the analysis and design of polymer electrolyte membrane water electrolyzers.

References

Abbasi, T. and Abbasi, S. A. 2011. Renewable' hydrogen: Prospects and challenges. *Renewable and Sustainable Energy Reviews*, 15: 3034–3040.

Agbli, K. S., Péra, M. C., Hissel, D., Rallières, O., Turpin, C., and Doumbia, I. 2011. Multiphysics simulation of a PEM electrolyser: Energetic macroscopic representation approach. *International Journal of Hydrogen Energy*, 36: 1382–1398.

Andrews, J. and Shabani, B. 2012. Re-envisioning the role of hydrogen in a sustainable energy economy. *International Journal of Hydrogen Energy*, 37: 1184–1203.

Baik, K. D., Hong, B. K., and Kim, M. S. 2013. Effects of operating parameters on hydrogen crossover rate through Nafion membranes in polymer electrolyte membrane fuel cells. *Renewable Energy*, 57: 234–239.

Barbir, F. 2005. PEM electrolysis for production of hydrogen from renewable energy sources. *Solar Energy*, 78: 661–669.

Bensmann, B., Hanke-Rauschenbach, R., Peña Arias, I. K., and Sundmacher, K. 2013. Energetic evaluation of high pressure PEM electrolyzer systems for intermediate storage of renewable energies. *Electrochimica Acta*, 110: 570–580.

Bensmann, B., Hanke-Rauschenbach, R., and Sundmacher, K. 2014. In-situ measurement of hydrogen crossover in polymer electrolyte membrane water electrolysis. *International Journal of Hydrogen Energy*, 39: 49–53.

Bockris, J. 1975. *Energy: The Solar-Hydrogen Alternative.* New York, Halsted Press.

Bockris, J. O. M. and Veziroglu, T. N. 2007. Estimates of the price of hydrogen as a medium for wind and solar sources. *International Journal of Hydrogen Energy*, 32: 1605–1610.

Busquet, S., Hubert, C. E., Labbé, J., Mayer, D., and Metkemeijer, R. 2004. A new approach to empirical electrical modelling of a fuel cell, an electrolyser or a regenerative fuel cell. *Journal of Power Sources*, 134: 41–48.

Carmo, M., Fritz, D. L., Mergel, J., and Stolten, D. 2013. A comprehensive review on PEM water electrolysis. *International Journal of Hydrogen Energy*, 38: 4901–4934.

Choi, P., Bessarabov, D. G., and Datta, R. 2004. A simple model for solid polymer electrolyte (SPE) water electrolysis. *Solid State Ionics*, 175: 535–539.

Choi, P., Haldar, P., and Datta, R. 2006. Proton-exchange membrane fuel cells. *Encyclopedia of Chemical Processing*, Taylor & Francis, Florida, pp. 2501–2530.

Choi, P., Jalani, N. H., and Datta, R. 2005. Thermodynamics and proton-transport in Nafion®. II. Proton diffusion mechanisms and conductivity. *Journal of Electrochemical Society*, 152: E123–E130.

Damjanovic, A., Dey, A., and Bockris, J. M. 1966. Electrode kinetics of oxygen evolution and dissolution on Rh, Ir, and Pt–Rh alloy electrodes. *Journal of the Electrochemical Society*, 113: 739–746.

Grigoriev, S. A., Millet, P., and Fateev, V. N. 2008. Evaluation of carbon-supported Pt and Pd nanoparticles for the hydrogen evolution reaction in PEM water electrolysers. *Journal of Power Sources*, 177: 281–285.

Grigoriev, S. A., Millet, P., Korobtsev, S. V., Porembskiy, V. I., Pepic, M., Etievant, C., Puyenchet, C., and Fateev, V. N. 2009. Hydrogen safety aspects related to high-pressure polymer electrolyte membrane water electrolysis. *International Journal of Hydrogen Energy*, 34: 5986–5991.

Grigoriev, S. A., Porembsky, V. I., and Fateev, V. N. 2006. Pure hydrogen production by PEM electrolysis for hydrogen energy. *International Journal of Hydrogen Energy*, 31: 171–175.

Grigoriev, S. A., Porembskiy, V. I., Korobtsev, S. V., Fateev, V. N., Auprêtre, F., and Millet, P. 2011. High-pressure PEM water electrolysis and corresponding safety issues. *International Journal of Hydrogen Energy*, 36: 2721–2728.

Hansen, H. A., Viswanathan, V., and Norskov, J. K. 2014. Unifying kinetic and thermodynamic analysis of 2e⁻ and 4e⁻ reduction of oxygen on metal surfaces. *The Journal of Physical Chemistry C*, 118: 6706–6718.

Hansen, M. K., Aili, D., Christensen, E., Pan, C., Eriksen, S., Jensen, J. O., von Barner, J. H., Li, Q., and Bjerrum, N. J. 2012. PEM steam electrolysis at 130°C using a phosphoric acid doped short side chain PFSA membrane. *International Journal of Hydrogen Energy*, 37: 10992–11000.

Holladay, J. D., Hu, J., King, D. L., and Wang, Y. 2009. An overview of hydrogen production technologies. *Catalysis Today*, 139: 244–260.

Ioroi, T., Yasuda, K., Siroma, Z., Fujiwara, N., and Miyazaki, Y. 2002. Thin film electrocatalyst layer for unitized regenerative polymer electrolyte fuel cells. *Journal of Power Sources*, 112: 583–587.

Ito, H., Maeda, T., Nakano, A., Hasegawa, Y., Yokoi, N., Hwang, C. M., Ishida, M., Kato, A., and Yoshida, T. 2010. Effect of flow regime of circulating water on a proton exchange membrane electrolyzer. *International Journal of Hydrogen Energy*, 35: 9550–9560.

Ito, H., Maeda, T., Nakano, A., Kato, A., and Yoshida, T. 2013. Influence of pore structural properties of current collectors on the performance of proton exchange membrane electrolyzer. *Electrochimica Acta*, 100: 242–248.

Ito, H., Maeda, T., Nakano, A., and Takenaka, H. 2011. Properties of Nafion membranes under PEM water electrolysis conditions. *International Journal of Hydrogen Energy*, 36: 10527–10540.

Ivy, J. 2004. Summary of electrolytic hydrogen production: Milestone completion report (No. NREL/MP-560-36734). Golden, CO, National Renewable Energy Lab.

Katsounaros, I., Cherevko, S., Zeradjanin, A. R., and Mayrhofer, K. J. 2014. Oxygen electrochemistry as a cornerstone for sustainable energy conversion. *Angewandte Chemie International Edition*, 53: 102–121.

Kocha, S. S., Yang, J. D., and Yi, J. S. 2006. Characterization of gas crossover and its implications in PEM fuel cells. *AIChE Journal*, 52: 1916–1925.

Kothari, R., Buddhi, D., and Sawhney, R. L. 2008. Comparison of environmental and economic aspects of various hydrogen production methods. *Renewable and Sustainable Energy Reviews*, 12: 553–563.

Leng, Y., Chen, G., Mendoza, A. J., Tighe, T. B., Hickner, M. A., and Wang, C. Y. 2012. Solid-state water electrolysis with an alkaline membrane. *Journal of the American Chemical Society*, 134: 9054–9057.

Levene, J. I., Mann, M. K., Margolis, R. M., and Milbrandt, A. 2007. An analysis of hydrogen production from renewable electricity sources. *Solar Energy*, 81: 773–780.

Li, X., Walsh, F. C., and Pletcher, D. 2011. Nickel based electrocatalysts for oxygen evolution in high current density, alkaline water electrolysers. *Physical Chemistry Chemical Physics*, 13: 1162–1167.

Makharia, R., Mathias, M. F., and Baker, D. R. 2005. Measurement of catalyst layer electrolyte resistance in PEFCs using electrochemical impedance spectroscopy. *Journal of the Electrochemical Society*, 152: A970–A977.

Marangio, F., Santarelli, M., and Calì, M. 2009. Theoretical model and experimental analysis of a high pressure PEM water electrolyser for hydrogen production. *International Journal of Hydrogen Energy*, 34: 1143–1158.

Marini, S., Salvi, P., Nelli, P., Pesenti, R., Villa, M., Berrettoni, M., and Kiros, Y. 2012. Advanced alkaline water electrolysis. *Electrochimica Acta*, 82: 384–391.

Marshall, A., Børresen, B., Hagen, G., Sunde, S., Tsypkin, M., and Tunold, R. 2006. Iridium oxide-based nanocrystalline particles as oxygen evolution electrocatalysts. *Russian Journal of Electrochemistry*, 42: 1134–1140.

Marshall, A. T., Sunde, S., Tsypkin, M., and Tunold, R. 2007. Performance of a PEM water electrolysis cell using $Ir_xRu_yTa_zO_2$ electrocatalysts for the oxygen evolution electrode. *International Journal of Hydrogen Energy*, 32: 2320–2324.

Meier, F. and Eigenberger, G. 2004. Transport parameters for the modelling of water transport in ionomer membranes for PEM-fuel cells. *Electrochimica Acta*, 49: 1731–1742.

Millet, P., Dragoe, D., Grigoriev, S., Fateev, V., and Etievant, C. 2009. GenHyPEM: A research program on PEM water electrolysis supported by the European Commission. *International Journal of Hydrogen Energy*, 34: 4974–4982.

Mitlitsky, F., Myers, B., and Weisberg, A. H. 1998. Regenerative fuel cell systems. *Energy & Fuels*, 12: 56.

Oliveira, L. F. L., Jallut, C., and Franco, A. A. 2013. A multi-scale physical model of a polymer electrolyte membrane water electrolyzer. *Electrochimica Acta*, 110: 363–374.

Onda, K., Ichihara, K., Nagahama, M., Minamoto, Y., and Araki, T. 2007. Separation and compression characteristics of hydrogen by use of proton exchange membrane. *Journal of Power Sources*, 164: 1–8.

Onda, K., Murakami, T., Hikosaka, T., Kobayashi, M., and Ito, K. 2002. Performance analysis of polymer-electrolyte water electrolysis cell at a small-unit test cell and performance prediction of large stacked cell. *Journal of the Electrochemical Society*, 149: A1069–A1078.

Rasten, E., Hagen, G., and Tunold, R. 2003. Electrocatalysis in water electrolysis with solid polymer electrolyte. *Electrochimica Acta*, 48: 3945–3952.

Ren, X., Springer, T. E., Zawodzinski, T. A., and Gottesfeld, S. 2000. Methanol transport through Nation membranes. Electro-osmotic drag effects on potential step measurements. *Journal of the Electrochemical Society*, 147: 466–474.

Rosenthal, N. S., Vilekar, S. A., and Datta, R. 2012. A comprehensive yet comprehensible analytical model for the direct methanol fuel cell. *Journal of Power Sources*, 206: 129–143.

Rozain, C. and Millet, P. 2014. Electrochemical characterization of polymer electrolyte membrane water electrolysis cells. *Electrochimica Acta*, 131: 160–167.

Schalenbach, M., Carmo, M., Fritz, D. L., Mergel, J., and Stolten, D. 2013. Pressurized PEM water electrolysis: Efficiency and gas crossover. *International Journal of Hydrogen Energy*, 38: 14921–14933.

Selamet, Ö. F., Acar, M. C., Mat, M. D., and Kaplan, Y. 2013. Effects of operating parameters on the performance of a high-pressure proton exchange membrane electrolyzer. *International Journal of Energy Research*, 37: 457–467.

Selamet, Ö. F., Becerikli, F., Mat, M. D., and Kaplan, Y. 2011. Development and testing of a highly efficient proton exchange membrane (PEM) electrolyzer stack. *International Journal of Hydrogen Energy*, 36: 11480–11487.

Siracusano, S., Baglio, V., Di Biasi, A., Briguglio, N., Stassi, A., Ornelas, R., Trifoni, E., Antonucci, V., and Arico, A.S. 2010. Electrochemical characterization of single cell and short stack PEM electrolyzers based on a nanosized IrO2 anode electrocatalyst. *International Journal of Hydrogen Energy*, 35: 5558–5568.

Thampan, T., Malhotra, S., Tang, H., and Datta, R. 2000. Modeling of conductive transport in proton-exchange membranes for fuel cells. *Journal of the Electrochemical Society*, 147: 3242–3250.

Thampan, T., Malhotra, S., Zhang, J., and Datta, R. 2001. PEM fuel cell as a membrane reactor. *Catalysis Today*, 67: 15–32.

Turner, J. A. 2004. Sustainable hydrogen production. *Science*, 305: 972–974.

van der Merwe, J., Uren, K., van Schoor, G., and Bessarabov, D. 2014. Characterisation tools development for PEM electrolysers. *International Journal of Hydrogen Energy*, 39: 14212–14221.

Vilekar, S. A. and Datta, R. 2010. The effect of hydrogen crossover on open-circuit voltage in polymer electrolyte membrane fuel cells. *Journal of Power Sources*, 195: 2241–2247.

Vilekar, S. A., Fishtik, I., and Datta, R. 2010. Kinetics of the hydrogen electrode reaction. *Journal of the Electrochemical Society*, 157: B1040–B1050.

Wang, M., Wang, Z., Gong, X., and Guo, Z. 2014. The intensification technologies to water electrolysis for hydrogen production—A review. *Renewable and Sustainable Energy Reviews*, 29: 573–588.

Weber, A. Z. and Newman, J. 2007. Chapter two: Macroscopic modeling of polymer-electrolyte membranes. *Advances in Fuel Cells*, 1: 47–117.

13

Fundamentals of Electrochemical Hydrogen Compression

Peter Bouwman

CONTENTS

13.1 Introduction

13.1.1 Electrochemical Hydrogen Pumping

Electrochemical hydrogen transport through a membrane using an ionic transport mechanism was known since the first precursors of Nafion® films were developed for fuel cells back in the 1960s, according to insiders (McElroy 1980, Eisman 1990, Grot 2011). Fuel cells and electrolyzers do employ hydrogen reduction or hydrogen oxidation, respectively, but these processes are markedly different from the process of pumping hydrogen because except for a small overvoltage at each electrode, the net chemical reaction itself does not give, or take, energy, or generate new chemical products. Even so, electrochemical hydrogen transport does bring additional benefits such as purification, compression, with silent operation, and (remote) flow control as surplus value features.

Basically, the working principle is simple: hydrogen is fed to the cell, oxidized to protons and electrons, driven through the membrane, and then reduced back to molecular hydrogen. The direction of the electrical current determines the direction of the hydrogen movement. Simple becomes sophisticated, when realizing that electrochemical compression is capable of transporting only hydrogen selectively against its own activity gradient—analogous to picking cherries from a mixed fruit basket. Thus, the transport mechanism of hydrogen gas combines an absolute purification step and actuates pressure differentials (i.e., compression), on the condition that the membrane and cell hardware integrity are maintained.

In this chapter, we present highlights of our fundamental understanding of the electrochemical hydrogen compression (EHC) to date with the ambition to inspire more research in this field. Admittedly, it may seem daunting to compress hydrogen up to high pressures employing only a thin membrane using only electrochemistry, and indeed it is a challenge which can be achieved with proper hardware and laboratory resources. We have encountered ourselves that the main obstacle in the development of the EHC technology is our limited mental picture of the physical processes occurring in the stack.

For example, our everyday life experiences make us associate compression with the tiresome effort of inflating tires, where even 5 bar is felt to be a lot. Stainless steel is viewed to be hard and dimensionally stable, while polymer films considered soft and easily torn. Most people cannot fathom the implications of the extreme forces involved with pressures up to 100 MPa (equivalent of ~10 km water column), and its long-term effect on materials needs to be investigated in more detail, before accurate models can simulate reality and cell hardware designs can be optimized. Therefore, the first step on this road is to add perspective to the mental picture by presenting to you a comprehensive set of results and to direct your attention to important matters and materials.

13.1.2 Existing Hydrogen Infrastructure

Hydrogen is gaining interest as an energy carrier (Grasman 2013), enabling the *conversion* from electrical power to chemical energy and back with reasonable efficiencies using devices such as electrolyzers and fuel cells. Contemporary reasons include growing concerns over diminishing fossil fuel reserves, the impact of our carbon footprint, and the increasing amount of solar and wind energy demanding active balancing of the power grid.

Electricity is a valuable form of energy, but neither be stored over long periods of time nor in large amounts, and hence energy conversion is required for storage. Figure 13.1 shows a graphical overview of the storage option considering capacities and timescales (ITM 2014), where high caloric gasses such as hydrogen and natural gas are ranked at the top of both scales. Vast amounts of energy could be stored in the existing gas pipeline distribution networks, even within the boundaries set by present pressure restrictions International Energy Agency (IEA) (2011).

Familiarity with methane gas and the ambition to save sustainable energy for later use drive "Power-to-Gas" initiatives to actively (re-)convert CO_2 to "green" methane gas, which can be injected into the gas grid if the composition complies with local standards. However, methanation is economically not viable when compared against the cost price of fossil methane gas (Stolten et al. 2013). Hydrogen is an essential reaction intermediate in this methanation sequence, and it would be logical to stop further energy conversion at this point and store the hydrogen gas for later use, instead of methane gas.

While the market for hydrogen as a fuel is in its nascent stages, as a commodity it is already used in large volumes. Hydrogen is the fundamental element in the production of ammonia, which is in turn used to make fertilizers. The other major market for hydrogen is the

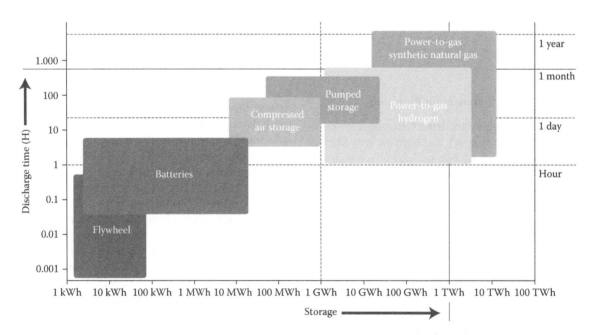

FIGURE 13.1
Overview of energy storage technologies categorized according to typical storage capacity and discharge timescale. (Courtesy of ITM Power Plc., London, U.K.)

oil refining industry, where hydrogen is used to process crude oil and make high-value distillate fuels such as gasoline and diesel. The global hydrogen consumption was estimated at 50 Mton year^{-1}, of which 2 Mton year^{-1} was produced from electrolysis; the majority was produced from hydrocarbon sources such as gas (24 Mton year^{-1}), oil (15 Mton year^{-1}), and coal (9 Mton year^{-1}) (Evers 2010). Hydrogen is a valuable chemical resource, where its cost price depends heavily on the source and the method of production and distribution (Raine et al. in Roads2HyCom 2009).

The Freedonia Group (2016) reports the global consumption of merchant and captive hydrogen is forecast to increase 3.5% annually through 2018 to 290 billion m^3 (ed. 26 Mton year^{-1}), of which 85% was generated by the petroleum refineries for hydro-treating and hydrocracking refining operations. Here, hydrogen is specifically used to remove impurities such as sulfur from the oil and upgrade the remaining heavy fractions to higher-value refined products. Hydrogen is also produced as an off-gas during the oil refining and petrochemical processes, which could potentially be recycled after purification to reduce merchant hydrogen costs and maintain greater margins. Considering the potential market for EHC in industrial hydrogen, there is virtually no upper limit in the demand for compression rate capability.

Continued growth is expected in the usage of hydrogen in the petrochemical refining industry for multiple reasons. Firstly, the demand for refined products is growing in especially emerging economies (not EU), driven by increasing vehicle ownership per capita. Secondly, more stringent sulfur-restricting regulations are being implemented and enforced in almost all of the countries in the world demanding lighter and cleaner fuels (Directive 2009/30/EC of the European Parliament and of the Council) to combat persistently difficult air quality issues. Thirdly, increased use of heavy oil is expected to be used as the feed in the refining process when sourcing from mature fields or shale deposits. Clearly, the automotive market is aiming for cleaner fossil fuels and is directly driving hydrogen demand.

Outside of refining, hydrogen is used at industrial scale in the production of fine chemicals, metals, electronics, and thin-film solar industries, edible oil processing. Although the chemical industry is the largest of these, more rapid growth is expected in several smaller markets. Additionally, the adoption of hydrogen energy technologies continues to proceed worldwide, and fuel cells will see greater mainstream adoption. Despite technical and other challenges, the emergence of a hydrogen market for fuel cell powered vehicles remains an option that is currently being developed by multiple manufactures within partnerships with gas suppliers such as the Clean Energy Partnership in Germany and national H$_2$ mobility initiatives.

In response to a worldwide demand ISO standards are being developed that address hydrogen gas vehicles and fueling stations including subcomponents (ISO 19880-4 for Compressed Hydrogen Compressors). According to the automotive SAE2601 standard for hydrogen refilling stations, the pressure at which hydrogen should be stored is set at 50 MPa for busses and 35 or 70 MPa for automotive vehicles. Hydrogen is compressed in stages into a series of buffer tanks, varying in size and pressure to be most cost-effective to refill empty tanks and only topping up from a high pressure storage tank at 85 MPa. Hydrogen storage at lower tank pressures is still under investigation using adsorption materials and metal alloys (Talaganis et al. 2009), but to date, no practical materials have been reported that store more than 10 wt.% (Schlapbach and Züttel 2001). Therefore, automotive demand will push the target compression capability for EHC technology to pressures exceeding 85 MPa.

Handling and storage of hydrogen gas is already done at small and large scale by mechanical compression using diaphragm pumps for capacities up to 1 kNm^3 h^{-1} and piston pumps for capacities up to 1 MNm^3 h^{-1} (not considering screw and turbo compression at very large scale). The compression process is adiabatic and divided into stages to reduce temperature excursions and reduce overall energy loss (i.e., heat). Additionally, the movement friction of seals and valves add 20% energy requirement, approximating 6 kWh kg^{-1} for 1–40 MPa compression. In comparison, the electrochemical compression process has the potential to be more energy efficient, down to 3 kWh kg^{-1} at very low pumping rates, whilst being practically isothermal and single stage. Note that the energy advantage of isothermal compression over adiabatic compression is limited to 17%, when restricting the output temperature to 120°C and input temperature to 0°C with an arbitrary number of stages. A major advantage of the EHC is that it can instantly change compression flow rates to match demand (lower is more efficient), hence adding energy flexibility to the already mentioned benefits of no moving parts (20%) and isothermal (17%) compared to mechanical compression. Note one must consider the entire system for a proper comparison, not just the stack; as a result the system may not be as efficient across the spectrum of pressures.

Although mechanical hydrogen compressors have established themselves as proven technology, many aspects could use upgrading. Electrochemical compression is not limited by any moving parts, noise levels, and there is no classic trade-off between pump capacity and pump output pressure as can be seen in traditional pump curve plots. Electrochemical compression could pick up any hydrogen in contact with the membrane, even near vacuum conditions, and compress the gas to output pressures up to 100 MPa output, without any dictated compression ratio. No power is consumed if there is no hydrogen present, while on the flipside, the quantity of hydrogen can be estimated directly from the Faradaic equivalent of the current passing through the EHC, thus providing a potential solution for the metering issue when making an allowance for any back diffusion of hydrogen gas.

The electrochemical compression principle can easily be scaled up by increasing the total active membrane area. Generally, this could mean increasing the active area per cell plate, stacking more cell plates. However, limitations apply for EHC applications due to the mechanical forces associated with pressures up to 100 MPa. The likely scenario for scaling up is placing more stacks in parallel to each other, each one conducting single-stage compression from the shared input to the shared output pressure. The stack costs increase nonlinearly with increasing output pressure (Moton et al. 2013). Typically, the EHC stack "building unit" is to be optimized for both cost and performance for each specific application, and in this book chapter we focus on all elements that need to be considered.

13.1.3 Prior Art, and Patents, and Projects

Before discussing the working principles of electrochemical compression in detail, we review work that has been published on this topic by relevant stakeholders. Hydrogen pumping is most frequently used as a diagnostic test for measuring the crossover in fuel cells, so let us start here.

Hydrogen crossover can be measured electrochemically as a diagnostic tool to gauge the integrity of the membrane in a polymer electrolyte membrane fuel cell (PEMFC) setup and to evaluate the presence of pin holes or membrane thinning is well summarized by Kocha et al. (2006). Earlier work has been reported by Yeo and McBreen (1979), Ogumi et al. (1985), and Tsou et al. (1992). Practically, a positive potential is imposed on the cathode side of a PEMFC with a H_2–N_2 gas atmosphere on the anode–cathode, and subsequently the measured Faradaic current is associated with the molecular hydrogen that has diffused through the polymer membrane from the hydrogen (anode) to the nitrogen side (cathode), where the hydrogen gas is converted to protons on the cathode electrode. The operating principle is identical to a hydrogen pump, only here the feed side has been starved of a partial hydrogen pressure.

First accurate publications on the working principle of EHC were published by Sedlak et al. (1981), electrochemically pumping hydrogen from a low to a high pressure and separation of hydrogen from an inert gas to provide high-purity hydrogen. To our best knowledge, the first patent registering hydrogen purification using a typical electrochemical cell configuration is R.G. Haldeman US3475302 dating back to 1969 followed by H.J.R. Maget, U.S. Patent 3,489,670 filed in 1964. Interestingly, patent

applications were filed later on this topic in 1985 (J.J. Zupancic US4664761) describing electrochemical separation of hydrogen from gaseous mixtures, and in 1996 (N. Nobuaki JPH08176873A; K. Kunimatsu JPH08193286A), where the protons are generated by the oxidation of water.

In 1998, Bessarabov (Bessarabov 1998) clearly elaborated on the possibilities of hydrogen pumping in a paper on electrochemically aided membrane separation, and in the same year Rohland et al. (1998) reported on application of the principle in a hydrogen compressor. The potential ability of EHC to purify and compress hydrogen was recognized, using the general structure of a fuel cell (Stroebel et al. 2002).

Two major solid electrolyte membrane types are distinguished, ceramic (Kurita et al. 2002, Iwahara et al. 2011) and cation exchange polymer type (Perry et al. 2008 and references therein). The latter type is widely used in proton exchange membrane fuel cells, giving the technology its name.

More recent literature on hydrogen pumping can be distinguished into two main application categories:

1. *Hydrogen compression/recirculation* (Kurita et al. 2002, Ströbel et al. 2002, Onda et al. 2007, Casati et al. 2008, Perry et al. 2008, Thomassen et al. 2010, Gregoriev et al. 2011, Moton et al. 2013).

2. *Hydrogen purification/extraction* (Lee et al. 2004, Granite and O'Brien 2005, Gardner and Ternan 2007, Ibeh et al. 2007, Onda et al. 2007, 2009, Casati et al. 2008, Abdullah et al. 2011, Thomassen et al. 2010, Gregoriev et al. 2011, Pasierb and Rekas 2011), reported in the last decade.

Haile et al. (2006) published an overview of work conducted on cesium dihydrogen phosphate as a proton conductor and fuel cell electrolyte. Hydrogen pumping was investigated using ceramic proton-conducting membranes in Japan (Takaaki et al. 2008) with niche applications in tritium enrichment monitoring in fusion reactor technology (Tanaka et al. 2006, 2011). Other types of applications of the hydrogen pump principle are in hydrogen sensor technology (Yamaguchi et al. 2008).

Multiple patent applications on hydrogen compression systems have been filed of which five U.S. patents have been accepted in the last 10 years. Table 13.1 gives an overview.

Pioneering research was often conducted with public funding. In United States, for example, the Department of Energy (DOE) has funded multiple projects through an SBIR award for ANALYTIC POWER LLC in 2005 (Mackenzie and Bloomfield 2006), FuelCell Energy together with Sustainable Innovations from 2010 till 2014 (Lipp 2014), Proton Energy, Nuvera. H2Pump demonstrated hydrogen purification and compression to a pressure of 150 psi and successfully engaged Phase I and Phase II SBIR DOE

TABLE 13.1

List of Patents on the Subject of Electrochemical Hydrogen Compression

First Inventor	Patent Number	Priority Date
R.G. Haldeman	US3475302 (A)	28/10/1969
J.J. Zupancic	US4664761 (A)	27/12/1985
N. Nobuaki	JPH08176873 (A)	09/07/1996
K. Kunimatsu	JPH08193286 (A)	30/07/1996
D. Bloomfield	US5900031 (A)	04/05/1999
W. Juda	US2003155252 (A1)	21/08/2003
	US7169281 (B2)	
T.Y.H. Wong	US20040211679 (A1)	7/03/2003
	WO 03/075379 (A2, A3)	12/09/2003
T. Molter	DE10307112 (A1)	30/10/2003
G.E. Benson	US2004058209 (A1)	25/03/2004
	US7011903 (B2)	
F. Barbir	US2004142215 (A1)	22/07/2004
	US6994929 (B2)	
A.W. Ballentine	US2005053813 (A1)	10/03/2005
	US7252900 (B2)	
A.D. Deptala	US8790506 (B2)	29/09/2011
	US2011233072 (A1)	

program funding to achieve 300 psi (differential) pressure using polybenzimidazoles (PBI) membranes, which was completed in 2012. Buelte et al. (2011) reported using this membrane for hydrogen pumping, studying the influence of the acid concentration on performance.

Most recently, in June 2014, the Energy Department (DOE 2014) announced $20 million for 10 new R&D projects to advance hydrogen production and delivery technologies. All of these projects strive to reach the hydrogen cost goal of under $4 per gallon gas equivalent, with six of these projects being directed toward producing and dispensing hydrogen, while the remaining four are directed toward hydrogen delivery, including compression, storage, and dispensing at the station. For example, Southwest Research Institute of San Antonio, Texas will receive $1.8 million to demonstrate a hydrogen compression system. Also, Nuvera Fuel Cells Inc. of Billerica, Massachusetts was announced to receive $1.5 million to design and demonstrate an integrated, intelligent high-pressure hydrogen dispenser for FCEV fueling.

In Europe HyET is involved in the Joint Technology Initiatie - Fuel Cells and Hydrogen Joint Undertaking (JTI-FCH-JU) "PHAEDRUS" project (JTI FCH-JU) which aims to realize in 2015 a scalable, commercial hydrogen refueling station featuring an optimal combination of mechanical and electro-chemical compression up to 100 MPa.

In the JTI-FCH "DON QUICHOTE" project (JTI FCH-JU), the target is to capture renewable wind and solar energy, and store this in the form of compressed hydrogen (60 kg H_2 day^{-1} at 40 MPa). This hydrogen can be dispensed on demand to refill fuel cell–powered forklift trucks or

alternatively feed an on-site, stationary fuel cell system to again make electricity. In this latter project, HyET will also investigate how much compression energy can be recovered during decompression, thus making the overall energy cycle even more efficient (Cownden et al. 2001).

HyET is also engaged in injection and selective removal of hydrogen from the natural gas grid and green methanated gas mixtures through our involvement in the Dutch two-year funded project "PurifHy" (PurifHy project [TKI-Gas]). We place the connection between the natural gas grid and topic of hydrogen compression and storage in greater contemporary context in the next section.

13.1.4 Visionary Applications

The EHC is now gaining interest, because hydrogen gas itself has been shortlisted as convenient energy carrier for storing renewable electricity as lightweight chemical compound. One could say that the disadvantageous properties of using hydrogen gas can be compensated to a large degree by the features and flexibility of applying electrochemical compression. Figure 13.2 depicts all visionary applications with infrastructural connections to the hydrogen gas supply and demand sides. In contrast with incumbent fuels, there is no deposit of hydrogen, like there is no deposit of electricity. Significant distribution losses could be avoided provided the harvested energy was stored locally.

We need to store electricity generated by renewable energy sources with inherently intermittent supply and no base "load." For example, solar and wind energy generate electricity in a random and daily, weekly, or season-bound patterns. Harvesting and buffering from these patterns is most efficiently resolved using pumped hydro for stationary grid applications, but few geographic locations present themselves as being suitable, and most are already being exploited (Stolten et al. 2013).

Water electrolysis appears less energy efficient, but is a versatile solution for grid balancing and storing surplus energy in hydrogen and oxygen. The hydrogen is extremely pure and needs no further purification. Simultaneous compression of the generated gases is feasible for membrane electrolyte-type electrolyzers, as done with electrochemical compression, but there are some key disadvantages in comparison: Firstly, compressing oxygen and hydrogen gasses equally leads to a highly reactive oxygen atmosphere, and potential hazards upon membrane rupture. Secondly, the electrolyzer hardware is made from expensive titanium cell plates, making it expensive to use bulkier structures to cope with higher pressures. A combination between an electrolyzer with a subsequent EHC could be safer and more economically attractive with optimization of the intermediate hydrogen handover pressure.

FIGURE 13.2
Schematic diagram showing how Electrochemical Hydrogen Compression (EHC) could connect renewable energy sources (solar, wind, biomass, biofuels) and even fossil fuels (natural gas) to the hydrogen demand in industrial markets, energy storage, and automotive applications by using hydrogen gas (arrows) as the carrier.

In most urban, residential areas hydrogen could be supplied locally through reforming of present chemical resources, such as natural gas. Gas compositions are strictly governed by codes and standards at a national level to maintain the exact gas composition and Wobbe index in the pipelines, so that end users avoid having malfunctioning equipment. When significant hydrogen content is already introduced in the natural gas mixture, it would simply flow through the reforming process and be picked up by the EHC unit. Rural areas could also source from biomass or biofuels to make reformate hydrogen (Ottinger 2009). Note that biocategories are notoriously ill-defined: their price and composition may vary significantly with seasons, along with the level and number of contaminants (Blease 2013). Considering all

reformate hydrogen sources, the EHC would benefit from its purification ability to deliver clean hydrogen.

On the demand side, the industrial consumption of hydrogen is the largest category and this application could even return a contaminated hydrogen stream back to the EHC for recycling. The automotive application will pay the highest cost price for hydrogen at pressure, but also has the highest purity standards. The output needs to be 5N pure hydrogen, as the gas quality has great impact on the fuel cell performance and lifetime (Zhang 2010).

The box indicating the category "energy storage" symbolizes energy harvesting in the form of hydrogen gas, ranging in scale depending on application and location. Renewable energy resources are huge, but distributed as mapped by the International Renewable Energy Agency (IRENA 2012). Around 885 million TWh worth of solar radiation reaches the Earth's surface each year (IEA 2011). The consumer energy demand pattern is also distributed in nature, but both distributions seldom overlap. A little energy storage could make a big difference, as shown by the following example.

In Europe, the average solar resource is around 1200 kWh m^{-2} year^{-1} (IRENA 2012). Assuming a solar panel conversion efficiency of 10%, and 60% conversion efficiency for electrolysis and the hydrogen lower heating value (LHV) of 33.3 kWh kg^{-1}, this could provide just over 2 kg of hydrogen year^{-1} m^{-2}. A photovoltaic area greater than 120 m^2 would be needed to make enough hydrogen to refill a FCEV every week with a tank size of 5 kg. Boldly put, the whole property would need to be covered in solar panels, although statistics need to be accumulated to correlate the dwelling size of FCEV owners. Fortunately, the average European household itself needs a fraction of this amount in case no energy conversion is required, only 34.5 m^2 would be sufficient—according to the Eurostat 2005–2009 data indicating the average household in Europe (EU-27) utilized an energy equivalent of 356 kilogram(s) of oil equivalent per year, or 80 kWh week^{-1}, with large differences between countries. Vice versa, one FCEV on the driveway with 166 kWh energy storage capacity integrated with the residential power supply could help survive a blackout period for about 2 weeks (and more *with* solar panels).

A home refueling system could be created by connecting the electrochemical compressor unit to a reformer to produce pure hydrogen from natural gas (see Figure 13.3), and/or have an electrolyzer generate hydrogen from cheap (renewable) electricity. As the pipeline system has been extensively developed over the last century, most households have a natural gas connection with a throughput capacity large enough to cover the automotive energy requirement of all residents. Also, the technology exists to produce hydrogen through electrolysis, compress (EHC), store (tank), and regenerate electricity (fuel cell) on a small scale using the FC car. It remains to be seen if installation of a hydrogen tank storage will be permitted in residential areas, or whether the tank in the FC vehicle will be sufficient. Together with the choice of new energy systems, the information sharing through internet and automated control provide a paradigm shift. Harvesting and storing energy is now supplemented by tracking, sharing, selling, and enabling SMART (micro-) grids with optimal user comfort and financial return.

Ideally, residents have the choice to use their solar panels and/or a good contract with the gas company to refuel their FCEV. Even better, the tank remains filled up and money is made by converting cheap gas into (temporarily) high-priced electricity and selling

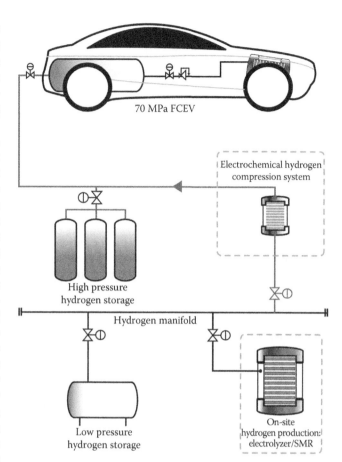

FIGURE 13.3
Schematic representation of the visionary home refueling system where the EHC could source hydrogen from either an electrolyzer or a reformer to refuel the FCEV as a means to store energy (additional stationary tanks are considered optional).

this back to the grid. Countries are also investigating whether surplus hydrogen can be reinserted into the local gas network, making good use of the distribution and inherent storage capacity of the pipeline volume. The options are vast.

Yes, there is an established market for hydrogen compressors. There is also a huge potential to extend the possibilities on integrating and revolutionizing the energy infrastructure with hydrogen and EHC. Key is its conversion ability, linking competing electricity and gas domains. The role of the electrochemical compressor plays is purifying and compressing all hydrogen from different (cheap) sources into a tank or a pipeline—if the technology can achieve 85 MPa, it can certainly do 8.5 MPa. Capital investment costs no longer restrict this technology to large businesses, and a swarm of distributed power plants could be most effective when synchronized appropriately. A leading role is waiting to be fulfilled by forward thinking gas and power utility companies, paving the way for the "average Joe" who may still be tempted to buy a diesel generator, and to provide a more attractive, sustainable option.

13.2 Theoretical Considerations

13.2.1 Working Principle behind Hydrogen Pumping

A cell capable of hydrogen pumping highly resembles the typical PEM-based membrane electrode assembly for fuel cells. It involves the anodic oxidation of molecular hydrogen to protons, identical to the anodic reaction of a hydrogen-fed fuel cell, and the subsequent migration of the resulting protons over a cation exchange membrane and the reduction of these protons back to molecular hydrogen at the cathode, all driven by an imposed cell voltage. The electrons are transferred via an external circuit to the opposite catalyst layer on the other side of the membrane. Effectively, this induces mass transport of hydrogen, and hydrogen only, enabling simultaneous purification. The take-home-message is: one electron equals one proton, equals half a hydrogen gas molecule. The schematic diagram is shown in Figure 13.4.

Current actively drives hydrogen transport in the EHC. Governed by overall charge neutrality, the amount of electrons traveling through the outer circuit is directly correlated to the amount of protons traversing the membrane. The pumping capacity is determined by the membrane active area multiplied with the current density, according to the following relationship:

$$J = \frac{I \cdot M_w}{N \cdot F} = \frac{j \cdot A \cdot M_w}{n \cdot F} \qquad (13.1)$$

where,

J is the mass flow (g s^{-1})
I is the current (A)
n is the no of electrons in reaction: $H_2 = 2H^+ + 2e^- = H_2$
F is the Faraday constant (C mol^{-1})
M_w is the Molar weight (g mol^{-1})
j is the current density (A cm^{-2})
A is the total active area (cm^2)

The resistive energy losses (heat) associated with passing a current through the electrochemical cell are predicted by Ohm's law. The derived equation clearly states the power consumed in order to transport hydrogen at given rate, irrespective of pressure differentials.

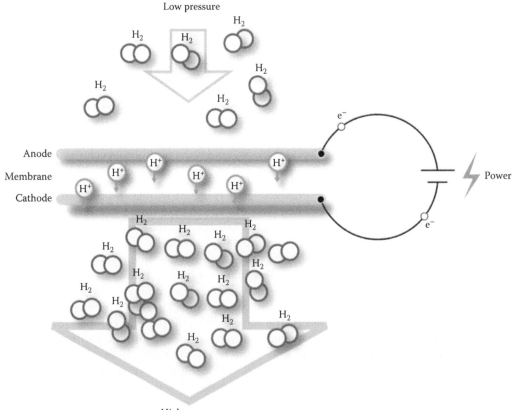

FIGURE 13.4

Principle of Electrochemical Hydrogen Compression using an external driving current to force hydrogen conversion on the first electrode into protons, which traverse through the solid, proton-conductive membrane before being reconverted back to hydrogen gas on the opposite electrode. The hydrogen flow is reversed when the direction of the current is reversed.

Doubling the current density, doubles the hydrogen flow, but quadruples the dissipated power according to Equation 13.2.

$$V = (I \cdot R) \, \text{Ohms Law}$$

$$\begin{aligned} P_{loss} &= I \cdot V = I^2 \cdot R \\ &= (j \cdot A)^2 \cdot R \\ &= \frac{j^2 \cdot A^2 \cdot r}{A} \\ &= j^2 \cdot A \cdot r \end{aligned} \qquad (13.2)$$

where
V is the polarization voltage (V)
I is the current (A)
R is the overall resistance (Ω)
P_{loss} is the power (W)
r is the specific resistance ($\Omega \, cm^2$)

The hydrogen transport energy is calculated per unit of mass (kWh kg^{-1}) by combining Equations 13.1 and 13.2. Combining both earlier mentioned equations gives the following expression, showing a linear relationship exists as a function of both current density and specific resistance.

$$\begin{aligned} \text{Energy requirement } \eta &= \frac{P_{loss}}{J} \\ &= \frac{3600 \cdot j \cdot r \cdot N \cdot F}{M_w} \end{aligned} \qquad (13.3)$$

where η is transport energy required to drive the oxidation of hydrogen, the proton transport, the reduction of hydrogen, and cover the electronic resistance (of the hardware) through the cell assembly, quantified here in the unit (kWh kg^{-1}). This value can also be expressed relative to its LHV (LHV = 120 MJ kg^{-1} or 33.33 kWh kg^{-1}). Note that so far only the displacement of hydrogen has been considered, without any pressure difference. The individual elements contributing to the specific resistance r are further discussed in Section 13.2.4.

13.2.2 Nernst Compression Energy

The energy required to compress gas under isothermal conditions is dictated through thermodynamic principles (work). When the EHC cell is in equilibrium, the Gibbs free energy difference between the uncompressed and compressed state translates itself to a voltage difference (Equation 13.5) following the Nernst–Einstein relationship (Equation 13.4). The Nernst voltage is a linear

relationship with regard to temperature, meaning the energy requirement for compression is lower at lower operating temperatures.

$$\Delta G = -n \cdot F \cdot V_{Nernst} \qquad (13.4)$$

$$V_{Nernst} = \left(\frac{R \cdot T}{n \cdot F} \right) \cdot \ln \left(\frac{p_1}{p_0} \right) \qquad (13.5)$$

where
G is the Gibbs free energy (work) (J kg^{-1})
V_{Nernst} is the pressure induced voltage (V)
R is the gas constant (J mol^{-1} K^{-1})
T is the temperature (K)
p_0 is the pressure on input (MPa)
p_1 is the pressure on output (MPa)

The open circuit potential (OCP) provides an accurate reading of the hydrogen pressure differential across the membrane, as shown in the graph in Figure 13.5 for a temperature of 150°C. Note that a positive pressure difference of +60 and +400 bar translates into an OCP of 30 and 60 mV, respectively. The voltage becomes negative exponentially in case of reverse pressure differential. Interestingly, in principle, the said compression energy can be recovered in an electrochemical cell upon decompression by utilizing the OCP generated by the pressure difference as a driving force to power the reverse current. The "self-decompression" is feasible at low current density, since the Nernst voltage is limited to typically tens of mV range. Boosting decompression costs more energy, when applying an additional voltage to draw higher currents, but we still see the benefit of the Nernst potential offset working in our favor. As such, compression energy is always regained. Disregarding back diffusion effects, the maximum roundtrip energy efficiency depends most on the hydrogen (de-)compression *rate* in combination with the total cell-to-cell resistance, than on the Nernst voltage.

13.2.3 Total Energy Requirement for Compression

The compression energy (work) discussed in Section 13.2.2 needs to be added to the previously discussed Ohmic losses (heat, Section 13.2.1) to obtain the total energy required.

The graph in Figure 13.6 displays the relative impact of Ohmic losses (heat) and compression energy (work) on the total energy required during operation. The compression energy (work) is displayed as a horizontal line, as it is determined by pressure difference and temperature, but it is independent of current density.

The specific resistance of the membrane has major impact on the relative efficiency of compression, scaling

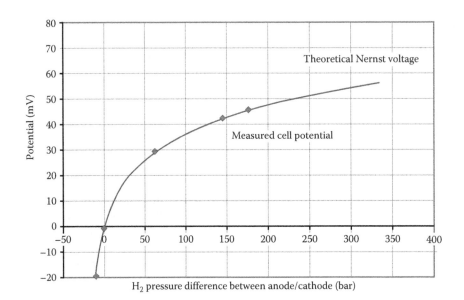

FIGURE 13.5
Graphic representation of Nernst voltage measured (diamonds) and calculated as a function of the pressure difference on either side of the membrane, both positive and negative for a temperature of 150°C.

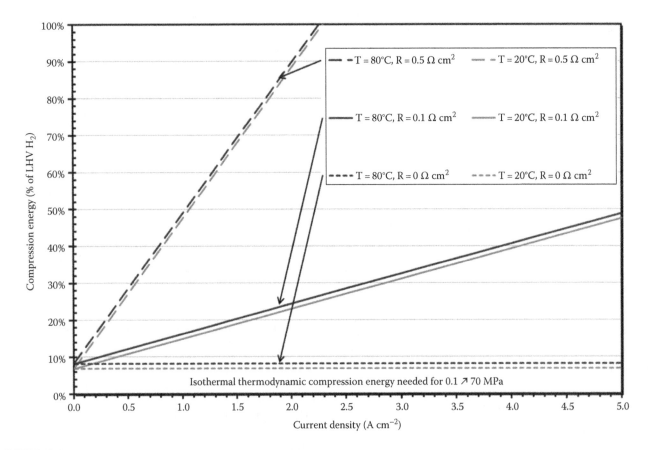

FIGURE 13.6
Graph showing the total energy required for hydrogen compression from 0.1 to 70 MPa taking into account the impact of two temperatures on work, and two selected cell-to-cell resistance values on heat losses as a function of current density.

linearly with the selected current density working point. For example, compression from 0.1 to 70 MPa using a membrane with "high" specific resistance of 500 mΩ cm^2 with a current density above 2 A cm^{-2} will consume almost the same amount of energy as contained in the hydrogen gas itself (i.e., 100% of LHV).

Clearly, the influence of temperature on the compression energy (work) is minimal. This suggests that the EHC should operate at the optimum temperature for MEA components, where overall resistance is minimal, which typically depends on the choice of ionomer membrane materials. Note that the overall resistance may be affected by other operating conditions such as humidity, pressure, gas purity, contact resistance, diffusion profiles, and at the selected current density. These contributing factors will be placed in context in the following section.

13.2.4 Equivalent Circuit

As mentioned before, current drives the proton pumping process: one proton inside equals one electron around the outside. This direct relationship is obvious from the reaction equation (see Figure 13.4) and provides a key to control and monitor the mass flow through our ability to manage electrical current. Here we take a look at the process from an electrical point of view by constructing an equivalent circuit to describe the individual contributing elements.

In Figure 13.7 the process of the EHC is represented as an equivalent circuit. The faradaic current

is equivalent to the flow of hydrogen as protons during forward compression or as hydrogen gas during backward diffusion. The phenomenon of hydrogen gas (H$_2$) dissolving in the membrane and diffusing toward lower concentrations is referred to as "back diffusion" (Kocha et al. 2006, Cheng et al. 2007). The back diffusion process can be viewed as a current source/sink which is dependent on the pressure difference (dP), the temperature (T), and the relative humidity (RH %) due to changing membrane properties. All current attributed to the back diffusion current source (product of current and voltage) is considered a loss when working against a positive pressure difference. The Faradaic equivalent of the back-diffused hydrogen gas can be pumped forward again once it hits the anode electrode and is reconverted to protons, but it is double work that goes at the expense of the (kW kg^{-1}) ratio. Brunetti et al. (2012) proposed introducing a new parameter called the "Transport Performance Index" defined as the ratio of hydrogen permeability and the proton conductivity, leading to an immediate idea of the whole mass transport performance of a membrane.

Vice versa, in case of a negative pressure difference (i.e., de-compression process), the back diffusion current works in the same direction by delivering additional hydrogen gas, but its compression energy is not recoverable through any Nernst voltage because the permeation process itself does not send electrons flowing through the external circuit.

The energy required to drive the faradaic current depends on the Nernst voltage, as discussed in the previous section, and the overall resistance of the system. In Figure 13.7, this is depicted as one general resistor with an observed dependency on the current density itself, the pressure difference, relative humidity, and time. Different processes contributing to this resistance can be distinguished as shown in Figure 13.8. The breakdown takes into account an Ohmic contact resistances on both sides of the membrane, the gas transport delivering hydrogen gas to the anode catalyst and away from the cathode catalyst, the reaction kinetics for the oxidation, hydrogen oxidation reaction (HOR) and the evolution, hydrogen electrode reaction (HER) on the anode and cathode respectively, and the transport of protons through the electrolyte phase. Under optimal conditions more than 90% of the resistance would be attributed to the latter process of proton transport, since the other processes have the potential to be extremely fast. This become apparent from the typical resistance fingerprints as observed with Electrochemical Impedance Spectroscopy (EIS).

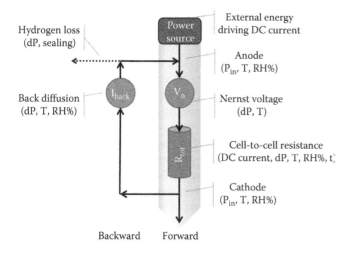

FIGURE 13.7
Equivalent circuit describing the faradaic current flow equivalent to the hydrogen transport during the process of the Electrochemical Hydrogen Compression.

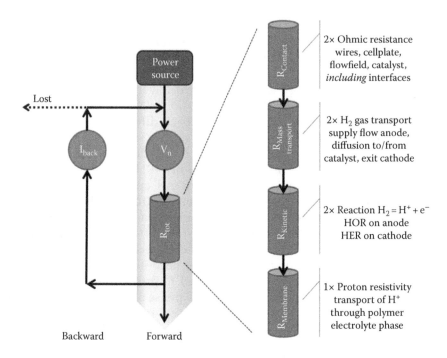

Lost

Backward   Forward

2× Ohmic resistance
wires, cellplate,
flowfield, catalyst,
including interfaces

2× H$_2$ gas transport
supply flow anode,
diffusion to/from
catalyst, exit cathode

2× Reaction H$_2$ = H$^+$ + e$^-$
HOR on anode
HER on cathode

1× Proton resistivity
transport of H$^+$
through polymer
electrolyte phase

FIGURE 13.8
Breakdown of processes contributing to the overall cell-to-cell resistance in the equivalent circuit of the electrochemical hydrogen pumping process.

13.3 Results and Discussion

13.3.1 Typical Impedance Spectra

An overview of the influences on the impedance characteristics is provided in Figure 13.9, where a number of spectra are presented to show the potential influence of different operating conditions.

First we point out the two impedance spectra positioned on the real axis around 150 mΩ cm^2 that almost overlap, also shown in the inset graph of Figure 13.9. These spectra correspond with the first and the last measurement done with no pressure difference at 1–1 MPa of pure hydrogen gas. The similarity between the (insignificant) features indicated that any resistance characteristics incurred during the pressure cycle up to 70 MPa were completely reversible.

In this example during compression the real impedance of the spectra increased with increasing pressure difference, showing an almost linear relationship for the high frequencies intersecting with the real axis. This cutoff value is typically attributed to the contact and the membrane electrolyte resistance (Yuan et al. 2010). The contribution of the membrane electrolyte resistance can be distinguished by measuring with EIS at this particular frequency as a function of temperature and comparing the Arrhenius behavior with the expected proton conductivity characteristics of the membrane as shown in Figure 13.10. The imaginary component was also

observed to increase with increasing pressure difference, giving rise to a semicircle pattern at 70 MPa that remained small in rest when no current was applied to the system.

The semicircle pattern became more prominent during operation when a current density of 500 mA cm^{-2} was applied to the cell to compress hydrogen against a positive pressure difference of 1–70 MPa. Typically, the semicircle pattern could be attributed to the reaction kinetics and described using an equivalent circuit including a parallel configuration of a real reaction resistance and an electrode interface capacitance. Any influence of diffusion has not been considered here yet. Interestingly, here we observe a decrease in the high frequency cutoff of the real axis, which was attributed to contact/membrane resistance in the previous paragraph. This implies the membrane (interface) changed for the better during operation when facing a pressure differential. However, the total resistance increased significantly, as the impedance measured at the lowest frequency of 1 Hz showed a much higher value compared to the situation at rest. We will not elaborate further on physical processes underlying the absolute increase of cell-to-cell resistance here, as we believe more study is required to understand the impact and interactions between multiple parameters controlling local operating conditions (Bockris and Reddy 1977). We have observed that (lack of) hydrogen gas supply played a minor role, as increasing the feed pressure in the range 1–20 bara had no significant effect on the described impedance spectrum.

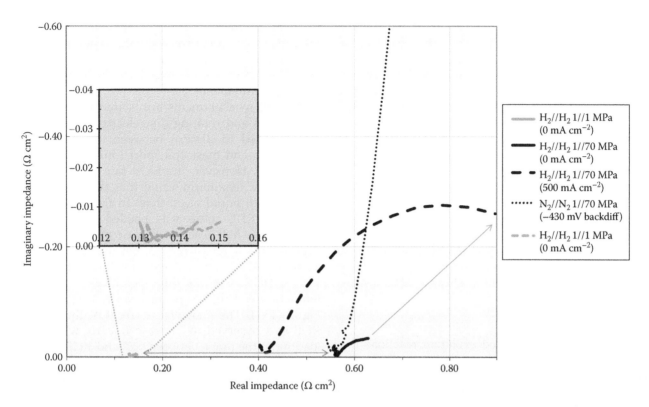

FIGURE 13.9
Overview of typical impedance spectra that may be observed during compression of hydrogen from 1 up to 70 MPa measured in a frequency range of 30 kHz to 1 Hz and a temperature around 60°C.

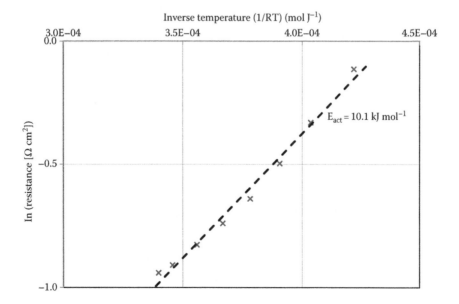

FIGURE 13.10
Temperature dependence of the cell-to-cell resistance measured with EIS at a frequency typical of the electrolyte resistance plotted against the inverse absolute temperature to show Arrhenius behavior in line with expected proton conductivity.

The impact of the hydrogen gas supply became evident when instead nitrogen gas was being fed to the cell anode input. In Figure 13.9, the dotted curve was measured in a specific operational mode* to measure the back diffusion caused by the pressure difference across the membrane. Only lack of hydrogen supply was now limiting the reaction rate and hence pumping capability; the only hydrogen gas available to the anode electrode was permeating through the membrane from the high pressure side. As can be expected, no change was observed of the real resistance cutoff attributed to membrane/contact resistance. However, the semicircle expanded greatly to mimic reaction kinetics with blocking electrodes (Yuan et al. 2010).

This particular example was selected to elaborate on the influence of operating condition, but in practice these effects should be insignificant in a healthy EHC system.

13.3.2 Kinetics

The hydrogen oxidation and evolution reactions (i.e., HOR and HER) have been studied extensively, but received relatively little attention due to their high reaction rate relative to the oxygen reduction reaction (ORR). The HOR has one of the fastest known specific rate constants in aqueous solutions. Superior catalyst performance is presented by platinum and platinum group metals, including Pt, Ru, Pd, Ir, Os, and Rh, thanks to their intermediate values of H adsorption and their ability to disassociate H_2 in the presence of H_2O (Li et al. 2008). Unlike in this reference, we will not review all proposed stepwise reaction mechanisms here, but merely state a consensus that the energetic state between the reactive intermediate $H_{adsorbed}$ and the catalytic Pt surface determines the kinetics of the HOR in aqueous solutions.

Hydrogen diffusion limitations often complicate the measurement of the actual kinetic reaction rate, for which many experimental setups have been devised and reported (Olender et al. 1982). Mass transport limitations quickly lead to concentration polarization across thick electrodes limiting the active thickness to approximately 5 μm at high current densities, regardless of the actual geometric thickness of the catalyst layer (Wilson and Gottesfeld 1992). We have found that for EHC applications the utilization of low platinum loadings down

to 10 μg Pt cm^{-2} are perfectly feasible, thus shifting the problem from electrode thickness and cost toward maintaining an even distribution of catalyst particulates. It is worthwhile to investigate the limit in catalyst loading requirement.

Half-cell measurements are geared to use thin catalyst layers and rotating disc electrode (RDE) setups are designed to discern between catalyst turnover frequencies and mass transport limitations (Garsany et al. 2010). However, the HOR rate is so fast that it exceeds the maximum value that can accurately be measured in liquid electrolyte. In fact, a special gas-breathing RDE setup was needed to be designed to study the half reaction (Zalitis et al. 2013), which experimentally measured geometric current densities as high as 497 A mg^{-1}_{Pt} at a low loading of 2.2 μg_{Pt} cm^{-2} for the HOR in 4 mol dm^{-3} perchloric acid at room temperature. This key result has been reproduced in Figure 13.11. The geometric current of the "floating" RDE was observed to increase linearly with catalyst loading in the range between 0.72 and 10.15 mg_{Pt} cm^{-2}, spanning sub-monolayer and multilayer coverage of catalyst, which translates to an average maximum specific current density of 600 ± 60 mA cm^{-2}_{spec}. In this configuration, mass flow limitation was reached at a current density of 5.7 A cm^{-2}_{Geo} at the highest loading.

The kinetic limitation of the HOR reaction is shown by the fine structure observed between 0.18 < V versus RHE < 0.36 and the performance decay above 0.36 V in Figure 13.11. The origin of these structures are still subject to debate and have cautiously been associated with changing reaction rates, pathways, surface (zero) charge or specific surface adsorption due to hysteresis effects. Further investigation is ongoing, but we can already comprehend two things for the EHC application: unless these features are observed in the current–voltage relationship (so-called IV curve), the performance of the EHC is not likely kinetically limited by the HOR. Secondly, there is little gain in raising the driving voltage beyond 0.5 V considering the optimum catalyst reaction rate.

At HyET, these fine structures have not been observed when testing a fuel cell or EHC using cyclic voltammetry with humidified hydrogen gas atmosphere on both sides of the membrane. The corresponding reaction rate at the catalyst surface area was orders of magnitude lower, than in the situation anticipated by Zalitis (2013). Typically, a generic cell-to-cell resistance value is observed and hence a linear relationship is expected between voltage and current density. This suggest the fast processes of HOR at the anode, and HER at the cathode, were overshadowed by other processes, for example, the limited proton conductivity in the membrane phase or within the membrane-electrode interfacial layer.

A clue was provided by a striking discrepancy observed when conducting IV measurements to evaluate

---

* In short, hydrogen gas on the anode input was purged with nitrogen gas at 1 MPa and the driving potential was increased to 430 mV to avoid adsorption effects on the platinum catalyst. All H_2 gas back-diffusing through the membrane was now reconverted at the anode electrode and pumped back to the cathode as protons. The back diffusion gas stream was quantified accurately by measuring the current, as there was no other hydrogen gas supplied to the cell. Ideally, under these circumstances the pressure on the cathode output remained the same, provided no hydrogen gas escaped through external leaks in the cell enclosure.

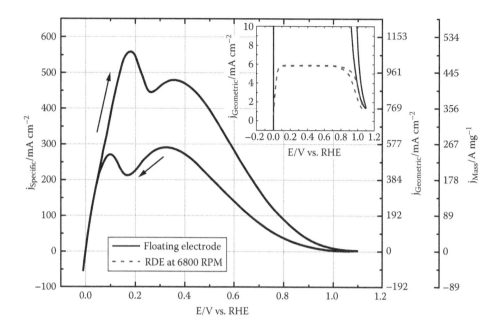

FIGURE 13.11

RDE electrode containing 2.2 μg_{Pt} cm^{-2} catalyst exposed to hydrogen, run in 4 mol dm^{-3} HClO$_4$ at 10 mV s^{-1} at 298 K. CE = Pt, RE = RHE. The ordinate axis corresponds to the specific current density (left), geometric current density (first right), and mass activity (second right). The inset shows the HOR of Pt on a RDE in comparison to the floating electrode in terms of geometric current density. The RDE was rotated at 6800 rpm, in 0.5 mol dm^{-3} HClO$_4$, at 10 mV s^{-1}. (From Zalitis, C.M., Kramer, D., and Kucernak, A.R., Electrocatalytic performance of fuel cell reactions at low catalyst loading and high mass transport, *Phys. Chem. Chem. Phys.*, 15, 4329–4340, 2013. Reproduced by permission of The Royal Society of Chemistry.)

the hydrogen pumping process on both our PEMFC and EHC setups without any pressure differential. Figure 13.12 shows a direct comparison between the two duplicate datasets for identical membrane and catalyst compositions. The expected linear trend was observed in the PEMFC setup, while the EHC setup appeared more restricted above a certain current density, and showed leveling off. Duplicate measurements confirmed this contrasting trend. Although some degree of performance stabilization may be expected when comparing results from cyclic voltammetry with current–voltage polarization measurements, here both MEAs should display similar behavior based on design. This suggests the limitation is caused by the MEA environment, that is, the local operating conditions (Figure 13.12).

There are numerous small differences between the typical EHC and the PEMFC cell setups, but both are expected have ample supply of hydrogen, thus ruling out "choking" of the catalyst. Probably the most important parameter influencing the performance behavior is the presence of sufficient humidity. The carbon PEMFC cell design has a serpentine structure and a built-in membrane humidifier operating at the same (dew point) temperature as the cell. The EHC cell setup may be more sensitive to the humidity changes, as it lacks such features guaranteeing full humidification, but the observed limitation is unlikely to be caused by simple membrane drying. We suspect water could play

an underestimated role in the reaction kinetics, which becomes visible in the situation of the EHC, because there is no constant resupply of product water from the cathode. We consider model studies in literature.

Wagner and Moylan together reported in 1987 and 1988 that the presence of liquid water in the electrolyte (and/or other component of the electrochemical double layer) had a major effect on the hydrogen coverage–potential relationship of the catalyst, and signaled that any attempt to model hydrogen electro-sorption should include at least water along with hydrogen. In contrast with this observation, the heat of water adsorption on metals is only little influenced by co-adsorbed hydrogen (Kizhakevarian 1992), in other words, it is not just presence of water that is influencing hydrogen adsorption.

Hydrogen electrosorption has been studied extensively on various single-crystal Pt surfaces and was found to be highly reversible, as discussed by Wagner and Moylan (1992), who observed pronounced differences between electrochemical and hydrogen-gas-only adsorption behavior. They suggested this discrepancy was caused by complex, potential-dependent interactions between co-adsorbed H and H$_3$O$^+$ ratios, which would also explain the unexpected breadth and flatness of the hydrogen regions in voltammograms on well-ordered surfaces. No hydronium formation was observed by Wagner in 1988 on the Pt catalyst surface for hydrogen coverages less than 20% of saturation.

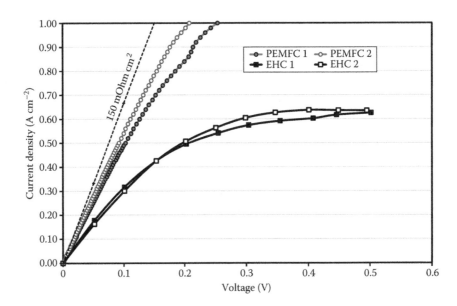

FIGURE 13.12
Graph showing exemplary IV curves of identical MEAs testing hydrogen pumping performance in a EHC compression cell setup (solid black line, squares) and in the PEMFC measurement setup (solid gray line, circles), relative to the linear relationship expected from a total resistance value of 150 mOhm cm² (dashed line).

Chen et al. (1999) stated the species identified as "hydronium," are not likely to be just H_3O^+, but more likely a mixture of hydrated intermediates such as $H_5O_2^+$, $H_7O_3^+$, and $H_9O_4^+$ based on calculated fits of more accurate measurement data. Evidently, some form of hydronium plays a pivotal role, provided sufficient hydrogen and water are present for its formation on the catalyst surface.

Water needs to be considered as an essential reactant in the HOR mechanism as pointed out by Wagner et al. (1988) and explained by Kizhakevarian et al. (1992) with the kinetic model studies on model Pt surfaces. While the standard hydrogen redox reaction involved all three states of matter and is noted as follows (also see Equation 13.1):

$$2\,H^+\left(aq\right)+2\,e^-\left(m\right)=H_2\left(gas\right) \qquad (13.6)$$

We actually need to consider the expression of the reversible elementary surface reaction step where all species are adsorbed:

$$H+H_2O=H_3O^++e^-\left(m\right) \qquad (13.7)$$

Isotope exchange experiments were conducted in high vacuum to confirm this reaction mechanism occurred on monocrystalline Pt surfaces in water hydrogen coadsorbate. Evidence was reported of hydronium formation and subsequent intra-layer isotope exchange observed between coadsorbed D and H_2O on monocrystalline Pt surface with high work

functions (5.1 eV for Pt(100)) by Kizhakevarian et al. (1992), Sass et al. (1991), and Lackey et al. (1991), but not observed on Cu (110) with a low work function (3.6 eV). Apparently, having a catalyst work function greater than 4.54 eV allowed this surface reaction in Equation 13.7 to occur reversibly by making formation of the hydronium ions thermodynamically favorable and enabling the isotope exchange; lower work functions favor chemisorbed hydrogen.

A Born–Haber cycle is used by Kizhakevarian et al. (1992) to explain the influence of the work function on the formation of hydronium ions as reproduced in Figure 13.13, leaving room for alignment of the vacuum versus electrochemistry potential scales. Large energies are involved with this four-step process: (1) desorption of atomic hydrogen (+2.7 eV), (2) ionization (+13.6 eV), (3) electron transfer to the metal (−5.1 eV), and (4) proton hydration to the aqueous layer (−11.3 eV). Although each step generally involves large energy changes, the entire reaction is essentially energetically neutral (i.e., the overall energy change is only −0.1 eV in the example provided) and thus reversible. The most important consideration for EHC is: as water desorbs or is carried away from the surface of the catalyst, the hydronium ions readily convert back into chemisorbed atomic hydrogen which cannot participate in the proton conduction process.

The EHC application benefits greatly from extremely fast hydrogen reaction kinetics on platinum group metal catalyst, enabling electrodes with extremely low catalyst loadings, and application of low driving potentials. The hydrogen-to-hydrogen conversion looks simple, as it

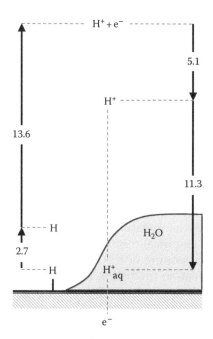

FIGURE 13.13
Born–Haber thermodynamic cycle illustrating the conversion of chemisorbed hydrogen to hydronium ion through the gas phase route. The scales on the left and right indicated energy difference in the units of eV. The central portion of the figure illustrates schematically the species involved. (Reprinted from *Surface Science*, 275, Kizhakevariam, N. and Stuve, E.M., Coadsorption of water and hydrogen on Pt(100): Formation of adsorbed hydronium ions, 223–236, Copyright (1992), with permission from Elsevier.)

is often over-simplified, but the subsequent HOR and HER processes appear quite complex and delicately balanced, involving potential-dependent coadsorbed "hydronium" complexes stabilized by matching catalyst work functions. The suggested explanations quoted here still need proper verification for the situation of EHC and with better understanding, alternative catalyst materials can be constructed. The message is this: when converting hydrogen (electro-) catalytically, we cannot ignore the role of water.

13.3.3 Thermal Control

In contrast to a fuel cell, the EHC generates no net chemical reaction energy and the electrical power supplied to the EHC is controlled by the operator. Any excess driving voltage (exceeding the overall Nernst voltage) is converted into heat. Considering previous discussion in Section 13.2.4, the heat is generated mainly within the membrane of the MEA. Note that the membrane in a fuel cell is also warmed through Ohmic losses as a function of the current density, but generally the reaction overpotential at the cathode ORR constitutes a much greater source of heat (Owejan 2014). Hotspots and nonuniform temperature distributions need to be avoided by having sufficient measures for heat spreading and dissipation.

An EHC stack generates 2 kWh kg^{-1} of heat when operated at 0.5 A cm^{-2} and assuming a cell-to-cell resistance of 150 mOhm cm^2. In other words, such stack would produce approximately 167 W of heat when scaled to a pumping capacity of 2 kg of hydrogen (disregarding compression work). Mathematically, a linear relation exists between the current density and the pump capacity and the heat output. Except in case of high current densities, the thermal cooling requirement remains low and could even be resolved using convective air cooling. This is commonly the preferred option, as it is difficult to integrate cooling channels, manifold structures, and gastight seals inside an electrochemical cell designed for high-pressure hydrogen.

We have to distinguish between thermal cooling load and temperature control. Note, the cell-to-cell resistance is a function of the temperature as a result of major influence of the membrane proton conductivity characteristics. Colder cells show higher resistance and therefore dissipate more heat with passing of current during operation. The proton conductivity at room temperature is sufficient to cold-start a full size stack and heat itself to operating temperature (subject to membrane material used). The temperature distribution depends on the heat dissipation through conductive, convective, radiative losses for all different stack elements such as flanges, plates, and cells in absence of interflowing cooling media for thermal control.

EHC stacks typically consist of two cylinder-shaped flanges, bolting together an alternating series of round cell plates and MEAs. When relying on internal heating during start-up, a parabolic temperature distribution develops with likely the hottest cell in the middle of the stack. This middle cell would be least capable of dissipating excess heat, relying mainly on conductive heat transport toward its neighboring cells. In contrast, when external flange heating is used to raise the temperature before starting operation, the temperature of the middle cell would be the lowest, being furthest away from the source of heat. Note that a combination with external flange heating cannot correct the mentioned parabolic temperature distribution during operation, because it will only affect a temperature offset on the whole profile and thus further increase the mid-cell temperature as well. Poor stack design could lead to extreme temperature differences reaching tens of °C between the hottest middle cell and the outer coolest cell in the situation of a passive cooling.

Temperature control of the EHC stack is important, because it directly impacts the cell-to-cell voltage distribution on more than one level. We mentioned membrane proton conductivity (see Figure 13.10), but note that water management, Nernst compression voltage (see Figure 13.6), and back diffusion (see Figure 13.23) are influenced by temperature as well. To our knowledge, we have not encountered a comprehensive mathematical model in open literature designed to simulate these aspects of the EHC simultaneously. Creating such model is recommended to predict the required level of thermal control and avoid unstable or even detrimental off-spec operating conditions.

13.3.4 Humidity and Water Management

Hydrogen dehumidification is another interesting effect observed during isothermal compression as shown in Figure 13.14, similar to air conditioning when the pressure is increased and subsequently released. Fully humidified hydrogen carries more vapor at *low* pressure than at *high* pressure per mass unit of hydrogen, because the water vapor pressure is determined by temperature and equal on both sides under isothermal conditions, by definition. Compression of hydrogen gas directly increases the pressure, but compression of water vapor leads to condensation of surplus water. In the situation of a positive pressure differential, not all water carried into the anode side can evaporate on the (over-) saturated cathode side, if both dew point temperatures equal operation temperature. Hence, less water vapor exits the EHC upon compression.

Upon re-expansion to atmospheric conditions, the dew point of the hydrogen gas effectively dropped from fully saturated (input) down to −40°C (output) after having been subjected to a pressure differential of 200 bar. Surplus moisture remains as liquid water within, or in contact with the membrane. Commonly, a PFSA-type membrane is permeable to water and able to equilibrate

between anode and cathode side (Motupally et al. 2000). Provided liquid water does not impede gas transport, it provides indispensable humidification for optimum proton conductivity in the membrane bulk and surface interfaces (Choi and Data 2003, Bautista-Rodríguez et al. 2009).

The parameter lambda is used to describe the quantity of water residing in the membrane, expressed as the number of water molecules per sulfonic acid. In 1991, both Zawodzinksi et al. and Springer et al. published a correlation between the equilibrium water vapor pressure and the lambda value ranging between 0 and 14 [H_2O/SO_3^-] for Nafion PFSA material at atmospheric pressure and a temperature of 30°C. Hinatsu et al. (1994) reported an empirical formula describing this relationship at a temperature of 80°C and Ge et al. (2005) provided an equation predicting the water content across the whole range of 30°C–80°C. Typically, decreased water content resulted from lower vapor adsorption at elevated temperatures.

Different water transport mechanisms play an important role in determining the water content and its distribution within the membrane. Water residing in the membrane is transported in three ways in a typical fuel cell operated on hydrogen–air: (1) electro-osmotic drag of water by protons transported from the anode to the cathode side, (2) back diffusion by the concentration gradient of water, and (3) convection generated by the pressure gradient, albeit limited. Modeling results from Yi and Nguyen (1998) showed that the performance of a PEMFC could be improved by anode humidification and positive differential pressure between the cathode

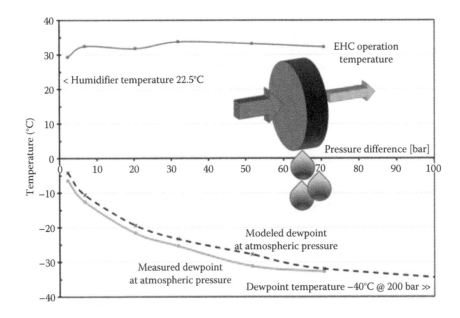

FIGURE 13.14
Graph showing dehumidification effect caused by pressure increment during isothermal compression of fully humidified hydrogen at room temperature.

and anode to increase the back transport rate of water across the membrane. Results also showed that effective heat removal was necessary to prevent excessive temperature excursions, which could lead to local membrane dehydration (Hinatsu et al. 1994). These findings could applicable to the EHC as well, but the scale of the applied pressure differential may exceed the implied validity of the physical models (Berg et al. 2004, Haddad et al. 2008, Misran et al. 2013, Mulyazmi et al. 2013).

Physical pressure has a marked effect on the membrane. Kusoglu et al. (2011) discussed the internal balance between chemical and mechanical forces determining the water content and the d-spacing of the water domains in Nafion PFSA membranes. Small-angle x-ray scattering (SAXS) measurements showed direct hydrostatic pressure confinement reduced the water content. Figure 13.15 shows the impact of pressure on the water content of soaked and humidified PFSA membranes with different thermal history. The model predictions were extended up to 25 MPa to show that the decrease in water content became less significant at higher pressures.

Unlike with uniaxial pressure, hydrostatic compression results in an isotropic change in the d-spacing, similar to that caused by thermal pretreatment. Apparently, both thermal history and (ex situ) applied pressure could change the mechanical energy of the structure in PFSA membranes (Kusoglu et al. 2011).

The water mass balance *around* the membrane needs to consider multiple phenomena: the so-called Schroeder's

paradox (Choi et al. 2003) for the amount of absorption from vapor versus liquid water, different mass transfer rates for water adsorption versus desorption to expand (or contract) the lattice (Ge et al. 2005), and dissimilar dew points of the hydrogen gas entering and leaving the EHC at incremental pressures, as was discussed earlier using the example of dehumidification.

The water balance *within* the membrane is more complex, as we need to take into account osmotic drag driven by the Faradaic protonic equivalent of the current density, diffusion based on activity, permeation of water driven by pressure gradient, and restricted evaporation from the membrane surface. We will discuss these points briefly in the following.

Electro-osmotic drag signifies the transport of water molecules with the migration of protons (i.e., hydronium ion complexes). Conflicting measurement results prompt the debate how many water molecules are associated with each proton, and whether an influence of humidity and temperature is significant (Jinnouchi et al. 2004). Springer et al. (1991) specifically distinguished between electro-osmotic drag at unit activity of water throughout the system (i.e., immersed), or in case a water gradient exists. In the latter situation, the drag is expected to be smaller, but commonly a value of unity ratio [H_2O/H^+] is utilized.

Diffusion of water is the product of an activity gradient across the membrane thickness, scaled with factors describing the concentration of acid groups and the water diffusion coefficient (Springer et al. 1991, Xing et al. 2013). The diffusion coefficient itself is also a function of the water content lambda, but its (nonlinear) relationship is still under discussion. Majsztrik et al. (2007) provide insight in measurement methods reported in literature for quantifying diffusion, whilst distinguishing sorption and desorption effects. Motupally et al. (2000) reported the quantity of water diffusion was affected by the absolute nitrogen gas pressure (1 and 5 atm tested) by indirectly influencing the water activity gradient across the membrane in their ex situ test setup.

Hydraulic permeation describes the major influence of pressure on the water transport through the connecting network of hydrophilic channels of submicroscopic dimensions (Xing et al. 2013). Typical values for Nafion are circa $4*10^{-16}$ cm^2 at 23°C (Duan et al. 2012). This water transport process was long ignored in mathematical models, because for gases no significant viscous flow permeation was observed in absence of pinholes. The water residing in these channels cannot be "blown out" easily by a large pressure gradient. Single pore capillary models predict huge internal liquid and osmotic pressures exceeding hundreds of bars, when taking into account discontinuities in the water distribution and swelling of the ionomer (Eikerling and

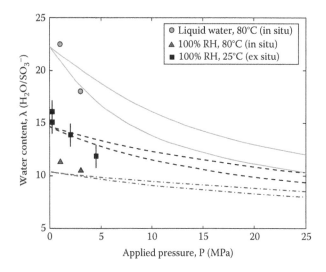

FIGURE 13.15
Model predictions for the effect of pressure on the water content of Nafion® membrane for liquid-equilibrated at 80°C and saturated vapor equilibrated at 25°C and 80°C. Measured water contents from in situ and ex situ neutron imaging are also shown for each condition. The lines are the upper and lower bound predictions of the model. (Reprinted from Kusoglu, A. et al., *Journal of the Electrochemical Society*, 15812, B1504, 2011. With permission from the Electrochemical Society.)

Berg 2011). Recent publications argue that liquid–vapor pervaporation fluxes were significantly greater than the liquid–liquid hydraulic permeation in case of an activity gradient and limited pressure difference (<10 kPa), but vice versa when there was liquid water on both sides (Duan et al. 2012).

Validation of these models is practically difficult, if not impossible, to reproduce quantitative values unambiguously due to the intricate interplay of membrane composition, microscopic structure, reorganization upon swelling, and effective materials properties (stress, strain, modulus) along with statistical fluctuations in pore ensembles. The key message for the modeling of EHC properties is that the internal driving forces for water fluxes and external conditions should be clearly distinguished.

13.3.5 Compression and Decompression

Estimations have been published of the anticipated maximum pressure that could be achieved using electrochemical compression. Sedlak et al. (1981) predicted pressure of 2 MPa would be achievable with proper support of the membrane. Grigoriev et al. (2008) achieved a pressure of 13 MPa with a dedicated cell, but recommended to use a cascade to reach higher pressures in 5 MPa steps. Godula-Jopek et al. (2012) discussed this technology and claimed a cascade would be required to reach high pressures, quoting that a conventional PEM cell with a membrane with a thickness of 180 μm achieved a maximum pressure of 40 MPa (at 90°C and 0.5 A cm⁻²).

We would argue that a cascade configuration is in principle the same as using *one* single stage with a cumulatively thicker membrane, considering the effect on back diffusion and proton conductivity is additive. From an economic and energy perspective, it makes sense to pull hydrogen through a membrane only once, and make this membrane as thin as possible.

HyET demonstrated single stage, EHC up to a pressure of 100 MPa in 2012 as shown in Figure 13.16, thus providing proof-of-concept on a laboratory scale. A membrane thickness less than 100 μm was sufficient to hold a pressure equivalent to a 10 km water column. The slight curvature of the line was due to the influence of the Nernst potential (see Figure 13.5) that was subtracted from the driving voltage under potentiostatic control. After reaching 100 MPa the applied potential was removed to stop the current and the active forward hydrogen transport. Usually, one would close the valve to the hydrogen storage tank at this point to confine the pressure in a dedicated enclosure.

Figure 13.16 shows the 100 MPa pressure decreased gradually due to permeation of hydrogen gas from the cathode chamber of the EHC through seals and back diffusion through the membrane itself, toward the anode chamber. This phenomenon was most visible at rest in the absence of current (hydrogen pumping), but it occurred continuously whenever there was a (partial) pressure or activity gradient across the membrane. Back diffusion can be actively countered by reconverting hydrogen gas to protons on the anode side, and driving these back to the cathode chamber with current at the expense of efficiency. The pressure of 100 MPa could be

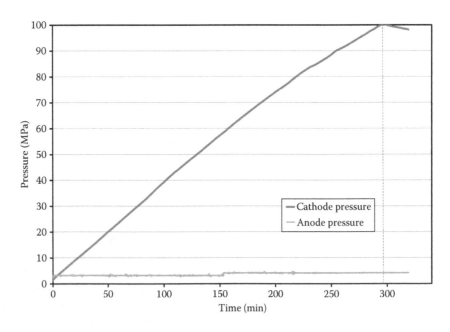

FIGURE 13.16
Proof of concept showing that electrochemical compression is capable of achieving 100 MPa output pressure using a single-stage configuration.

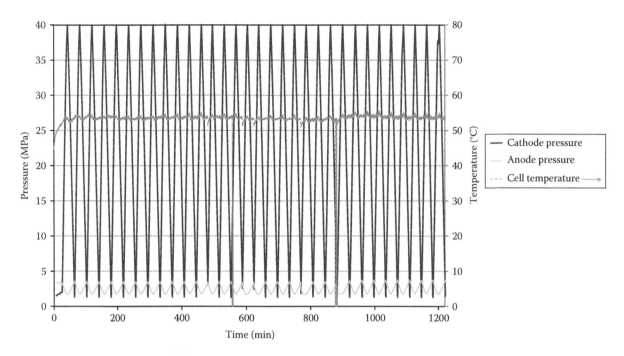

FIGURE 13.17
Repeated electrochemical pressure cycling of hydrogen from anode input of 2–40 MPa under automated potentiostatic control at a stack temperature between 50°C and 55°C (32 cycles shown here).

achieved here, because the forward H^+ proton transport rate far outweighed the H_2 gas back diffusion process. Back diffusion will be discussed separately later on in Section 13.3.6.

The electrochemical process is isothermal and fully reversible by changing the direction of the external current. In Figure 13.17 the hydrogen supplied to the anode input was actively compressed, and then actively decompressed under potentiostatic control at a temperature around 50°C–55°C. Here, the Nernst potential acted as a superimposed driving force during decompression in addition to the applied voltage, which explains the slightly convex–concave deformation of the curves in Figure 13.17 during repeated cycling to 40 MPa on the cathode (32× shown here). The anode pressure increased during decompression, as hydrogen is fed back into the supply line. Clearly, the pattern is reproducible and with the exception of the hydrogen escaping through external gaskets (estimated <1%), all pumped hydrogen is recovered during the decompression cycle.

Here, the major contribution to the cell-to-cell resistance was attributed to consistent membrane proton conductivity, due to a constant cell temperature. In a multi-cell stack configuration, the contribution of the contact resistance was less sensitive to pressure differential than projected by the typical impedance spectra shown Figure 13.9. The influence of current density was investigated by repeating the compression sequence to 40 MPa in a single cell configuration, stepwise increasing the current density from 0.1,

0.2, ..., 0.5 A cm². The cell-to-cell resistance was corrected for the Nernst voltage and plotted as a function of pressure in Figure 13.18. Further research and development always strives to reduce the absolute cell-to-cell resistance level. The separate curves all overlapped and showed the same trend as a function as pressure, indicating in this current density range the hydrogen pump rate could be varied freely as suggested in Figure 13.6 without any change in cell-to-cell resistance.

The cell-to-cell resistance did experience some reversible influence of the absolute pressure differential across the membrane, evident from a slight kickup above 150 bara cathode pressure. This effect was attributed to increasing contact resistance between the hardware and the MEA, as was confirmed using continuous EIS measurements at 1 kHz frequency. Movement of the MEA within cell hardware should be restricted as much as possible to avoid cell-to-cell variations and premature failure upon cycling. We continue to stress this point with the following extreme example.

Degrees of movement are not necessarily equal for all MEAs in a stack configuration when subjected to increasing–decreasing pressure differential. A typical example is shown in Figure 13.19, where the cell-to-cell resistance variation was measured continuously as a function of pressure in an exemplary three-cell EHC stack. Upon compression, the cell voltage monitoring showed their individual energy requirement was different between cells, but more interesting, also changed

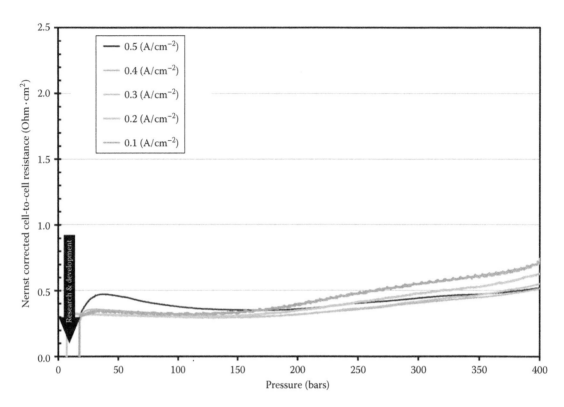

FIGURE 13.18
Nernst corrected cell-to-cell resistance plotted as a function of the cathode pressure for a number of consecutive galvalnostatic compression cycles up to 40 MPa using a range of current densities between 0.1 and 0.5 A cm^{-2}.

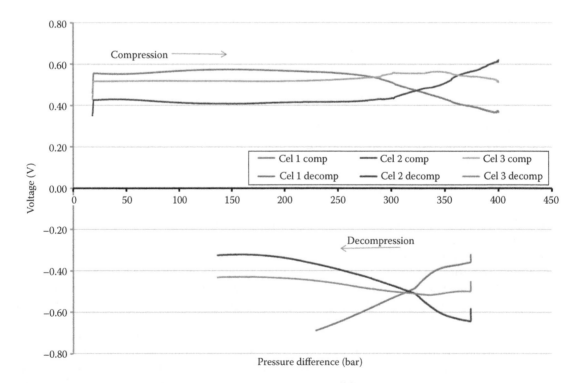

FIGURE 13.19
Extreme example of a varying voltage distribution within a three-cell stack measured during potentiostatic compression of hydrogen to 40 MPa and subsequent decompression, plotted in one graph to show the reverse similarity between both patterns in this exemplary observation.

relative to each other as a function of pressure. Note, all cells exhibited a similar configuration and experienced the same operating conditions and pressure differential across the membrane. The trend appeared to be reversible upon decompression, as the reverse change pattern was observed at the same cathode pressure point. Likely, movement of MEAs within hardware caused a redistribution of the cell-to-cell contact resistance as a function of pressure. Such variability needs to be controlled to ensure scalability of the EHC technology in large stack configuration.

The membrane is subjected to extreme pressure differentials in the EHC application and more research is required to investigate any detrimental influence of high pressures on its durability and effective lifetime. Commercial PFSA membranes like Nafion 117® were subjected to 3–5 MPa pressure showed increased fluoride release by 2–6 fold according to Kusoglu et al. (2014) and 2–4.9 fold according to Yoon and Huang (2010), compared to the uncompressed condition. It was hypothesized that the energy of the mechanical load induced a polymer deformation and thereby changed the activation energy for any decomposition reactions. Note that in water, pressures greater than 3 MPa could already induce plastic deformation (Kusoglu et al. 2007, 2010, 2012).

More durability data is being accumulated with realistic drive cycles. Lipp and Patel (2013) and Lipp (2014) reported achieving 10,000 h durability testing compressing from 30 to 3000 psig (0.2–20.4 MPa) and set a U.S. compression record of 12,000 psig (81.6 MPa) with an energy requirement of approximately 20 kWh kg^{-1}. It was not disclosed which membrane material was used. It is worth investigating whether a correlation exists between observed degradation behavior caused by local mechanical stress experienced in hydrogen–air fuel cells and sustaining exposure to high isostatic pressures in hydrogen–hydrogen compression cells. The lifetime target for EHC applications is aiming well beyond the average lifetime of the fuel cell in the FCEV.

The membrane is the key to EHC. The membrane material and its geometric dimensions largely determine the potential window of operation and system specifications, in terms of temperature, pump rate capability, maximum output pressure, and energy requirement. Boundaries exist, since the mechanical forces associated with the high pressure impose restrictions on the active geometric area of each cell, and may cause variations between cells in a stack during compression such as discussed earlier (see Figure 13.19).

Three main membrane characteristics that need to be considered are schematically depicted in Figure 13.20 and can be mutually conflicting in their practical realization. These are mechanical integrity (i.e., strength, durability), proton conductivity, and back diffusion.

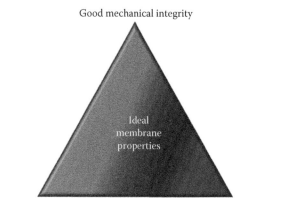

FIGURE 13.20
This schematic triangle illustrates the playing field to find the optimum balance between three main membrane characteristics that largely determine the specifications of the electrochemical compressor system.

For example, a thicker membrane is stronger and has lower back diffusion, but lower proton conductivity as well. The optimum balance between these characteristics can be different for various applications, whilst considering the overall economic feasibility.

13.3.6 Back Diffusion

The back diffusion of molecular hydrogen gas occurs when there exists a partial pressure difference (activity gradient) across the membrane between anode and cathode (Kocha et al. 2006, Cheng et al. 2007). The degree of hydrogen back diffusion depends on the membrane thickness, material composition, porosity, and integrity. As discussed previously in the first Section 13.3.1 (see footnote), the back diffusion quantity can be measured electrochemically by polarizing with at least 430 mV (direction of forward pumping) whilst supplying nitrogen gas to the feed anode and retaining pressurized hydrogen on the cathode side. Two typical datasets are shown in Figure 13.21, where the resulting current density caused by back diffusion was plotted for a commercial LT membrane (PFSA, 50 μm) and a proprietary HT membrane (100 μm) tested at their common operating temperatures as a function of the cathode hydrogen pressure. Clearly, our HT membrane has superior resistance against hydrogen permeation making it most suitable for EHC application.

The current response appeared linear with cathode pressure, indicating the diffusion mechanism does not change significantly in this specific pressure regime. Constant values are expected when the current density values are converted to gas permeability units normalized for time, pressure difference, and thickness (Stern 1968). These permeability coefficients are plotted in Figure 13.22 for both tested membranes.

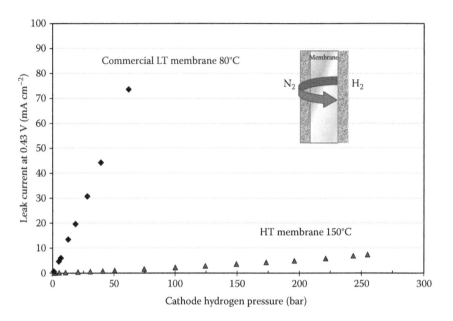

FIGURE 13.21

Faradaic equivalent of hydrogen back diffusion measured as a function of pressure for two different types of membranes at their preferred operating temperatures. The inset shows a schematic representation of hydrogen gas permeating from the cathode through the membrane and being reconverted into protons at the anode electrode.

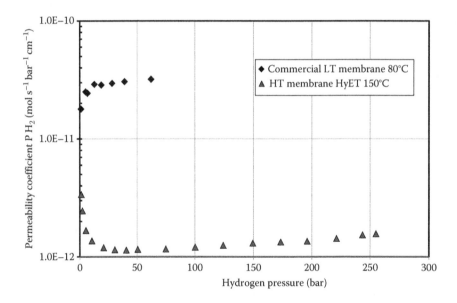

FIGURE 13.22

The permeability coefficient calculated from the data shown in Figure 13.21 plotted as a function of the cathode pressure for both types of membranes.

Indeed this graph shows stabilized values at higher cathode pressures greater than >30 bara cathode pressure and confirmed the HT membrane had up to 30 × lower hydrogen permeation than the commercial LT membrane material. Note that we here presume that the absolute hydrogen permeation scales linearly with the thickness of the membrane, because we have not yet observed discriminating evidence showing the skin has any different influence than the bulk, as is the case with liquid water permeability (Kusoglu et al. 2011). Note the strong tangent deviations observed at low cathode pressures below <30 bara were identified to be an artefact caused by the measurement method, when measuring in a downward direction from an initial high cathode pressure through an intended small external leak. Caution is advised with detailed physical interpretation, since there could be an effect of local pressure distributions in the cell and the escaping hydrogen gas may

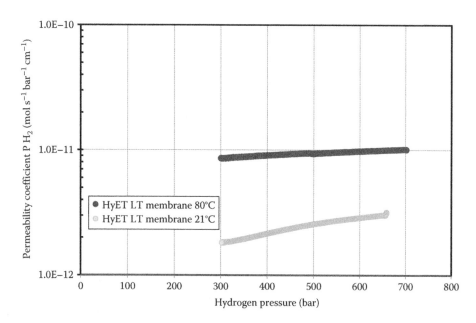

FIGURE 13.23

The permeability coefficient plotted as a function of the cathode pressure up to 70 MPa measured using a proprietary low temperature membrane to show the significant influence of operating temperature between 21°C and 80°C.

have changed the operating conditions with the removal of moisture. Nonetheless, this is the first reported back diffusion data measured in an electrochemical cell up to extremely high pressure differentials. The commercial LT membrane was clearly limited in its maximum pressure capability.

HyET has conducted many experiments to qualify various membrane materials for the application in EHC. Figure 13.23 shows two datasets measuring back diffusion on one single proprietary LT membrane material at difference temperatures, going from an extreme cathode pressure of 70 MPa down to 30 MPa. We aim to demonstrate the sensitivity of membrane's hydrogen gas permeability to the operating conditions within the high pressure cell. Between 80°C and 21°C the hydrogen gas permeability decreased 3× and 5× fold when compared at 70 and 30 MPa cathode pressure, respectively. Reducing the cell operating temperature helps mitigate back diffusion, particularly in situations where the quantity of back diffusion of hydrogen gas is significant such as encountered with high-pressure differentials, thin or commercial membrane materials with greater porosity.

The optimum membrane operating temperature for EHC applications is a compromise between high enough proton conductivity and low enough hydrogen gas back diffusion to minimize the overall energy requirement for compression.

HyET foresees a growing academic interest in studying the correlation between membrane nanostructure and physical properties such as gas permeability and proton mobility for EHC applications, once dedicated

high-pressure EHC cells are made available. Little is understood at this time how the isostatic pressure influences the polymer chain behavior, including their spacing, domain segregation or crystallization, also in the presence of water and electric fields. For example, considering at 5 bara pressure water boils at approximately 150°C (instead of at 100°C in atmosphere), this suggests even PFSA-type membranes could be operated at higher temperatures well beyond 100°C without loss of proton conductivity, as experienced with fuel cells (de Bruijn et al. 2008). However, based on the data shown earlier, there could be little gain from pushing proton conductivity to higher levels, if greater sacrifice is made on the hydrogen gas back diffusion and residual membrane strength under targeted conditions.

It makes sense to always minimize the hydrogen gas back diffusion, irrespective from the ongoing academic discussion and characterization of the back diffusion mechanisms through acidic, polymeric membrane materials. Potentially, this could be a viable reason to pursue research on ceramic-type, proton-conducting membrane materials, where hydrogen gas permeation could be less sensitive to the temperature compared to polymers. Combinations of both commercial ionomer and ceramic filler material could prove synergistic, and various, multi-component composite structures continue to be evaluated (Grigoriev et al. 2011).

13.3.7 Total Energy Requirement

Interest in EHC technology is generally motivated by the perception of decreased energy requirement versus mechanical compressors, taking advantage from more favorable process principles like: no moving parts (i.e., no friction), single-stage compression, and more isothermal compression conditions. Indeed, lower energy requirements are feasible under specific conditions, but not guaranteed.

In this section, we will project the total energy requirement for hydrogen compression, taking into account the three key membrane properties that need to be balanced: mechanical strength, proton conductivity, and back diffusion of hydrogen gas. The appropriate membrane design is based on the anticipated operation condition specified by the customer application. Specification as input and output pressure, duty cycle, lifetime, OPEX, and CAPEX first define the total active area membrane and then translate to operating current density, compression ratio, and optimum temperature for the selected material. We have attempted to visualize the energy requirement for EHC as a function of current density and compression ratio in a 3D graph, shown in Figure 13.24, based on the properties of a proprietary LT-membrane.

The 3D compression plot shows that at zero compression ratio, (i.e., Pin = Pout = 10 bara), the energy requirement follows a linear relationship as a function of the current density according to Ohm's law. This energy is converted entirely into heat. Additional work is done when the hydrogen is compressed from 10 up to 700 bara, following Nernst law of thermodynamics. At elevated cathode pressures, the energy requirement shows a vertical tangent at low current densities, because here the back diffusion almost fully compensates the Faradaic forward compression rate. Note that these inter-relationships hold for any given membrane material and thickness, but the specific energy requirement can be optimized by tuning the membrane accordingly.

A similar 3D compression plot has been calculated for a HT-membrane in Figure 13.25, which suffers less hydrogen gas back diffusion than the LT-membrane version discussed previously. As a result, lower current density operation provides lower compression energy requirement, and hence this particular membrane is most suitable for high compression ratios, proven up to 100 MPa (see Figure 13.16).

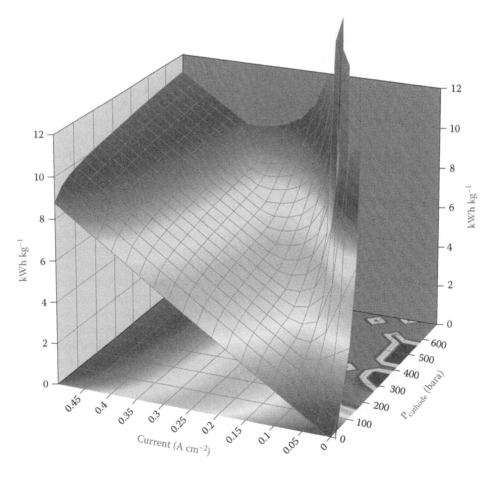

FIGURE 13.24
Typical 3D graph showing calculated hydrogen compression energy as a function of the current density and the cathode pressure, based on an input pressure of 1 MPa and the properties of a proprietary low temperature membrane.

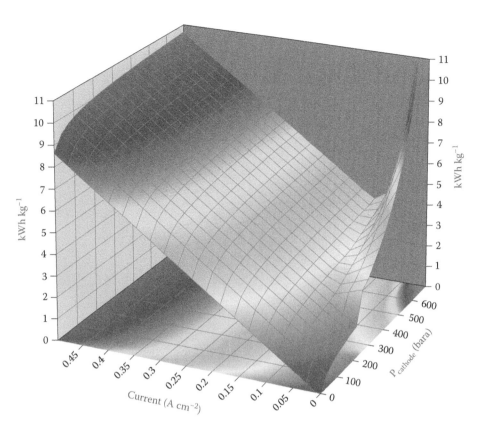

FIGURE 13.25
Typical 3D graph showing calculated hydrogen compression energy as a function of the current density and the cathode pressure, based on an input pressure of 1 MPa and the properties of a proprietary high temperature membrane.

In principle, the thermodynamic work energy can be regained using the EHC in reverse operation; or more practically said, the Nernst voltage is used to power the proton conductivity, drive a current density and deliver hydrogen gas at lower pressure together with the residual electrical energy. The Nernst voltage itself typically measures only several tens of mV, for example, 51.4 mV for 1–40 MPa at 50°C. In compression mode, 50 mV of driving voltage typically delivered a current density of 0.33 A cm^{-2} (assuming a cell-to-cell resistance of 0.15 Ohm cm^2). Therefore, in decompression mode similar or lower current densities would be conceivable, except when additional driving voltage is applied to assist decompression.

Two complications are encountered in self-sustaining decompression mode that determine the upper and lower limit current density, respectively. Firstly, if we consume the full driving voltage, we do not gain any residual energy from the thermodynamic compression energy and it would be easier to open a separate valve to deflate the storage tank pressure. Secondly, the hydrogen gas back diffusion continues to be present in case of pressure difference across the EHC membranes and thus we need to set a sufficient decompression current density in order to minimize the relative parasitic drain on the total

stored energy. In the situation of the LT-membrane, the calculated 3D plot displayed in Figure 13.26 estimated no more than 0.45 kWh kg^{-1} energy could be regained during decompression. In the situation of the HT-membrane, the calculated 3D plot shown in Figure 13.27 shows more than 1 kWh kg^{-1} could be recovered. In both situations the recoverable energy quantity appeared fairly independent on the cathode pressure (differential), due of the shape of the Nernst potential curve. These examples clearly show the impact of the back diffusion on the required/recoverable compression energy and underline the one major reason why commercial fuel cell membranes often prove unsuitable for EHC applications.

A final comment to consider is that the cell-to-cell resistance is assumed only a function of temperature and pressure in this projection, based on the characterization test results during compression of hydrogen. However, during decompression, the direction of the proton conductivity is reversed and thus the drag of water molecules is in the same direction as the water permeation driven by the physical pressure difference. The shift in the balanced water management could result in drying out of the cathode electrode and hence increase of the total cell-to-cell resistance, particularly when dry hydrogen gas is supplied from a high pressure storage vessel

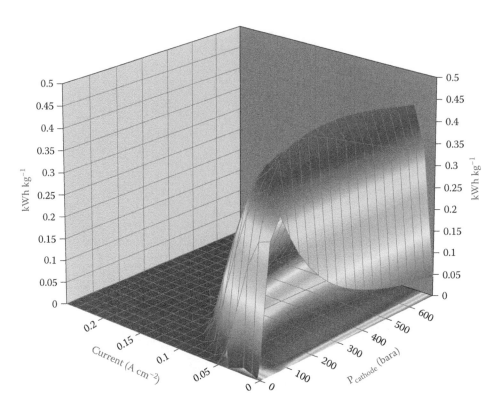

FIGURE 13.26
Typical 3D graph showing calculated hydrogen decompression energy that could be regained as a function of the current density and the cathode pressure, based on an input pressure of 1 MPa and the properties of a proprietary low temperature membrane.

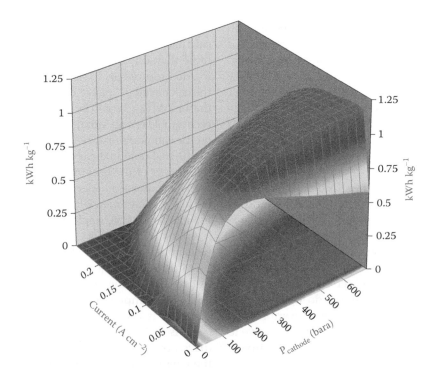

FIGURE 13.27
Typical 3D graph showing calculated hydrogen decompression energy that could be regained as a function of the current density and the cathode pressure, based on an input pressure of 1 MPa and the properties of a proprietary high temperature membrane.

with a dew point temperature of −40°C. Such practical complications should be considered and risk being overlooked when people see how easy it is to reverse the hydrogen flow direction with the reversal of the current.

13.3.8 Implicit Hydrogen Purification

Hydrogen gas supplied to the EHC is implicitly purified, since other gaseous impurities cannot conform to the working principle of subsequent oxidation, protonic diffusion, and reduction at low voltages. The possibility exists that these impurities could build up on the anode input, if the EHC system uses a dead-end flow configuration. Intermittent purging may prove necessary to maintain the overall performance. On the positive side, this purification principle could also be applied successfully to separation of gas mixtures.

Commonly, the membrane separation capability of polymer systems is governed mostly by the gas diffusion coefficient; the gas solubility characteristics play a more important role in the perm-selectivity properties for more polar gases and at higher pressures (Robeson 1991). Different solubility–selectivity relationships have been measured for perfluorinated polymers and compared to hydrocarbon–aromatic polymers (Robeson 2008).

Particularly the separation of H_2 from CO_2 is surprisingly difficult in micro-porous membranes (Koros and Chern 1987). Although H_2 has a high diffusivity, because of its low condensability, it has a very low solubility in membranes. A more soluble, lower diffusivity gas such as CO_2 may have a steady-state permeability comparable to that of H_2. Rowe et al. (2010) predicted the H_2/CO_2 separation selectivity at fixed permeability to increase with increasing temperature, as a result of the strong solubility decrease of the more condensable gas (CO_2) with increasing temperature, which led to an overall enhancement of the selectivity toward H_2, despite the decreasing diffusivity selectivity with increasing temperature.

Typically, solubility differences dominate the predicted behavior for gas pairs with large differences in condensability such as H_2/CO_2, while diffusivity differences control behavior when the gas molecules differ greatly in size such as H_2/N_2.

PBIs are a class of heterocyclic polymers which possess extremely high thermal stability, excellent chemical, and moisture resistance (Hosseini et al. 2010), with a relatively high glass transition temperature of ∼418°C, high H_2/N_2, and H_2/CO_2 gas selectivity, but extremely low gas permeability due to a high packing density of the polymer matrix. Polymer membranes based on PBI-based materials have been reported to outperform the "Robeson upper bound" prediction (Robeson 1991, Dal-Cin et al. 2008) and are promising materials for high

temperature H_2 separation from syngas (Li et al. 2014). As the polymer molecular spacing becomes tighter, the permeability decreases due to decreasing diffusion coefficients, but the separation characteristics are enhanced. However, the low hydrogen gas permeability of PBI dictates only ultrathin selective top layers have commercially attractive H_2-fluxes.

Much to our advantage, Robeson upper bound rules do not apply in case of electrochemical hydrogen purification (EHP), because protons do not classify as gas molecules and have an inherently different migration mechanism. We pump the atomic hydrogen forward against its activity gradient to the high pressure cathode, taking advantage of the low hydrogen gas permeability of our selected proton conducting membrane. Any back diffusion of hydrogen gas is still regarded a parasitic energy loss, but it may help drive back other contaminant gas species diffusing across the membrane. Our proprietary HT-membrane has the advantage that the solubility of gas species (e.g., CO_2) is reduced at elevated temperatures, making these inherently more suitable for H_2/CO_2 separations than LT-membranes.

Note that no sweep gas needs to be used here on the permeate side in order to reduce the activity of hydrogen gas on the membrane surface and maintain high permeability performance, as is the case with microporous membranes. Here, the permeate side contains practically pure hydrogen and the Nernst voltage reflects the activity gradient, which is easily overcompensated by the driving voltage. Only if the partial hydrogen pressure becomes very low in the feed pressure, does the Nernst voltage rise to significant values, for example, EHP from 0.001 mbar to 100 bar takes 210 mV Nernst voltage at 150°C. There are some realistic applications where such purification of low pressure hydrogen (isotopes) could be economically feasible and fortunately, this Nernst energy toll is to be paid only once the hydrogen molecule is actually recovered.

The most likely application for EHP is the purification of hydrogen from reformate gas streams fuelled by natural gas, oil, coal, or biomass sources. HyET has tested multiple gas compositions for compatibility with the catalyst and membrane system and here we show the purification results from a premixed gas mixture after typical Water Gas Shift reaction. Figures 13.28 shows a schematic representation of the EHP process together with a table listing typical gas compositions analysis results taken from the source, retentate, and permeate gas samples.

Clearly, purification was achieved successfully delivering almost pure hydrogen with 188 ppm of CO_2, 14 ppm CO, and no detectable CH_4 in the permeate sample. The concentration reduction was estimated to be a factor 1000× for CO_2, factor 5000× for CO and "infinite" for CH_4. This perm-selectivity is orders

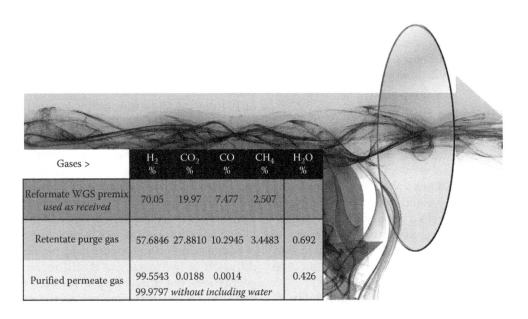

Gases >	H_2 %	CO_2 %	CO %	CH_4 %	H_2O %
Reformate WGS premix *used as received*	70.05	19.97	7.477	2.507	
Retentate purge gas	57.6846	27.8810	10.2945	3.4483	0.692
Purified permeate gas	99.5543 99.9797 *without including water*	0.0188	0.0014		0.426

FIGURE 13.28
Schematic representation of the Electrochemical Hydrogen Purification (EHP) process together with gas composition analysis results measured on a typical reformate stream after Water Gas Shift reaction. (1% = 10,000 ppm.)

of magnitude above the upper bound estimations and matched only by metallic palladium membranes (Godula-Jopek et al. 2012).

Note a significant water content is measured in the retentate and the permeate stream, attributed to the presumed equilibrium vapor pressure at the surface of the proton conducting membrane. Water content was also observed in the reference measurement, where only pure hydrogen was supplied to the system operated in identical manner. As discussed earlier, the water management is controlled through the EHC operating conditions and less water is carried through to the permeate side when a high pressure differential is maintained, that is, purifying and compressing simultaneously, in line with the graphs shown in Figure 13.14.

13.4 Summary and Outlook

The physical principles behind EHC have been discussed elaborately, providing insight in the key abilities and limitations in displacing hydrogen against its activity gradient.

Even at room temperature, the overall process is fast and reversible thanks to the superior hydrogen reaction kinetics on platinum group metal catalyst and the fast diffusion coefficients of hydrogen gas. The EHC principle and associated technology have been studied for some time already. Fortunately, hydrogen gas is gaining attention as a carrier for storing renewable energy. This application calls for an efficient pump featuring

instant response and remote control through internet. Now is the time that EHC technology may have a window of opportunity to help fulfil a potential new market demand.

Contrary to general expectations, the EHC technology is most competitive when operated at low to medium current densities (e.g., between 0.1 and 0.5 A cm^{-2}), whilst having the option to operate high current densities ($\gg$1 A cm^{-2}) at the expense of energy requirements due to Joule heating. Compression itself is easy and consumes relatively little energy, which in theory even could be recovered as Nernst voltage. Compared to mechanical compressors, the EHC has the potential to reduce energy consumption in kWh kg^{-1} to some degree, based on isothermal principles and no moving parts that induce friction. However, its flexibility is likely to deliver the major surplus value of the EHC, as mechanical compressors suffer from input pressures limitations, fixed compression ratios, and a trade-off between compression and throughput. Ironically, we do not wish to rule out that the "best" solution may well be a combination of mechanical and EHC.

The Achilles' heel of the EHC technology is the complex interactions within the electrochemical system, where water management plays a critical role in maintaining optimum conditions for hydrogen oxidation–reduction on the electrodes and hydrated proton conductivity in the membrane phase. Most hydrogen sources, being electrolyzers or reformers, generate a gas stream that is oversaturated with water and can be matched up perfectly with EHC moisture requirements that depend on the pressure difference and the operating temperature. The hydrogen gas exiting the EHC at high pressure contains some water vapor due to evaporation from the membrane, and less at higher pressure differentials. Still posttreatment may be required prior to storage and dispensing, depending on the specific protocol standard (Schneider 2012). The hydrogen quality from the EHC is bound to meet the stringent fuel cell standards.

Purification of the supplied hydrogen is implicit to the EHC working principle. Only hydrogen gas can be (dis-)associated and transported against its partial pressure gradient. This is contrary to porous membranes, where the different permeability resistance governs the separation of gas mixtures diffusion from high to low (partial) pressures. The solid membranes of the EHC are designed to hold back high pressure hydrogen gas, so are well capable to resist permeation of larger gas molecules. Perm-selectivity ratios between H_2/CO and H_2/CO_2 exceeding factor 1000 have been observed, making EHC technology also suitable for inherently difficult purification steps.

New membrane materials need to be developed to further optimize overall strength, proton conductivity and reduce the hydrogen gas permeability. Although the EHC technology benefits strongly from progress in fuel cells and electrolyzers, this application is less demanding on chemical stability as to the reducing hydrogen environment and the lower voltages applied during operation (<0.5 V per cell) avoid generating detrimental oxygen and hydroxyl radicals. Low-cost membrane materials with matching catalyst systems would help enable economically feasibility and allow using large active areas to optimize rated current density.

The key advantages of the EHC technology are its scalability, the cost-down potential of the electrochemical cell stacks and the instant, remote control-ability of the device. Eventually, its superior, maintenance-free lifetime will become a convincing argument to prefer EHC over incumbent solutions, once the EHC technology has proven itself in the field toward a wide range of applications.

Acknowledgments

This work was prepared as book chapter by invitation of Dr. D. Bessarabov with valuable contributions from the author's colleagues Dr. M.J.J. Mulder, Dr. A. Bos, J. Konink, M. Bosch, D. Semerel, L. Raymakers, M. Koeman, W. Dalhuijsen, E. Milacic, and we thank Dr. G.A. Eisman for discussion and comments on the manuscript.

References

Abdulla, A., Laney, K., Padilla, M. et al. 2011. Efficiency of hydrogen recovery from reformate with a polymer electrolyte hydrogen pump. *AIChE Journal* 57(7):1767–1779.

Bautista-Rodríguez, C.M., Rosas-Paleta, A., Rivera-Márquez, J.A. et al. 2009. Study of electrical resistance on the surface of Nafion 115® membrane used as electrolyte in PEMFC technology, Part I: Statistical inference and Part II: Surface response methodology. *International Journal of Electrochemical Science* 4:43–59 and 4:60–76.

Berg, P., Promislow, K., St. Pierre, J. et al. 2004. Water management in PEM fuel cells. *Journal of the Electrochemical Society* 151:A341–A353.

Bessarabov, D.G. 1998. Electrochemically-aided membrane separation and catalytic processes. *Membrane Technology: An International Newsletter* 93:7–11.

Blease, G. 2013. What happened to biofuels? *The Economist—Technology Quarterly* Q3:2013. http://www.economist.com/news/technology-quarterly/21584452 (accessed on December 2014).

Bockris, J.O'M. and Reddy, A.K.N. 1977. *Modern Electrochemistry.* Plenum, New York.

Brunetti, A., Fontananova, E., Donnadio, A. et al. 2012. New approach for the evaluation of the membranes transport properties for polymer electrolyte membrane fuel cells. *Journal of Power Sources* 205:222–230.

Buelte, S.J., Lewis, D.J., and Eisman, G.A. 2011. Effects of phosphoric acid concentration on platinum catalyst and phosphoric acid hydrogen pump performance. *ECS Transactions* 41:1955–1966.

Casati, C., Longhi, P., Zanderighi, L. et al. 2008. Some fundamental aspects in electrochemical hydrogen purification/compression. *Journal of Power Sources* 180:103–113.

Chen, N., Blowers, P., and Masel, R.I. 1999. Formation of hydronium and water–hydronium complexes during coadsorption of hydrogen and water on (2 × 1) Pt (110). *Surface Science* 419:150–157.

Cheng, X., Zhang, J., Tang, Y. et al. 2007. Hydrogen crossover in high-temperature PEM fuel cells. *Journal of Power Sources* 167:25–31.

Choi, P. and Datta, R. 2003. An explanation of Schroeder's Paradox. *Journal of the Electrochemical Society* 150:E601–E607.

Cownden, R., Nahon, M., and Rosen, M.A. 2001. Modelling and analysis of a solid polymer fuel cell system for transportation applications. *Hydrogen Energy* 26:615–623.

Dal-Cin, M.M., Kumar, A., and Layton, L. 2008. Revisiting the experimental and theoretical upper bounds of light pure gas selectivity–permeability for polymeric membranes. *Journal of Membrane Science* 323:299–308.

de Bruijn, F.A., Dam, V.A.T., and Janssen, G.J.M. 2008. Review: Durability and degradation issues of PEM fuel cell components. *Fuel Cells* 8:3–22.

Department Of Energy (DOE). 2014. Energy Department Invests $20 Million to Advance Hydrogen Production and Delivery Technologies. http://energy.gov/eere/articles/energy-department-invests-20-million-advance-hydrogen-production-and-delivery (accessed on June 2015).

Don Quichote project (JTI FCH-JU) 2012–2017. Demonstration of new qualitative innovative concept of hydrogen out of wind turbine electricity. www.don-quichote.eu.

Duan, Q., Wang, H., and Benziger, J. 2012. Transport of liquid water through Nafion membranes. *Journal of Membrane Science* 392–393:88–94.

Eikerling, M. and Berg, P. 2011. Poroelectroelastic theory of water sorption and welling in polymer electrolyte membranes. *Soft Matter* 7:5976–5990.

Eisman, G.A. 1990. The application of Dow Chemical's perfluorinated membranes in proton-exchange membrane fuel cells. *Journal of Power Sources* 29:389–398.

Eurostat. 2005–2009. Sustainable development—Consumption and production Electricity consumption per household, by country, 2005–2009 Figure 21. http://epp.eurostat.ec.europa.eu (accessed on December 2014).

Evers, A.A. 2010. *The Hydrogen Society: More Than Just a Vision.* Oberkraemer, Marwitz, Germany: Hydrogeit Verlag.

Gardner, C.L. and Ternan, M. 2007. Electrochemical separation of hydrogen from reformate using PEM fuel cell technology. *Journal of Power Sources* 171:835–841.

Garsany, Y., Baturina, O., Swider-Lyons, K.E. et al. 2010. Experimental methods for quantifying the activity of platinum electrocatalysts for the oxygen reduction reaction. *Analytical Chemistry* 82:6321–6328.

Ge, S., Li, X., Yi, B. et al. 2005. Adsorption, desorption, and transport of water in polymer electrolyte membranes for fuel cells. *Journal of the Electrochemical Society* 152:A1149–A1157.

Godula-Jopek, A., Jehle, W., and Wellnitz, J. 2012. *Hydrogen Storage Technologies: New Materials, Transport, and Infrastructure.* Weinheim, Germany: Wiley-VCH.

Granite, E.J. and O'Brien, T. 2005. Review of novel methods for carbon dioxide separation from flue and fuel gases. *Fuel Processing Technology* 86:1423–1434.

Grasman, S.E. (Ed.) 2013. Hydrogen and Electricity: Parallels, Interactions, and Convergence. In: *Hydrogen Energy and Vehicle Systems.* Boca Raton, FL: CRC Press, Taylor & Francis Group, p. 1–22.

Grigoriev, S.A., Shtatniy, I.G., Millet, P. et al. 2011. Description and characterization of an electrochemical hydrogen compressor/concentrator based on solid polymer electrolyte technology. *International Journal of Hydrogen Energy* 36:4148–4155.

Grot, W. 2011. *Fluorinated Ionomers.* Norwich, NY: William Andrew.

Haddad, A., Bouyekhf, R., and El Moudni, A. 2008. Dynamic modelling and water management in proton exchange membrane fuel cell. *International Journal of Hydrogen Energy* 33:6239–6252.

Haile, S.M., Chisholm, C.R.I., Sasaki, K. et al. 2006. Solid acid proton conductors: From laboratory curiosities to fuel cell electrolytes. *Faraday Discussions* 134:17–39.

Hinatsu, J.T., Mizuhata, M., and Hiroyasu Takenaka, J. 1994. Water uptake of perfluorosulfonic acid membranes from liquid water and water vapor. *Journal of the Electrochemical Society* 141:1493–1498.

Hosseini, S.S., Peng, N., and Chung, T.-S. 2010. Gas separation membranes developed through integration of polymer blending and dual-layer hollow fiber spinning process for hydrogen and natural gas enrichments. *Journal of Membrane Science* 349:156–166.

Ibeh, B., Gardner, C., and Ternan, M. 2007. Separation of hydrogen from a hydrogen/methane mixture using a PEM fuel cell. *International Journal of Hydrogen Energy* 32:908–914.

International Energy Agency (IEA). 2011. *Solar Energy Perspectives.* Paris, France: IEA/OECD.

International Renewable Energy Agency (IRENA). 2012. *Renewable Energy Technologies: Cost Analysis Series. Solar Photovoltaics,* Volume 1: Power Sector. Issue 4/5. Abu Dhabi, UAE: International Renewable Energy Agency (IRENA), pp. 8–9.

ITM Power Plc. 2014. ITM Power Plc is a public limited company registered in England and Wales (company number 5059407). http://www.itm-power.com (accessed on May 2015).

Iwahara, H., Asakurab, Y., Katahirac, K. et al. 2011. Prospect of hydrogen technology using proton-conducting ceramics. *Solid State Ionics* 16:299–310.

Jinnouchi, R., Yamada, H., and Morimoto, Y. 2004. Measurement of electro-osmotic drag coefficient of Nafion using a concentration cell. *14th International Conference on the Properties of Water and Steam in Kyoto.* http://www.iapws.jp/Proceedings/…/403Jinnouchi.pdf (accessed on December 2014).

Joint Technology Initiative - Fuel Cells and Hydrogen - Joint Undertaking (JTI-FCH-JU). http://www.fch.europa.eu/page/governance (accessed on June 2015).

Kizhakevariam, N. and Stuve, E.M. 1992. Coadsorption of water and hydrogen on Pt(100): Formation of adsorbed hydronium ions. *Surface Science* 275:223–236.

Kocha, S.S., Yang, J.D., and Yi, J.S. 2006. Characterization of gas crossover and its implications in PEM fuel cells. *AIChE Journal* 52:1916–1925.

Koros, W.J. and Chern, R.T. 1987. Separation of gaseous mixtures using polymer membranes. In: R.W. Rousseau (Ed.), *Handbook of Separation Process Technology.* New York: John Wiley & Sons, p. 862.

Kurita, N., Fukatsy, N., Otsuka, H. et al. 2002. Measurements of hydrogen permeation through fused silica and borosilicate glass by electrochemical pumping using oxide protonic conductor. *Solid State Ionics* 146:101–111.

Kusoglu, A., Calabrese, M., and Weber, A.Z. 2014. Effect of mechanical compression on chemical degradation of Nafion membranes. Role of mechanical factors in controlling the structure–function relationship of PFSA ionomers. *ECS Electrochemistry Letters* 3:F33–F36.

Kusoglu, A., Karlsson, A.M., Santare, M.H. et al. 2007. Mechanical behavior of fuel cell membranes under humidity cycles and effect of swelling anisotropy on the fatigue stresses. *Journal of Power Sources* 170:345–358.

Kusoglu, A., Kienitz, B.L., and Weber, A.Z. 2011. Understanding the effect of compression and constraints on water uptake of fuel-cell membranes. *Journal of the Electrochemical Society* 158:B1504–B1514.

Kusoglu, A., Savagatrup, S., Clark K.T. et al. 2012. Role of mechanical factors in controlling the structure–function relationship of PFSA ionomers. *Macromolecules* 45:7467–7476.

Kusoglu, A., Tang, Y.J., Lugo, et al. 2010. Constitutive response and mechanical properties of PFSA membranes in liquid water. *Journal of Power Sources* 195:48–92.

Lackey, D., Schott, J., Sass, J.K. et al. 1991. Surface-science simulation study of the electrochemical charge-transfer reaction $(H)_{ad} + (H_2O)_{ad} \rightarrow (H_3O^+)_{ad} + e^-$ metal on Pt(111) and Cu(110). *Chemical Physics Letters* 184:277–281.

Lee, H.K., Choi, H.Y., Choi, K.H. et al. 2004. Hydrogen separation using electrochemical method. *Journal of Power Sources* 132:92–98.

Li, H., Lee, K., and Zhang, J. 2008. Electrocatalytic H_2 oxidation reaction. In: J. Zhang (Ed.), *PEM Fuel cell Electrocatalysts and Catalyst Layers—Fundamentals and Applications.* London, U.K.: Springer, pp. 135–159.

Li, X., Singh, R.P., Dudeck, K.W. et al. 2014. Influence of polybenzimidazole main chain structure on H_2/CO_2 separation at elevated temperatures. *Journal of Membrane Science* 461:59–68.

Lipp, L. 2014. Electrochemical hydrogen compressor. DOE presentation. FuelCell Energy, Inc., June 17, 2014. http://www.hydrogen.energy.gov/pdfs/review14/pd048_lipp_2014_o.pdf (accessed on September 2014).

Lipp, L. and Patel, P. 2013. Electrochemical hydrogen compressor. FY 2013 Annual Progress Report III.6. FuelCell Energy, Inc. http://www.hydrogen.energy.gov/pdfs/progress13/iii_6_lipp_2013.pdf (accessed on September 2014).

Mackenzie, B.S. and Bloomfield, D.P. 2006. Electrochemical hydrogen compressor. DOE Grant DE-FG02-05ER84220 to Analytic Power Corp. Final report. http://www.osti.gov/scitech/servlets/purl/883089 (accessed on September 2014).

Majsztrik, P.W., Satterfield, M.B., Bocarly, A. et al. 2007. Water sorption, desorption and transport in Nafion membranes. *Journal of Membrane Science* 301:93–106.

McElroy, J. 1980. Solid polymer electrolyte fuel cell technology program, NASA contract #Nas 9-15286. Final report. General Electric Co. http://ntrs.nasa.gov/archive/nasa/casi.ntrs.nasa.gov/19900011150.pdf (accessed on December 2014).

Misran, E., Hassan, N.S.M., Daud, W.R.W. et al 2013. Water transport characteristics of a PEM fuel cell at various operating pressures and temperatures. *International Journal of Hydrogen Energy* 38:9401–9408.

Moton, J., James, B., and Colella, W. 2013. Design for manufacturing and assembly (DFMA) analysis of electrochemica hydrogen compression (EHC) systems. *Proceedings of EFC2013: 5th European Fuel Cell Technology and Applications Conference,* EFC13-179, Rome, Italy, pp. 297–298.

Motupally, S., Becker, A.J., and Weidner, J.W. 2000. Diffusion of water in Nafion 115 membranes. *Journal of the Electrochemical Society* 147:3171–3177.

Mulyazmi, Daud, W.R.W., Majlan, E.H., and Rosli, M.I. 2013. Water balance for the design of a PEM fuel cell system. *International Journal of Hydrogen Energy* 38:9409–9420.

Ogumi, Z., Kuroe, T., and Takehara, Z. 1985. Gas permeation in SPE method II. Oxygen and hydrogen permeation through Nafion. *Journal of the Electrochemical Society* 132:2601–2605.

Olender, H., McBreen, J., O'Grady, W.E. et al. 1982. Design of a cell for electrode kinetic investigations of fuel cell reactions. *Journal of Electrochemical Society* 129:135–137.

Onda, K., Araki, T., Ichihara, K. et al. 2009. Treatment of low concentration hydrogen by electrochemical pump or proton exchange membrane fuel cell. *Journal of Power Sources* 188:1–7.

Onda, K., Ichihara, K., Nagahama, M. et al. 2007. Separation and compression characteristics of hydrogen by use of proton exchange membrane. *Journal of Power Sources* 164:1–8.

Ottinger, R.L. 2009. Biofuels: Potentials, problems & solutions. Pace Law Faculty Publications. http://digitalcommons.pace.edu (accessed on December 2014).

Owejan, J.P. 2014. Transport resistance in polymer electrolyte fuel cells. PhD dissertation. University of Tennessee. http://trace.tennessee.edu/utk_graddiss/2721 (accessed on December 2014).

Pasierb, P. and Rekas, M. 2011. High-temperature electrochemical hydrogen pumps and separators. *International Journal of Electrochemistry* 2011(9): 1–10.

Perry, K., Eisman, G.A., and Benicewicz, B.C. 2008. Electrochemical hydrogen pumping using a high-temperature polybenzimidazole (PBI) membrane. *Journal of Power Sources* 177:478–484.

PHAEDRUS project (JTI FCH-JU). 2012–2015. High Pressure Hydrogen All Electrochemical Decentralized RefUeling Station. www.phaedrus-project.eu (accessed on June 2015).

PurifHy project (TKI - Technology Knowledge Initiative). 2014–2015. Hydrogen extraction from green methane mixtures enabling Power-2-Gas and subsequent injection in pipeline network.

Raine, D., Williams, R., Strøm et al. 2009. Roads2HyCom: Deliverable D2.4—Analysis of the current hydrogen cost structure. http://www.Roads2Hy.com (accessed on September 14, 2014).

Robeson, L.M. 1991. Correlation of separation factor versus permeability for polymeric membranes. *Journal of Membrane Science* 62:165–185.

Robeson, L.M. 2008. The upper bound revisited. *Journal of Membrane Science* 320:390–400.

Rohland, B., Eberle, K., Ströbel, R. et al. 1998. Electrochemical hydrogen compressor. *Electrochimica Acta* 43:3841–3846.

Rowe, B.W., Robeson, L.M., Freeman B.D. et al. 2010. Influence of temperature on the upper bound: Theoretical considerations and comparison with experimental results. *Journal of Membrane Science* 360:58–69.

Sass, J.K., Lackey, D., Schott, J. et al. 1991. Electrochemical double layer simulations by halogen, alkali and hydrogen coadsorption with water on metal surfaces. *Surface Science Letters* 247:A239–A247.

Schlapbach, L. and Züttel, A. 2001. Review article hydrogen-storage materials for mobile applications. *Nature* 414:353–358.

Schneider, J. 2012. SAE J2601—Worldwide. Hydrogen fueling protocol: Status, standardization & implementation. SAE Fuel Cell Interface Group Chair. www.energy.ca.gov (accessed on November 2014).

Sedlak, J.M., Austin, J.F., and LaConti, A.B. 1981. Hydrogen recovery and purification using the solid polymer electrolyte electrolysis cell. *International Journal of Hydrogen Energy* 6:45–51.

Springer, T.E., Zawodzinski, T.A., and Gottesfeld, S. 1991. Polymer electrolyte fuel cell model. *Journal of the Electrochemical Society* 138:2334–2342.

Stern, S.A. 1968. The "Barrer" permeability unit. *Journal of Polymer Science* Part A-2, 6:1933–1934.

Stolten, D., Grube, T., and Schiebahn, S. 2013. Hydrogen and fuel cells for transportation becoming a major trend for the future? In: D. Stolten and V. Scherer (Eds.), *Transition to Renewable Energy Systems*. Weinheim, Germany: Wiley-VCH, pp. 195–214.

Stroebel, R., Oszcipok, M., Fasil, M. et al. 2002. The compression of hydrogen in an electrochemical cell based on a PE fuel cell design. *Journal of Power Sources* 105:208–215.

Takaaki, S., Matsumoto, H., Kudo, T. et al. 2008. High performance of electroless-plated platinum electrode for electrochemical hydrogen pumps using strontium-zirconate-based proton conductors. *Electrochimica Acta* 53:8172–8177.

Talaganis, B.A., Esquivel, M.R., Meyer, G. et al. 2009. A two-stage hydrogen compressor based on (La,Ce,Nd,Pr)Ni$_5$ intermetallics obtained by low energy mechanical alloying—Low temperature annealing treatment. *International Journal of Hydrogen Energy* 34:2062–2068.

Tanaka, M., Y. Asakura, Uda, T. et al. 2006. Hydrogen enrichment by means of electrochemical hydrogen pump using proton-conducting ceramics for a tritium stack monitor. *Fusion Engineering and Design* 81:1371–1377.

Tanaka, M., Sugiyama, T., Ohshima, T. et al. 2011. Extraction of hydrogen and tritium using high-temperature proton conductor for tritium monitoring. *Fusion Science and Technology* 60:1391–1394.

The Freedonia Group. 2016. World hydrogen report, 331 pages—SKU: FG5292199. http://www.marketresearch.com/Freedonia-Group-Inc-v1247/Hydrogen-8274798/ (accessed on December 2014).

Thomassen, M., Sheridan, E., and Kvello, J. 2010. Electrochemical hydrogen separation and compression using polybenzimidazole (PBI) fuel cell technology. *Journal of Natural Gas Science and Engineering* 2:229–234.

Tsou, Y., Kimble, M.C., and White, R.E. 1992. Hydrogen diffusion, solubility, and water uptake in Dow's short-side chain perfluorocarbon membranes. *Journal of the Electrochemical Society* 139:1913–1917.

Wagner, F.T. and Moylan, T.E. 1987. Identification of surface hydronium: Coadsorption of hydrogen fluoride and water on platinum (111). *Surface Science* 182:125–149.

Wagner, F.T. and Moylan, T.E. 1988. Generation of surface hydronium from water and hydrogen coadsorbed on Pt(111). *Surface Science* 206:187–202.

Wilson, M.S. and Gottesfeld, S. 1992. This-film catalyst layers for polymer electrolyte fuel cell electrodes. *Journal of Applied Electrochemistry* 22:1–7.

Xing, L., Mamlouk, M., and Scott, K. 2013. A two dimensional agglomerate model for a proton exchange membrane fuel cell. *Energy* 61:196–210.

Yamaguchi, T., Takisawa, M., Kiwa, T. et al. 2008. Analysis of response mechanism of a proton-pumping gate FET hydrogen gas sensor in air. *Sensors and Actuators B—Chemical* 133:538–542.

Yeo, R.S. and McBreen, J. 1979. Transport properties of Nafion membranes in electrochemically regenerative hydrogen/halogen cells. *Journal of the Electrochemical Society* 126:1682–1687.

Yi, J.S. and Nguyen, T.V. 1998. An along-the-channel model for proton exchange membrane fuel cells. *Journal of the Electrochemical Society* 145:1149–1159.

Yoon, W. and Huang, X. 2010, Acceleration of chemical degradation of perfluorosulfonic acid ionomer membrane by mechanical stress: Experimental evidence. *ECS Transactions* 33:907–911.

Yuan, X.-Z., Song, C., Wang, H. et al. 2010. *Electrochemical Impedance Spectroscopy in PEM Fuel Cells—Fundamentals and Applications*. London, U.K.: Springer.

Zalitis, C.M., Kramer, D., and Kucernak, A.R. 2013. Electrocatalytic performance of fuel cell reactions at low catalyst loading and high mass transport. *Physical Chemistry Chemical Physics* 15:4329–4340.

Zawodzinksi, T.A., Neeman, M., Sillerud, O., and Gottesfeld, S. 1991. Determination of water diffusion coefficients in perfluorosufonate ionomeric membranes. *Journal of Physical Chemistry* 95:6040–6044.

Zhang, X. 2010. Effect of fuel impurities on polymer electrolyte fuel cells. PhD dissertation. University of Connecticut. http://digitalcommons.uconn.edu/dissertations/AAI3429227 (accessed on November 2014).

14

Large-Scale Water Electrolysis for Power-to-Gas

Rob Harvey, Rami Abouatallah, and Joseph Cargnelli

CONTENTS

14.1 Introduction

Over a century ago, Alexander T. Stuart began to take an interest in hydrogen energy while studying chemistry and mineralogy at the University of Toronto. At the time, Niagara Falls' hydroelectric generating capacity was being utilized at only 30%–40%. The question was: How could such surplus capacity is converted to useable energy? The obvious answer was electrolysis of water. In 1948, father and son founded the Electrolyzer Company and it became a leading designer and manufacturer of electrolytic hydrogen and oxygen generation plants for markets around the world. By the 1990s, the company had built several hundred installations in over 80 countries and 5 continents. With its 2004 acquisition of the renamed company, Stuart Energy, Hydrogenics Corporation entered the electrolytic hydrogen generation market and today it has developed a megawatt-scale proton exchange membrane (PEM) electrolyzer stack technology that will be the building block platform for Power-to-Gas, a revolutionary approach to energy conversion and storage using hydrogen.

Around the world, the development of clean sources of power generation has increased rapidly over the past 30 years. In many jurisdictions today, such as Germany and California, over 20% of the electricity consumed comes from renewable energy sources, predominately wind and solar. The challenge with these renewable sources is that they are intermittent. Unlike a thermal power plant such as a gas-turbine-driven power plant, it is not possible to dispatch the wind or the sun and turn it on to match the demand profile of when electricity is needed. When renewable generation makes up a small portion of the generation grid mix, the task of integrating their output is relatively simple. However, when the proportion of renewable generation reaches a critical mass, there may be frequent periods during the day when demand is lower than production, and even consecutive days during the late spring and early fall where there is surplus power generation. Today, the default solution is simply to curtail, or waste the wind or solar generation. This is truly a waste of a resource. Although the electrical power grid allows for easy distribution of energy, the grid itself has no storage capacity. To store electricity, it usually has to be converted into other forms of energy. A variety of storage technologies currently exist such as electric energy storage (supercapacitors), potential energy storage (traditional pumped storage), mechanical energy storage (compressed air reservoirs, flywheels), electrochemical energy storage (batteries), and finally chemical energy storage (hydrogen, synthetic natural gas, methanol). The rated power, energy storage capacity, charge and discharge time and frequency, and capital and operation costs characteristics of these storage systems vary greatly.

The future potential for renewables is higher for electricity generation than it is in gas (biogas, landfill gas, etc.) and transport (biofuel) combined, but the addition of even more renewable generation capacity will result in even greater surpluses of energy—it is a case of diminishing returns. However, if we can better use the surplus renewable power to help achieve higher overall efficiencies and environmental goals across the entire energy spectrum, including gas and transportation

sectors, then the overall impact and environmental benefit would be significantly higher. Hydrogen as an energy carrier will make this possible.

Power-to-Gas is the term that describes the use of water electrolysis to convert surplus renewable electrical power generation into renewable hydrogen gas. Applications of Power-to-Gas include hydrogen fuel for fuel cell electric vehicles (FCEVs), as a substitute for natural gas in pipelines, as renewable natural gas or as a renewable hydrogen feedstock in the refining of conventional liquid fuels. A key element of the value proposition of Power-to-Gas is that it can also provide the power grid operator with grid ancillary services such as frequency regulation. The concept of Power-to-Gas has been developed in Europe and to date, 18 Power-to-Gas pilot projects representing all of the different applications have been launched with the majority of these projects being commissioned in Germany. Across the rest of Europe, there are another dozen projects and in July 2014, the first Power-to-Gas project in North America was announced (Figure 14.1).

The scale of the Power-to-Gas projects built to date range from 100 kW to 6 MW of capacity. For the majority of these projects, existing industrial electrolyzers have been deployed. Water electrolysis has long been used to supply hydrogen in industrial applications such

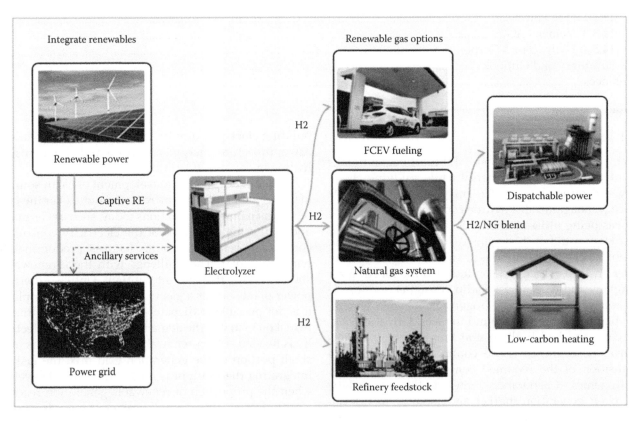

FIGURE 14.1
Power-to-Gas solution. (Courtesy of Hydrogenics Corporation, Ontario, Canada.)

as for laboratories, float glass manufacturing, electronics production, metallurgy, food processing, and high efficiency hydrogen cooled power plant generators. The most prolific industrial electrolysis technology is alkaline based utilizing an aqueous solution of potassium hydroxide (KOH) as the electrolyte. Alkaline electrolysis is proven to be a safe, cost-effective, and a very reliable solution; many installations have been in service for more than 25 years.

While alkaline water electrolysis is well suited for Power-to-Gas applications for smaller-scale deployments, there are some distinctive requirements for larger-scale commercial projects. For jurisdictions with a high renewable generation grid mix, the scale of energy storage required is not measured in MWh but in GWh. Typical Power-to-Gas projects will likely be anywhere from 40 MW to over 100 MW in size. The amount of space needed for a 100 MW alkaline electrolyzer plant is very large and may be impractical for many potential sites. Other requirements include partial load range, rapid response, scalability, and a low capital and operating cost. A less mature alternative electrolyzer technology that has the potential to deliver much higher power densities and hence a smaller plant footprint and lower cost is PEM water electrolysis.

Today, four companies—Hydrogenics Corporation, ITM Power, Proton OnSite, and Siemens AG—are developing large-scale PEM electrolyzers for Power-to-Gas applications. Policy goals for cleaner energy systems in electricity, gas and transportation around the world will drive the demand for Power-to-Gas projects. There is a strong focus on improvements in electrolyzer performance and capital cost as this nascent Power-to-Gas market is developing. Hydrogenics Corporation is the first company to develop a megawatt-scale PEM electrolyzer stack; the first installation utilizing this megawatt-scale PEM technology will be commissioned in 2015 in Reitbrook, a suburb of Hamburg with E.ON SE, one of the world's largest gas and electric utilities. However, the full story of large-scale electrolysis and Power-to-Gas is only just beginning.

14.2 Water Electrolysis Market Overview and Growth

Hydrogen is the first element of the periodic table of elements. It is a colorless, odorless and tasteless gas. The global annual production of hydrogen is approximately 56 million metric tons. About 96% of all hydrogen is obtained from fossil fuels, primarily by steam methane reforming (SMR). In this process, a reformer reacts steam at high temperature with natural gas to liberate hydrogen. The by-product of the reforming process is carbon dioxide. Petroleum refining and ammonia production account for over 80% of the demand for hydrogen. For these major uses, the refinery or fertilizer manufacturer typically produces their own hydrogen with a dedicated SMR at its facilities. Merchant gas companies also produce hydrogen using SMR and deliver it to their customers. However, in many parts of the world, the best alternative for industrial hydrogen applications is to install an electrolyzer plant to generate hydrogen on site due to the higher cost of delivered hydrogen or the purity of the hydrogen required. Electrolysis is used to meet about 4% of hydrogen demand today.

14.2.1 Industrial Electrolysis

Water electrolysis has been used for hydrogen production for industrial applications for more than 125 years. The typical size of electrolyzer plants for the prominent industrial applications is shown in Table 14.1.

The industrial water electrolyzer market is a mature and stable market.

14.2.2 Hydrogen Refueling Stations

Over the past decade there has been a steady increase in the number of hydrogen fueling stations built for the test fleets of automotive FCEVs and for fuel cell bus terminal fueling. In anticipation of the announced commercial launch of FCEVs in 2015 and 2016 by several different automobile manufacturers, many jurisdictions have hydrogen mobility initiatives in place to help fund and build out a critical mass of hydrogen fueling stations, particularly California, Germany, Japan, and South Korea. As of March 2013, there were 208 hydrogen refueling stations worldwide—80 in Europe, 76 in North America, and 49 in Asia (Carter and Wing 2013). It is estimated that over 40 of these hydrogen refueling stations are electrolysis based. These stations have an electrolyzer on site that eliminates the need to frequently transport hydrogen as many of these stations are located in dense urban settings. In addition to the

TABLE 14.1

Common Size of Water Electrolysis for Industrial Applications

Industrial Applications	Typical Plant Size (Nm³ h⁻¹)
Power Plant Generator Cooling	10–60
Laboratory	0.1–10
Float Glass Manufacturing	120–360
Metallurgy	60–480
Food Industry	120–480
Electronics Manufacturing	10–120

130 kg day^{-1} (Stuttgart, Germany)

65 kg day^{-1} (Santa Monica, CA)

780 kg day^{-1} (Hamburg, Germany)

FIGURE 14.2
Electrolysis-based hydrogen fueling stations.

electrolyzer, the hydrogen refueling station includes compression, hydrogen storage and dispensing at 700 bar pressure. Figure 14.2 shows two examples.

14.2.3 Power-to-Gas

The concept of Power-to-Gas was first established in 2009 by Dr. Michael Sterner who investigated a new concept for storing and balancing surplus renewable power for his dissertation at Kassel University in Germany.

He recognized that the full potential of renewable generation goes beyond the electrical power grid to impact the heat and transport energy systems as well and that hydrogen gas derived from electrolysis was the ideal energy carrier. At higher levels of renewable generation penetration, the need for very large GWh scale energy storage emerges. One of the inherent characteristics of the natural gas energy system is seasonal storage. The natural gas system often includes underground gas storage, which provides the capacity to store energy for months at a time during the off-season period when it is charged with gas to the winter heating season when it is discharged. The original concept of Power-to-Gas to produce renewable natural gas by processing electrolytic hydrogen through a methanation process has been

expanded to include other renewable hydrogen options as follows: hydrogen for refueling FCEVs, direct injection of hydrogen into the natural gas system, and hydrogen as a feedstock for the refining of conventional liquid fuels such as gasoline and diesel.

Simply stated, Power-to-Gas provides a service enabling the power grid operator to integrate large amounts of surplus renewable generation when it is not needed and provides options to store and transport this energy as renewable hydrogen. Power-to-Gas provides unparalleled scale of energy storage and enhances system flexibility. In its comprehensive review of Power-to-Gas, the SBC Energy Institute examined the technical and economic basis for hydrogen-based energy conversion and storage. "Exploiting hydrogen's versatility, chemical energy storage opens up alternatives to the usual approach to electricity storage in three ways: first, its volumetric energy density is superior to all other bulk-electricity-storage technologies; second, hydrogen-based technologies could reduce infrastructure investments required for integrating intermittent generators to the grid; and third, the versatility of hydrogen-based solutions is better than other energy storage technologies because it is not restricted to providing electricity back to the grid" (Decourt et al. 2014).

As mentioned earlier, electrolyzers have been utilized for hydrogen refueling stations for many years. Several of the Power-to-Gas pilot projects launched recently have the same application of hydrogen refueling. The primary difference between past electrolysis based fueling stations and recent Power-to-Gas fueling stations is in the way that the facility is intended to operate. Earlier electrolysis based hydrogen fueling stations where run as on-site hydrogen generators in the same fashion as an electrolyzer installed on an industrial plant. The electrolyzer would be run each day as long as needed to meet the hydrogen fueling demand. In contrast, the operation of a Power-to-Gas facility will help integrate renewables by absorbing surplus power (at low prices) and providing grid stabilization services to the electrical grid operator. The dynamic response of an electrolyzer can be measured in milliseconds. As a result, a Power-to-Gas facility can respond to dispatch signals from the electrical grid operator to provide frequency regulation services, which the grid operator uses to fine-tune the balance between generation supply and load demand throughout the day.

A second Power-to-Gas application is the direct injection of hydrogen into the natural gas grid. In a study by NREL, it was determined that "blending hydrogen into the existing natural gas pipeline network has been proposed as a means of increasing the output of renewable energy systems such as large wind farms. If implemented with relatively low concentrations, less than 5%–15% hydrogen by volume, this strategy of storing, and delivering renewable energy to markets appears to be viable without significantly increasing risks associated with utilization of the gas blend in end-use devices (such as household appliances), overall public safety, or the durability and integrity of the existing natural gas pipeline network" (Melaina et al. 2013). The study also identified that further research, testing, and modifications to pipeline monitoring practices are needed for large-scale deployment. Today, Germany has set an initial pipeline blend concentration of 2% hydrogen. It is important to recognize that even a very low concentration of 2% represents an enormous amount of storage—for example, in the province of Ontario, Canada, 2% of the total natural gas storage capacity is equivalent to 1600 GWh of energy. No other bulk energy storage solution including pumped hydro and compressed air energy storage (CAES) can match this scale of energy storage. Another major advantage of utilizing the natural gas pipeline network for storage is that the infrastructure already exists and there is no incremental capital cost associated with the storage of energy.

A third application of Power-to-Gas involves methanation, a process used to generate methane.

Instead of injecting hydrogen gas directly into the natural gas network, methanation involves combining hydrogen with carbon dioxide to produce synthetic natural gas or methane. The advantage of this approach is that there is no blending limitation of injecting synthetic methane into the natural gas network compared with hydrogen gas. There are two methods of methanation as follows: the first method, a well known process, is referred to as the Sabatier process where hydrogen is reacted with carbon dioxide at elevated temperatures in the presence of a catalyst, and the second method is biological, referred to as biogas methanation. In this second method, the carbon dioxide in biogas from anaerobic digestion is upgraded to biomethane by combining the carbon dioxide containing biogas with hydrogen gas in a reactor using special microbes.

The fourth application of Power-to-Gas involving electrolytic hydrogen from renewable sources for petroleum oil refining is attracting growing interest. Historically, refineries have been net producers of hydrogen because it is a by-product of the catalytic reforming process in crude oil refining. However, as more stringent sulfur oxide levels in fuel standards have been enacted and with increasing heavy oil production, refineries have had to source additional supplies of hydrogen, usually by SMR production. A game changer in this application would be the recognition of renewable electrolytic hydrogen under the various renewable fuel standards. Today, refiners can meet the renewable fuel standard by using biofuels such as ethanol or bio-diesel, but both of these feedstocks have some disadvantages: biofuels have a lower energy output than traditional fuels, the process to produce biofuels can have a hefty carbon emissions footprint; large quantities of water are required and valuable cropland to grow fuel crops could have an impact on the cost of food. Renewable electrolytic hydrogen could provide an attractive alternative for refiners.

The range of Power-to-Gas options for renewable hydrogen applications is indicative of its versatility as an energy carrier that can create a bridge between the electricity energy system and the gas heating and transportation energy systems. Today we are at the pilot project stage for Power-to-Gas and at the threshold of commercial-scale projects. There are many drivers that influence the rate of adoption of Power-to-Gas, principally the regulatory environment and how quickly large-scale electrolyzers move down the capital cost curve. While it is a challenging task to estimate the size of the Power-to-Gas market in the future, Navigant Research estimates it will be on the order of $500 million to $600 million market in the 2022–2023 timeframe (Dehamna 2013).

14.3 Power-to-Gas Demonstration Projects

In its 2013 study, DNV GL identified 30 Power-to-Gas demonstration projects in Europe (Grond et al. 2013). Together they cover all of the applications of Power-to-Gas and range in size from 6 MW to under 100 kW; eight of the projects are 1 MW or larger and Hydrogenics Corporation has supplied the electrolyzers for seven of the MW projects.

A brief overview of five prominent Power-to-Gas pilot projects is provided: the E.ON Gas Storage project, the Audi E-Gas project, the Energie Park Mainz project, the Thüga Group project, and the WindGas project. As well, the first North American project has recently been announced by Hydrogenics in Ontario, Canada.

14.3.1 E.ON Gas Storage (Falkenhagen, Germany)

E.ON Gas Storage inaugurated the first Power-to-Gas facility in Falkenhagen, Germany, to inject hydrogen gas directly into the natural gas grid in August 2013. It is a two megawatt facility connected to the local electrical distribution grid deploying six Hydrogenics' alkaline electrolyzer units producing a total of 360 Nm³ h⁻¹ of hydrogen gas. As of late 2014, the E.ON facility had injected more than two million kilowatt-hours of hydrogen gas into the natural gas distribution network. A partner on the project is SwissGas AG, which has an off-take contract for a portion of the renewable hydrogen produced. E.ON has also introduced a premium gas product for its residential customers called WindGas, which is marketed with 10% hydrogen content (Figure 14.3).

FIGURE 14.4
6 MW Power-to-Gas facility. (Courtesy of Audi.)

14.3.2 Audi E-Gas (Wertle, Germany)

In the fall of 2013, Audi commissioned a new 6 MW Power-to-Gas facility in Wertle, Germany to produce E-Gas, renewable synthetic methane for fueling cars. It is designed to provide enough fuel to power 1,500 Audi A3 vehicles for 15,000 km a year. The facility combines electrolysis with a catalytic methanation process and the renewable methane is injected into the natural gas grid and dispensed at an E-gas fueling station (Figure 14.4).

14.3.3 Thüga Group (Frankfurt am Main, Germany)

In December 2013, ITM Power commissioned a 315 kW PEM electrolyzer facility with the Thüga Group. The facility injects hydrogen gas directly into the Frankfurt am Main natural gas distribution network (Figure 14.5).

14.3.4 Energie Park Mainz (Mainz, Germany)

This project will be connected to a 10 MW wind farm and will have 6 MW (peak output) of electrolyzer capacity using Siemens' PEM technology. It will integrate the fluctuating wind output and provide ancillary services into the grid. It is designed to both inject into the municipal gas grid and also have a hydrogen trailer filling station. The project consortium includes the municipal energy supplier in Mainz, Linde, Siemens, and RheinMain University of Applied Sciences. Planned commissioning is in 2015.

14.3.5 WindGas (Hamburg, Germany)

A new project in Hamburg led by E.ON SE with partners Hydrogenics Corporation, Solvicore GmbH & Co. KG, DLR, and Fraunhofer ISE will deploy the world's first PEM megawatt electrolyzer stack developed by Hydrogenics Corporation. The project will feed renewable hydrogen into the local natural gas distribution

FIGURE 14.3
E.ON gas storage 2 MW direct injection Power-to-Gas facility. (Courtesy of E.ON Gas Storage, Germany.)

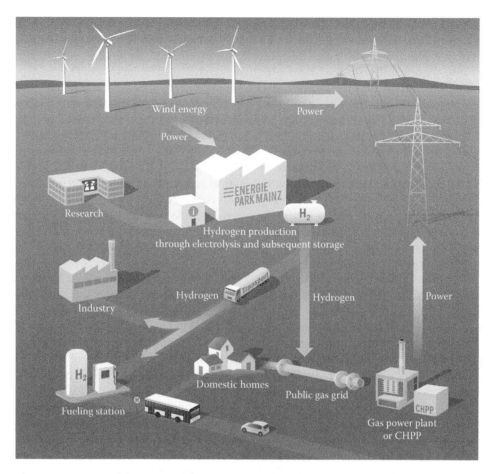

FIGURE 14.5
Energie Park Mainz project overview. (Courtesy of Siemens.)

network in Hamburg. Commissioning of this plant is planned for early 2015.

14.3.6 Ontario Power-to-Gas (Toronto, Canada)

The first Power-to-Gas project in North America was announced in July 2014 when Hydrogenics Corporation was awarded a regulation service contract by the Independent Electricity System Operator (IESO) for a 2 MW facility in Toronto. Enbridge Inc. is a partner in this project.

14.4 Preferred Technology Platform for Power-to-Gas

While the scale of pilot Power-to-Gas projects built to date range from 100 kW to 6 MW (with the majority less than 1 MW), the capacity required for commercial projects will likely be 40–100 MW in size. The reason for this is the massive scale of energy storage required.

For jurisdictions with a high proportion of renewable generation in the mix, the energy storage requirement is not measured in MWh but GWh, and the size of fluctuations of renewable output will be in the order of hundreds of megawatts. Both alkaline and PEM electrolyzer technologies have the ability to deliver on site and on demand hydrogen (load following), pressurized hydrogen without the need of a compressor, and high purity and carbon-free hydrogen. Although alkaline technology is well suited to Power-to-Gas for smaller-scale projects, there are some distinctive advantages for using PEM technology for large-scale commercial projects.

14.4.1 Megawatt-Scale PEM Technology Is the Essential Enabler

Although PEM electrolyzer technology is less mature, it has several advantages over classical alkaline electrolysis as it relates to Power-to-Gas applications. PEM electrolyzers operate at significantly higher current densities, which if leveraged in stack designs, can lead to extremely high-stack power densities. As an example, Hydrogenics' recently developed MW PEM electrolyzer stack is only 5% of its equivalent industrial alkaline electrolyzer stack volume, which requires 12 stacks. The significant stack volume reduction with PEM technology can translate into significantly smaller Power-to-Gas plant footprints. PEM electrolyzer technology also has excellent gas permeability characteristics and good mechanical stability allowing for stacks designs that are able to operate reliably with a differential pressure from the cathode to anode side allowing the anode circuit to be designed for near ambient pressure. The absence of corrosive electrolytes is a positive feature combined with the fact that PEM electrolyzers are able to achieve very high-quality hydrogen production. Properly designed PEM electrolyzer systems also have a simpler system design compared with alkaline systems allowing for a reduced number of components. Combine the earlier mentioned characteristics with the demonstrated potential for higher operating efficiencies, excellent partial load operating range, and millisecond response to fluctuating electrical power inputs and you have the essential enabling technology for Power-to-Gas applications.

Today, the capital cost per MW and the efficiency (percentage of electrical energy converted to hydrogen) of most PEM electrolyzers is similar to that of alkaline electrolyzers. While alkaline electrolysis is a mature technology that has reached a performance plateau, PEM electrolysis technology has plenty of runway left to increase power density, efficiency, and of course cost. We are at the early stages of large-scale megawatt PEM development. The primary components of a PEM electrolyzer and a PEM fuel cell—the bipolar plates and the membrane electrode assembly—have different design specifications but the supply chain for these components and materials will likely be identical.

There has been a dramatic drop in PEM fuel cell production costs over the past decade primarily due to volume production as well as early large investments in research and development. This is expected to accelerate even further as automotive manufacturers ramp up FCEV production beginning in 2015. Since PEM electrolysis has the ability to share many of the key material and component suppliers of PEM fuel cells, the cost learning curve for PEM electrolyzer components will be much shorter than it was for fuel cells and it has the potential to reduce capital expenditures (capex) in PEM electrolyzers significantly over the next 3–5 years. There is considerable runway for capital cost reductions on PEM electrolyzer technology as a result. Over the last several years, PEM electrolyzer technology has been able to share in the material improvements made for PEM fuel cells. Hydrogenics has achieved significant improvements in PEM electrolyzer stack efficiency between 2001 and 2010 while expanding the operating range as shown in the exhibit in Figure 14.6. As more material and membrane suppliers put their design efforts to optimizing performance for PEM electrolysis, it is anticipated that there will be further breakthroughs on the efficiency of PEM electrolyzers.

The capital cost and performance of electrolyzers is one critical driver to the speed of adoption of Power-to-Gas. The PEM electrolyzer technology platform holds the highest promise for the lowest capital cost along with higher power densities, smaller footprint, larger dynamic range, and a scalable design. There are a number of electrolyzer manufacturers including Hydrogenics Corporation, ITM Power, Proton OnSite, and Siemens AG who are actively pursuing this PEM development path for Power-to-Gas.

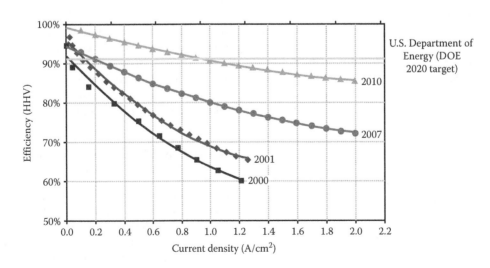

FIGURE 14.6
Hydrogenics' PEM stack efficiency. (Courtesy of Hydrogenics.)

14.5 PEM Electrolyzer Competitive Landscape

A handful of companies are focused on developing PEM electrolysis products. Proton OnSite, an established U.S. based company, manufactures PEM electrolyzers for on-site hydrogen generators for a wide range of applications. ITM Power, a UK-based company, manufactures and markets PEM electrolyzers for hydrogen refueling stations and other applications. Siemens AG is a global powerhouse in both industrial electrical system equipment and environmental products. Siemens has announced very ambitious development plans for its electrolyzers. Hydrogenics Corporation is the only company who manufactures both alkaline and PEM electrolyzers and recently completed development of the first megawatt sized PEM electrolyzer stack. All of these companies are working hard to address the emerging Power-to-Gas market.

14.5.1 Proton OnSite

Proton OnSite provides equipment for on-site gas generation with a product line covering hydrogen, nitrogen, and oxygen. Since its inception in 1996, it has developed and markets a range of turnkey electrolytic hydrogen generation systems ranging from 2 to 65 kg day^{-1} for laboratories, semiconductor manufacturers, and power station turbine generator cooling. Its largest commercial product, the HOGEN® C30 has a maximum capacity of 30 Nm^3 h^{-1} of hydrogen (65 kg day^{-1}) with an output pressure of 30 bar. In this configuration, it deploys three PEM electrolyzer stacks as shown in Figure 14.7.

Proton OnSite has adapted its C30 platform for use in fully-integrated, packaged hydrogen fueling stations. Proton began collaborating with the U.S. Department of Energy in February 2010 to develop a PEM electrolyzer that can produce hydrogen gas at the pressure required to fuel a vehicle without the need for a compressor. The result is the new FuelGen® hydrogen fueling station

launched in August 2012 utilizing a high-differential pressure PEM stack that can safely generate hydrogen at 350 bar while releasing outgoing oxygen gas at atmospheric pressure. This design feature also has an application for Power-to-Gas because it will permit the direct injection of hydrogen virtually anywhere on the natural gas system without the need for mechanical compression.

Proton OnSite is also scaling up its PEM electrolyzer as described in technical presentations by Everett Anderson, Vice President of Electrochemical Technology, at conferences in Hannover, Germany in April 2012 (Hannover Masse Conference [2012]) and in Copenhagen, Denmark in May 2012. The 10 Nm^3 h^{-1} stack is its largest commercial stack today. Work is underway to develop a 90 Nm^3 h^{-1} stack as shown in Figure 14.8.

In November 2012, Proton OnSite announced that it is developing a 1 MW PEM electrolyzer system targeted at renewable energy storage systems.

14.5.2 ITM Power

ITM Power has developed a suite of integrated PEM electrolyzer products for a range of applications from laboratory-based analytical systems to clean fuel provision for hydrogen-powered vehicles. ITM Power delivered the electrolyzer system for the Thüga Power-to-Gas demonstration project in Germany.

ITM Power is part of a larger consortium for the PHAEDRUS project with a €3.6 million grant for the development of an advanced refueling system using a high-pressure hydrogen electrolysis technology. Other partners include Hydrogen Efficiency Technologies, Daimler AG, and Shell Global Solutions. ITM's role in the three-year project will be to develop an electrolyzer design capable of delivering 200 kg day^{-1} of hydrogen at a pressure of 200 bar and a higher current density of 3 A cm^{-2}.

ITM Power recently announced that it will be launching a new 350 kW single stack where three stacks will be used to achieve 1 MW power levels (Figure 14.9).

FIGURE 14.7
Proton OnSite HOGEN C30 Electrolyzer. (Courtesy of Proton.)

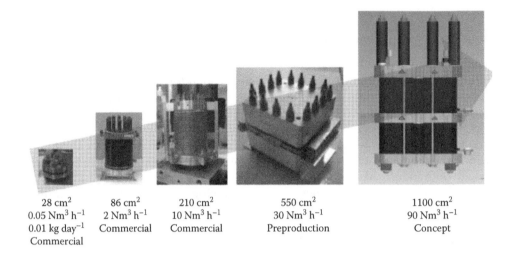

28 cm^2	86 cm^2	210 cm^2	550 cm^2	1100 cm^2
0.05 Nm^3 h^{-1}	2 Nm^3 h^{-1}	10 Nm^3 h^{-1}	30 Nm^3 h^{-1}	90 Nm^3 h^{-1}
0.01 kg day^{-1}	Commercial	Commercial	Preproduction	Concept
Commercial				

FIGURE 14.8
Proton OnSite PEM stack development. (Courtesy of Proton, U.K.)

FIGURE 14.9
315 kW PEM electrolyzer facility. (Courtesy of ITM Power, London, U.K.)

14.5.3 Siemens AG

Siemens AG has been developing and testing PEM electrolyzers in their Corporate Technology Laboratory in Erlangen, Germany since the mid-2000s. Siemens has a very good understanding of the Power-to-Gas requirements from a capacity perspective and has decades of experience designing and maintaining power equipment to meet the expectations and technical requirements of their power utility customers. Siemens has a focused development program in place with the goal of "developing 'three-digit megawatt' facilities." Siemens is developing MW size PEM electrolyzer stacks with peak power of 2 MW and hydrogen output pressure of 3.5 MPa (Figure 14.10).

Siemens and partners are currently working on the Energie Park Mainz project located in Mainz-Hechtsheim, Germany where a 10 MW wind farm will

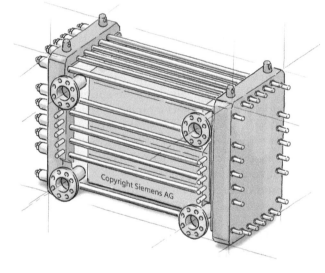

FIGURE 14.10
Siemens 2 MW PEM electrolyzer stack. (Courtesy of Siemens.)

be connected to a 6 MW peak electrolyzer along with 1000 kg of hydrogen storage. The Energie Park Mainz project overview is shown in the following.

14.5.4 Hydrogenics Corporation

Hydrogenics has been developing a large active area PEM electrolyzer cell stack platform since 2010. In 2014, it built the world's first megawatt-scale PEM electrolyzer cell stack. The cell stack shown in the picture in Figure 14.11 is capable of absorbing 1.5 MW of electrical input power in a Power-to-Gas application, and can produce hydrogen gas at an output of 285 Nm^3 h^{-1} and at a pressure of up to 40 bar. Hydrogenics has also received CE certification according to the Pressure Equipment

FIGURE 14.11
Hydrogenics' 1.5 MW PEM electrolyzer cell stack. (Courtesy of Hydrogenics Corporation, Ontario, Canada.)

Directive (PED) 97/23/EC for its Series 1500E PEM electrolyzer stack design. Hydrogenics' MW cell stack design will provide the essential building block for large-scale Power-to-Gas projects going forward.

14.6 Summary and Outlook

Energy storage is on the edge of becoming an important building block of the electricity infrastructure of the future. During the last century, the modern world has developed energy silos: the transportation energy silo made up of mainly gasoline and diesel for vehicles, the heating energy silo made up of the natural gas distribution network and the electricity energy silo for powering motors, appliances, and electronics.

The result today is a melange of energy technologies and infrastructures that do not connect and interact with each other and that by and large are not able to efficiently capitalize on the benefits of clean renewable energy sources. Power-to-Gas will act as the bridge between these energy silos and megawatt-scale PEM electrolysis has the ability to be the enabling technology ushering in a new era of renewable energy penetration. The Power-to-Gas opportunity is multifaceted and involves potentially rich value propositions for numerous stakeholders and interests in different industries. Large-scale PEM electrolysis is expected to have improved performance and lower cost, which will help unlock the benefits associated with Power-to-Gas discussed in this chapter.

References

Carter, D. and J. Wing. 2013. The Fuel Cell Industry Review 2013. Fuel Cell Today. http://www.fuelcelltoday.com/media/1889744/fct_review_2013.pdf.

Decourt, B., B. Lajoie, R. Debarre, and O. Soupa. 2014. Hydrogen-Based Energy Conversion More than Storage: System Flexibility. SBC Energy Institute. http://bit.ly/1yb1me8.

Dehamna, A. 2013. In Energy Storage, Power-to-Gas Seeks a Market. Navigant Research Blog. http://www.navigantresearch.com/blog/in-energy-storage-power-to-gas-seeks-a-market.

Grond, L., P. Schulze, and J. Holstein. 2013. Systems analyses Power to Gas: A technology review. DNV KEMA Energy & Sustainability. http://bit.ly/1vLdhfl.

Hannover Masse Conference, 2012. *Hydrogen/Fuel Cell Technical Forum and Symposium on Water Electrolysis and Hydrogen as Part of the Future Renewable Energy System,* April 23–27. Hannover Messe, Messegelände, 30521 Hannover, Germany.

Melaina, M.W., O. Antonia, and M. Penev. 2013. Blending Hydrogen into Natural Gas Pipelines: A Review of Key Issues. National Renewable Energy Laboratory. http://www.nrel.gov/docs/fy13osti/51995.pdf.

15

Depolarized Proton Exchange Membrane Water Electrolysis: Coupled Anodic Reactions

Sergey Grigoriev, Irina Pushkareva, and Artem Pushkarev

CONTENTS

15.1 Introduction

Proton exchange membrane (PEM) water electrolysis requires energy consumption of about >4 kWh per 1 Nm³ hydrogen, which is a relatively high cost of electrolytic hydrogen. In this regard, PEM electrolysis with anode depolarization has a definite prospect. When the main product of electrolysis is only hydrogen (oxygen is not required), it is reasonable to apply the anode depolarization, which is realized as a partial or complete replacement of the most energy-consuming anodic oxygen evolution reaction (OER) by the electrochemical decomposition of other compounds—so-called

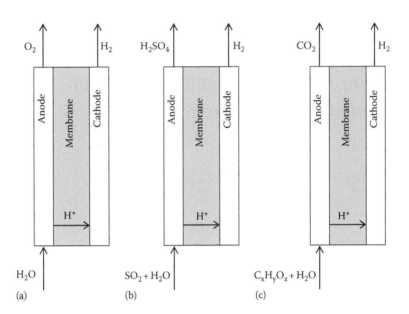

FIGURE 15.1
Schematic diagrams of PEM electrolysis of water (a), sulfuric anhydride (b), and organic compounds (c) water solutions.

depolarizers (sulfur dioxide, methanol, ethanol, glycerol, formic acid, and other reducing agents, specified in Figure 15.1 and Table 15.1). Upon decomposition of the compounds mentioned or their aqueous solutions, the anode electrochemical process will proceed at a lower potential than the potential of the water

decomposition reaction (Table 15.1), offering a considerable reduction of energy consumption required for the production of hydrogen and the cost thereof. Thus, depolarization of the anode can reduce the energy consumption for the production of electrolytic hydrogen by about half.

TABLE 15.1

Main Parameters of Electrolysis of Water and Various Depolarizers, Including Their Aqueous Solutions

Compound Decomposed at the Anode	Anodic Reaction[a]	Anode Potential, V	Cathodic Reaction	Cathode Potential, V	Overall Reaction	Cell Potential, V	Specific Power Consumption, kWh Nm^{-3} H$_2$
Water	$H_2O \rightarrow 2H^+ + 1/2O_2$ $+ 2e^-$	1.229	$2H^+ + 2e^- \rightarrow$ H_2	0	$H_2O \rightarrow 1/2O_2 + H_2$	1.229	≥ 4
Sulfur dioxide + water	$SO_2 + 2H_2O - 2e^- \rightarrow$ $H_2SO_4 + 2H^+$	0.158	$2H^+ + 2e^- \rightarrow$ H_2	0	$SO_2 + 2H_2O \rightarrow$ $H_2SO_4 + H_2$	0.158	2.1 at 0.4 A cm^{-2}
Methanol + water	$CH_3OH + H_2O \rightarrow$ $CO_2 + 6H^+ + 6e^-$	0.016	$6H^+ + 6e^- \rightarrow$ $3H_2$	0	$CH_3OH + H_2O \rightarrow$ $3H_2 + CO_2$	0.016	1.48 at 0.5 A cm^{-2}
Ethanol + water	$C_2H_5OH + 3H_2O \rightarrow$ $2CO_2 + 12e^- +$ $12H^+$	0.09	$12H^+ + 12e^- \rightarrow$ $6H_2$	0	$C_2H_5OH + 3H_2O \rightarrow$ $6H_2 + 2CO_2$	0.090	2.63 at 0.25 A cm^{-2}
Glycerol + water	$C_3H_8O_3 + 3H_2O \rightarrow$ $3CO_2 + 14H^+ + 14e^-$	0.22	$14H^+ + 14e^- \rightarrow$ $7H_2$	0	$C_3H_8O_3 + 3H_2O \rightarrow$ $3CO_2 + 7H_2$	0.220	1.1–1.67 at 0.001–0.01 A cm^{-2}
Formic acid	$HCOOH \rightarrow$ $CO_2 + 2H^+ + 2e^-$	0.17	$2H^+ + 2e^- \rightarrow H_2$	0	$CH_2O_2 \rightarrow CO_2 + H_2$	0.170	≥ 2 at 0.2 A cm^{-2}
Hydrogen bromide + water	$2HBr \rightarrow Br_2 +$ $2H^+ + 2e^-$	0.58	$2H^+ + 2e^- \rightarrow H_2$	0	$2HBr \rightarrow Br_2 + H_2$	0.580	≥ 4.3–5.1 at 0.4 A cm^{-2}
Anhydrous hydrogen chloride	$2HCl_{(l)} \rightarrow Cl_{2(g)} +$ $2H^+ + 2e^-$	1.36	$2H^+ + 2e^- \rightarrow H_2$	0	$2HCl_{(l)} \rightarrow Cl_{2(g)} +$ $2H^+ + 2e^-$	1.360	3.3 (theoretical)
	$2HCl_{(g)} \rightarrow Cl_{2(g)} +$ $2H^+ + 2e^-$	0.987	$2H^+ + 2e^- \rightarrow H_2$	0	$2HCl_{(g)} \rightarrow 2H_{2(g)} +$ $2Cl_{2(g)}$	0.987	3.5–4.3 (real) at 0.4–1.0 A cm^{-2}

Source: Corgnale and Summers (2011), Marshall and Haverkamp (2008), Kostin et al. (2008), Sasikumar et al. (2008), Gorensek et al. (2009b), Sivasubramanian et al. (2007), Cloutier and Wilkinson (2010), Lamy et al. (2011), Caravaca et al. (2012), Pham et al. (2013a,b), Kostin et al. (2014), and Grigoriev and Bessarabov (2012).

[a] Additional reactions are possible.

The current–voltage characteristic of the electrolysis cell can be described by the following equation:

$$U = E + \eta_\kappa + \eta_a + iR \qquad (15.1)$$

where
  U is the voltage of electrolysis
  E is the value of the back electromotive force (EMF) of the electrolysis cell at given temperature and pressure (the difference between the thermodynamic equilibrium potentials of the anodic and cathodic reactions)
  η_κ is the cathodic overpotential (polarization)
  η_a is the anodic overpotential (polarization)
  iR is the ohmic losses

In the electrolysis of water, the theoretical back electromotive force (EMF) (counter-EMF) of the cell is 1.229 V (at 25°C and atmospheric pressure), and is determined only by the potential of the anode reaction (the potential of the cathode hydrogen evolution reaction is 0, see Table 15.1). When specific reducing agents (depolarizers) are added to water supplied to the anode, the anode reaction will proceed at a lower potential than the potential of the water decomposition reaction. Specific energy consumption for the production of electrolytic hydrogen is determined by the following equation:

$$W = \frac{IUt}{V} \qquad (15.2)$$

where
  I is the current intensity (A)
  U is the voltage of the electrolysis cell (V)
  t is the time (h)
  V is the volume of the produced gas (m^3)

Therefore, energy consumption for the hydrogen production W is reduced by reducing the electrolysis voltage U.

The following sections review the features of electrolysis with anode depolarization by various compounds in more detail.

15.2 Electrochemical Oxidation of Sulfur Dioxide

15.2.1 Hybrid Sulfur Cycle

Among the various anode depolarizers, sulfur dioxide (SO_2) is very interesting for practical applications. In the case of SO_2 depolarization, the integration of the electrolysis process in the sulfuric acid cycle is important. This hybrid thermochemical cycle (so-called Mark-11 or Westinghouse cycle), developed in the 1970s, uses the heat from high-temperature gas-cooled nuclear reactors (HTGRs) (Sivasubramanian et al. 2007, Kostin et al. 2008, Jung and Jeong 2010, Leybros et al. 2010, Kostin et al. 2014) or from solar concentrators (Corgnale and Summers 2011) in the temperature range 800°C–900°C. A schematic diagram of the hybrid sulfur (Hy-S) cycle is shown in Figure 15.2.

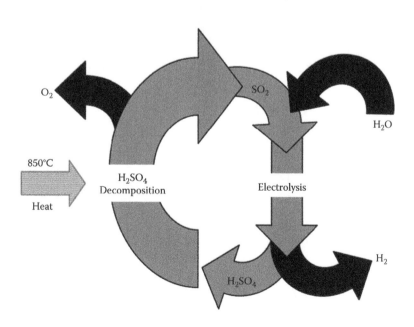

FIGURE 15.2
A schematic diagram of the hybrid sulfur cycle. (Reprinted from Sivasubramanian, P.K. et al., *Int. J. Hydrogen Energy*, 32, 463, 2007. With permission from International Journal for Hydrogen Energy.)

The first, high-temperature step of this hybrid cycle is carried out by the reaction

$$H_2SO_4 \rightarrow H_2O + SO_2 + \frac{1}{2O_2} \qquad (15.3)$$

The second, low-temperature (80°C–100°C) step proceeds by PEM electrolysis with SO_2 anode depolarization

Anodic reaction: $SO_2 + 2H_2O \rightarrow H_2SO_4 + 2H^+ + 2e^-$

$E^0 = 0.158$ V versus standard hydrogen electrode (SHE)

Cathodic reaction: $2H^+ + 2e^- = H_2$

$E^0 = 0$ V vs. SHE

Overall reaction: $2H_2O + SO_2 = H_2 + H_2SO_4$

Figure 15.3 shows a schematic diagram of a PEM electrolyzer with SO_2 depolarization (Sivasubramanian et al. 2007).

15.2.2 Limitations and Performances

As is evident from the aforementioned reactions, the HyS cycle is closed and only water is consumed as a reactant to produce hydrogen. Water required for the anodic reaction is supplied by the crossover through the membrane from the cathode (Staser and Weidner (2008)). The flux of water across the membrane decreases with increasing current density because electroosmotic drag pulls water from the anode to the cathode to counter the diffusion of water from the cathode to the anode. No additional water is required to be fed with the SO_2 at current densities <0.7 A cm^{-2}. It is not possible to increase the rate of the anodic reaction by increasing the voltage or the SO_2 flow rate. Rather, the reaction rate is limited by the rate of water transport across the membrane. Hence, it should be possible to increase the limiting current by using thinner membranes or by humidifying the SO_2 before it enters the electrolyzer (Sivasubramanian et al. 2007).

It is notable that energy consumption for electrolysis at a low concentration of sulfuric acid is also low.

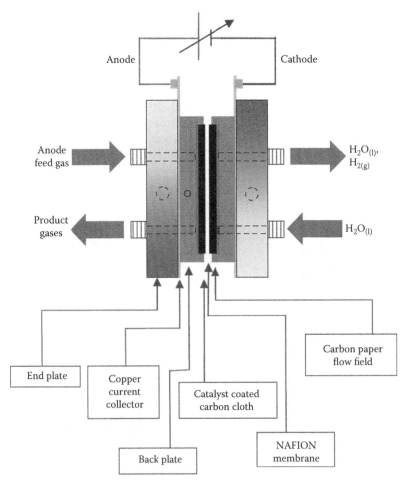

FIGURE 15.3
Schematic diagram of PEM electrolyzer assemble. (Reprinted from Sivasubramanian, P.K. et al., *Int. J. Hydrogen Energy*, 32, 463, 2007. With permission from International Journal for Hydrogen Energy.)

However, more concentrated sulfuric acid is preferable for the high-temperature acid decomposition reaction that regenerates SO_2. Thus, the optimal value of the sulfuric acid concentration determines the maximum efficiency of the entire cycle (Gorensek et al. 2009b).

The decisive contribution of the sulfuric acid concentration has been confirmed (Staser and Weidner (2008)). There is a trade-off between low voltages (fast water transport) and high sulfuric acid concentrations (slow water transport); a higher sulfuric acid concentration is desirable for downstream decomposition, but a high sulfuric acid concentration increases the cell voltage and hence the power required for the electrolysis.

The pressure difference between anode and cathode significantly affects the water transport and acid concentration (Kozolii and Kostin 2009, Staser et al. 2009, Xue et al. 2013). Kozolii and Kostin (2009) attempted to increase the SO_2 concentration in the electrolyte solution by increasing the pressure in the anodic chamber of the cell. The results of comparative tests of a three-cell electrolyzer at different pressures of SO_2 are presented in Figure 15.4.

Figure 15.4 shows that with increasing current density the difference between the current–voltage curves for the cell with a depolarized (1, 2) and a nondepolarized (3) anode decreases. Curve (2) for the cell with depolarization by SO_2 at $p = 0$ MPa is approaching curve (3) for the cell with water electrolysis at a current density of 700 mA cm^{-2}. Increasing the pressure to 0.3 MPa leads to the effect of "convergence" of the current–voltage curves, shifted to higher current

densities (1700 mA m^{-2}). Therefore, the mass-transfer limitations for SO_2 can be significantly reduced by increasing the pressure of SO_2 in the anodic compartment of the electrolyzer.

15.2.3 Membrane Materials

One of the most important components of the electrolytic cell with anode depolarization is the membrane. Several investigations into and the modeling of properties of different membranes (e.g., Nafion, polybenzimidazole [PBI]-based, etc.) to identify the most effective membrane for use in the HyS cycle have been carried out. The following have been identified as some of the critical characteristics: high conductivity, stability (Schoeman et al. 2012, Peach et al. 2014), and SO_2 crossover (Sethuraman et al. 2009, Elvington et al. 2010, Lokkiluoto and Gasik 2013, Opperman et al. 2014).

Several commercially available and experimental PEMs, including Nafion, have been investigated for application in a SO_2-depolarized electrolyzer (Elvington et al. 2010). The following have been measured: chemical stability at 80°C, SO_2 transport, ionic conductivity, and electrolyzer performance at 1 V. All of the membranes exhibited excellent chemical stability in hot, concentrated sulfuric acid solutions. PEMs that exhibit reduced transport of SO_2 were identified. Of the perfluorinated sulfonic acid (PFSA) type membranes, the DuPont bilayer, the 1500 EW membrane, the two treated PFSA membranes from DuPont (DuPont 112/pvp46 and DuPont 1135/pvp48), and the fluorinated ethylene propylene PFSA blends from Case Western Reserve University all showed reduced SO_2 transport relative to the baseline membrane Nafion 115. Of the non-PFSA membranes, BPVE and BPVE-6F from Clemson University, and the Celtec-V PBI membrane from BASF also showed reduced SO_2 transport. Only the BPVE, BPVE-6F, and PBI membranes exhibited increased electrolyzer performance coupled with lower SO_2 transport. The PBI-based Celtec-V membrane exhibited the best combination of performance and SO_2 transport, with a 27% increase in current density and a 67% decrease in SO_2 transport, compared to the baseline membrane Nafion 115.

Pure, as well as blended, PBI, partially fluorinated poly(arylene ether) (sFS) and nonfluorinated poly(arylene ethersulfone) (sPSU) membranes have been investigated in terms of their acid stability as a function of acid concentration (Schoeman et al. 2012). Five samples were studied: two pure sulfonated arylene main-chain polymer membranes (one sPSU and [sFS]), a pure PBI membrane, and two blend membranes composed of sPSU and PBI, and sFS and PBI, respectively. The nonfluorinated membranes (PBI, sPSU, and sPSU-PBI) showed excellent stability at <90 wt% H_2SO_4. However, sulfonation

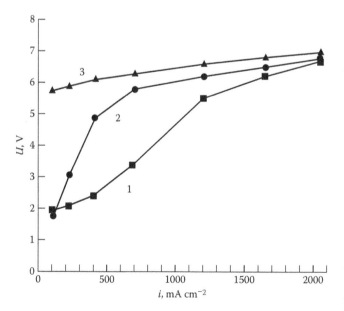

FIGURE 15.4
Current–voltages curves of electrolyzer at 25°C and different pressures in the anodic chamber: 1—$p = 0.3$ MPa; 2—$p = 0$ MPa; 3—without depolarization by SO_2.

occurred at 90 wt% H_2SO_4, resulting in the dissolution of the membranes. Fluorination of the membrane (sFS and sFS-PBI) seemed to improve the membrane stability in sulfuric acid in view of the improved resistance observed, especially at 90 wt% H_2SO_4. This is of little concern, however, because in electrolyzers concentrations of 30–40 wt% sulfuric acid are rarely exceeded. Indications are that future focus will be on base-excess PBI-sulfonated ionomer blend membranes.

Acid-excess blend membranes containing the partially fluorinated F_6PBI with different cross-linking and base components, and the resulting performance of SO_2 electrolyzers have been investigated (Peach et al. 2014). Results clearly showed that the blend membranes yielded comparable, and at low currents even better, *I–V* curves than the commercial Nafion 115 did. These results are similar to what was reported previously (Schoeman et al. 2012). The improved stability noted for the sFS-PBI blend membrane due to the additional partially fluorinated PBI (F_6-PBI) component is note worthy.

It has been shown that the type of membrane as well as different reaction conditions play a significant role in the SO_2 flux characteristics of a polymer membrane (Opperman et al. 2014). Permeability characteristics such as the diffusion and solubility of the gas in different membranes are affected differently under different reaction conditions. The SO_2 flux across the membrane is strongly influenced by the solubility of the SO_2 in the bulk solution. The flux is most likely purely a concentration gradient-driven process. Varying membrane compositions also affect these permeation characteristics. These sFS-PBI membranes showed a lower SO_2 flux than the Nafion membranes. The optimal conditions for the electrolysis were high temperature, low trans-membrane partial pressure, and high sulfuric acid concentration; while membranes that were thicker new-generation membranes, as well as membranes with a more rigid backbone structure were more suitable for restricting SO_2 crossover. All these factors should therefore be borne in mind when optimizing the HyS cycle with regards to ohmic losses due to the SO_2 crossover phenomenon.

15.2.4 Catalytic Materials

The anode and cathode catalytic materials are key components of SO_2 electrolyzers, ensuring that the electrolyzers operate at maximum energy efficiency and with long lifetimes. The kinetics of the electrochemical oxidation of SO_2 at the anode is very slow compared to at the cathode. Consequently, most of the inefficiencies of electrolyzers arise from the anode reaction kinetics. Chemical stability of the anodic catalyst for the oxidation of SO_2 is also important.

Platinum (Pt) has received the most attention, although palladium (Pd) is reported to be an excellent catalyst for the oxidation of SO_2. Although carbon is favored as the catalyst supporting medium, it has the disadvantage of being oxidized to carbon dioxide at low voltages, leading to the degradation of gas diffusion layers.

Pt- and Pd-based catalysts for the electrooxidation of SO_2 have been investigated (Colorn-Mercado and Hobbs 2007). The catalytic activity of $Pt^{0.45}/C$ and $Pd^{0.4}/C$ for SO_2 oxidation reaction were studied in 3.5–10.4 M H_2SO_4 solutions and at temperatures 30°C–70°C. Pt/C demonstrates very good stability and activity, than Pd/C, and instability in very high sulfuric acid concentrations (10.4 M) at temperatures of $\geq$50°C. In view of the fact that the exchange current density on Pt/C is much higher than on Pd/C (about 1000 times) and it has greater stability, the authors consider the Pt/C catalyst to be preferable.

The operation of a SO_2-depolarized PEM electrolyzer was studied by Sivasubramanian et al. (2007). They hypothesized that the water needed for the anodic reaction (1) could be provided by transport across the membrane from the cathode. Figure 15.5 shows the current–voltage response of the PEM electrolyzer for the oxidation of SO_2 to H_2SO_4 and the reduction

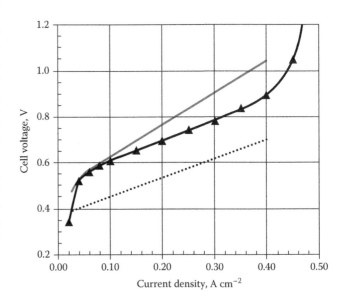

FIGURE 15.5
The current–voltage response for SO_2 electrolysis. The black triangles are data from the PEM electrolyzer at 80°C and 1.0 bar with a Pt loading on the anode and cathode of 0.66 and 0.7 mg cm⁻², respectively. The solid gray line is data from literature in an aqueous phase electrolyzer at 50°C and 1 bar, with a Pt loading on the anode and cathode of 7 and 10 mg cm⁻², respectively. The dotted line is the targeted cell performance data from literature at 100°C and 5–20 bar, which was not achieved. (Reprinted from Sivasubramanian, P.K. et al., *Int. J. Hydrogen Energy*, 32, 463, 2007. With permission from International Journal for Hydrogen Energy.)

of protons to H_2. Data (black triangles) were collected at 80°C, 1.0 bar and *ca.* 5%–40% conversion of SO_2. Results indicated that the water transported across the membrane was sufficient to sustain reaction (1) up to current densities of 0.4 A cm^{-2}. A mass transfer limiting current density is observed beyond this point. It is most likely that the limiting reactant in reaction (1) is water and not SO_2. Therefore, it is not possible to increase the rate of reaction (1) by increasing the voltage or the SO_2 flow rate. Rather, the reaction rate is limited by the rate of water transport across the membrane. It should be possible to extend the limiting current by using thinner membranes or by humidifying the SO_2 before it enters the electrolyzer.

An improvement of >150 mV in cell voltage at one-tenth of the Pt loading has been observed at 0.4 A cm^{-2}, but the temperature was 30°C higher than at (Sivasubramanian et al. 2007), the PEM electrolyzer shows promise as a means of carrying out reaction (15.3). This improvement is most likely due to improved gas-phase mass transfer and high membrane conductivity.

15.2.5 Recent Advances and Perspectives

Two HyS cycle process flow sheets intended for use with HTGRs have been reported by Gorensek and Summers (2009). The flow sheets were developed for the Next Generation Nuclear Plant (NGNP) program, and couple a PEM electrolyzer for the SO_2-depolarized electrolysis step with a silicon carbide bayonet reactor for the high-temperature decomposition step. One presumes an HTGR reactor outlet temperature (ROT) of 950°C, the other 750°C. Performance was improved (comparing with performance achieved in earlier flow sheets) by assuming that use of a more acid-tolerant PEM, like PBI, instead of Nafion, would allow higher anolyte acid concentrations. Lower ROT was accommodated by adding a direct contact exchange–quench column upstream from the bayonet reactor and reducing the decomposition pressure. Aspen Plus was used to develop material and energy balances. A net thermal efficiency is 44%–47.6%.

A HyS flow sheet using an ionic liquid absorbent to separate SO_2 from oxygen has been developed. The key chemical reactors in the HyS flow sheet are a sulfuric acid distillation column, sulfuric acid vaporization and decomposition reactor, a knockout drum to separate SO_2–O_2 mixed gas from the SO_3–H_2O recombined, and condensed aqueous phase, a separation part of SO_2, and a SO_2-depolarized electrolysis cell. The separation part of SO_2 consists of SO_2 absorption and desorption columns adopting a temperature swing methodology. A hydrogen production thermal efficiency of 39% can be expected from this HyS process (Shin et al. 2012).

Currently, there are developments of industrial facilities for hydrogen generation by the HyS cycle, where the solar towers are used as sources of thermal energy. Alumina sand (particle size 600 μm) is used as the high-temperature heat carrier. Heat exchange with the reactor is carried out through an intermediate heat-transfer agent (air or helium). The cost of the generated H_2 is estimated at 4\$ kg^{-1} for the year 2015 and 2.85\$ kg^{-1} for the year 2025 (Corgnale and Summers 2011).

Outotec (Rauhalanpuisto 9, 02230 Espoo, Finland) has developed an Outotec open cycle (OOC), linking together the production of hydrogen and sulfuric acid. A case is described where the OOC is connected to a pyrometallurgical plant (Lokkiluoto et al. 2012). SO_2 and water are the raw materials for the production of sulfuric acid and hydrogen. SO_2 can be obtained from sulfides of nonferrous metals after treatment (roasting, melting), regeneration of spent acid, or the combustion of elemental sulfur. Unlike the Westinghouse cycle, the OOC is an open process in which acid is a commercial product and it is not destroyed. The incoming SO_2 gas stream is divided in two sub streams: the first sub-stream is routed to the SO_2-depolarized electrolyzer or Sulfur dioxide depolarized electrolysis (SDE) and the second is routed to the acid plant to produce SO_3. The SO_3 is used for concentrating the sulfuric acid produced in the electrolyzer. The process is schematically shown in Figure 15.6 (Lokkiluoto et al. 2012).

The once-through hybrid sulfur (Ot-HyS) process, proposed by Jung and Jeong (2010), produces hydrogen by the same SDE process as in the original HyS. In the process proposed here, the sulfuric acid decomposition process in the HyS procedure is replaced with the

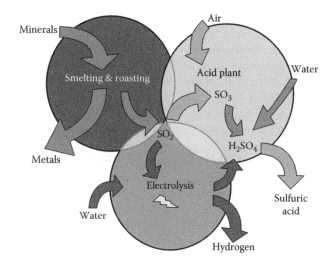

FIGURE 15.6
Schematic diagram of the Outotec open cycle process. (With kind permission from Springer+Business Media: *Environ. Dev. Sustain.* Novel process concept for the production of H_2 and H_2SO_4 by SO_2-depolarized electrolysis, 14, 2012, 529–540, Lokkiluoto, A., Taskinen, P.A., Gasik, M., Kojo, I.V., Peltola, H., Barker, M. H., and Kleifges, K.-H.)

well-established sulfur combustion process. A flow sheet for the Ot-HyS process was developed by referring to existing facilities and to the work done by the Savannah River National Laboratory (Savannah River Site near Jackson, South Carolina). The process was simulated with appropriate thermodynamic models. It was demonstrated that the Ot-HyS process has higher net thermal efficiency, as well as other advantages, over competing benchmark processes. The net thermal efficiency of the Ot-HyS process is 47.1% (based on lower heating value of hydrogen) and 55.7% (based on higher heating value of hydrogen) assuming 33.3% thermal-to-electric conversion efficiency of a nuclear power plant with no consideration given to the work for the air separation. Hydrogen produced through the Ot-HyS process would be used as off-peak electricity storage, to relieve the burden of load following, and could help to expand applications of nuclear energy, which is regarded as a sustainable development technology.

In conclusion, application of electrolysis with anode depolarization by SO_2 allows a significant (on average up to 50%) reduction in the power consumption for hydrogen production, with an estimated cost of 6\$–2.3\$ kg^{-1}, depending on the heat source and specific features of the cycle (Gorensek and Summers 2009, Jung and Jeong 2010, Leybros et al. 2010, Corgnale and Summers 2011).

15.3 Electrolysis of Aqueous Methanol Solutions

15.3.1 Mechanism of Methanol Electrooxidation

There has been some interest in the use of alcohols, specifically methanol, as anode depolarizers (Liu et al. 2006, Sasikumar et al. 2008, Cloutier and Wilkinson 2010). Although methanol is a valuable raw material and a fuel, in some cases the problem of its recovery has arisen. For example, a particular problem is the recycling of methanol produced as a by-product in a number of industrial processes, such as the production of dimethyl ether from syn-gas, and the processing of low-quality coals, and others (Take et al. 2007, Uhm et al. 2012, Kostin et al. 2014). In these cases, methanol can be processed at the anode of an electrolytic cell, and hydrogen will be released at the cathode (see Table 15.1).

The following reactions proceed on the electrodes in electrolysis of a methanol–water solution:

Anode: $CH_3OH + H_2O - 6e^- \rightarrow CO_2 + 6H^+$
Cathode: $2H^+ + 2H_2O \rightarrow H_2 \uparrow$
Overall: $CH_3OH + H_2O \rightarrow 3H_2 + CO_2$

The main problem associated with using alcohols (ethanol, *n*-propanol, isopropanol, ethylene glycol, etc.) as fuels in PEM fuel cells (Lamy et al. 2001, Cao and Bergens 2003, Peled et al. 2002), and hence as depolarizers in PEM electrolysis cells include difficulty in the breaking of the C–C bond and the arrest of the decomposition reactions in intermediate generation (acetic acid, etc.). However, these problems do not arise in the case of the electrochemical conversion of methanol.

It is currently assumed that the reaction mechanism has been identified with sufficient accuracy. The reaction of methanol oxidation takes place in several stages, each of which may have multiple paths (Figure 15.7).

In Figure 15.7, methanol appears in the left top corner, and the product of complete oxidation (CO_2) is located in the right bottom corner. The forward steps are associated with the process of hydrogen abstraction and formation of a proton–electron pair. The downward steps are also associated with hydrogen abstraction and formation of a proton–electron pair, but are accompanied by the joining or abstraction of an OH group. In this scheme, any path is possible; there are three forward and three downward steps that lead to the formation of CO_2, six protons, and six electrons.

Because it has the same anode reaction, the setup of the PEM methanol electrolyzer cell is quite similar to that of a direct methanol fuel cell (DMFC). It is well known that the sluggish kinetics of the methanol oxidation reaction, the permeation of methanol from the anode to the cathode (i.e., methanol crossover), and cathode flooding are key factors that have hampered the development of DMFC technology (Liu et al. 2006, Casalegno and Marchesi 2008, Zhao et al. 2010). It is particularly methanol crossover that causes great problems, including the creation of a mixed potential on the cathode (i.e., decreasing the cathode potential), low fuel utilization efficiency, etc.

Unlike in the DMFC, water is not produced in the cathode of the PEM methanol electrolysis cell and the

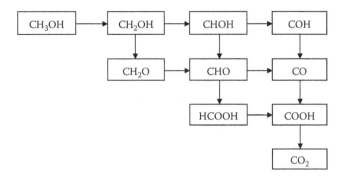

FIGURE 15.7
Schematic diagram of electrochemical methanol oxidation process. (With kind permission from Springer+Business Media: *Chem. Petrol. Eng.* Production of hydrogen by solid polymer electrolysis with anode depolarization, 49, 2014, 575–578, Kostin, V.I., Grigor'ev, S.A., Bessarabov, D.G., and Kryuger, A.)

flooding is not a problem. Therefore, the major challenges to achieving better performance in the PEM methanol electrolysis cell exist only on the anode side. Besides improving the sluggish kinetics of the methanol oxidation reaction, the mass transport related to supply of aqueous methanol and removal of CO_2 is also highly important, as in the DMFC. An appropriate flow-field design and gas diffusion layer are key in satisfying high hydrogen production performance (Liu et al. 2006, Pham et al. 2013a,b). The efficiency of the PEM methanol electrolysis cell was largely improved with a porous flow field made of sintered spherical metal powder, compared to when a conventional groove-type flow field was used (Pham et al. 2013b). This is attributed to an increase in the effective electrode area by using the porous material, which enables the flow field to supply reactant evenly to the electrode and remove carbon dioxide smoothly (Pham et al. 2013a).

15.3.2 Electrocatalytic Materials

Platinum, as monometallic catalyst, does not demonstrate sufficient activity for the direct oxidation of alcohol. Hence, much attention has been devoted to R&D on binary and ternary Pt-containing catalysts (Liu et al. 2006, Sasikumar et al. 2008, Hsieh and Lin 2009, Cloutier and Wilkinson 2010, Tarasevich and Kuzov 2010). Within the bifunctional mechanism, the promoter metal should provide adsorption of oxygenate particles at low potentials. Furthermore, the promoter metal should be sufficiently stable and should not inhibit the adsorption rate and the dehydrogenation of methanol.

Currently, it is generally recognized that the most appropriate electrocatalytic system for electrochemical methanol oxidation is Pt–Ru (Liu et al. 2006, Cloutier and Wilkinson 2010, Tarasevich and Kuzov 2010). Various types of Pt–Ru systems (alloys, mixtures, etc.) have a catalytic effect. Despite the ongoing debate about the optimal structure of these systems, the most important condition for the electrocatalytic effect is a well-developed Pt–Ru contact boundary. In efforts to form more complicated systems, some attempts have been made to address more complicated systems using Nb, W, Mo, or their oxides.

Sn, Bi, Mo, Ru, Os, and Ir were investigated as components of Pt–M binary systems (Tarasevich and Kuzov 2010). The optimum content of Ru in Pt–Ru was found to be 20%–50% of Ru (Figure 15.8), depending on both the method of synthesis and the experimental conditions (alcohol concentration of the electrolyte, electrode potential, and temperature). An increase in temperature leads to a shift in the activity maximum towards higher Ru content.

A bifunctional mechanism of methanol oxidation on the Pt–Ru system is described in detail by Tarasevich

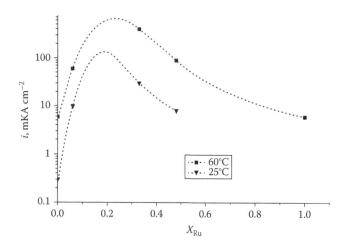

FIGURE 15.8
The dependence of the rate of 0.5 M methanol oxidation in 0.5 M H_2SO_4 upon Pt–Ru alloy composition at various temperatures. (Reprinted from Tarasevich, M.R. and Kuzov, A.V., Int. Sci. J. *Altern. Energy Ecol.*, 87, 86, 2010. With permission from Scientific Technical Centre "TATA.")

and Kuzov (2010). Adsorption and dehydration of methanol occurs on Pt, while the active oxygen species required for the final oxidation CO_{ads} to CO_2 are formed on Ru. This simple scheme is suitable for a relatively low overvoltage ($E < 0.4$ V) and temperature. When the potential shifts to the anode side, oxygen particles begin to form on Pt, while at higher temperatures (60°C–80°C) Ru begins to adsorb and dehydrogenate alcohol. The first step of the reaction is adsorption and the dehydrogenation of methanol

$$CH_3OH \rightarrow (CO)_{ads} + 4H^+ + 4e^-$$

with the formation of $Pt(CO)_{ads}$ and $Ru(CO)_{ads}$.

Simultaneously, the anodic dissociation of water takes place

$$H_2O \rightarrow (OH)_{ads} + H^+ + e^-$$

with the formation of $Pt(OH)_{ads}$ and $Ru(OH)_{ads}$, which reacts with $(CO)_{ads}$

$$(CO)_{ads} + (OH)_{ads} \rightarrow CO_2 + H^+ + e^-$$

General acceleration of this reaction takes place if it occurs between the particles of CO adsorbed on Pt and OH particles adsorbed on Ru:

$$Pt(CO)_{ads} + Ru(OH)_{ads} \rightarrow CO_2 + H^+ + e^-$$

and, thus, is possible with high-mobility CO on the surface.

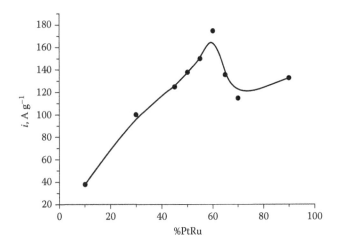

FIGURE 15.9

Dependence of specific activity at $E = 0.65$ V upon the content of the PtRu in the catalyst. (Reprinted from Tarasevich, M.R. and Kuzov, A.V., *Int. Sci. J. Altern. Energy Ecol.*, 87, 86, 2010. With permission from Scientific Technical Centre "TATA.")

The compensation effect was discovered later: OH_{ads} transfer from Ru to Pt offset by the transfer of energy with water transfer from Pt to Ru. Electrocatalysts with high metal content should be used in the active layer in order to reduce the transport limitations.

Figure 15.9 shows the dependence of specific activity ($E = 0.65$ V) upon the content of the Pt–Ru in the catalyst. Here, the optimal activity is achieved when the content of the metal phase is about 60%.

15.3.3 Performances

Research on methanol decomposition in a 25 cm^2 PEM cell using 40% PtRu/C on the anode and Pt/C on the cathode, and Nafion 117 as membrane, was carried out by Sasikumar et al. (2008). The effect of the methanol concentration was studied at 1 V and 30°C. Results for dependence of methanol concentration on current density and hydrogen production rate are presented in Figure 15.10. A sharp increase in current density was observed up to about 4 M of methanol concentration, from 4 to 6 M the increase was only marginal. Highest current density (*ca.* 680 mA cm^{-2} at 1 V) was obtained using 6 M methanol. Similar results (increase in current density with increase of methanol concentration) have been obtained by other research groups (Take et al. 2007, Uhm et al. 2012).

The decrease in current density at methanol concentrations > 6 M may be attributed to the dissolution of Nafion ionomer in the catalytic layer at the higher concentrations of methanol, or a decrease in membrane conductivity. When membrane thickness decreased, cell performance increased; but methanol crossover to the cathode also increased significantly. Due to the lower methanol crossover and better mechanical stability, rather thicker Nafion 117 is thus preferred for hydrogen production by aqueous methanol electrolysis.

An optimum temperature of 80°C has been recommended for high efficiency. Energy consumption is 1.48 kWh Nm^{-3} at 0.5 A cm^{-2} and 0.62 V (4 M methanol

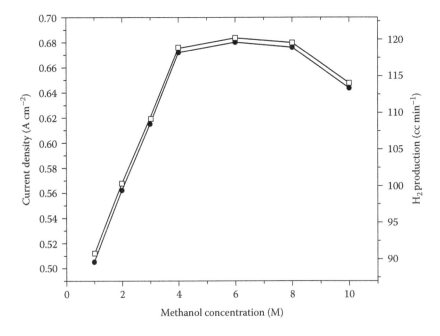

FIGURE 15.10

Effect of methanol concentration on current density and hydrogen production rate for a methanol electrolyzer at 1.0 V and 30°C with a Nafion 117 membrane. (□) current density, (●) hydrogen productivity. (Reprinted from Sasikumar, G. et al., *Int. J. Hydrogen Energy*, 33, 5905, 2008. With permission from International Journal for Hydrogen Energy.)

solution), which is much lower than the energy consumption of water electrolysis under similar conditions. The increase in temperature leads to an increase in current density at the same voltage values. This has also been confirmed by other researches (Take et al. 2007, Cloutier and Wilkinson 2010, Uhm et al. 2012).

The comparative results for water and methanol–water solution electrolysis are presented in Figure 15.11 (Uhm et al. 2012).

After analyzing the dependence of cell voltage and current efficiency on the operating conditions and properties of the electrocatalysts, the energy requirements necessary to produce the same amounts of hydrogen were compared. On the basis of 300 mA cm^{-2}

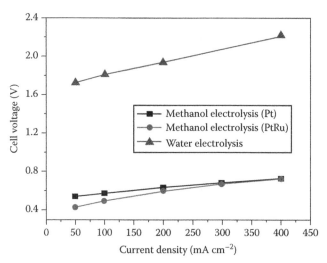

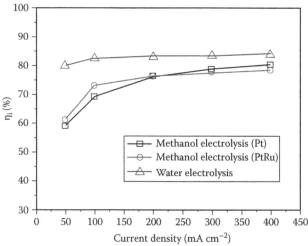

FIGURE 15.11

Cell voltage variations as a function of applied current and corresponding current efficiency. All experiments were carried out at 70°C with 1.0 M methanol. (Reprinted from *J. Power Sources*, 198, Uhm, S., Jeon, H., Kim, T.J., and Lee, J., Clean hydrogen production from methanol–water solutions via power-saved electrolytic reforming process, 218–222, Copyright 2012, with permission from Elsevier.)

operation with Pt–Ru or Pt catalysts, the methanol electrolysis can proceed at an effective potential of *ca.* 0.65 V with an energy consumption of 1.95 kWh Nm^{-3}. On the other hand, 5.5–6.0 kWh is required to produce 1 Nm^3 of hydrogen by water electrolysis, thereby consuming 65% more energy than a methanol electrolysis (Uhm et al. 2012).

15.4 Electrolysis of Aqueous Ethanol Solution

15.4.1 Mechanism of Ethanol Electrooxidation

The following reactions occur at the electrodes:

Anode:　$C_2H_5OH + 3H_2O - 12e^- \rightarrow 2CO_2 + 12H^+$

Cathode:　$2H^+ + 2H_2O \rightarrow H_2 \uparrow$

Overall:　$C_2H_5OH + 3H_2O \rightarrow 2CO_2 + 6H_2$

with $\Delta H° = -347.95$ kJ mol^{-1}, $\Delta S° = -839.1$ J K^{-1}, $\Delta E° = 0.085$ V,

where

　$\Delta H°$ is a thermal effect of the reaction

　$\Delta S°$ is entropy change

　$\Delta E°$ is decomposition voltage

The overall ethanol electrooxidation mechanism has been considered in details by Tarasevich and Kuzov (2010). The reaction involves the release of 12 electrons per 1 alcohol molecule, and is more complicated in comparison with the oxidation of methanol. Its parallel–series nature is presented schematically in Figure 15.12.

It was found that the main products of ethanol electrooxidation on Pt are acetic acid, acetaldehyde, and some CO$_2$ (Figure 15.12a). Acetic acid is the main product of ethanol electrooxidation at potentials in the range 0.5–0.8 V (Figure 15.12b). Oxidation of acetic acid commences at a significant rate on Pt only at 180°C–200°C. It can be assumed that the reaction through the C–C link gap after the partial dehydrogenation of adsorbed alcohol molecules is proceeded (Figure 15.12c).

A decrease in the overall reaction rate (resulting in a reduction of the specific activity) is indirectly related to the rate of the ethanol oxidation, which depends on the relative yield of carbon dioxide.

Increasing the adsorption energy of ethanol required to break C–C bond leads to the formation of M–CH$_2$ and M–CHOH particles, which oxidation rate to CO$_2$ is less than the oxidation rate of ethanol to acetaldehyde

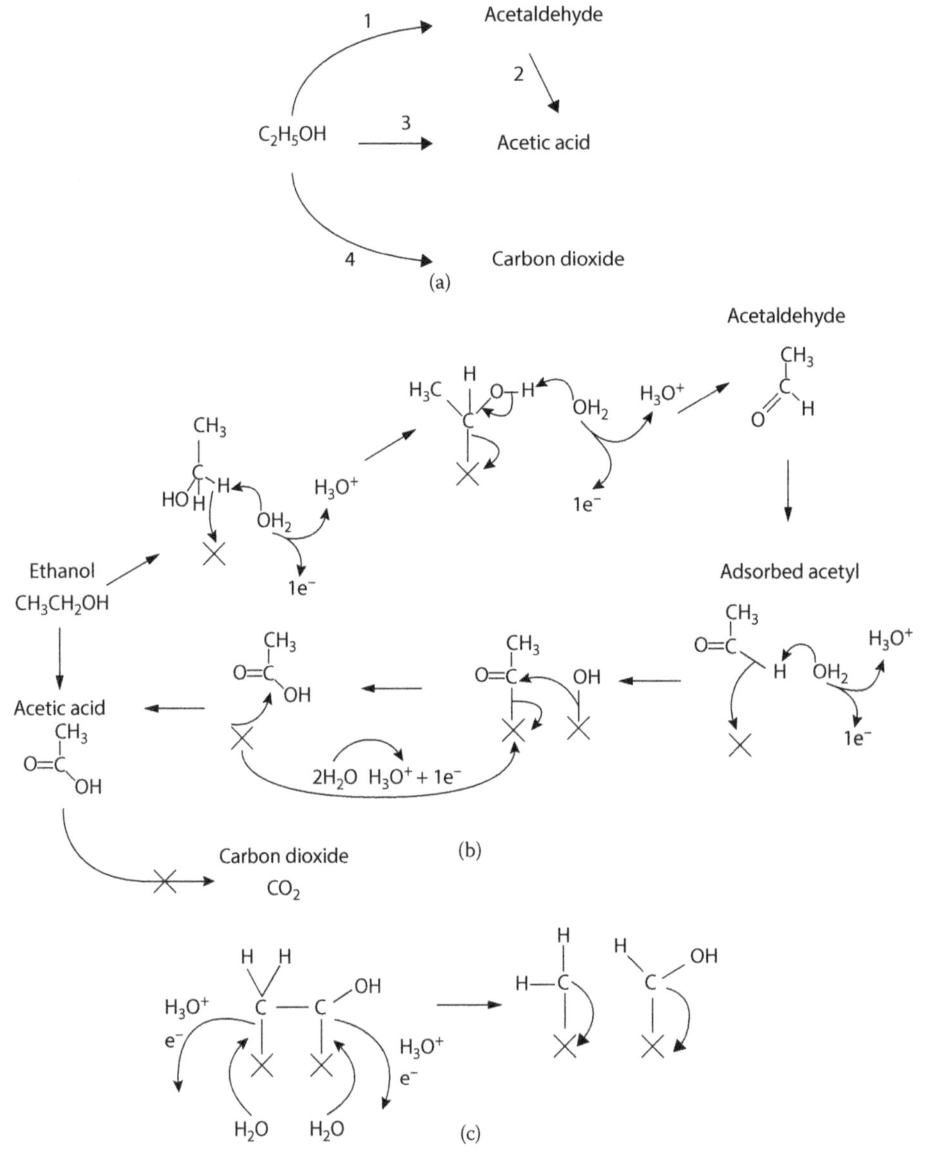

FIGURE 15.12
Schematic diagram of ethanol oxidation mechanism. a: Main products ; b: specific mechanism; c: reaction through C–C link gap (see Section 15.4.1 for more details). (Reprinted from Tarasevich, M.R. and Kuzov, A.V., *Int. Sci. J. Altern. Energy Ecol.*, 87, 86, 2010. With permission from Scientific Technical Centre "TATA.")

and acetic acid. The quantities of acetic acid and carbon dioxide have maximum values at low concentrations of ethanol (<0.1 M).

The yield of acetaldehyde is low because it undergoes further oxidation to acetic acid. Upon increasing the initial concentration of ethanol, the yield of carbon dioxide and acetic acid is reduced; however, the yield of acetaldehyde is increased significantly (Figure 15.13). This is apparently due to blocking of the active sites required for the formation of oxygen-containing particles involved in oxidative desorption of the products of incomplete ethanol oxidation.

Thus, in the scheme plotted in Figure 15.12, a, the reactions 1–3 are dead-ended, while reaction 4 proceeds

with breaking of a C–C bond, and only affords the final product via reaction 1 with transfer of 12 electrons. It was found that Pt may cause C–C bond breaking, but it does not provide an adequate degree of adsorption of oxygen-containing particles required for the final oxidation of adsorbates.

15.4.2 Electrocatalytic Materials

The main systems currently used as anode electrocatalysts in ethanol oxidation systems are the Pt-based ones modified by transition metals and other metals, and their oxides, supported on finely divided carbon carriers (Tsiakaras 2007, Kuzov et al. 2008, Korchagin

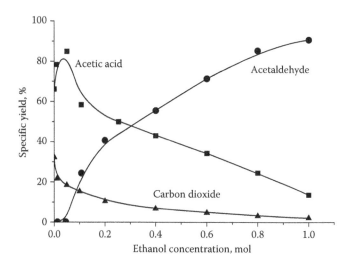

FIGURE 15.13

Specific yields of carbon dioxide, acetic acid, and acetaldehyde depending on the concentration of ethanol. (Reprinted from Tarasevich, M.R. and Kuzov, A.V., *Int. Sci. J. Altern. Energy Ecol.*, 87, 86, 2010. With permission from Scientific Technical Centre "TATA.")

et al. 2010, Tarasevich and Kuzov 2010, Lamy et al. 2014). The main objective of electrocatalysis of ethanol oxidation is to cause rupture of the C–C bond at temperatures <100°C. Furthermore, to complete the reaction to CO_2, electrocatalysts should provide adsorption of oxygenate particles. The following were studied as anode catalysts materials: Pt, Ru, Ir, Au, Co, and alloys of Pt/X (X = Ru, Mo, Co, Cr, Sn). However, very few of them (only Ru, Sn, Mo) lead to a noticeable positive effect. Table 15.2 and Figure 15.14 show comparisons of the various systems for which the catalytic activity of the electrochemical reaction of ethanol was studied.

Currently, the main catalysts for anode ethanol oxidation are Pt-based systems modified by transition and other metals, their oxides, supported on finely dispersed carbon carriers (Tsiakaras et al. 2007, Kuzov et al. 2008, Korchagin et al. 2010, Tarasevich and Kuzov 2010, Lamy et al. 2014). The main objective of electrocatalysis

of ethanol oxidation is breaking of C–C bond at temperatures below 100°C. Furthermore, to complete the reaction to CO_2, electrocatalysts should provide adsorption of oxygenate particles. Anode catalysts materials were studied for Pt, Ru, Ir, Au, Co, and alloys of Pt/X (X = Ru, Mo, Co, Cr, Sn). However, very few of them (Ru, Sn, Mo) lead to a noticeable positive effect. Table 15.2 and Figure 15.14 provide comparisons of the various catalytic systems.

Many researchers have noted the highest acceleration of ethanol oxidation with Pt–Sn under model conditions (Kuzov et al. 2008, Korchagin et al. 2010). Pt–M (M = W, Pd, Rh, Re, Mo, Ti, Se) catalysts have higher activity in the electrochemical reaction of ethanol electrooxidation compared to Pt, but lower than Pt–Ru and Pt–Sn. Some increase in activity of the Pt–Sn system can be achieved by introduction of a third component (Kuzov et al. 2008, Tarasevich and Kuzov 2010). An important feature in the effective anode catalyst development is an increase in metal loading on the support from 20 to 60 wt%, thereby reducing the thickness of the anode active layer.

A new approach in the preparation of ethanol oxidation catalysts is the use of nanostructured carbon supports such as carbon nanotubes (CNTs) and carbon nanofibers (CNFs). One of the benefits of the use of this approach is the ability to form the mesoporous structure of an anode active layer of such catalysts (Tarasevich and Kuzov 2010).

It was demonstrated that the anode electrode could be regenerated by the application of high potentials. This was attributed to the activation of water molecules to produce OH species, which leads to the oxidation of adsorbed intermediates (Caravaca et al. 2012).

15.4.3 Performances

A thermodynamic evaluation of ethanol decomposition voltage in a variety of media, under standard conditions,

TABLE 15.2

The Parameters of Ethanol Electrooxidation on Different Catalysts under Model Conditions (0.5 M H_2SO_4 + 1 M C_2H_5OH)

Catalyst	Carbon Support	Temperature, °C	Scan Rate, mV s^{-1}	Current, mA cm^{-2} at	
				$E = 0.4$ V	$E = 0.6$ V
PtRu	Vulcan XC72R	25	10	25	23
PtW				30	45
PtPd				20	25
PtSn				40	55
Pt				17	22
PtSn	Vulcan XC72	60	5	36	82
PtSn	CNT	60		57	215

Source: Tarasevich, M.R. and Kuzov, A.V., *Int. Sci. J. Altern. Energy Ecol.*, 87, 86, 2010.

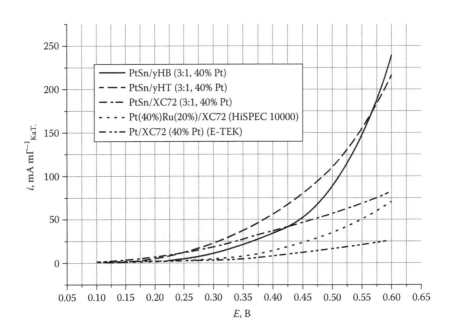

FIGURE 15.14
Polarization curves of ethanol electrooxidation on different catalysts in 0.5 M H_2SO_4 + 1 M C_2H_5OH at 60°C. (Reprinted from Tarasevich, M.R. and Kuzov, A.V., *Int. Sci. J. Altern. Energy Ecol.*, 87, 86, 2010. With permission from Scientific Technical Centre "TATA.")

including the decomposition to intermediates (acetaldehyde, acetic acid, ethylene glycol, etc.) indicates that in all processes, including the conversion of ethanol to the final products in aqueous solutions, its value does not exceed 0.1–0.7 V. This value is more than two times lower than the decomposition voltage of water into hydrogen and oxygen, which is equal to 1.23 V (Kuzov et al. 2008, Korchagin et al. 2010, Tarasevich and Kuzov 2010, Caravaca et al. 2012, Lamy et al. 2014).

Caravaca et al. (2012) have reported on the electrochemical reforming of ethanol–water solutions for pure H_2 production in a PEM electrolysis cell. Maximum current density values of up to 0.25 A cm^{-2} were measured at 80°C under application of 1.1 V, which corresponds to an energy consumption of 29.17 kWh kg^{-1} hydrogen or 2.62 kWh Nm^{-3} hydrogen. In earlier studies, higher values of current densities of up to 2 A cm^{-2} (Sasikumar et al. 2008) and 0.35 A cm^{-2} (Take et al. 2007) have been reported for the electrochemical reforming of methanol–water solutions under similar reaction conditions. However, the lower values obtained in Caravaca et al. (2012) for ethanol–water solutions could be offset by a number of advantages of ethanol versus methanol (e.g., availability, nontoxicity, and ease of storage), as mentioned in Section 15.1. It should be noted that optimization of the ethanol–water electro-reforming process (e.g., by the use of other catalysts) may, moreover, enhance the performance of the system for its further practical development (Caravaca et al. 2012, Lamy et al. 2014).

On the other hand, the ethanol electrolysis process seems to be much more efficient than the glycerol one, since much higher current densities (one order of magnitude larger) were obtained here in comparison to what is reported by Marshall and Haverkamp (2008) for glycerol reforming. This effect is probably related to the complexity of the molecules as a function of the C–C bonds, since their rupture represents one of the most difficult steps in the electrooxidation of alcohols (Caravaca et al. 2012).

The performance of the system was almost unaffected by the ethanol concentration at the low polarization range from 0.4 to 0.7 V. However, an important ethanol concentration effect was observed at potentials >0.7 V (Figure 15.15). Thus, the higher the ethanol concentration was in the feed, the higher was the current obtained under the application of a fixed potential. This behavior could be explained by considering the limiting step at each potential and concentration range. Under low polarization conditions, the electrolysis process may be controlled by the activation process and hence by the kinetics of the electrochemical reactions that occur in the system (Alzate et al. 2011), leading to a negligible effect of the ethanol composition in the feed. However, at higher potentials and higher current densities, the system started to be limited by ohmic losses and by surface phenomena (mass transfer), and hence the important effect of ethanol concentration was observed (Caravaca et al. 2012).

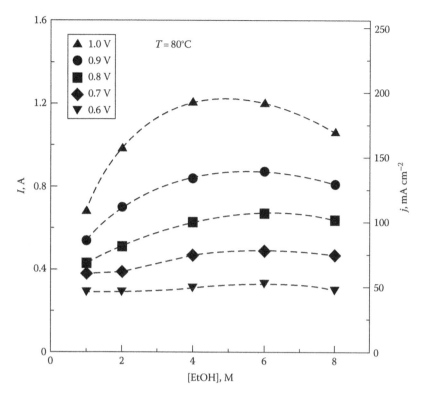

FIGURE 15.15

Effect of the ethanol concentration on the current and current density under the application of different cell potentials. Conditions: 1–8 M C_2H_5OH, 80°C. (Reprinted from Caravaca, A. et al., *Int. J. Hydrogen Energy*, 37, 9504, 2012. With permission from International Journal for Hydrogen Energy.)

The effect of the ethanol concentration from 1 to 8 M on the obtained current at different potentials is shown in Figure 15.15. As already discussed, at low polarizations (e.g., <0.8 V), the obtained current is almost independent of the ethanol concentration (activation process control step), whereas at higher potentials (>0.8 V) the current is strongly influenced by the ethanol concentration (surface phenomenon control step). In addition, at high polarizations ($V_{cell} > 0.8$ V), an optimum ethanol composition value of 6 M was obtained. This optimum composition value, which maximized the obtained current density (e.g., 0.2 A cm^{-2} for 1 V), clearly determined the change of the limiting process. Above an ethanol concentration of 6 M, the system started to be limited by ohmic losses because the conductivity of the membrane may decrease at high ethanol concentrations. A similar optimum methanol concentration of 6 M has been reported by Sasikumar et al. (2008) for the case of methanol electro-reforming, which has been attributed to changes in the membrane conductivity with the methanol concentration (Caravaca et al. 2012).

The voltage of a PEM electrolysis cell (5 cm² surface area) with ethanol depolarization did not exceed 0.9 V at 100 mA cm^{-2}, which corresponds to the electrical energy consumption of <2.3 kWh Nm^{-3}, that is, less than one half of the energy needed for water electrolysis (4.7 kWh Nm^{-3} at 2 V) (Lamy et al. 2014).

15.5 Electrochemical Reforming of Glycerol–Water Solutions

15.5.1 Electrochemistry of the Process

Using electrochemical reforming, it is possible to convert a biologically or synthetically produced glycerol to hydrogen (Marshall and Haverkamp 2008, Selembo et al. 2009). In this case, an aqueous solution of glycerol is supplied to the anode. Glycerol and water are oxidized to protons, which migrate through the membrane, carbon dioxide, and electrons. Pure hydrogen is evolved at the cathode.

The overall cell reaction for the electrochemical reforming of a glycerol–water solution is

$$C_3H_8O_3 + 3H_2O \rightarrow 3CO_2 + 7H_2$$

The anode and cathode reactions are given by

$$\text{Anode:} \quad C_3H_8O_3 + 3H_2O \leftrightarrow 3CO_2 + 14H^+ + 14e^-$$

$$\text{Cathode:} \quad 14H^+ + 14e^- \rightarrow 7H_2$$

Glycerol is a by-product of some industrial processes, in particular, soap manufacturing and biodiesel production. Given that biodiesel production rates are increasing considerably, a glut in the global glycerol market has developed, and there has been a reduction in the price of glycerol.

Glycerol is renewable, relatively safe, low toxic, non-flammable, and has a high boiling point. Furthermore, unlike other alcohols such as methanol and ethanol, glycerol does not swell the Nafion PEM. These factors make glycerol a very suitable compound for hydrogen production by PEM-based electrochemical reforming (Marshall and Haverkamp 2008, Selembo et al. 2009, Hunsom and Saila 2013).

15.5.2 Electrocatalysts

Electrochemical decomposition of glycerol–water solutions in a PEM electrolyzer was investigated by Marshall and Haverkamp (2008). $Pt^{20}/RuO_2–IrO_2$ (1:1) with 10% Nafion was used as anode catalyst and Pt^{20}/C with 10% Nafion as cathode catalyst. A membrane electrode assembly was prepared by spraying an ink-containing electrocatalyst and Nafion ionomer solution onto Nafion 212 membranes. The total catalyst loading was about 3 mg cm^{-2} per electrode.

Cyclic voltammetry of the prepared Pt^{20} on Ru–Ir oxide with Nafion in 0.5 mol/1 H_2SO_4 without glycerol revealed a large pseudocapacitive charging behavior from the oxide support, along with the characteristic peaks associated with Pt (Figure 15.16a).

When glycerol was added, the voltammograms changed dramatically, as expected, with large anodic currents due to the oxidation of glycerol (Figure 15.16b). On the forward scan, oxidation of glycerol commenced from about 0.5 V, with the current peaking around 0.65 V before decreasing, due to the formation of Pt oxide. On the reverse scan, the Pt oxide was reduced, allowing the glycerol oxidation to recommence.

15.5.3 Performances

The cell voltage required to drive an electrochemical reaction is about 0.22 V (Marshall and Haverkamp 2008). This is significantly lower than what is required

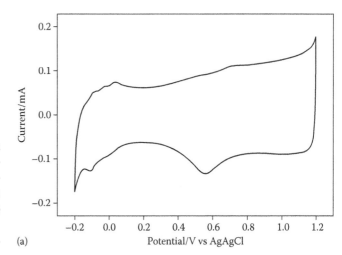

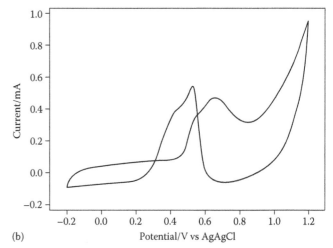

FIGURE 15.16

Cyclic voltammogram of glassy carbon electrode with 20% Pt/Ru–Ir (Ox) with 10% Nafion at 20 mV s^{-1} in (a) 0.5 mol/1 H_2SO_4, and (b) 0.5 mol/1 H_2SO_4 with 1 mol/1 glycerol. (Reprinted from Marshall, A.T. and Haverkamp, R.G., *Int. J. Hydrogen Energy*, 337, 4649, 2008. With permission from International Journal for Hydrogen Energy.)

for water electrolysis (1.23 V). Therefore, 18% of the energy carried by the hydrogen produced by this method is provided by electricity and 82% is provided by glycerol.

The performance of a PEM cell for glycerol reforming to produce hydrogen was assessed by recording polarization curves (Figure 15.17). Cell voltages in the range 0.48–0.7 V were obtained, compared with 1.33–1.40 V for water in the same cell. Results also show that the performance of the cell depends on the glycerol concentration.

The maximum hydrogen productivity of 0.37 g h^{-1} per 1 m^2 of electrode area has been achieved at a current density of 10 A m^{-2} and power cost of 1.1 kWh Nm^{-3}. Compared with water electrolysis, in the same cell and at the same H_2 production rate, electrochemical glycerol reforming saves around 2.1 kWh Nm^{-3} of hydrogen.

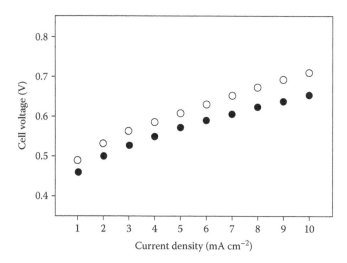

FIGURE 15.17

Polarization curves of PEM electrolysis cell at 70°C with (●) 2 mol L^{-1} and (○) 8.5 mol L^{-1} of glycerol. (Reprinted from Marshall, A.T. and Haverkamp, R.G., *Int. J. Hydrogen Energy*, 337, 4649, 2008. With permission from International Journal for Hydrogen Energy.)

The energy efficiency of conversion of glycerol to hydrogen is around 44%.

A possible explanation for the cell voltage increase, and therefore loss in performance, is poisoning of the catalyst by glycerol or its oxidation products. It was confirmed that the hydrogen evolution reaction is not inhibited by glycerol (in the case of glycerol crossover). As the reaction kinetics is slow due to poisoning of the anode electrocatalyst, development of more stable and efficient catalytic materials is required.

15.6 Electrochemical Decomposition of Formic Acid

15.6.1 Mechanism of the Process

Using biomass feedstocks is very promising for low-cost hydrogen production. The reason for this is that the theoretical cell voltage for the electrochemical decomposition of such compounds for hydrogen production is lower than the theoretical cell voltage of water electrolysis (1.23 V under standard conditions) (Kilic et al. 2009, Wang et al. 2009, Lamy et al. 2011).

Formic acid, a reactive hazardous pollution agent, is generally produced from oxidation processes of various compounds. Formic acid has to be removed from waste water before it can be discharged. Electrolysis with anode depolarization by formic acid allows, simultaneously, reduced pollution of waste water and extraction of clean energy in the form of hydrogen.

The schematic principle of electrochemical decomposition of water and formic acid in a PEM electrolysis cell is presented in Figure 15.18. Formic acid is supplied to the anodic compartment where it is oxidized, producing carbon dioxide and protons, as follows:

$$HCOOH + \frac{1}{2}O_2 \rightarrow CO_2 + 2H^+ + 2e^-$$

Protons crossover through the membrane and reach the cathodic compartment, where they are reduced to hydrogen

$$2H^+ + 2e^- \rightarrow H_2$$

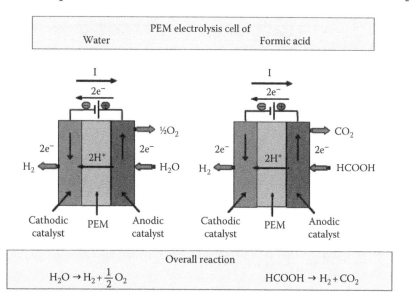

FIGURE 15.18

Schematic principle of electrochemical decomposition of water and formic acid in a PEM electrolysis cell. (Reprinted from *J. Power Sources*, 245, Lamy, C., Jaubert, T., Baranton, S., and Coutanceau, C. Clean hydrogen generation through the electrocatalytic oxidation of ethanol in a Proton Exchange Membrane Electrolysis Cell (PEMEC): Effect of the nature and structure of the catalytic anode, 927–936, Copyright 2012, with permission from Elsevier.)

This corresponds to the overall electrochemical decomposition of formic acid into hydrogen and carbon dioxide

$$HCOOH \rightarrow CO_2 + H_2$$

15.6.2 Thermodynamics

The thermodynamic characteristics under standard conditions are: $\Delta H = 31.5$ kJ mol^{-1}, $\Delta G = -33$ kJ mol^{-1}. Formic acid has a low decomposition energy compared to that required for water decomposition ($\Delta H \sim 286$ kJ mol^{-1}). Since $\Delta G < 0$, the decomposition proceeds spontaneously. The corresponding theoretical cell voltage can be calculated from ΔG, that is, $E_0 = \Delta G/2F$. Thus, $E_0 = -0.17$ V for formic acid decomposition and $E_0 = 1.23$ V for water electrolysis. However, the relatively slow kinetics of the anodic reaction in both processes leads to high anodic overvoltages, >1.8 V in the case of water electrolysis, where high current densities (<1 A cm^{-2}) are required for high hydrogen production rates.

To compare the cell voltages required for water electrolysis and formic acid electrolysis, current–voltage dependencies corresponding to the water oxidation reaction, formic acid reaction, hydrogen oxidation reaction (positive current density), oxygen reduction reaction, and proton reduction reaction (negative current density) are presented in Figure 15.19 (Lamy et al. 2011).

In order to obtain a competitive cost of energy for the production of hydrogen, the overvoltages have to be reduced to acceptable values. The development of new electrocatalysts is thus urgently needed. This has been

the subject of many studies in recent years (Yi et al. 2007, Wang et al. 2009, Lamy et al. 2011, Habibi 2013).

15.6.3 Catalysts and Reaction Mechanisms

Two parallel pathways for the electrooxidation of formic acid to CO_2 are proposed. The first reaction (direct reaction pathway) occurs without formation of adsorbed CO as intermediate product, whereas the second reaction (indirect reaction pathway) involves the reaction of formic acid on a Pt catalytic site to form a CO intermediate, which is further oxidized to CO_2.

Direct reaction pathway:

$$Pt_s + HCOOH \rightarrow Pt_s + CO_2 + 2H^+ + 2e^-$$

Indirect reaction pathway:

$$Pt_s + HCOOH \rightarrow Pt - CO + H_2O$$

$$Pt_s + H_2O \rightarrow Pt - OH + H^+ + e^-$$

$$Pt - CO + Pt - OH \rightarrow 2Pt_s + CO_2 + H^+ + e^-$$

where Pt_s represents a Pt catalytic site.

Such mechanisms are expected to also occur on Pd-based catalysts. The addition of Pd to Pt has led to the enhancement of the direct reaction pathway. Pd-based catalysts are less poisoned by CO particles (Lamy et al. 2011).

Bimetallic, Ti-supported nanoporous network Pt–Ir electrocatalysts fabricated by the hydrothermal process

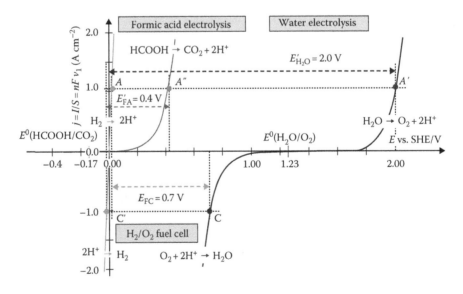

FIGURE 15.19
Comparison of the theoretical current–voltage curves obtained by the Butler–Volmer kinetics law for water oxidation, formic acid oxidation, oxygen reduction, and proton reduction. (Reprinted from *J. Power Sources*, 245, Lamy, C., Jaubert, T., Baranton, S., and Coutanceau, C., Clean hydrogen generation through the electrocatalytic oxidation of ethanol in a Proton Exchange Membrane Electrolysis Cell (PEMEC): Effect of the nature and structure of the catalytic anode, 927–936, Copyright 2012, with permission from Elsevier.)

were investigated for formic acid oxidation in (Yi et al. 2007). Three bimetallic compositions were studied: $Pt_{59}Ir_{41}/Ti$, $Pt_{44}Ir_{56}/Ti$, and $Pt_{22}Ir_{78}/Ti$. Experiments carried out in 0.5 M H_2SO_4 and 0.5 M HCOOH solutions showed that all three catalysts exhibited significant current densities in formic acid oxidation compared to a pure Pt catalyst. The highest anodic current density was achieved on $Pt_{44}Ir_{56}/Ti$.

Formic acid oxidation on these Pt–Ir electrocatalysts follows a "CO pathway," illustrated by the following reactions:

$$HCOOH + M \rightarrow M - CO + H_2O \qquad (15.4)$$

$$M + H_2O \rightarrow M - OH + H^+ + e^- \qquad (15.5)$$

$$M - CO + M - OH \rightarrow 2M + CO_2 + H^+ + e^- \qquad (15.6)$$

The formation of M–CO results in a decrease of active sites of the electrode surface, leading to a decrease in current density. The regeneration of the sites shown in reaction (Equation 15.6) leads to an increase in current density. High steady-state current densities indicate that Ir addition to a Pt electrocatalyst enhances the reaction (Equation 15.6).

Electrocatalysts based on Pd nanoparticles on an Al electrode has been investigated (Habibi 2013). Pd/Al has a higher surface area and was found to be more active than pure Pd catalysts.

15.6.4 Performances

Current–voltage curves of a water electrolyzer with anode depolarization by formic acid solution were measured using a test bench (Kostin et al. 2014), as shown in Figure 15.20.

Comparison of the current–voltage curves obtained for electrolysis of formic acid of various concentrations (Figure 15.21) allows us to conclude that the effect of electrolysis voltage reduction takes place at low formic acid concentrations (up to 0.1 M). It could be explained by insufficient efficiency of the Pt/C electrocatalyst applied for formic acid oxidation.

A series of Pd-based electrocatalysts (Pd/C, Pd_xAu_{1-x}/C, and Pd_xPt_{1-x}/C) were investigated by Lamy et al. (2011). Experiments in a single electrolysis cell with a 5 cm² geometric surface area were carried out by feeding the anode with 5 M formic acid solutions and the cathode with an acidic solution (0.5 M H_2SO_4) at room temperature. In all the experiments of Lamy et al. (2011), the specific amount of electrical energy was <2 kWh Nm⁻³ (since $U_{cell} < 0.9$ V), which is twice as low as the energy

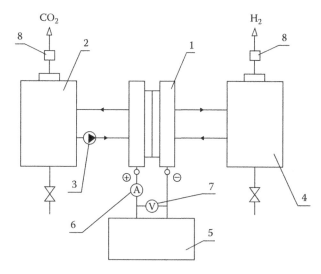

FIGURE 15.20
Schematic diagram of the test bench for investigating water electrolysis with anodic depolarization. 1. PEM electrolysis cell; 2. feed tank; 3. circulation pump; 4. separator; 5. direct current source; 6. ammeter; 7. voltmeter; 8. flowmeter. (With kind permission from Springer+Business Media: *Chem. Petrol. Eng.*, Hydrogen and sulfuric acid production by electrolysis with anodic depolarization by sulfurous anhydride, 44, 2014, 121–127, Kostin, V.I., Fateev, V.N., Bokach, D.A., and Korobtsev, S.V.)

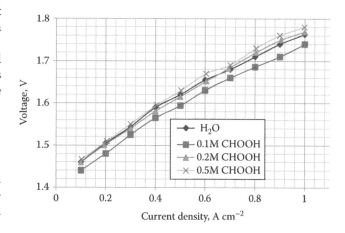

FIGURE 15.21
Current–voltage curves of water electrolysis with anodic depolarization in formic acid at different concentrations. (With kind permission from Springer+Business Media: *Chem. Petrol. Eng.*, Hydrogen and sulfuric acid production by electrolysis with anodic depolarization by sulfurous anhydride, 44, 2014, 121–127, Kostin, V.I., Fateev, V.N., Bokach, D.A., and Korobtsev, S.V.)

consumed for water electrolysis. For example, the power cost of 1.8 kWh Nm⁻³ of hydrogen was obtained at 0.2 A cm⁻² for $Pd_{0.8}Au_{0.2}/C$ as anodic catalyst.

Pd-based electrocatalysts investigated in Lamy et al. (2011) must be improved in terms of performance and stability. Development of more active and stable catalysts for the electrooxidation of organics is needed. The working current density has to be increased above 0.5 A cm⁻², keeping low cell voltages, in order to make the system competitive with the water electrolysis process. The stability of the catalyst (in particular, the tolerance to CO) needs to be improved.

15.7 Electrochemical Decomposition of Anhydrous Hydrogen Bromide

15.7.1 Hydrogen Bromide–Based Thermochemical Cycles

The Mark-13 cycle has been developed by General Atomics. The following reactions take place:

$$2HBr \rightarrow H_2 + Br_2 \quad \text{(electrolysis, } 80°C - 200°C\text{)}$$

$$Br_2 + SO_2 + 2H_2O \rightarrow 2HBr + H_2SO_4 \quad (20°C - 100°C)$$

$$H_2SO_4 \rightarrow H_2O + SO_2 + \frac{1}{2}O_2$$

$$\text{(thermolysis, } 700°C - 900°C\text{)}$$

Thermal decomposition of acid bromide in this cycle is replaced by the electrolytic decomposition, in which the following reactions proced on the electrodes with the theoretical potentials:

$$2Br^- - 2e^- \rightarrow Br_{2(l)} \quad E° = 1.065 \text{ V vs. SHE}$$

$$2H^+ + 2e^- \rightarrow H_2 \quad E° = 0 \text{ V vs. SHE}$$

Thermal decomposition of acid bromide could be replaced by the electrolytic decomposition in the Ca–Br thermochemical cycle (Figure 15.22) (Weidner 2005, Sivasubramanian et al. 2007, Weidner and Holland 2009).

The following reactions take place in the Ca–Br cycle:

$$CaO + Br_2 \rightarrow CaBr_2 + \frac{1}{2}O_2 \quad \text{at } 550°C$$

$$CaBr_2 + H_2O \rightarrow CaO + 2HBr \quad \text{at } 730°C$$

$$2HBr \rightarrow Br_2 + H_2$$

The last reaction in the Ca–Br cycle is carried out in PEM electrolyzer at 80°C, as follows:

Anode: $2HBr \rightarrow Br_2 + 2H^+ + 2e$ $E° = 0.58$ V vs. SHE
Cathode: $2H^+ + 2e^- = H_2$ $E° = 0$ V vs. SHE

Only water and energy are consumed in the Ca–Br thermochemical cycle. The liquid-phase electrolysis of HBr suffers mainly from low current densities due to mass

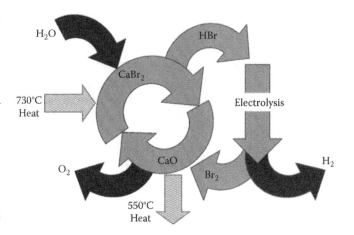

FIGURE 15.22
Schematic representation of the modified Ca–Br cycle. (Reprinted from Sivasubramanian, P.K. et al., *Int. J. Hydrogen Energy*, 32, 463, 2007. With permission from International Journal for Hydrogen Energy.)

transfer limitations and complicated product separation due to dissolution of Br_2 in solution. In efforts to exclude these limitations, gas-phase electrolysis has been attempted in phosphoric acid and molten salt cells (Sivasubramanian et al. 2007).

15.7.2 Performances

Figure 15.23 shows the current–voltage responses for the electrolysis of SO_2 and HBr.

Current–voltage dependence for HBr electrolysis shown in Figure 15.23 (grey rhombus) was obtained at 80°C, 1 bar, 2 mg cm^{-2} RuO_2 loading on the anode, and corresponds to 50% conversion efficiency. Current–voltage dependence for SO_2 electrolysis (black triangles) was measured at 80°C, 1 bar, and 0.6–0.7 mg cm^{-2}

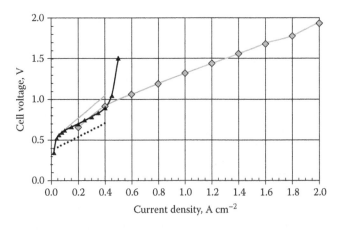

FIGURE 15.23
Current–voltage responses for SO_2 (black triangles) and HBr (grey rhombus) electrolysis in a PEM electrolyzer. (Reprinted from Sivasubramanian, P.K. et al., *Int. J. Hydrogen Energy*, 32, 463, 2007. With permission from International Journal for Hydrogen Energy.)

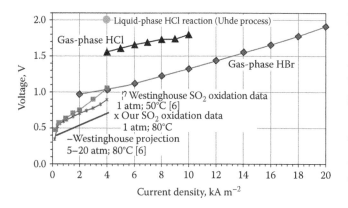

FIGURE 15.24
The current–voltage performances of SO₂, HCl, and HBr PEM electrolysis. (From Weidner, Low temperature electrolytic hydrogen production, DOE Hydrogen Program, Annual Progress Report, IV. I3, 2005, pp. 392–395.)

of Pt. The solid gray line comprises data from Sivasubramanian et al. (2007) in a liquid-phase electrolyzer at 50°C, at 1 bar, and with Pt loading on the anode and cathode 7 and 10 mg cm⁻², respectively. The dotted line is the targeted cell performance at 100°C and 5–20 bar (Sivasubramanian et al. 2007), which was not achieved. The main difference between the HBr and SO₂ electrolysis is that HBr electrolysis can be realized at significantly higher current densities.

Current–voltage curves of HBr electrochemical decomposition in a Ca–Br cycle as reported by Weidner (2005) are presented in Figure 15.24. The voltage of gas-phase HBr electrolysis was about 0.5 V below that of the gas-phase HCl reaction, and a current density of 2 A cm⁻² was achieved with no signs of oxygen evolution.

According to Figures 15.23 and 15.24, the HBr cycle looks rather promising because of its wider operating window (i.e., large current densities), lower cell voltage, requirement for a less costly catalyst (RuO₂ instead of Pt), and more stable operation. Energy consumption for HBr electrochemical decomposition is 4.3–5.1 for 1 m³ h⁻¹ of hydrogen at 0.4 A cm⁻² (Weidner 2005).

15.8 Electrochemical Decomposition of Anhydrous Hydrogen Chloride

15.8.1 Application of Hydrogen Chloride Electrolysis

More than 50% of the chlorine used in chemical industry ends up as hydrogen chloride. For example, in the poly(vinyl chloride) and polyurethane industries, millions of tons of HCl gas are produced as by-product.

Recovery of chlorine from by-product HCl could be mainly by either catalytic oxidation or electrolysis.

For catalytic oxidation, commercial processes under trade names such as "Kel-Chlor," "Shell-Chlor," or "MT-Chlor" have been implemented. For electrolysis, the only commercial electrolysis technology is offered by Uhde GmbH (Germany), and is based on the following process:

Cathode: $2H^+ + 2e^- \rightarrow H_{2(g)}$ $E = 0.00$ V vs. SHE

Anode: $2HCl_{(l)} \rightarrow Cl_{2(g)} + 2H^+ + 2e^-$ $E = 1.36$ V

Overall: $2HCl \rightarrow 2H_{2(g)} + 2Cl_{2(g)}$ $E = 1.36$ V

(Ziegelbauer et al. 2007, Gulla et al. 2007, Twardowski et al. 2008).

A gas-phase HCl electrolyzer (Figure 15.25) was developed by DuPont in 1996. The following reactions proceed:

Cathode: $2H^+ + 2e^- \rightarrow H_{2(g)}$ $E = 0.00$ V vs. SHE

Anode: $2HCl_{(g)} \rightarrow Cl_{2(g)} + 2H^+ + 2e^-$ $E = 0.987$ V vs. SHE

Overall: $2HCl_{(g)} \rightarrow 2H_{2(g)} + 2Cl_{2(g)}$ $E = 0.987$ V vs. SHE

Whereas only sufficiently hydrated Nafion membrane is effective as an ionic conductor, water circulation through the cathode chamber of the electrolyzer is required. Conductivity of Nafion equilibrated with acid is lower than the conductivity of water-equilibrated Nafion. Furthermore, Nafion conductivity, separation efficiency, chemical, and mechanical stability are influenced by membrane pretreatment (thermal or electrochemical). In the case of electrolysis of an aqueous solution of HCl, the acid concentration has an influence on proton conductivity (Motupally et al. 2002, Vidakovic-Koch et al. 2012).

15.8.2 Performances

An HCl electrolysis cell voltage of 1.55–1.80 V at current density of 0.4–1.0 A cm⁻² was achieved using RuO–IrO₂ electrocatalyst and Nafion as PEM (Weidner 2005) (Figure 15.25). It is further reported that the specific energy consumption is about 3.5 kWh Nm⁻³ of hydrogen if the current efficiency is 1.

As a result of a 5-year collaboration (1995–2000), DuPont, together with US Berkley, has increased the operating current density to 1.3 A cm⁻². Pt, RuO₂, and SnO₂ were used as anode catalysts, and Pt was applied as the cathode catalyst. Efficiency at high currents was <50%. Despite water diffusion from the hydrated H₂ to the dry Cl₂ electrode, due to the high current density, the Cl₂ product remains relatively dry (Motupally et al. 2002, Tolmachev 2013).

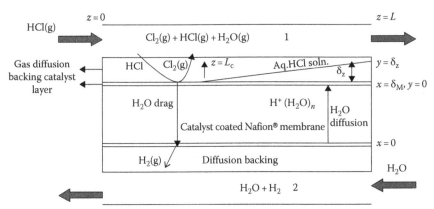

FIGURE 15.25
Schematic representation of the side view of the experimental cell of Motupally et al. (2002). Dry HCl and water are fed to flow channels 1 and 2, respectively. Water diffuses across the Nafion membrane from the liquid to the gas side due to a gradient in the activity of water. After the condensation of water, the diffusion across the membrane is governed by the concentration of HCl in the condensed phase. (Reprinted from Motupally, S. et al., *J. Electrochem. Soc.*, 149, D63, 2002. With permission from The Electrochemical Society.)

15.9 Summary

Anode depolarization (electrooxidation of various reducing agents instead of water) in PEM electrolysis reduces the energy consumption required for the production of electrolytic hydrogen by about 50%, on average.

Electrolysis with anode depolarization by sulfur dioxide, hydrogen bromide, and hydrochloric acid is appropriate as a low-temperature stage of hybrid thermochemical cycles. Hydrogen bromide electrolysis in the Ca–Br cycle is more efficient than other cycles due to a cheaper catalyst (RuO_2 instead of Pt), a wide range of current densities, low voltage, etc. However, SO_2 electrolysis has significant potential for developments in the field involving cheaper catalysts and reducing the voltage of electrolysis.

Anodic oxidation of methanol is a well-studied process because the same reaction proceeds in methanol PEM fuel cells. At the same time, low energy consumption for the production of hydrogen by electrolysis with depolarization by methanol makes this process feasible and promising for a number of industrial applications where utilization of methanol (as a by-product) is required.

The use of organic compounds as the depolarizers is limited by the low activity of conventional anode electrocatalysts because strong C–C bonds need to be broken. However, organic depolarizers (formic acid, glycerol, ethanol, etc.) are quite promising as potential raw materials.

Basically, the process of HCl electrolysis is considered as a source of chlorine (chlorine regeneration from by-products of chemical industry). R&D efforts have focused on reducing the energy consumption for the production of chlorine, including the use of cathode depolarization by oxygen. The energy consumption for the production of hydrogen in this process is high and comparable to the energy consumption for the electrolysis of water.

Of particular importance is the successful integration between a PEM electrolyzer with anode depolarization and power plants based on renewable and nuclear energy sources. In the latter case, the concept of so-called nuclear-hydrogen energy has been translated into action. The large-scale implementation of electrolysis with anode depolarization including hydrogen infrastructure facilities can provide additional impact to the commercialization of environmentally friendly hydrogen vehicles and stationary power plants for decentralized power supply.

Acknowledgment

This review was done at the expense of the Russian Scientific Foundation (project number 14-29-00111).

References

Alzate, V., Fatih, K., and Wang, H. 2011. Effect of operating parameters and anode diffusion layer on the direct ethanol fuel cell performance. *Journal of Power Sources* 196:10625–10631.

Cao, D. and Bergens S. H. 2003. A direct 2-propanol polymer electrolyte fuel cell. *Journal of Power Sources* 124:12–17.

Caravaca, A., Sapountzi, F. M., de Lucas-Consuegra, A., Molina-Mora, C., Dorado, F. and Valverde, J. L. 2012. Electrochemical reforming of ethanol-water solutions for pure H_2 production in a PEM electrolysis cell. *International Journal of Hydrogen Energy* 37:9504–9513.

Casalegno, A. and Marchesi, R. 2008. DMFC performance and methanol cross-over: Experimental analysis and model validation. *Journal of Power Sources* 185:318–330.

Cloutier, C. R. and Wilkinson, D. P. 2010. Electrolytic production of hydrogen from aqueous acidic methanol solutions. *International Journal of Hydrogen Energy* 35:3967–3984.

Colorn-Mercado, H. R. and Hobbs, D. T. 2007. Catalyst evaluation for a sulfur dioxide-depolarized electrolyzer. *Electrochemistry Communications* 9:2649–2653.

Corgnale, C. and Summers, W. A. 2011. Solar hydrogen production by the hybrid sulfur process. *International Journal of Hydrogen Energy* 36:11604–11619.

Elvington, M. C., Colon-Mercado, H., McCatty, S., Stone, S. G., and Hobbs, D. T. 2010. Evaluation of proton-conducting membranes for use in a sulfur dioxide depolarized electrolyzer. *Journal of Power Sources* 195:2823–2829.

Gorensek, M. B., Staser, J. A., Stanford, T. G., and Weidner, J. W. 2009. A thermodynamic analysis of the SO_2/H_2SO_4 system in SO_2-depolarized electrolysis. *International Journal of Hydrogen Energy* 34:6089–6095.

Gorensek, M. B. and Summers, W. A. 2009. Hybrid sulfur flowsheets using PEM electrolysis and a bayonet decomposition reactor. *International Journal of Hydrogen Energy* 34:4097–4114.

Grigoriev, S. A. and Bessarabov, D. G. 2012. PEM electrolysis with anode depolarization for production of hydrogen. *Alternative Fuel Transport* 6(30):69–71.

Gulla, A. F., Gancs, L., Allen, R. J., and Mukerjee, S. 2007. Carbon-supported low-loading rhodium sulfide electrocatalysts for oxygen depolarized cathode applications. *Applied Catalysis A*: General 326:227–235.

Habibi, B. 2013. Aluminum supported palladium nanoparticles: Preparation, characterization and application for formic acid electrooxidation. *International Journal of Hydrogen Energy* 38:5464–5472.

Hsieh, C. T. and Lin, J. Y. 2009. Fabrication of bimetallic Pt–M (M = Fe, Co, and Ni) nanoparticle/carbon nanotube electrocatalysts for direct methanol fuel cells. *Journal of Power Sources* 188:347–352.

Hunsom, M. and Saila, P. 2013. Product distribution of electrochemical conversion of glycerol via Pt electrode: Effect of initial pH. *International Journal of Electrochemical Science* 8:11288–11300.

Jung, H. J. and Jeong, Y. H. 2010. Development of the once-through hybrid sulfur process for nuclear hydrogen production for nuclear hydrogen production. *International Journal of Hydrogen Energy* 35:12255–12267.

Kilic, E. O., Koparal, A. S., and Ogutveren, U. B. 2009. Hydrogen production by electrochemical decomposition of formic acid via solid polymer electrolyte. *Fuel Processing Technology* 90:158–163.

Korchagin, O. V., Kuzov, A. V., Novikov, V. T., Bogdanovskaya, V. A., and Tarasevich, M. R. 2010. Optimized catalysts for a fuel cell with direct oxidation of ethanol. *Electrochemical Energy* 10(1):11–18 (in Russian).

Kostin, V. I., Fateev, V. N., Bokach, D. A., and Korobtsev, S. V. 2008. Hydrogen and sulfuric acid production by electrolysis with anodic depolarization by sulfurous anhydride. *Chemical and Petroleum Engineering* 44:121–127.

Kostin, V. I., Grigor'ev, S. A., Bessarabov, D. G., and Kryuger, A. 2014. Production of hydrogen by solid polymer electrolysis with anode depolarization. *Chemical and Petroleum Engineering* 49:575–578.

Kozolii, A. V. and Kostin, V. I. 2009. Pressure effect on water electrolysis with anode depolarization sulfur dioxide. *Electrochemical Energy* 10(1):34–37 (in Russian).

Kuzov, A. V., Tarasevich, M. R., and Korchagin, O. V. 2008. Anode catalysts for direct ethanol electrooxidation. *Fifth Baltic Conference on Electrochemistry*, April 30–May 3, 2008, Tartu, Estonia.

Lamy, C., Devadas, A., Simoes, M., and Coutanceau, C. 2011. Clean hydrogen generation through the electrocatalytic oxidation of formic acid in a Proton Exchange Membrane Electrolysis Cell (PEMEC). *Electrochimica Acta* 60:112–120.

Lamy, C., Jaubert, T., Baranton, S., and Coutanceau, C. 2014. Clean hydrogen generation through the electrocatalytic oxidation of ethanol in a Proton Exchange Membrane Electrolysis Cell (PEMEC): Effect of the nature and structure of the catalytic anode. *Journal of Power Sources* 245:927–936.

Lamy, C., Belgsir, E. M., and Leger, J. M. 2001. Electrocatalytic oxidation of aliphatic alcohols: Application to the direct alcohol fuel cell (DAFC). *Journal of Applied Electrochemistry* 31:799–809.

Leybros, J., Saturnin, A., Mansilla, C., Gilardi, T., and Carles, P. 2010. Plant sizing and evaluation of hydrogen production costs from advanced processes coupled to a nuclear heat source: Part II: Hybrid-sulphur cycle. *International Journal of Hydrogen Energy* 35:1019–1028.

Liu, H., Song, C., Zhang, L., Zhang, J., Wang, H., and Wilkinson, D. P. 2006. A review of anode catalysis in the direct methanol fuel cell. *Journal of Power Sources* 155:95–110.

Lokkiluoto, A. and Gasik, M. M. 2013. Modeling and experimental assessment of Nafion membrane properties used in SO_2 depolarized water electrolysis for hydrogen production. *International Journal of Hydrogen Energy* 38:10–19.

Lokkiluoto, A., Taskinen, P. A., Gasik, M., Kojo, I. V., Peltola, H., Barker, M. H., and Kleifges, K.-H. 2012. Novel process concept for the production of H_2 and H_2SO_4 by SO_2-depolarized electrolysis. *Environment, Development and Sustainability* 14:529–540.

Marshall, A. T. and Haverkamp, R. G. 2008. Production of hydrogen by the electrochemical reforming of glycerol-water solutions in a PEM electrolysis cell. *International Journal of Hydrogen Energy* 337:4649–4654.

Motupally, S., Becker, A. J., and Weidner, J. W. 2002. Water transport in polymer electrolyte membrane electrolyzers used to recycle anhydrous HCl. *Journal of the Electrochemical Society* 149:D63–D71.

Opperman, H., Kerres, J., and Krieg, H. 2014. SO_2 crossover flux of Nafion® and sFS-PBI membranes using a chronocoulometric (CC) monitoring technique. *Journal of Membrane Science* 415–416:842–849.

Peach, R., Krieg, H. M., Kruger, A. J., van der Westhuizen, D., Bessarabov, D., and Kerres, J. 2014. Comparison of ionically and ionical-covalently cross-linked polyaromatic membranes for SO_2 electrolysis. *International Journal of Hydrogen Energy* 39:28–40.

Peled, E., Livshits, V., and Duvdevani, T. 2002. High-power direct ethylene glycol fuel cell (DEGFC) based on nanoporous proton-conducting membrane (NP-PCM). *Journal of Power Sources* 106:245–248.

Pham, A. T., Baba, T., and Shudo, T. 2013b. Efficient hydrogen production from aqueous methanol in a PEM electrolyzer with porous metal flow field: Influence of change in grain diameter and material of porous metal flow field. *International Journal of Hydrogen Energy* 38:9945–9953.

Pham, A. T., Baba, T., Sugiyama, T., and Shudo, T. 2013a. Efficient hydrogen production from aqueous methanol in a PEM electrolyzer with porous metal flow field: Influence of PTFE treatment of the anode gas diffusion layer. *International Journal of Hydrogen Energy* 38:73–81.

Sasikumar, G., Muthumeenal, A., Pethaiah, S. S., Nachiappan, N., and Balaji, R. 2008. Aqueous methanol electrolysis using proton conducting membrane for hydrogen production. *International Journal of Hydrogen Energy* 33:5905–5910.

Schoeman, H., Krieg, H. M., Kruger, A. J., Chromik, A., Krajinovic, K., and Kerres, J. 2012. H_2SO_4 stability of PBI-blend membranes for SO_2 electrolysis. *International Journal of Hydrogen Energy* 37:603–614.

Selembo, P. A., Perez, J. M., Lloyd, W. A., and Logan, B. E. 2009. High hydrogen production from glycerol or glucose by electrohydrogenesis using microbial electrolysis cells. *International Journal of Hydrogen Energy* 34:5373–5381.

Sethuraman, A. V., Khan, S., Jur, J. S., Haug, A. T., and Weidner, J. W. 2009. Measuring oxygen, carbon monoxide and hydrogen sulfide diffusion coefficient and solubility in Nafion membranes. *Electrochimica Acta* 54:6850–6860.

Shin, Y., Shin, H., Lee, J., and Kim, Y. 2012. A Hybrid-sulfur flowsheet using an ionic liquid absorbent to separate sulfur dioxide from oxygen. *Energy Procedia* 29:576–584.

Sivasubramanian, P. K., Ramasamy, R. P., Freire, F. J., Holland, C. E., and Weidner, J. W. 2007. Electrochemical hydrogen production from thermochemical cycles using a proton exchange membrane electrolyzer. *International Journal of Hydrogen Energy* 32:463–468.

Staser, J. A. and Weidner, J. W. 2008. Effect of water transport on the production of hydrogen and sulfuric acid in a PEM electrolyzer. *Journal of the Electrochemical Society* 159:B16–B21.

Take, T., Tsurutani, K., and Umeda, M. 2007. Hydrogen production by methanol–water solution electrolysis. *Journal of Power Sources* 164:9–16.

Tarasevich, M. R. and Kuzov, A. V. 2010. Direct alcohol fuel cells. *International Scientific Journal for Alternative Energy and Ecology* 87:86–108.

Tolmachev, Y. V. 2013. Hydrogen-halogen electrochemical cells: A review of applications and technologies. *Russian Journal of Electrochemistry* 50(4):301–316. DOI: 10.1134/S1023193513120069.

Tsiakaras, P. E. 2007. PtM/C (M = Sn, Ru, Pd, W) based anode direct ethanol–PEMFCs: Structural characteristics and cell performance. *Journal of Power Sources* 171:107–112.

Twardowski, Z., Drackett, T., and Harper, S. R. 2008. Mediated hydrohalic acid electrolysis. U.S. Patent No. 7,341,654 B2.

Uhm, S., Jeon, H., Kim, T. J., and Lee, J. 2012. Clean hydrogen production from methanol–water solutions via power-saved electrolytic reforming process. *Journal of Power Sources* 198:218–222.

Vidakovic-Koch, T., Martinez, I. G., Kuwerts, R., Kunz, U., Turek, T., and Sundmacher, K. 2012. Electrochemical membrane reactors for sustainable chlorine recycling. *Membranes* 2:510–528.

Wang, J., Yin, G., Chen, Y., Li, R., and Sun, X. 2009. Pd nanoparticles deposited on vertically aligned carbon nanotubes grown on carbon paper for formic acid oxidation. *International Journal of Hydrogen Energy* 34:8270–8275.

Weidner, J. W. 2005. Low temperature electrolytic hydrogen production. DOE Hydrogen Program, 2005 Annual Progress Report, IV. I3, pp. 392–395.

Weidner, J. W. and Holland, C. E. 2009. Production of low temperature electrolytic hydrogen. U.S. Patent No. 2009/0000956 A1.

Xue, L., Zhang, P., Chen, S., Wang, L., and Wang, J. 2013. Sensitivity study of process parameters in membrane electrode assembly preparation and SO_2 depolarized electrolysis. *International Journal of Hydrogen Energy* 38:11017–11022.

Yi, Q., Chen, A., Huang, W., Zhang, J., Liu, X., Xu, G., and Zhou, Z. 2007. Titanium-supported nanoporous bimetallic Pt-Ir electrocatalysts for formic acid oxidation. *Electrochemistry Communications* 9:1513–1518.

Zhao, T. S., Yang, W. W., Chen, R., and Wu, Q. X. 2010. Towards operating direct methanol fuel cells with highly concentrated fuel. *Journal of Power Source* 195:3451–3462.

Ziegelbauer, J. M., Gulla, A. F., O'Laoire, C., Urgeghe, C., Allen, R. J., and Mukerjee, S. 2007. Chalcogenide electrocatalysts for oxygen-depolarized aqueous hydrochloric acid electrolysis. *Electrochimica Acta* 52:6282–6294.

16

Generation of Ozone and Hydrogen in a PEM Electrolyzer

Dmitri Bessarabov

CONTENTS

16.1 Introduction

16.1.1 Brief Review of Ozone Generation Methods

Ozone has been used for almost 100 years in water treatment in Europe, where it is largely used for disinfection, the control of taste and odor, and the removal of color. The role of ozone in the treatment of potable water supplies and waste water is as a disinfectant and a powerful oxidant. As a disinfectant, ozone successfully inactivates enteric bacteria, viruses, amoebic cysts, and spores. As an oxidant, ozone oxidizes many inorganic materials completely and rapidly, for example, sulfides to sulphates, nitrites to nitrates, etc. Ozone also oxidizes organic materials such as unsaturated and aromatic compounds, which are oxidized and cleaved at the double bonds; humates and fulvates, which are commonly found in potable water supplies are effectively bleached; and foul-tasting phenol materials are readily destroyed. In addition to the direct reactions of molecular ozone described earlier, ozone can also react indirectly via the radical species formed when ozone decomposes in water. The production of hydroxyl radicals (•OH) can be enhanced by an increase in pH, addition of hydrogen peroxide, or irradiation with ultraviolet (UV) light. Ozone competes with chlorine-based technologies on the water treatment market. The safety of ozone treatment plants is an advantage over chlorine plants.

Ozone can be produced using pure oxygen or air by means of the corona discharge method or via UV irradiation. Ozone cannot be transported; it has to be produced on site. Ozone production by means of the corona discharge method requires a permanent source of oxygen. An additional requirement is a low dew point of the oxygen to be used for ozone generation. Technological developments over the past decade have led to corona discharge generators capable of producing ozone in concentrations of up to 15% w/w when oxygen is used as the feed gas. More recent developments in electrochemical ozone generation have led to high-concentration ozone (HCO), with concentrations as high as 40% w/w.

Ozone, the triatomic form of oxygen (O_3), is an unstable compound that decomposes spontaneously, or by contacting with oxidizable chemicals, producing oxygen, hydroxyl radicals, and other free radical species. Ozone is a very powerful oxidant (E = +2.07 V) that can react with numerous organics present in water. Table 16.1 shows a list of a few selected chemical reagents known as strong oxidants, including ozone (Langlais et al. 1991).

There are numerous applications for ozonation in waste water treatment and a few new promising applications have recently been identified (Langlais et al. 1991). These include, for example, ozonation of pulp and paper mill waste water, municipal waste water, and waste water contaminated with pesticides. One of the main disadvantages of using ozone for water disinfection, however, is the lack of a residual in water distribution systems. Ozone has also shown promise to inactivate cryptosporidium oocysts in water (Rennecker et al. 1999).

TABLE 16.1

Some Strong Oxidants

F_2 (gas) + 2e$^-$ → 2F$^-$ (aq)	E_o = +2.87 V
OH• + H$^+$ + e$^-$ → H$_2$O	E_o = +2.42 V
O_3 + 2H$^+$ + 2e$^-$ → O_2 + H$_2$O	E_o = +2.07 V
MnO$_4^-$ + 8H$^+$ + 5e$^-$ → Mn$^+$ + 4H$_2$O	E_o = +1.51 V
Cl$_2$ (gas) + 2e$^-$ → 2Cl$^-$ (aq)	E_o = +1.36 V

Source: Data from Langlais, B. et al., Ozone in water treatment. Application and engineering, Cooperative Research Report, AWWA Research Foundation, Lewis Publishers, Boca Raton, FL.

The benefits of using ozone in drinking water treatment and in the food industry are also well known. They include the following (Kim et al. 1999): improvement of organoleptic properties of water, enhancement of coagulation and filtration processes, oxidation of color-causing compounds and some inorganic chemicals, control and reduction of microbiological growth potential of the water and its biological stabilization, and a superior disinfection performance when compared to chlorine. It is further expected that the use of ozone-based technologies in the food industry should be boosted by a recent formal approval of the use of ozone as an antimicrobial agent for the treatment, storage, and processing of foods in gas and aqueous phases by the U.S. Food and Drug Administration (Rice and Graham 2001). The earlier mentioned benefits and new processes have been the reasons for an increase in the number of ozone installations worldwide.

The formation of ozone from oxygen is presented as follows:

$$3O_2 \rightarrow 2O_3 \qquad (16.1)$$

It is an endothermic reaction, with a large and unfavorable entropy:

$$\Delta H^\circ_{1atm} = +284.5 \text{ kJ mol}^{-1}, \quad \Delta S^\circ_{1atm} = -69.9 \text{ J mol}^{-1} \text{ K}^{-1}$$

Liquefying ozone by compression may result in spontaneous explosions. Transportation of ozone is potentially hazardous, hence ozone is generally manufactured on site, where it is used.

Ozone is currently most widely commercially produced by means of the silent electric discharge process, wherein air or oxygen is passed through an intensive, high-frequency alternating-current electric field. However, the conventional corona discharge ozone generation method has some shortcomings. These include high capital costs, requirement for an external source of dry oxygen, and only a relatively low ozone concentration is obtained. Typical values of ozone concentration obtained are 3 wt.% for corona discharge with an air feed, and up to 15 wt.% with an oxygen feed (Lang et al. 1993).

For smaller-scale generators, the practicalities and economics of operation mean that air is generally used, and then extensive drying is required for stable operation, which, in turn, introduces additional costs and concerns about reliability. In case of the corona discharge method with an air feed (which naturally contains nitrogen), the following side reactions may occur:

$$O_2 + N_2 \rightarrow 2NO \qquad (16.2)$$

$$O_2 + N_2 \rightarrow 2NO \qquad (16.3)$$

$$N_2O_5 + H_2O \rightarrow 2HNO_3 \qquad (16.4)$$

These side reactions limit the use of air in ozone production for water treatment. Another issue for consideration in some circumstances is the safety of working with high-voltage equipment.

Ozone can also be produced photochemically. In this case, the formation of ozone takes place when oxygen is exposed to light at 140–190 nm. The shortcomings of a UV-based method, however, include low quantum yield of ozone formation from oxygen in comparison to relatively high quantum yield of photolysis of ozone, resulting in low concentrations of ozone obtained. The typical value for UV ozone generation is 0.2 wt.%. Except for small-scale uses or synergetic effects, the UV-photochemical generation of ozone has not yet reached maturity.

In recent years, the electrochemical generation of ozone by the electrolysis process has received increasing interest; it is an environmentally friendly process (ozone can be directly dispensed in water in the electrolysis cell), and the by-product of the electrolysis, hydrogen, can be utilized as a fuel in fuel cell systems (Christensen et al. 2009, 2013, Wang and Chen 2013).

16.1.2 Overview of Concepts and Challenges for Production of Ozone by Electrolysis

It has been known for over a century that ozone can be synthesized by electrolysis. Ozone can be generated electrochemically by the anodic oxidation of water (Foller 1979, Foller and Tobias 1982a,b, Menth and Stucki 1982, Foller and Goodwin 1984, Foller 1985, Stucki et al. 1985, Shimamune et al. 1991, Couper and Bullen 1992, Nakamatsu et al. 1993, Tatapudi and Fenton 1993, Shimamune and Sawamoto 1995, 1997, Murphy and Hitchens 1999).

$$9H_2O \rightarrow 6H_3O^+ + 6e^- + O_3 \qquad (16.5)$$

Although the use of this electrochemical reaction for ozone generation has been known for a long time, no entirely suitable system that has the potential for being sufficiently attractive as the basis of a commercial generator of gaseous ozone of high concentration has yet been identified. Thermodynamically, the preferred reaction is the generation of oxygen.

$$6H_2O \rightarrow 4H_3O^+ + 4e^- + O_2 \qquad (16.6)$$

General conditions for how ozone can be generated commercially were developed during the early 1980s

by Foller and Tobias at the University of California, Berkeley, CA (Foller and Tobias 1982a, Foller 1985). Further major commercialization steps were taken by OxyTech, Inc. and ICI Chemicals & Polymers Ltd. (Couper and Bullen 1992).

The advantages of using proton exchange membranes (PEMs) in electrochemical reactors for ozone generation were later clearly demonstrated (Stucki et al. 1985, Tatapudi and Fenton 1993). In particular, it was shown that the performance of an ozone reactor strongly correlated with properties of the membrane and its interface with the anode.

Table 16.2 summarizes the methods of ozone generation described earlier in terms of typical ozone concentration values that can be achieved by each method (Bessarabov 2000, Christensen et al. 2013, Wang and Chen 2013).

Historically, several companies have been involved in the development of commercially available electrochemical ozone generators (Stucki et al. 1985, Tatapudi and Fenton 1993) under various trade names, such as: MEMBREL™, FISHER, LynnTech Inc., D-Ozone, "ECHO" generator by ICI Watercare, etc. for specific small-scale applications, such as, for example, sterilization in military field hospitals and water treatment.

Most military field hospitals require fast sterilization capability with minimal logistical support for cleaning surgical tools and other medical devices. Most U.S. hospitals and medical device manufacturers, for example, use ethylene oxide as the primary gas for sterilization. This presents a problem in the field hospital applications because ethylene oxide-based sterilization is a lengthy process. However, the commercial success was very limited, if any, in spite of the fact that ozone as a sterilant can avoid problems with toxicity, flammability, air pollution, sterilant residue.

There is a need for further optimization of the electrochemical ozone generation system, aiming at the reduction of power consumption, through better anode catalyst materials and design of a catalyst layer. Additional research should focus on membranes and their interactions with ozone and catalysts that are used in such systems in order to achieve better performance and durability of the entire

TABLE 16.2

Ozone Generation Methods

Method of Ozone Generation	Typical Concentration of Ozone (wt.%)	Typical Power Consumption (kWh kg^{-1} O$_3$)[a]
UV	0.01–0.2	20–300
Air-fed corona discharge	0.5–4	15–20
O$_2$-fed corona discharge	4–15	9–20
Anodic oxidation of water	15–40	50–175

[a] Power consumption depends on the ozone concentrations.

ozone generation system. There are two approaches in the electrochemical generation of ozone.

The first approach includes the use of a PEM electrolyzer system with deionized water feed, a suitable PEM, and specially selected anodic and cathodic catalysts as electrodes. This approach is very similar to the standard PEM water electrolysis technology.

The second approach includes the use of a suitable electrolyte dissolved in water and suitable inert electrodes resistant to the conditions of strong electrical polarization occurring during ozone generation. Several relevant original patents are available (Foller and Tobias 1982a, Foller 1985). Some liquid electrolytes suitable for ozone generation include fluoro anions such as F^-, BF_4^-, BF_6^-, etc. (Bessarabov 2000, Christensen et al. 2009, 2013, Wang and Chen 2013). Earlier investigations of the anodic oxidation of water to ozone almost exclusively employed H_2SO_4 or $HClO_4$ as an electrolyte. Platinum (Pt) anodes were used.

The following fundamental criteria are applied to the selection of a suitable electrolyte:

- The anion must be resistant to further oxidation
- The cation should undergo no cathodic reactions, nor be influenced by pH changes near the electrodes
- The salt combination must be of sufficient solubility
- No reactions should take place with the ozone produced

Although both approaches are mentioned in this chapter, the focus will be on the PEM-based technology, similar to standard PEM water electrolysis. For example, U.S. Patent 4,416,747 (Menth and Stucki 1982) relates to the process for the synthetic production of ozone by electrolysis, wherein water saturated with oxygen is used as raw material and, as the electrolyte, a solid electrolyte is used, which is coated with Pt on the cathode side and with PbO_2 on the anode side and is in the form of a thin cation exchange membrane. The membrane is made of the perfluorinated cation-exchange material. In the process, the oxygen-saturated water is supplied both on the cathode side and on the anode side, parallel to the solid electrolyte membrane, and the electrical current is supplied to the coatings of the solid electrolyte, which serve as electrodes.

U.S. Patent No. 5,203,972 (Shimamune et al. 1991) describes an electrolytic ozone generator comprising a perfluorinated cation-exchange (proton exchange solid electrolyte) membrane with a catalyst embedded on the membrane and thus separating the anode chamber and the cathode chamber from each other. An anode catalyst is usually the β-crystalline form of PbO_2. The water is fed from the cathode side of the electrolytic cell only.

U.S. Patent No. 5,326,444 (Nakamatsu et al. 1993) describes an electrolytic ozone generating apparatus comprising an anode, a perfluorinated cation-exchange membrane, and a cathode. Oxygen-containing gas is supplied to the cathode (oxygen reduction electrode) to minimize electrical power consumption. Water is used as a raw material.

U.S. Patent No. 5,407,550 (Shimamune and Sawamoto 1995) describes a process of manufacturing electrodes for ozone production. A perfluorinated cation-exchange membrane is used along with a beta-crystalline form of lead dioxide as the anode catalyst.

U.S. Patent No. 5,607,562 (Shimamune and Sawamoto 1997) describes an improved electrolytic ozone generator containing a perfluorinated cation-exchange membrane between the anode and cathode. Deionized water is used as a raw material for ozone production. Lead dioxide anodes are used in the system.

U.S. Patent No. 5,972,196 (Murphy and Hitchens 1999) describes a method of electrochemical production of ozone and hydrogen peroxide. The method includes anodic oxidation of water on the lead-dioxide-based anodes separated from pyrolyzed cobalt-porphyrin-containing cathodes by perfluorinated cation-exchange membranes.

Potential advantages of the ozone generating processes described earlier include the following: use of low-voltage DC current, no feed-gas preparation, no pressure-swing adsorption or oxygen cylinders, reduced equipment size, generation of ozone in high concentrations, generation of ozone directly in water (hence eliminating ozone-to-water contacting processes), greater transfer efficiencies in contacting systems due to higher ozone concentration, higher ozone residual, and the possibility of simultaneous generation of ozone and H_2O_2 in a single reactor to increase oxidative efficiency of the system (Bessarabov 2000).

Possible disadvantages include the following: corrosion and erosion of membranes and electrodes, thermal overloading due to high current densities, a need for special electrolytes or water with low conductivity, and higher electrical consumption required than in corona discharge systems (Bessarabov 2000).

16.2 Description of the Process: Electrochemical Ozone Generation

16.2.1 Reaction Mechanisms

The exact mechanisms of electrochemical ozone generation are still under discussion because it was demonstrated that the ozone evolution reaction cannot be optimized by raising the anode potential alone to its highest limits (Foller 1979). The fact is that, in many cases, anionic additives are required to improve the electrochemical efficiency of the ozone generation. It has been shown that both electrolyte pH and anionic strength have great influence on the ozone yields (Foller 1979, Bessarabov 2000). For example, one of the highest ozone current efficiencies (>50%) was demonstrated with using β-PbO_2 at 0°C in 7.3 M HBF_6 by Foller and Tobias (1982b). The general consensus is that the following reactions to produce hydroxyl radicals are key for ozone generation (Christensen et al. 2013, Wang and Chen 2013):

$$H_2O \rightarrow OH^{\bullet}_{ads} + H^+ + e^- \quad (16.7)$$

$$2OH^{\bullet}_{ads} \rightarrow O_{2ads} + 2H^+ + 2e^- \quad (16.8)$$

$$OH^{\bullet}_{ads} + O_{2ads} \rightarrow {}^{\bullet}HO_{3ads} \quad (16.9)$$

$${}^{\bullet}HO_{3ads} \rightarrow H^+ + O_3 + e^- \quad (16.10)$$

16.2.2 Role of the PEM

The key phenomenon in the electrocatalytic method of ozone production is the anodic oxidation of deionized water on the surface of a PEM which contains a suitable electrocatalyst with high overpotential for oxygen evolution.

Figure 16.1 shows a schematic representation of a perfluorosulfonic acid (PFSA) cation-exchange membrane with two different catalysts in contact with it (Grubb 1959, Nutall 1977, Bessarabov 1998, 1999).

An electric direct current (DC) is applied to the catalyst layers of the membrane. The anode is marked as "+" and the cathode side is marked as "−," and deionized water is fed to the cathode side of the membrane. The electrochemical reactions taking place on the anode and cathode are also shown in the figure. Formation of hydrogen (at least 99.99%, as confirmed by gas chromatography or GC) takes place on the cathodic catalyst.

Due to high water sorption and relatively high permeability of PFSA membranes, water molecules diffuse to the anode side of the reactor, where formation of humidified oxygen and ozone in gaseous form takes place. Feeding a PEM reactor with water from the anode side could also be an option when ozone must be generated directly into deionized water.

Membranes that are used in such a process are generally manufactured from perfluorinated cation-exchange materials. These materials have certain advantages such as: high ionic conductivity, outstanding chemical and thermal stability, and good mechanical strength.

Reduction of protons to hydrogen gas takes place on the cathode side of the membrane. It is also possible to arrange an operation of the electrochemical ozone

$$4H_3O^+ + 4e^- \longrightarrow 4H_2O + 2H_2$$

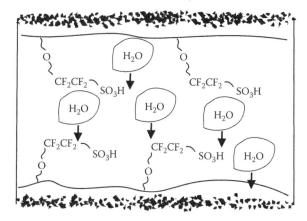

Thin layer of catalyst

Thin layer of catalyst

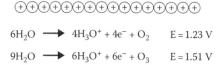

$$6H_2O \longrightarrow 4H_3O^+ + 4e^- + O_2 \qquad E = 1.23\ V$$

$$9H_2O \longrightarrow 6H_3O^+ + 6e^- + O_3 \qquad E = 1.51\ V$$

FIGURE 16.1

Schematic representation of the principle of water electrolysis on a perfluorinated cation-exchange membrane containing cathodic (−) and anodic (+) electrocatalysts. Cathodic and anodic catalysts are embedded into the PEM. (Reprinted from *Membrane Technology*, 114, Bessarabov, D.G., Membranes help to produce high-concentration ozone: New challenges, 5–8, Copyright 1999, with permission from Elsevier.)

generator in such a way that cathodic reduction of oxygen (or air) to water or hydrogen peroxide may take place. These are fuel-cell-type reactions, which reduce the cell voltage, maintain the water balance of the system, and increase the oxidation efficiency of ozone. The current fuel cell and gas diffusion electrode technology can beneficiate R&D in the field of membrane-based ozone generation.

In the electrochemical ozone generation process, PEMs function as follows:

- A charge carrier, providing proton transfer in the system, which finally makes it possible to use deionized water as a feed without any conductivity at all

- A very efficient separation barrier between a mixture of gaseous oxygen with ozone and hydrogen gas

- A catalytic membrane surface for embedding an appropriate catalyst

In addition to its good stability, an electrocatalyst, which is in contact with a PEM, deposited directly on the membrane (catalyst coated membrane or CCM) or on the electrode (gas diffusion electrode) should provide large active areas and a suitable structure to ensure the best possible performance of the entire system.

Despite recent progress made in solid polymer electrolyte(SPE)-based electrocatalytic membrane technology, there are still several aspects that require further detailed studies. For example, there is the question of how the catalytic activity of both the catalyst particles and functional groups of an SPE polymeric matrix determine the resulting catalytic properties of the entire PEM system (Bessarabov et al. 2001). Another fundamental problem that requires further investigation is the effect of membrane surface modification on the morphology of a catalyst deposited on the surface of a SPE membrane (Bessarabov et al. 2000). The development of membrane-based SPE composite catalysts with optimal performance for heterogeneous reactions taking place under conditions of electrical polarization requires knowledge and understanding of the structure, morphology, concentration, and energy distribution of active sites on the surface of a membrane (Wieckowski 1999).

It is important to acknowledge that the state of a membrane surface can contribute to various important characteristics of membranes, such as the electrocatalytic activity, interface formation, and film growth. This also includes nonuniformity in distribution of hydrophobicity on a membrane surface, which may determine the hydrodynamic instability of the membrane in various membrane processes (e.g., electrodialysis, ultrafiltration, microfiltration, etc.) or fluctuation in electrical resistance of a membrane surface (Timashev et al. 2000).

Two factors that must be considered for understanding the electrochemical behavior of solid electrodes include the following: the influence of the substrate crystallography and the topography of the electrode surface area on the kinetics of the electrocatalytic processes (Salvarezza and Arvia 1996).

The observed electrocatalytic activity of metal particles is an aggregate effect of morphological factors (particle size and distribution), geometry (inter-atomic spacing, crystal structure), thermodynamic properties (surface energy, heat of adsorption), as well as the electronic state of the reactive site. It has been shown that enhancing the roughness factor of a PEM catalytic system improves the electrocatalytic activity of the catalyst system (Delime et al. 1998, 1999). Thus, the quantification of the surface profile of electrocatalytic membranes is important. There are various approaches to quantifying the surface roughness of a membrane; a few relevant references are provided in this Section 16.2 for further reading (Bessarabov and Michaels 2001b).

16.2.3 Role of Anode and Cathode Electrocatalysts

An inert anode with high oxygen overvoltage is required for efficient electrochemical ozone evolution. The electrode material must be stable under conditions of strong anodic polarization, that is, it must be in its highest oxidation state, or be kinetically resistant to further oxidation. The anodic material must possess high electronic conductivity, as high current densities may be needed to achieve a sufficient anodic potential for electrochemical ozone generation. The anodic material must also be stable at high interfacial acid concentrations produced by the anodic discharge of water.

The noble metals are known for the type of kinetic stability required. Various highest valance state oxides such as PbO_2 and SnO_2 are also of interest. Glassy carbon materials are also promising. Antimony-doped tin oxide electrodes as well as boron-doped diamond electrodes are also under consideration (Honda et al. 2013).

The use of PbO_2 anodes in ozone generation has been a subject of several recent investigations (Wang and Chen 2013). In its highest oxidation state, PbO_2 is only slightly subject to chemical corrosion, which results from high hydrogen-ion concentration produced by anodic water discharge. PbO_2 has two crystalline forms, denoted alpha (α) and beta (β). Three main variables control the ratio between the two forms of PbO_2 during its formation by anodic polarization, that is, the pH of a solution, temperature, and electric current density in the deposition process.

The beta form of PbO_2 is known to be an "active" crystalline form in the process of electrolytic ozone generation. The alpha form has lower oxygen overvoltage than the beta form at high current densities. For this reason, it is believed that the beta crystalline form is a more suitable anode material for the evolution of ozone.

The hydrated PEM used in a reactor is highly acidic, with a pH equivalent of a 10 wt.% H_2SO_4 solution. During the process of electrocatalytic ozone generation using PEMs, the hydrogen evolution reaction occurs from the discharge of hydroxonium ions. This is referred to as the cathodic process. The mechanism and kinetics of the electrocatalytic reaction occurring on the electrode generally includes the discharge of hydroxonium ions leading to the formation of adsorbed atomic hydrogen, followed by their removal by either recombination (Equation 16.11) or electrochemical desorption (Equation 16.12), as given by the following equations:

$$2S-H \rightarrow 2S + H_2\left(gas\right) \qquad (16.11)$$

$$S-H + H_3O^+ + e^- \rightarrow H_2O\left(liquid\right) + H_2\left(gas\right) + S \quad (16.12)$$

where S is an active site on the cathode surface.

In order to minimize power consumption during electrocatalytic ozone generation, one needs an efficient cathodic catalyst for hydrogen evolution with fast kinetics of hydroxonium ions discharge.

16.2.4 PEM Electrolyzer Designs for Ozone Generation: Example

The design of PEM electrolyzer stacks for ozone generation is very similar to that of the conventional hydrogen PEM generators–electrolyzers described in this book. In this section, a brief description is given of some key components (excluding balance of plant) of one of the most common zero-gap PEM electrolyzers for ozone production developed by the South African company Dinax Technologies (Bessarabov 1999, 2002, Offringa 2000).

An electrochemical membrane reactor with a variable active anode area was constructed to study ozone generation (Bessarabov 2002). The active anodic area of the reactor varied from 50 to 200 cm². Figure 16.2 shows some of the typical key electrolyzer components. From left to right: Ti bipolar plate with flow field; separator ring; permeable, porous sintered Ti electrode; PEM with Pt catalyst, CCM (shown from cathode side); and separator ring.

Figure 16.3 shows a schematic diagram of a PEM ozone generating cell containing one membrane. The main components are the following: 1—sintered Ti electrodes; 2—proton exchange membrane (PEM); 3—isolators between Ti electrodes and reactor; 4—outlet (for water

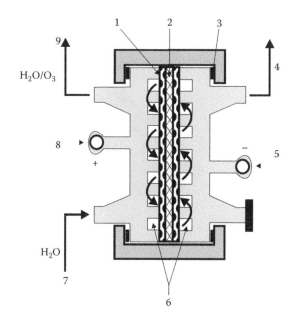

FIGURE 16.3
Schematic representation of the ozone generator described by Bessarabov (2002). (Reprinted from Bessarabov, D.G., Electrochemical generation of high-concentration ozone in compact integrated membrane systems, Final Report to the Water Research Commission of South Africa, South African Water Research Commission, Pretoria, South Africa, 2002. With permission.)

and hydrogen); 5—cathode side of a reactor; 6—gas and liquid distribution channels; 7—anodic inlet for water; 8—anode side of the reactor; and 9—anodic outlet for ozone and water.

FIGURE 16.2
Some key typical electrolyzer components. From left to right: Ti bipolar plate with a flow field, separator ring; Ti sintered current collector; PEM with cathodic platinum catalyst (cathode CCM); separator ring. (Reprinted from Bessarabov, D.G., Electrochemical generation of high-concentration ozone in compact integrated membrane systems, Final Report to the Water Research Commission of South Africa, South African Water Research Commission, Pretoria, South Africa, 2002. With permission.)

16.3 Preparation and Characterization Methods for Catalyst Layers

16.3.1 Anode Materials and Preparations: PbO$_2$ Anodes for Ozone Generation

Lead dioxide–based composite anodes were prepared using anodic polarization of specially pretreated sintered porous Ti electrodes in acidic solutions of lead nitrate at above room temperatures at predetermined time intervals. The current density during anodic polarization ranged from 0.01 to 0.03 A cm⁻². The density of loading of active PbO$_2$ on Ti anodes was measured to be approximately 0.025 g cm⁻².

Figure 16.4 shows typical morphology of PbO$_2$ crystals deposited on porous Ti supports. The image was obtained by means of atomic force microscopy (AFM) in the noncontact mode. It is seen that the microcrystals of PbO$_2$ have a layered structure with a very rough surface. Standard parameters of the surface were obtained

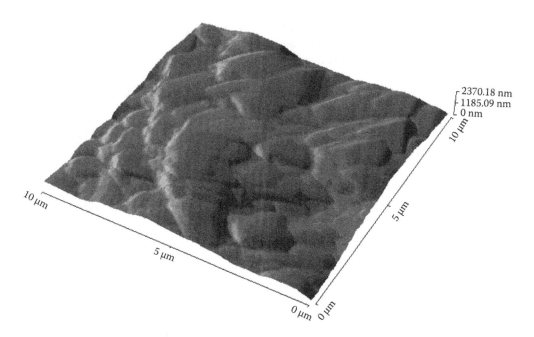

FIGURE 16.4

Typical morphology of PbO$_2$ layer in the manufactured anodes. (Reprinted from Bessarabov, D.G., Electrochemical generation of high-concentration ozone in compact integrated membrane systems, Final Report to the Water Research Commission of South Africa, South African Water Research Commission, Pretoria, South Africa, 2002. With permission.)

using software attached to the AFM. Standard parameters of a PbO$_2$–Ti electrode surface are the following:

Area Ra (nm) 382; Area RMS (nm) 451; Av height (nm): 1239; Max range (nm) 2370.

The composition of the crystalline phase in the PbO$_2$ layers deposited on the Ti disks was studied by means of the x-ray diffraction (XRD) technique, using a Philips PW 2273/20 diffractometer. Results confirmed that the deposition procedure used led to the formation of PbO$_2$ enriched mainly with the β-PbO$_2$ crystalline

form. Figure 16.5 shows the typical composition of PbO$_2$ obtained in this study; the predominant phase is β-PbO$_2$.

16.3.2 Cathode Catalyst Layer

Due to the corrosive environment of ozone generation in the SPE configuration, it is desirable to avoid carbon-supported cathode catalysts. There are various methods that can be considered to prepare cathode catalyst layers

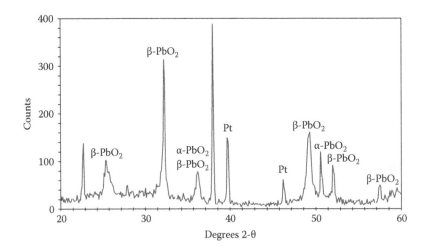

FIGURE 16.5

Typical composition of PbO$_2$ coatings produced. (Reprinted from Bessarabov, D.G., Electrochemical generation of high-concentration ozone in compact integrated membrane systems, Final Report to the Water Research Commission of South Africa, South African Water Research Commission, Pretoria, South Africa, 2002. With permission.)

for both PEM water electrolysis and the electrochemical ozone generation: hot press methods, electroless plating, and many other methods known from the fuel cell industry (Han et al. 2002).

The electroless method is a suitable method for the preparation of cathode CCMs for ozone generation. Electroless deposition methods will now be discussed in more detail.

Electroless deposition is an autocatalytic method of metal plating on a membrane surface without a supply of external electrons. The advantages of the technique include the following:

- A uniform metal deposition thickness can be obtained on a substrate of any shape
- A strong adhesive bonding between the metal film and the substrate can be obtained

The counter-diffusion deposition method and the impregnation–reduction method are both electroless deposition methods and will be briefly discussed further later in the chapter.

In optimizing the morphology of a catalyst deposited onto a SPE membrane for electrocatalytic ozone generation, the following aims are typically set:

- To improve electrical conductivity within a membrane electrode assembly (MEA) by continuous and uniform catalyst deposition (lower IR drop)
- To achieve higher catalytic activity and/or selectivity by deposition of small and specifically shaped catalyst particles (higher product yields, selectivity)
- To achieve a low-cost MEA by low catalyst loading density (cost effective systems)

A combination of the following experimental variables can be used (applicable to both of the earlier mentioned methods (Bessarabov 2000).

- Pretreatment of the membrane surface
- Chemical modification of the membrane surface
- Nature of chemical reducing agents
- Anionic–cationic nature of Pt complex ions in a solution
- Conditions of catalyst deposition (pH, temperature, concentration of Pt-ion-containing solution and/or reducing solution, time of the deposition process, etc.)
- External activation applied
- Use of specific additives

16.3.3 Counter-Diffusion Deposition (Takenaka–Torikai Method)

Takenaka et al. (1982) developed a method for depositing noble metals and their alloys onto a SPE. This method does not require using a binder or attaching the metal directly to the SPE. In the counter-diffusion deposition process, one side of a cation exchange membrane is in contact with a reducing agent, while the other is in contact with an anionic metal-ion solution (e.g., a platinic acid solution). The metal-ion solution is reduced by the reducing agent, resulting in a layer of the metal being deposited on the membrane surface. Pt particles deposited on perfluorinated cation-exchange membranes by means of the Takenaka–Torikai method typically result in spherical or flake-like shapes. The chemical reaction for the counter-diffusion deposition process can be represented as follows:

$$PtCl_6^{2-} + N_2H_4 \rightarrow N_2 + Pt + 6Cl^- + 4H^+ \quad (16.13)$$

It is possible to deposit Pt metal on the other free surface of the membrane using the same procedure. As the metal layer formed on the membrane has a microscopically porous structure, the reducing agent can diffuse through the metal layer and the membrane to reduce the metal-ion solution. This deposition process was modified by Fedkiw and Her (1989) and Fedkiw et al. (1990).

16.3.4 Impregnation–Reduction Deposition (Fedkiw and Her Method)

The impregnation–reduction method of Fedkiw and coworkers is a two-step procedure that involves the impregnation (ion-exchange) of a SPE with a cationic metal complex, followed by the reduction of the cationic metal complex. (Ion-exchanged) Nafion was impregnated with $Pt(NH_3)_4Cl_2$ before reducing the metal salt with a reducing agent, $NaBH_4$ (Fedkiw and Her 1989). It was also shown that greater depth of Pt particles within the SPE could be achieved by increasing the time of impregnation.

16.3.5 Pretreatment of a Membrane

Pretreatment of a membrane prior to the actual deposition of a catalyst does not involve any specific chemical modification of the membrane or its surface. Pretreatment is aimed at the improvement of the morphology of the catalyst. This includes achieving, for example, continuous layers of a catalyst, even thickness, and catalyst loading density, etc. The membrane pretreatment procedure should be chosen on the basis of their nature and basically involves softening of the

---

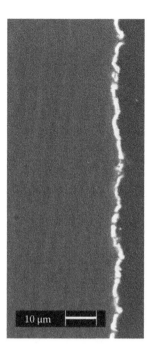

membranes prior to the catalyst deposition. An example of pretreatment is simply boiling membranes in water or a water–alcohol mixture before plating with a catalyst.

16.3.6 Cathode CCM: Experimental Example

The deposition of thin layers of highly dispersed Pt on PEM is achieved by autocatalytic reduction of $[PtCl_6]^{2-}$ complex anions using various reducing agents according to a technique known as the Takenaka–Torikai method.

According to this method, one specially pretreated surface of a PEM is in contact with a chemical reducing agent, while the other surface is in contact with an anionic metal-ion solution (e.g., an H_2PtCl_6 solution). The reducing agent reduces the metal-ion solution to form a metal layer on the membrane surface. This occurs when the reducing agent penetrates into the membrane and comes into contact with the metal-ion solution. This method does not involve the use of a binder nor does it involve an attempt to attach the metal directly to the SPE. The chemical deposition of Pt using, for example, hydrazine as a reducing agent is based on the Reaction (16.13).

The range of Pt loading density onto the membrane (mg cm^{-2}), depending on the time of deposition under certain temperature, pH and concentrations of both reducing and Pt-containing solutions, was determined for further optimization studies. Figure 16.6 shows the relationship between the Pt loading density and the time of deposition obtained experimentally using a specific reducing agent. Membranes containing approximately 1 mg cm^{-2} of a Pt catalyst were used in the tests described further in the following.

Figures 16.7 and 16.8 show SEM (a back-scattering detector was used) images of typical cross section of membranes containing a Pt catalyst (seen as white lines) deposited by means of the Takenaka–Torikai (Bessarabov and Michaels 2001a). Figure 16.7 represents a membrane that was boiled in water prior to the catalyst deposition. Figure 16.8 represents a membrane that was boiled in a mixture of water and an alcohol. All other conditions of the catalyst deposition were the same (temperature, concentration of the platinic acid and the reducing agent, time of deposition, etc). Pretreatment of the membranes resulted in an increase in the thickness of the Pt layer and improved the morphology of the catalyst coating. The Pt loading density was 0.31 and 2.42 mg cm^{-2} in each of the cases, respectively. Figure 16.8 shows a typical cross section of the Pt/PEM CCM produced in this study and used for ozone generation. The image was obtained using a SEM with a back-scattering detector. Pt layers are seen as white lines on the sides.

Figures 16.9 and 16.10 show that the morphology of the Pt catalyst can differ, depending on whether the membrane was pretreated or not.

16.3.7 Chemical Modification of Membrane Surface

Chemical surface modification of various ion-exchange membranes has been considered as a powerful method to change membrane properties (electrical, separation,

FIGURE 16.6
The relationship between the Pt loading density and the time of deposition established experimentally using a specific reducing agent. (Reprinted from Bessarabov, D.G., Electrochemical generation of high-concentration ozone in compact integrated membrane systems, Final Report to the Water Research Commission of South Africa, South African Water Research Commission, Pretoria, South Africa, 2002. With permission.)

FIGURE 16.7
SEM image, using a back-scattering detector, of the profile of Pt deposition on a membrane (seen as white spots) that was pretreated with boiling water prior to the plating. (Reprinted from *Membrane Technology*, 139, Bessarabov, D.G. and Michaels, W.C., Morphological diversity of platinum dusters deposited on proton exchange, perfluorinated membranes, 5–9, Copyright 2001, with permission from Elsevier.)

FIGURE 16.8
SEM image, using a back-scattering detector, of the profile of Pt deposition on a membrane (seen as a white line) that was pretreated in a boiling mixture of water and an alcohol prior to the plating. (Reprinted from *Membrane Technology*, 139, Bessarabov, D.G. and Michaels, W.C., Morphological diversity of platinum dusters deposited on proton exchange, perfluorinated membranes, 5–9, Copyright 2001, with permission from Elsevier.)

catalytic, etc.). The modification of the surface of ion-exchange membranes provides a method for controlling the structure and size of the Pt catalyst to achieve optimization of Pt deposition (Nidola and Martelli 1982, Bessarabov et al. 2000).

A new method to control the morphology of a catalytic layer by means of chemical modification of a membrane surface with a cationic surfactant was developed by Bessarabov and Michaels (the Bessarabov and Michaels method) in 2001 (Bessarabov and Michaels 2001b). Perfluorinated cation-exchange flat-sheet membranes were modified with $CH_3(CH_2)_{15}N(CH_3)_3Br$ ionic surfactant. The formation of a complex between the SPE membranes and the surfactant was verified by means of infrared spectroscopy.

The deposition of a Pt catalyst on the membranes treated with the surfactant was achieved by means of the Takenaka–Torikai method. Modification of the membranes resulted in a change in the shape of the Pt particles obtained; the Pt particles were small in size and pyramidally textured (Figure 16.11) (Bessarabov and Michaels 2001b). The shape of the Pt particles obtained in this study is unique and differs from that previously reported (Takenaka 1982, DeWulf and Bard 1988, Fedkiw and Her 1989, Sheppard et al. 1998, Liu et al.

1992, Liu and Fedkiw 1992, Fournier et al. 1997, Hirano et al. 1997); the Pt particles deposited on the surfactant-treated membranes were quite small. The average size of the particles now (measured as an average height) was approximately 10 nm, and the average length was approximately 0.3 μm.

Such a regular structure of the Pt particles could be attributed to the lyotropic phase of the surfactant–polyelectrolyte structure, which acts as a structure-forming medium during the deposition process. However, the change in the shape of the Pt catalyst particles could also be attributed to a lower transfer rate of the solution of the reducing agent through the membranes due to the membrane modification. Further characterization work carried out on membrane/Pt composites has recently been reported (Ingle et al. 2014).

16.3.8 Role of Chemical Reducing Agents

Autocatalytic electroless deposition of a metal catalyst onto ion-exchange membranes is thermodynamically possible if the following condition is valid:

$$\Delta E = E_{ME} - E_{RED} > 0 \qquad (16.14)$$

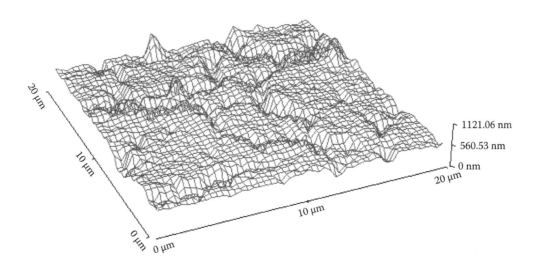

FIGURE 16.9

Typical AFM image (3D mesh view) of the surface of the membrane with platinum catalyst (seen as "flakes"). The membrane was pretreated with boiling water prior to the catalyst deposition. (Reprinted from *Membrane Technology*, 139, Bessarabov, D.G. and Michaels, W.C., Morphological diversity of platinum clusters deposited on proton exchange, perfluorinated membranes, 5–9, Copyright 2001, with permission from Elsevier.)

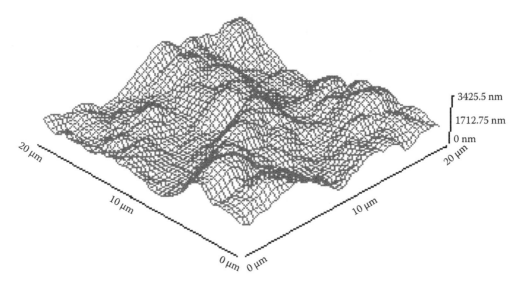

FIGURE 16.10

Typical AFM image (3D mesh view) of the surface of a membrane with platinum catalyst (seen as unevenly distributed semi-spherical particles). The membrane was pretreated by boiling in a water–alcohol mixture prior to the catalyst deposition. (Reprinted from *Membrane Technology*, 139, Bessarabov, D.G. and Michaels, W.C., Morphological diversity of platinum clusters deposited on proton exchange, perfluorinated membranes, 5–9, Copyright 2001, with permission from Elsevier.)

where E_{ME} and E_{RED} are the potentials of a metal in a solution containing its ions and a reducing agent, respectively (Bessarabov and Michaels 2001a). Many chemicals are suitable as reducing agents; for example, formaldehyde, hypophosphite, borohydride, hydrazine, and others.

The statistical probability (ω) of the formation of a metal catalytic nucleus during the autocatalytic process is proportional to the following Equation 16.15:

$$\omega \approx \exp\left[\left(\frac{\Delta E}{RT/F}\right)^{-2}\right] \qquad (16.15)$$

$$\rho = \frac{2\sigma V}{zF\Delta E} \qquad (16.16)$$

An equation for the critical size of a stable nucleus (ρ) also includes the variable ΔE: where σ is the surface tension between the catalytic particle and the solution, V is a metal molar volume, z is a number of electrons, and F is the Faradic number.

It is seen from these equations that the nature of the reducing agent allows control of the morphology (e.g., size, distribution profile, etc.) of Pt particles, resulting

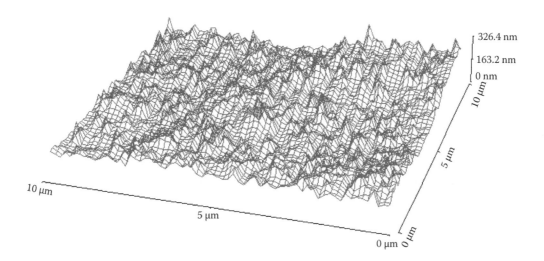

FIGURE 16.11
Typical 3D AFM image of a textured Pt catalyst deposited on the membrane. The platinum particles are small in size and pyramidally textured. (Reprinted from *Membrane Technology*, 139, Bessarabov, D.G. and Michaels, W.C., Morphological diversity of platinum dusters deposited on proton exchange, perfluorinated membranes, 5–9, Copyright 2001, with permission from Elsevier.)

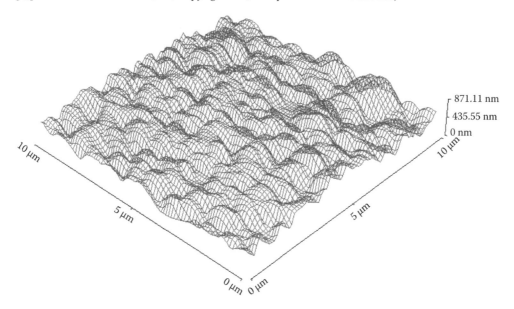

FIGURE 16.12
Typical 3D AFM surface image of a platinum catalyst seen as relatively small in size and evenly distributed semispherical particles. The catalyst was deposited onto the perfluorinated cation-exchange membrane using ultrasonic activation of the platinum-containing solution. (Reprinted from *Membrane Technology*, 139, Bessarabov, D.G. and Michaels, W.C., Morphological diversity of platinum dusters deposited on proton exchange, perfluorinated membranes, 5–9, Copyright 2001, with permission from Elsevier.)

from several variables, such as solution concentration, pH, temperature, etc. All these variables affect the rate of the formation of the nucleus and the rate of the autocatalytic process itself. Some information regarding an effect of the nature of reducing agents on the morphology of the Pt catalyst is available in the literature (e.g., Sheppard et al. 1998).

16.3.9 Effect of External Activation

Various types of external activation may be applied to the autocatalytic process of metal deposition onto membranes.

An example is the application of ultrasound to the solution containing Pt ions during the deposition process.

Figure 16.12 demonstrates an effect of ultrasound on the morphology of a Pt catalyst. Hydrazine was used as a reducing agent and ultrasound was applied to a solution of platinic acid. Application of the ultrasound resulted in the formation of evenly distributed semispherical Pt particles.

16.3.10 Characterization Methods

Key technological aspects of PEM-based electrocatalytic ozone generation membrane systems include manufacturing, characterization, optimization, and ultimate improvement of a thin porous anode and cathode catalytic layer attached to and partially embedded into the membrane polymeric matrix, which consists of clusters of metals or particles of metal oxides. Due to the high cost of the noble metals used in PEM-based membrane electrocatalytic systems, it is important to achieve low catalyst loading on the membranes for the PEM technology to be economically viable. Some of the strategies include modification of the chemical composition of the catalyst as well as optimization of the morphology of the catalyst (Delime et al. 1998, 1999). In practice, both approaches should be considered simultaneously. It is therefore important to employ such characterization methods that provide fundamental understanding on both chemical–electrochemical and structural levels. Since the characterization methods are well described elsewhere, only the related references are given here (Sheppard et al. 1998, Ingle et al. 2014).

16.4 Performance Testing

16.4.1 Cathode Water Feed Test

Deionized thermostated water was fed into the inlet of the electrolyzer (cathode). The DC was applied to the reactor. The formation of hydrogen (up to 99.99%) as measured by GC took place on the "cathodic" catalyst embedded into the PEM. Due to the high water permeability of the PEM, water molecules diffused to the anode side of the electrolyzer where the formation of oxygen and ozone in gaseous form took place. Ozone was collected from the outlet and its concentration was monitored by means of a BMT-963 ozone analyzer. The experimental setup is shown in Figure 16.13 (Bessarabov 2002). Considerable evolution of ozone was detected at the water temperature of 40°C (measured at the inlet of the reactor). The typical potential across the membrane cell containing a single membrane reached 3.1–3.2 V at the current density of 0.10 A cm^{-2}. The ozone concentration was as high as 6 wt.%.

The current efficiency of ozone evolution, η, was calculated according to the Faraday's law of electrolysis. It is defined as follows:

$$\eta = \frac{\Delta g}{K \times i \times t} \times 100\% \tag{16.17}$$

where
Δg is the mass of ozone obtained
K is the electrochemical ozone equivalent
i is the electric current (A)
t is the time (h)

K is defined as follows:

$$K = \frac{3600 M_{ozone}}{Fz} \tag{16.18}$$

where
M_{ozone} is the molecular weight of ozone
F is the Faraday constant (96,500 C/eq)
z is the number of electrons necessary for the formation of 1 mol of ozone ($z = 6$)

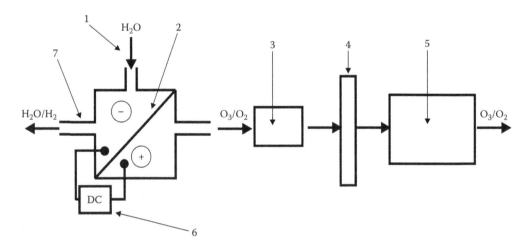

FIGURE 16.13
Schematic of the experimental set-up. 1—Deionized water inlet, 2—PEM with embedded catalysts, 3—BMT DH3 ozone dehumidifier, 4—flow meter, 5—BMT 963 ozone analyzer, 6—DC supply, 7—cathodic reactor outlet. (Reprinted from Bessarabov, D.G., Electrochemical generation of high-concentration ozone in compact integrated membrane systems, Final Report to the Water Research Commission of South Africa, South African Water Research Commission, Pretoria, South Africa, 2002. With permission.)

Typical values of η ranged between 27% and 36%. Under the given experimental conditions (400–500 cumulative hours of operation), no destruction of separator rings in the reactor was observed. However, the concentration of ozone decreased to 0.5 wt.% after the cumulative experimental time of 500 h. The membrane reactor was disassembled and membranes and electrodes examined and investigated further in terms of the anodic catalysts chemical composition.

16.4.2 Durability Challenges

Experimental runs with a PbO_2-based O_3 generator carried out for relatively long periods of time (up to 500 cumulative hours of operation) have demonstrated a decline in ozone generation efficiency. Measurements of ion content in water subjected to anodic oxidation were carried out using the inductively coupled plasma (ICP) technique. Table 16.3 shows the results of the ICP measurements of water that was subjected to anodic oxidation in the membrane cell for 400 h. Degradation of the PbO_2-based anode and the cell material (Ti) took place, resulting in the formation of the corresponding ions in

TABLE 16.3

Results of Water Analysis by ICP

	Concentration (ppm)	
Ion	Pb	Ti
Initial distilled water	0.0	0.0
Recycled water used for ozone generation	5.3	3.5

the water. The decline in the efficiency of ozone generation is explained by the degradation of the PbO_2 anodes under strong anodic polarization conditions. The cell voltage across the PbO_2-based cell decreased and the electrical current withdrawn from the cell increased, hence overheating the system and decreasing the yield of the ozone produced. Such a phenomenon could be explained by "poisoning" of the cation-exchange membrane by cations accumulating in the feed water and by degradation of the anodic catalyst (PbO_2 in this case). However, as it will be shown further in the following, the main problem is not the membrane "poisoning" by cations (the membrane can be regenerated by ion exchange), but rather the irreversible degradation of the anode coatings.

A MEA, as shown in Figure 16.2, comprising a perfluorinated cation-exchange membrane coated with a Pt catalyst on one (cathode) side was removed from the cell after 400 cumulative hours of operation. The perfluorinated cation-exchange membrane was then placed alone in an aqueous solution of nitric acid and held at 70°C for 48 h. This procedure allowed for the ion-exchange of various cations (if any) present in the membrane to form corresponding nitrates in the aqueous solution. The aqueous solution was then removed into a glass container and evaporated. The resulting crystals were slightly yellowish. The crystals were then analyzed by means of the XRD method, using a Philips PW 2273/20 diffractometer. Figure 16.14 shows the results of the XRD analysis. The crystal phase mostly consisted of lead nitrate (there was only one peak that was not identified as lead nitrate).

The results are significant as they again clearly demonstrated that the process of degradation of the PbO_2

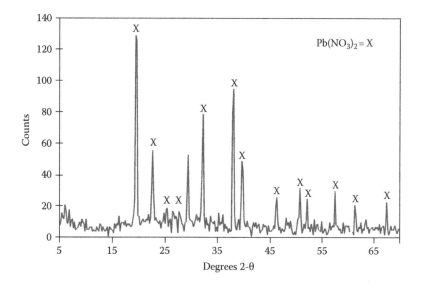

FIGURE 16.14

Results of the XRD analysis of nitrate crystals obtained by evaporation of the solution of aqueous nitric acid. It is seen that the chemical composition of the crystals corresponds to lead nitrate. (Reprinted from Bessarabov, D.G., Electrochemical generation of high-concentration ozone in compact integrated membrane systems, Final Report to the Water Research Commission of South Africa, South African Water Research Commission, Pretoria, South Africa, 2002. With permission.)

coating under the conditions of anodic ozone generation indeed took place. This loss of lead from the PbO₂ coatings is consistent with the slight solubility of lead (IV) in strong acid (Pourbaix 1974).

The mechanism of the degradation could be described as follows. During the first stage, under the conditions of strong anodic polarization, the thermodynamically possible dissolution of PbO₂ occurs at the interface of the cation-exchange perfluorinated membrane and Ti porous electrode (Foller 1979). During the second stage, lead ions migrate into the bulk of the cation-exchange membrane as countercations. When the membrane is treated with nitric acid, the corresponding lead nitrate is formed. Lead nitrate can then be crystallized by evaporation of the nitric acid, and analyzed.

$$PbO_2 + 4H^+ = Pb^{4+} + 2H_2O \qquad (16.19)$$

$$Pb^{4+} + 2e^- = Pb^{2+} \qquad (16.20)$$

Thus, irreversible degradation of the PbO₂ electrodes was observed. Some experimental runs included operation of the PEM-based electrolytic ozone generator when deionized water was fed to the anode side of the generator. It was demonstrated that the anodic water supply allows a longer lifespan of membranes and the entire system. However, irreversible degradation of the PbO₂ electrodes also occurred in this case.

16.5 Examples of Applications of Electrochemical Ozone Systems

16.5.1 Laboratory Evaluation of Phenol Removal by Means of Direct Anodic Oxidation and Low-Concentration Ozone Generated Electrochemically

There is an increasing interest in the use of electrochemical methods in the treatment of phenol in waste streams (Smith de Sucre and Watkinson 1981, Sharifan and Kirk 1986).

There are at least two approaches to designing such electrochemical systems. The first approach includes the use of anodes with high oxygen overvoltage (e.g., PbO₂) for the direct oxidation of refractory organic chemicals (Comninellis and Plattner 1988, Comninellis 1990, Kotz et al. 1991). The second approach includes electrochemical generation of strong oxidants, such as ozone, to be used for the oxidation of organic pollutants (Stucki et al. 1985, Foller and Tobias 1982b).

In practice, the anodic oxidation of organic pollutants on high-overvoltage anodes involves various oxidation mechanisms. One of the mechanisms, for example,

involves the formation of ozone on PbO₂ anodes through the formation of hydroxyl radicals (·OH) [8]:

$$OH^{\bullet}_{(ads)} + O_{2(ads)} \rightarrow HO^{\bullet}_{3(ads)} \qquad (16.21)$$

$$HO^{\bullet}_{3(ads)} \rightarrow HO^+_{3(ads)} + e^- \qquad (16.22)$$

$$HO^+_{3(ads)} \rightarrow O_3 + H^+ \qquad (16.23)$$

Equations 16.21 through 16.23 show that it is not only gaseous ozone that is involved in the oxidation of organics, but also hydroxyl radicals. A singlet oxygen, ¹O₂, was also reported to be formed on the PbO₂ anodes during the anodic oxidation of (Wabner and Grambow 1985).

The direct anodic oxidation of phenol has been studied by various researchers. It was found that phenol is readily oxidized in aqueous solutions but the oxidation of by-products is difficult to achieve. The reaction sequence transforming phenol to hydroquinone, benzoquinone, and other products, including carbon dioxide, is not well understood (Sharifan and Kirk 1986). Various by-products have been reported, including hydroquinone, p-benzoquinone, catechol, maleic acid, carbon monoxide, o-benzoquinone, formic and racemic acids, and others (Smith de Sucre and Watkinson 1981, Sharifan and Kirk 1986).

As mentioned earlier, the mechanism of direct anodic oxidation of organics (phenol, in particular) is a complicated process. As the direct electrochemical anodic oxidation of phenol (and organic pollutants in waste water in general) is seen as a potentially feasible process, there is a need for further investigation of by-products formed during the process.

16.5.2 Anodic Oxidation of Phenol on PbO₂ Anodes

Ozone can be generated electrochemically on the surface of PbO₂ anodes, as was described in Section 16.4. The use of PbO₂ anodes allows generation of ozone at relatively low concentration (up to 5–6 wt.%). Deionized water is required as raw material for ozone generation. In this case, no corrosive electrolyte is needed, making it a potentially useful application in medicine and food production. The efficiency of ozone generation depends on the cell voltage, electrical current density applied, and temperature of the cell.

It was observed that ozone is formed in measurable amounts when the cell voltage exceeded 3.1 V for one membrane stack. In these electrochemical experiments, an aqueous solution of phenol (concentration 80–100 mg L⁻¹, without a buffer) was fed into the membrane system along the PbO₂-based anode designed to generate low-concentration ozone (LCO) directly into the water stream, as shown in Figure 16.15. The efficiency of phenol removal was analyzed by means of the high-performance liquid chromatography (HPLC) technique.

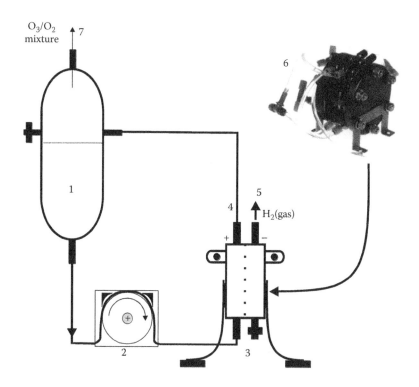

FIGURE 16.15
Schematic of the electrochemical set-up. 1—glass vessel containing phenol solution, 2—peristaltic pump, 3—electrochemical cell, 4—anode side of the electrochemical cell, 5—cathode side of the electrochemical cell and outlet for hydrogen evolution, 6—picture of actual electrochemical cell, 7—outlet for ozone and/or oxygen. (Reprinted from Bessarabov, D.G., Electrochemical generation of high-concentration ozone in compact integrated membrane systems, Final Report to the Water Research Commission of South Africa, South African Water Research Commission, Pretoria, South Africa, 2002. With permission.)

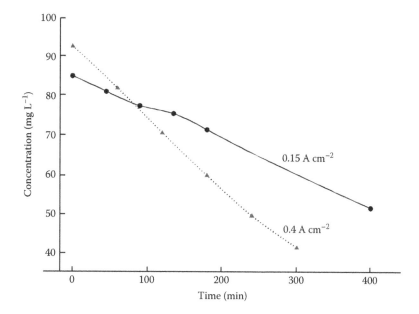

FIGURE 16.16
Effect of the electrical current density applied to the membrane/PbO$_2$ anode system on phenol removal. (Reprinted from Bessarabov, D.G., Electrochemical generation of high-concentration ozone in compact integrated membrane systems, Final Report to the Water Research Commission of South Africa, South African Water Research Commission, Pretoria, South Africa, 2002. With permission.)

TABLE 16.4

Conditions of the Experiments (Current Density 0.15 A cm^{-2})

Time (min)	Amps (A)	Volts (V)	Temperature (°C)	Phenol Conc. (mg L^{-1})
0	7.5	2.754	25	85.5
45	7.5	2.757	29	81.0
90	7.5	2.777	30	77.3
135	7.5	2.798	31	75.3
180	7.5	2.827	31	71.2
400	7.5	2.893	31	51.1

TABLE 16.5

Conditions of the Experiments (Current Density 0.4 A cm^{-2})

Time (min)	Amps (A)	Volts (V)	Temperature (°C)	Phenol Conc. (mg L^{-1})
0	20.0	3.530	24	92.6
60	20.0	3.127	39	81.9
120	20.0	3.142	39	70.4
180	20.0	3.156	39	59.4
240	20.0	3.152	40	49.1
300	20.0	3.149	40	40.9

Figure 16.16 shows the effect of the electrical current density applied to the membrane on the phenol removal. An increase in the current density resulted in a decrease of the time of the phenol removal. Tables 16.4 and 16.5 show conditions of the experiments. It is important to note that the evolution of gaseous ozone was observed when the cell voltage was >3.1 V. Under the given current density applied (0.15 A cm^{-2}) (see Table 16.4), the evolution of ozone was not observed and the oxidation of phenol took place according to the so-called "direct anodic oxidation" mechanism. However, in case of the current density being 0.4 A cm^{-2} (see Table 16.5), anodic evolution of gaseous ozone was observed and the concentration of ozone was measured at ≈2.5 wt.%.

Figure 16.17 shows the efficiency of phenol removal from water versus the electrical power consumed.

Figure 16.18 shows the results of the HPLC analysis of the phenol solution corresponding to the curve marked as "0.4 A cm^{-2}" in Figure 16.17. It is seen that the phenol removal is incomplete and the intermediate products formed are not easily degradable under the experimental conditions.

Figure 16.19 shows the efficiency of phenol removal from water versus power consumption in case of direct anodic oxidation and direct ozone treatment (both with HCO and LCO). It is seen that almost 100% removal of phenol could be achieved at much lower cost when using electrochemically generated HCO.

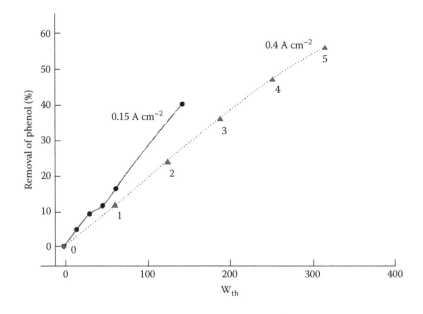

FIGURE 16.17

Efficiency of the phenol removal from water versus the electrical power consumed. Numbers "0," "1," "2," "3," "4," and "5" correspond to the HPLC analysis made for each sample, respectively. (Reprinted from Bessarabov, D.G., Electrochemical generation of high-concentration ozone in compact integrated membrane systems, Final Report to the Water Research Commission of South Africa, South African Water Research Commission, Pretoria, South Africa, 2002. With permission.)

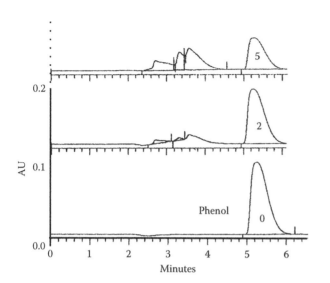

FIGURE 16.18
Results of the HPLC analysis of the phenol solution. "0" corresponds to the initial analysis where phenol only was detected. (Reprinted from Bessarabov, D.G., *Electrochemical generation of high-concentration ozone in compact integrated membrane systems*, Final Report to the Water Research Commission of South Africa, South African Water Research Commission, Pretoria, South Africa, 2002. With permission.)

It was shown in this section (Figure 16.18) that the use of electrochemically generated LCO for phenol removal, which was carried out within certain time intervals, results in only partial oxidation of phenol and occurrence of several peaks on the HPLC analysis.

Further HPLC analysis, using advanced software, was carried out to identify these compounds. These results are reported in the following section.

16.6 Detailed HPLC Analysis of Products of Phenol Oxidation by Means of Ozone Generated Electrochemically on PbO$_2$ Anodes

HPLC analysis of the products of phenol oxidation by means of ozone generated electrochemically on PbO$_2$ anodes was carried out. The following experimental details applied: HPLC instrument—with Chromquest Chromatographyä software; column—PL HiPlex H (cationic); detector: UV-based, absorbency measurement at 210 nm; temperature: at 55°C; injection volume: 80 μL; eluent: 0.03 N H$_2$SO$_4$. A solution corresponding to the fifth point in Figure 16.17 was analyzed in detail. Use of the HiPlex H (cationic) column allowed detailed analysis of the phenol solution.

Figure 16.20 shows the results of the HPLC analysis. Intermediate products were formed. The main component was oxalic acid. Traces of maleic and fumaric acids as well as catechol were also observed. Oxidation of phenol by means of HCO is a faster process, which results in the almost complete oxidation of the phenol as analyzed by HPLC.

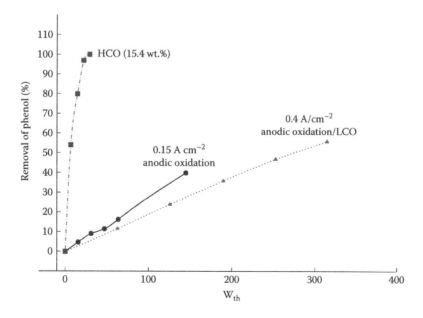

FIGURE 16.19
Efficiency of the phenol removal from water versus power consumption in cases of direct anodic oxidation and direct ozone treatment. (Reprinted from Bessarabov, D.G., *Electrochemical generation of high-concentration ozone in compact integrated membrane systems*, Final Report to the Water Research Commission of South Africa, South African Water Research Commission, Pretoria, South Africa, 2002. With permission.)

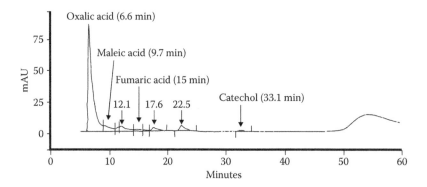

FIGURE 16.20
HPLC analysis of an aqueous solution of phenol that was treated in the ozone electrochemical cell, at 0.4 A cm^{-2} electrical current density. (Reprinted from Bessarabov, D.G., Electrochemical generation of high-concentration ozone in compact integrated membrane systems, Final Report to the Water Research Commission of South Africa, South African Water Research Commission, Pretoria, South Africa, 2002. With permission.)

16.7 Conclusions

The electrochemical generation of ozone and hydrogen in a PEM-based electrolyzer cell was discussed in this chapter. In theory, the electrochemical processes which are used for ozone generation by means of PEM-based cells are capable of producing ozone very efficiently. However, in practice, the theoretical efficiency cannot be easily achieved, and the electrical current consumption in the generators is higher than that achieved in large corona discharge units. It is for this reason that it is expected that such a system could be used for small-scale applications.

Experimental data related to electrochemical ozone generators comprising PEMs, platinum cathodes, and lead dioxide anodes were also described. The morphological diversity of platinum clusters deposited onto a PEM was demonstrated.

It was further demonstrated that the anodic generation of ozone using a PEM-supported lead dioxide catalyst suffers from the relatively fast and irreversible degradation of the catalyst. The method could, nonetheless, be suitable for the generation of ozone of low concentration when relatively low-current densities are used (<0.1 A cm^{-2}), and compete with conventional technologies (such as UV-based generation) for the production of LCO on small scale.

Further research is required to optimize operational conditions of PEM-based electrochemical ozone generation and to improve the stability of the metal-oxide-based anodic catalysts and/or other type of materials, as well as the efficiency of the cathodic catalyst. Hydrogen, as a by-product in this process, can potentially be used in fuel cells and ultimately contribute to the partial recovery of the high costs associated with the required high power consumption.

References

Bessarabov, D. G. 1998. Electrochemically-aided membrane separation and catalytic processes. *Membrane Technology/International Newsletter* 93:8–11.

Bessarabov, D. G. 1999. Membranes help to produce high-concentration ozone: New challenges. *Membrane Technology/International Newsletter* 114:5–8.

Bessarabov, D. G. 2000. Electrochemical generation of high-concentration ozone for water treatment. In: *Proceedings of the Water Institute of South Africa (WISA), Biennial Conference and Exhibition*, Sun City, South Africa, 28 May–1 June 2000. http://www.ewisa.co.za/literature/files/99bessarov.pdf.

Bessarabov, D. G. 2002. Electrochemical generation of high-concentration ozone in compact integrated membrane systems. Final Report to the Water Research Commission of South Africa. Pretoria, South Africa: South African Water Research Commission.

Bessarabov, D. G. and W. C. Michaels. 2001a. Morphological diversity of platinum dusters deposited on proton-exchange, perfluorinated membranes. *Membrane Technology* 139:5–9.

Bessarabov, D. G. and W. C. Michaels. 2001b. Solid polyelectrolyte (SPE) membranes containing a textured platinum catalyst. *Journal of Membrane Science* 194:135–140.

Bessarabov, D. G., W. C. Michaels, and Y. M. Popkov. 2001. Galvanodynamic study of the electrochemical switching effect in perfluorinated cation-exchange membranes modified by ethylenediamine. *Journal of Membrane Science* 194:81–90.

Bessarabov, D. G., W. C. Michaels, and R. D. Sanderson. 2000. Preparation and characterisation of chemically-modified perfluorinated cation-exchange platinum-containing membranes. *Journal of Membrane Science* 179:221–229.

Christensen, P. A., W. F. Lin, H. Christensen, A. Imkum, J. M. Jin, G. Li, and C. M. Dyson. 2009. Room temperature, electrochemical generation of ozone with 50% current efficiency in 0.5 M sulfuric acid at cell voltages < 3V. *Ozone: Science & Engineering. The Journal of the International Ozone Association* 31:287–293.

Christensen, P. A., Yonar, T., and Zakaria, K. 2013. The electrochemical generation of ozone: A review. *Ozone: Science & Engineering. The Journal of the International Ozone Association* 35:149–167.

Comninellis, C. 1990. Electrochemical treatment of waste water containing organic pollutants. In: *Electrochemical Engineering and Small Scale Electrolytic Processing*, eds. C. W. Walton, J. W. Van Zee, and R. D. Varjian, pp. 71–87. Pennington, NJ: Electrochemical Society Inc.

Comninellis, C. and E. Plattner. 1988. Electrochemical waste water treatment. *CHIMIA* 42:250–252.

Couper, A. M. and S. Bullen. 1992. The electrochemical generation of ozone at high concentrations. In: *Proceedings of the Electrochemical Engineering and the Environment Symposium*, Loughborough University of Technology, Loughborough, U.K., 7–9 April, 1992. ICHEME Symposium Series No 127, pp. 49–58.

Delime, F., J. M. Léger, and C. Lamy. 1998. Optimization of platinum dispersion in Pt-PEM electrodes: Application to the electrooxidation of ethanol. *Journal of Applied Electrochemistry* 28:27–35.

Delime, F., J. M. Léger and C. Lamy. 1999. Enhancement of the electrooxidation of ethanol on a Pt-PEM electrode modified by tin. Part I: Half cell study. *Journal of Applied Electrochemistry* 29:1249–1254.

DeWulf D. W. and A. J. Bard. 1988. Application of Nafion/platinum electrodes (solid polymer electrolyte structures) to voltammetric investigations of highly resistive solutions. *Journal of the Electrochemical Society* 135:1977–1985.

Fedkiw, P. S. and W. Her. 1989. An impregnation-reduction method to prepare electrodes on Nafion SPE. *Journal of the Electrochemical Society* 136:899–900.

Fedkiw, P. S., J. M. Potente, and W. H. Her. 1990. Electroreduction of gaseous ethylene on a platinized Nafion membrane. *Journal of the Electrochemical Society* 137:1451–1460.

Foller, P. C. 1979. The kinetics and mechanism of the formation of ozone by the anodic oxidation of water. PhD dissertation, Berkeley, CA: University of California.

Foller, P. C. 1985. Process and device for the generation of ozone via the anodic oxidation of water. US Patent 4,541,989.

Foller, P. C. and M. L. Goodwin. 1984. The electrochemical generation of high concentration ozone for small-scale applications. *Ozone: Science and Engineering* 6:29–36.

Foller, P. C. and C. W. Tobias. 1982a. Electrolytic process for the production of ozone. US Patent 4,316,782.

Foller, P. C. and C. W. Tobias. 1982b. The anodic evolution of ozone. *Journal of the Electrochemical Society* 129:506–515.

Fournier, J., G. Faubert, J. Y. Tilquin, R. Côte, D. Guay, and J. P. Dodelet. 1997. High performance, low Pt content catalysis for electroreduction of oxygen in polymer-electrolyte fuel cells. *Journal of the Electrochemical Society* 144:145–154.

Grubb, W. T. 1959. Batteries with solid ion exchange electrolytes. I. Second cells employing metal electrodes. *Journal of the Electrochemical Society* 106:275–278.

Han, S.-D., K.-B. Park, R. Rana, and K. C. Singh. 2002. Developments of water electrolysis technology by solid polymer electrolyte. *Indian Journal of Chemistry* 41A:245–253.

Hirano, S., J. Kim, and S. Srinivasan. 1997. High performance proton exchange membrane fuel cells with sputter-deposited Pt layer electrodes. *Electrochimica Acta* 42:1587–1593.

Honda, Y., A. I. Tribidasari, T. Watanabe, K. Murata, and Y. Einaga. 2013. An electrolyte-free system for ozone generation using heavily boron-doped diamond electrodes. *Diamond and Related Materials* 40:7–11.

Ingle, N. J. C., A. Sode, I. Martens, E. Gyenge, D. P. Wilkinson, and D. Bizzotto. 2014. Synthesis and characterization of diverse Pt nanostructures in nafion. *Langmuir* 30:1871–1879.

Kim, G., A. E. Yousef, and S. Dave. 1999. Application of ozone for enhancing the microbiological safety and quality of foods: A review. *Journal of Food Protection* 62:1071–1087.

Kotz, R., S. Stucki, and B. Carcer. 1991 Electrochemical waste water treatment using high overvoltage anodes. Part I: Physical and electrochemical properties of SnO_2 anodes. *Journal of Applied Electrochemistry* 21:14–20.

Lang, H. V., E. Erni, and P. A. Liechti. 1993. Advanced ozone generation technology to solve the oxidation problems of today. In: *Proceedings of the 11th Ozone World Congress*, San Francisco, CA, Vol. 1. Pan American Group, International Ozone Association, Stamford, CT, pp. S-4-1–S-4-10.

Langlais, B., D. A. Reckhow, and D. A. Brink. 1991. Ozone in water treatment. Application and engineering. Cooperative Research Report, AWWA Research Foundation. Boca Raton, FL: D.C. Lewis Publishers.

Liu, R. and P. S. Fedkiw. 1992. Partial oxidation of methanol on a metallized Nafion polymer electrolyte membrane. *Journal of the Electrochemical Society* 139:3514–3523.

Liu, R., W.-H. Her, and P. S. Fedkiw. 1992. In situ electrode formation on a Nafion membrane by chemical platinization. *Journal of the Electrochemical Society* 139:15–23.

Menth, A. and S. Stucki. 1982. Process for the synthetic production of ozone by electrolysis and use there of. U.S. Patent 4,416,747.

Murphy, O. J. and G. D. Hitchens. 1999. Electrochemical production of ozone and hydrogen peroxide. U.S. Patent 5,972,196.

Nakamatsu, S., Y. Nishiki, and M. Katoh. 1993. Apparatus for electrolytic ozone generation. U.S. Patent 5,326,444.

Nidola, A. and G. N. Martelli. 1982. Deposition of catalytic electrodes onion-exchange membranes. U.S. Patent 4,364,803.

Nutall, L. J. 1977. Conceptual design of large scale water electrolysis plant using solid polymer electrolyte technology. *International Journal of Hydrogen Energy* 2:395–403.

Offringa, G. 2000. Membrane development in South Africa. *Membrane Technology* 119:4–7.

Pourbaix, M. 1974. *Atlas of Electrochemical Equilibria in Aqueous Solutions*, 2nd edn. Houston, TX: NACE International.

Rennecker, J. L., B. J. Marinas, J. H. Owens, and E. W. Rice. 1999. Inactivation of cryptosporodium parvum oocysts with ozone. *Water Research* 33:2481–2488.

Rice, R. G. and D. M. Graham. 2001. U.S. FDA regulatory approval of ozone as an antimicrobial agent—What is allowed and what needs to be understood. *Ozone News: The Newsletter of the International Ozone Association* 29:22.

Salvarezza, R. C. and A. J. Arvia. 1996. A modern approach to surface roughness applied to electrochemical systems. In: *Modern Aspects of Electrochemistry*, eds. B. E. Conway, J. O'M. Bockris, and R. E. White. New York: Plenum Press.

Sharifan, H. and D. W. Kirk. 1986. Electrochemical oxidation of phenol. *Journal of the Electrochemical Society* 15:921–924.

Sheppard, S. A., S. A. Campbell, J. R. Smith, G. W. Lloyd, T. R. Ralph, and F. C. Walsh. 1998. Electrochemical and microscopic characterisation of platinum-coated perfluorosulfonic acid (Nafion 117) materials. *Analyst* 123:1923–1929.

Shimamune, T. and Sawamoto I. 1995. Perfluorocarbon sulfonic acid ion exchange membrane as solid electrolyte, anode placed on one side, and lead oxide as electrode catalyst. U.S. Patent 5,407,550.

Shimamune T. and Sawamoto I. 1997. Electrolytic ozone generator. U.S. Patent 5,607,562.

Shimamune, T., I. Sawamoto, and Y. Nishiki. 1991. Method for electrolytic ozone generation and an apparatus therefor. U.S. Patent 5,203,972.

Smith de Sucre, V. and A. P. Watkinson. 1981. Anodic oxidation of phenol for waste water treatment. *Canadian Journal of Chemical Engineering* 59:52–59.

Stucki, S., G. Theis, R. Kotz, H. Devantay, and H. J. Christen. S. 1985. In situ production of ozone in water using a Membrel electrolyzer. *Journal of the Electrochemical Society* 132:367–371.

Takenaka, H., E. Torikai, Y. Kawami and N. Wakabayashi. 1982. Solid polymer electrolyte water electrolysis. *International Journal of Hydrogen Energy* 7:397–403.

Tatapudi, P. and J. M. Fenton. 1993. Synthesis of ozone in a proton exchange membrane electrochemical reactor. *Journal of the Electrochemical Society* 140:3527–3530.

Timashev, S. F., D. G. Bessarabov, R. D. Sanderson, S. Marais, and S. G. Lakeev. 2000. Description of non-regular membrane structures: A novel phenomenological approach. *Journal of Membrane Science* 170:191–203.

Wabner, D. and C. Grambow. 1985. Reactive intermediates during oxidation of water at lead dioxide and platinum electrodes. *Journal of Electroanalytical Chemistry* 195:95–108.

Wang, Y.-H. and Q.-Y. Chen. 2013. Anodic materials for electrocatalytic ozone generation. *International Journal of Electrochemistry* 2013:1–7.

Wieckowski, A. 1999. *Interfacial Electrochemistry: Theory, Experiment, and Applications*. New York: Marcel Dekker.

17

Isotope Separation Using PEM Electrochemical Systems

Mikhail Rozenkevich and Irina Pushkareva

CONTENTS

17.1 Introduction

In 1932, as a result of the spectral analysis of hydrogen remaining in 1 cm³ after equilibrium evaporation of 4 l of liquid hydrogen, the heavy isotope of hydrogen— deuterium ^2_1D—was discovered (Urey et al. 1932). In the following year, using stepwise reduction, by electrolysis, of the alkaline water solution from an initial 20 l to a final volume of 0.5 cm³, Lewis and MacDonald obtained water with a deuterium concentration of 67.5% for the first time (Lewis and MacDonald 1933). In 1934, the first plant for the production of heavy water by electrolysis began operation in Ryukane (Norway). In 1938, it produced 40 kg of the product, in 1939 twice as much, and the maximum capacity of the plant in the last year of its existence (1942) was 1.54 tons of heavy water (Benedict 1956). Thus, until 1943, all the world's stocks of heavy water were produced by electrolysis.

Commencing in the atomic era of the 1940s, the demand for heavy water increased. Due to its physical and physicochemical properties, it emerged as the best moderator of fast neutrons, which explains its large scale use in nuclear power reactors operated on natural uranium (containing only 0.7% $^{235}_{92}U$) as a fuel.

For example, one Canadian CANDU-type nuclear reactor with a capacity of approximately 750 MW requires about 80 tons of heavy water with a deuterium concentration of not less than 99.75 at.%. Heavy water is also used in industrial and research nuclear reactors intended for radioisotopes production or obtaining powerful neutron flux. The required amount of heavy water for one reactor depends on the reactor power; it ranges from 10 to 150 tons. In addition to these large-tonnage applications of heavy water, today it is also widely used in scientific research in the form of various labeled compounds and deuterated solvents, such as in nuclear magnetic resonance (NMR) studies. In the future it could also be used as a component of fuel for fusion reactors, where the fusion reaction of deuterium and tritium is required. All these factors led to the fact that in 60th–80th years of the last century the production of heavy water for isotope production reached the incredibly large volume of 10–15,000 tons per year. The greatest volume was produced in Canada and the United States, where, after the accumulation of substantial reserves, several large industries were later dismantled or mothballed. Today, heavy water is produced in countries where the use of nuclear energy has begun to develop.

Historically, the first use of electrolysis for the heavy water production has its own reasons. It is enough to compare two figures. In order to obtain just 1 g of water with a 10 times larger deuterium concentration (compared to on natural water) by reduction, (through evaporation), the water residue it is necessary to evaporate is approximately 3×10^{33} tons of water (this value is 500 million times greater than the mass of the Earth!). The same problem solved by electrolysis of water requires only 18–30 times reduction of the electrolyte volume. The reason is the large difference between the single separation effect under evaporation of water and water electrolysis; the value of separation effect under electrolysis is much greater. Nevertheless, nowadays, the process of water rectification under vacuum is widely used for producing heavy water at the final concentration stage (Andreev et al. 2007). Electrolysis, as an independent method of heavy water production, is not used now due to the high energy effort compared to other modern methods of separating hydrogen isotopes (hundreds of megawatts per 1 kg of heavy water) (Andreev 1999).

Over the last 20 years, interest in the use of water electrolysis as a part of hydrogen isotope separation facilities increased again. There are two reasons for this. First, despite the much less demand for producing heavy water (the world's reserves are currently sufficient), there is an increasingly serious problem in many countries that have nuclear technology of dealing with the tritium-containing waste. Second, it is now generally accepted that the most promising technology for this process is based on catalytic isotope exchange between water and hydrogen (Boniface et al. 2013).

At the present development level, this technology requires using the electrolyzer as an assembly in which the entire flow of water coming from the separation column should be converted into hydrogen and oxygen. It is possible to use two types of electrolyzers: an alkali electrolyzer or an electrolyzer with a proton exchange membrane (PEM).

The material presented in the previous chapters of this book provides good reason to focus on the benefits of using PEM electrolyzers compared to alkaline electrolyzers as part of isotopes separation plants.

Some advantages of using PEM electrolyzers are the following:

- Substantially lower specific energy consumption for water decomposition
- Minimal amount of electrolyte in the device with equal hydrogen performance
- Absence of any impurities (except oxygen and water vapor) in the generated hydrogen

The significance of the first benefit is clear without any further explanation. As for the amount of electrolyte, this factor in the separation units determines the parameter called the period of accumulation (Andreev et al. 2000)—the time required to reach the steady-state operation of the separation unit, when the desired profile of isotopic concentrations at sections of the column along its entire height is achieved.

The chemical purity of the generated hydrogen in the electrolyzer is important in terms of the absence of possible catalyst poisons for a heterogeneous catalyst isotopic exchange between water and hydrogen.

However, the use of PEM electrolyzers as part of a hydrogen isotope separation plant does present some problems; for example, reliability, and service life during continuous operation (service life should be at least tens of thousands of hours).

It should be noted that in any embodiment of an electrolyzer, the value of the single separation effect in the decomposition of water is an important parameter required for the calculations of separation installation size and its design.

17.2 Brief Introduction to the Theory of Isotope Separation

17.2.1 Reversible and Irreversible Separation Processes

Figure 17.1 shows a schematic diagram of any single process of a binary mixture separation.

Initial mixture of components A and B in an amount of G_0 using a physical or physical–chemical process is divided into the enriched in target component A fraction in an amount of G_1 and depleted in component A fraction in an amount G_2. The main characteristics of this separation process are the following:

- The material balance equation

$$G_0 = G_1 + G_2, \tag{17.1}$$

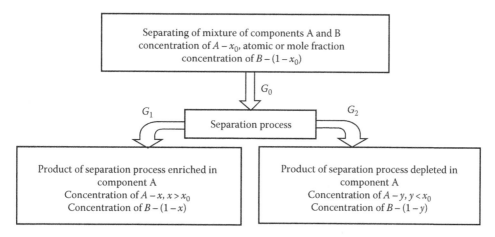

FIGURE 17.1
Schematic diagram of the separation of components in a binary mixture.

$$G_0 \cdot x_0 = G_1 \cdot x + G_2 \cdot y \quad \text{(balance for component A)},$$

$$(17.2)$$

$$G_0 \cdot (1 - x_0) = G_1 \cdot (1 - x) + G_2 \cdot (1 - y)$$

(balance for component B). $\quad$ (17.3)

- The quantitative characteristics of the efficiency of the separation process, separation factor α

$$\alpha = \frac{x/(1-x)}{y/(1-y)} = \frac{x \cdot (1-y)}{y \cdot (1-x)}, \quad (17.4)$$

which represents the ratio of the relative (dimensionless) concentrations of $x/(1-x)$, $y/(1-y)$ the target component in the enriched and depleted fractions, and *the enrichment factor* $\varepsilon = \alpha - 1$

$$\varepsilon = \frac{x - y}{y \cdot (1 - x)}. \quad (17.5)$$

From the definition of the separation factor and Equation 17.4, it follows that its absolute value is ≥ 1. In the case that the separation effect is absent, $\alpha = 1$. The greater the separation effect is, the greater the α value is. The absolute value of the separation effect, which is determined by the difference in target component concentrations in the enriched and depleted fractions, characterizes the enrichment factor ε.

Separation processes can be divided into two groups: irreversible (nonequilibrium) and reversible (equilibrium). Electrolysis along with other processes such as gas diffusion and thermal diffusion, centrifugation, photochemical and laser methods, and mass separation are some of the irreversible processes used for isotope separation. Distinctive features of these methods are that energy must be expended at each stage of the separation process and, moreover, the chemical form of the target component can differ in the initial mixture and the depleted or enriched phases. The cause of the isotope effects (exceeding α over the 1) in these processes is the difference in mass of isotopomers, on which, for example, the diffusion coefficients depend; the value of the centrifugal force; and the force of interaction of charged particles with electric and magnetic fields.

The simplest example of a reversible separation process is distillation. Due to equality of the heat of evaporation and condensation of substances under adiabatic conditions, the single isotopic equilibration between the vapor and liquid phases does not require additional energy consumption, and leads to obtaining enriched (usually liquid) and depleted (usually steam) fractions. The chemical nature of the substance in the enriched and depleted fractions in such processes, called phase isotope exchange (PhIE), remains unchanged.

Another group of reversible separation processes include the chemical isotope exchange reaction (ChIE), in which the isotopic equilibrium in the separation process is established between substances that differ in their nature, for example, isotopic exchange between water and hydrogen. In ChIE reactions, the real concentration of exchanging substances remains constant, only the distribution of isotopes between them changes.

The cause of the isotope effect in reversible processes has a thermodynamic nature. To concentrate a target isotope in any substance (in a ChIE reaction) or in any phase of a substance (in a PhIE reaction), it is necessary that the exchange be energetically favorable. When the separation process occurs, distribution of the target isotope in the substances should not be equiprobable because the total amount of free energy of the products is below the free energy of the starting materials during the determination of isotopic equilibrium. The nature of this isotope effect is the difference in ground state vibration levels for substance isotopomer molecules:

$$E_{\text{vibr},0} = \frac{hc\omega}{2},$$ (17.6)

where
  h is the Planck constant
  c is the speed of light
  ω is the natural molecule frequency (the wave number, cm^{-1})

Natural molecule frequency is expressed by the following equation:

$$\omega = \frac{1}{2}\pi c \cdot \sqrt{\frac{K}{\mu}},$$ (17.7)

where
  K is the molecule force constant (for given molecule does not depend on the isotopic substitutions)
  μ is its reduced mass

For a diatomic molecule AX, the reduced mass is calculated as follows:

$$\mu = \frac{m_A \cdot m_X}{m_A + m_X},$$ (17.8)

where m_A and m_X are the mass of the A and X atoms, respectively.

For example, we can calculate the ratio of the reduced mass of a diatomic water molecule fragment –OH with different isotopic substitutions: $-{}^{16}_{8}O^{1}_{1}H$, $-{}^{16}_{8}O^{2}_{1}H$, $-{}^{18}_{8}O^{1}_{1}H$. Reduced masses of these fragments are 16/17, 32/18, 18/19, respectively. Then the ratio of the vibration frequency, and hence the energy ground state vibration levels for these fragments, are as follows: $\omega^{16}_{8}O^{1}_{1}O / \omega^{16}_{8}O^{2}_{1}O = \sqrt{(32 \times 17)/(18 \times 16)} = 1.37$, $\omega^{16}_{8}O^{16}_{8}O / \omega^{18}_{8}O^{1}_{1}O = \sqrt{(18 \times 17)/(19 \times 16)} = 1.0033$.

An important conclusion follows from these calculations is as follows: the reversible hydrogen isotope separation processes will always be accompanied by a much larger isotope effect than the exchange processes with the participation of isotopes of other elements.

17.2.2 Definition of Separating Elements, Their Types, and Regularities of Work

A separating element is a device in which a physical, chemical, or physical–chemical process occurs that leads

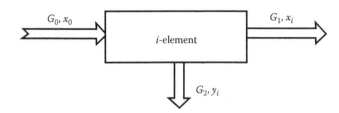

FIGURE 17.2
Separation elements that realize an irreversible separation process (separating element of the first type).

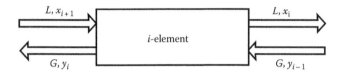

FIGURE 17.3
Separation elements that realize a reversible separation process (separating element of the second type).

to the separation of the initial mixture in an enriched and a depleted fraction of the target isotope. The ratio of relative concentration the target isotope in these fractions is equal to the value of a single separation factor α.

Figures 17.2 and 17.3 show schematics of separating elements that realize an irreversible (see Figure 17.2, separating element of the first type) or reversible (see Figure 17.3, separating element of the second type) separation process.

A typical example of the first type of separating element is the electrolyzer. The incoming (to electrolysis) flow of water (G_0) is partially decomposed into hydrogen (stream G_1) and oxygen (not indicated in the diagram), while the remaining water flow is removed from the electrolyzer (flow G_2). It is known that the concentration of heavy hydrogen isotopes in the resulting hydrogen is less than in water (Yakimenko et al. 1970; Fateev et al. 2005), so the exiting electrolyzer water flow is enriched with these isotopes. Some characteristics of the separating element of the first type are given in the following.

By definition of a separating element

$$\alpha = \frac{x_i \cdot (1 - y_i)}{y_i \cdot (1 - x_i)}.$$ (17.9)

Introducing the notation $\gamma = G_{1,i}/G_{0,i}$ (degree of subdivision of flow), we can write a consequence of the material balance equation as follows:

$$x_0 = \gamma \cdot x_i + (1 - \gamma) \cdot \left\{ \frac{x_i}{[\alpha - x_i \cdot (\alpha - 1)]} \right\}.$$ (17.10)

The following two important characteristics of the separating element are the degree of recovery of the target component Γ_i and the degree of separation K_i:

$$\Gamma_i = \frac{G_{1,i} \cdot x_i}{G_0 \cdot x_0}, \tag{17.11}$$

$$K_i = \frac{x_i \cdot (1 - x_0)}{x_0 \cdot (1 - x_i)}. \tag{17.12}$$

Both of these values depend on the degree of subdivision of flow in the separating element:

$$K_i - 1 = \frac{(\alpha - 1) \cdot (1 - \gamma)}{1 + \gamma \cdot (1 - x) \cdot (\alpha - 1)}. \tag{17.13}$$

From Equation 17.13 it follows that at $\gamma \to 1$, K_i tends to 1, that is, the separation effect is absent; and conversely, when $\gamma \to 0$, $K_i \to a$, but the value of withdrawn of the enriched phase tends to 0. With regard to the recovery degree, its maximum value is reached at the value of γ, depending on α.

The work of a separating element of the second type is characterized by the following equation:

- equilibrium equation

$$\alpha = \frac{x_i \cdot (1 - y_i)}{[y_i \cdot (1 - x_i)]} \quad \text{(see above)}, \tag{17.9}$$

- material balance equation

$$L \cdot x_{i+1} - G \cdot y_i = L \cdot x_i - G \cdot y_{i-1} = B \cdot x_B, \tag{17.14}$$

where $B \cdot x_B$ is a given quantity of the target isotope concentrate and its concentration.

Denoting the value of the flux ratio as $G/L = \lambda$, we obtain from Equation 17.14

$$x_i - \lambda \cdot y_{i-1} = \frac{B}{L} \cdot x_B. \tag{17.15}$$

From Equation 17.15, it is implied that the ratio of flows and its value depends on the task of separation.

17.2.3 Cascades of Separating Elements: Ideal and Square Cascades

Figures 17.4 and 17.5 show the schematic flow diagrams of cascades of separating elements of the first and second types for concentrating the target isotope.

In the cascade of separating elements of the second type (Figure 17.4), the feed flow on each element of the cascade decreases from G_0 to B, and thus for the separating of the entering on each element flow for enriched and depleted fractions, one needs to expend energy. In the case of return depleted fraction with concentration of target isotope $> x_0$ from i-stage to one of the preceding cascade stages, it must be either subjected to chemical conversion (for electrolysis—target isotope from hydrogen has to be transferred into the water form) or for his return it is necessary to expend energy (in the diffusion separation process—to increase the gas pressure).

In the cascade of separating elements of the second type, shown in Figure 17.5, L and G flows at all elements of the cascade do not change, but at the end of the cascade a device for the reversion of flow (DRF) is installed. Thus, L and G flows are many times greater than the separation process product flow B. To describe the process of separation throughout entire cascade, *the fundamental equation of enrichment* on the one stage of the cascade is valid (Rosen 1960):

$$\Delta x = x_i - x_{i+1} = \frac{\acute{\varepsilon} \cdot x_i}{1 - \acute{\varepsilon} \cdot x_i} (1 - x_i) - \frac{B}{L} \cdot \left[x_B - \frac{x_i}{\alpha \cdot (1 - \acute{\varepsilon} \cdot x_i)} \right], \tag{17.16}$$

where $\acute{\varepsilon} = (\alpha - 1)/\alpha$.

From this equation, it follows that the maximum enrichment for the separation stage $\Delta x_{max} = (\acute{\varepsilon} \cdot x_i)/(1 - \acute{\varepsilon} \cdot x_i)(1 - x_i)$ is reached at $B = 0$ or $L \to \infty$. If over the entire range of concentrations in the cascade the condition $\acute{\varepsilon} \cdot x_i \ll 1$ is valid, then Equation 17.16 transforms to the *equation of enrichment on the one stage of the fine separation cascade*:

$$\frac{dx}{dn} = \acute{\varepsilon} \cdot x(1 - x) - \frac{B}{L} \cdot (x_B - x), \tag{17.17}$$

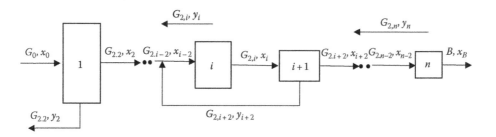

FIGURE 17.4
Schematic of flow diagram in concentrating cascade of separating elements of the first type.

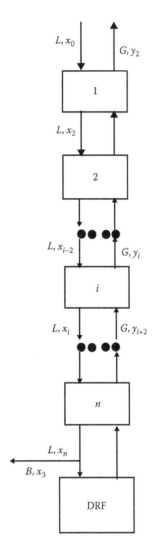

FIGURE 17.5
Schematic of flow diagram in concentrating cascade of separating elements of the second type.

From which it follows that the value of minimum flow entering the cascade is determined with $dx/dn \to 0$:

$$L_{\min} = \frac{B \cdot (x_B - x)}{\acute{\varepsilon} \cdot x \cdot (1 - x)}, \qquad (17.18)$$

and the minimum number of separation stages in the cascade is determined with a maximum enrichment of one stage:

$$\frac{dx}{dn} = \acute{\varepsilon} \cdot x \cdot (1 - x) \qquad (17.19)$$

and $B \to 0$

$$\text{as} \quad n_{\min} = \frac{1}{\acute{\varepsilon}} \cdot \ln \left[\frac{x_B}{x_0} \cdot \frac{(1 - x_0)}{(1 - x_B)} \right] = \frac{1}{\acute{\varepsilon}} \cdot \ln K, \quad (17.20)$$

where K is the degree of separation.

The cascade, where the condition $x_{i-1} = y_{1+1}$ is valid (see Figure 17.4), is *called an ideal one* (Rosen 1960). Maximum enrichment at this cascade stage is expressed by the equation

$$\frac{dx}{dn} = \frac{1}{2} \cdot \acute{\varepsilon} \cdot x \cdot (1 - x). \qquad (17.21)$$

From a comparison of Equations 17.19 and 17.21, the following relationship between the minimum flows and the number of stages in an ideal cascade and a fine separation cascade follows:

$$L_{\text{ideal}} = 2L_{\min} \quad \text{and} \quad n_{\text{ideal}} = 2n_{\min}.$$

A quantitative diagram that illustrates flows in a cascade of separating elements of the first and second types is presented in Figure 17.6.

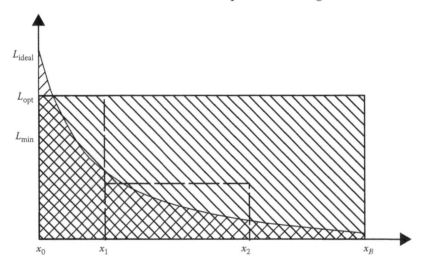

FIGURE 17.6
Quantitative diagram illustrating flows in a cascade of separating elements of the first and second types (curved line corresponds to the change at flow L in an ideal separating elements cascade of the first type with a change in the target isotope concentration from x_0 to x_B).

TABLE 17.1

Dependence of the Form Efficiency from a Given Separation Degree

Separation degree, K	2	3	5	10	100
Form efficiency, η	0.96	0.93	0.85	0.73	0.52

The curved line corresponds to the change at flow L in an ideal cascade of separating elements of the first type with a change in the target isotope concentration from x_0 to x_B. The total flow in the cascade is determined by the shaded area under the curve. Solid line bounds rectangle whose area corresponds to the total flow L in a second-type separation elements cascade, where the concentration change is achieved in one step. Obviously, the total flow in the cascade of the second-type elements is greater than in an ideal cascade. The ratio of these flows is characterized by a parameter called form efficiencyη: $\eta = \Sigma L_{ideal} / \Sigma L_{square}$.

Table 17.1 shows the values of this parameter (η), depending on the separation degree in the cascade (K).

For small separation degrees, the total flows in the ideal and rectangular cascades differ little, but with increasing K this difference becomes significant. Therefore, in practice, second-type separating elements cascades at practically significant values of K^* are divided into several successive stages: $K_\Sigma = K_1 \cdot K_2 \cdots K_N$. For example, in Figure 17.6, the dotted lines represent flows in a rectangular three-stage cascade, concentrating the target isotope from x_0 to x_B. The figure shows that the total flux in the three-stage cascade is much closer in magnitude to the flow in an ideal cascade.

17.3 Use of PEM Electrolyzers in Hydrogen Isotope Separation Facilities

17.3.1 Preliminary Considerations of Applications and Problems

In the introduction to this chapter, it was noted that in conventional systems designed for the separation of hydrogen isotopes, PEM electrolyzers can find their application as DRFs in the technology, which in literature carries the abbreviation combined electrolysis and catalytic exchange (CECE) or electrolysis and exchange (ELEX) processes (Andreev et al. 2007). The technology is based on the ChIE process between water and hydrogen:

$$ABO_l + AB_g \leftrightarrow B_2O_l + A_{2,g}, \qquad (17.22)$$

where A and B are the hydrogen isotopes (B is the heavier isotope, which is concentrated in the water), the symbols g and l represent the gas and liquid phases, respectively.

The process is a catalytic one; it occurs using hydrophobic platinum catalysts in two stages and it is realized in a countercurrent multistage column (see Figure 17.5).

$$ABO_{vapor} + AB_g \leftrightarrow B_2O_{vapor} + A_{2,g} \qquad (17.23)$$

$$B_2O_{vapor} + ABO_l \leftrightarrow ABO_{vapor} + B_2O_l \qquad (17.24)$$

The column is filled with a catalyst and a hydrophilic packing with the developed surface (for efficient reaction (17.24)). An electrolyzer is installed at the bottom of the column. The entire exiting from column water flow is converted into a directed to the column stream of the hydrogen.

Let us now draw a semi-quantitative comparison of using water electrolysis and the CECE process to solve the following task: getting 200 kg h^{-1} deuterium-enriched water with a concentration of 1.5 at.% from natural water. Using electrolyzers with separation factor $\alpha_{el.} = 5$, with $\gamma = 0.5$ in the technological scheme shown in Figure 17.4, and returning depleted fraction for feeding electrolyzer with the number $i-2$, 44.5 tH$_2$O h^{-1} must be fed to the cascade. The recovery degree is then 0.45 and the power consumption is 310 MWh. In a two-stage rectangular cascade that implements CECE technology with the separation factor of reaction 17.22 $\alpha_{ChIE} = 3.5$, with the same value of $K_\Sigma = 100$ ($x_B/x_0 = 1.5/0.015$), but broken on the stages of cascade with $K_1 = 5$ and $K_2 = 20$, the feed flow must be 33.0 tH$_2$O h^{-1}. The recovery degree will then be 0.61, and the power consumption for the water electrolysis in the DRFs is about 200 MWh.

It should be noted that the capital cost of an electrolysis separation cascade creation will greatly exceed the cost of a ChIE two-stage cascade column. Technologically, the scheme of the electrolysis cascade will consist of nine steps, including 221 electrolyzers (if their performance is 250 m^3 H$_2$ h^{-1}), with all attendant connections between them.

An important point related to the use of electrolyzers as DRFs should be noted. The separation units in CECE technology is a combination of the first-type separating element (electrolyzer) and ChIE columns, consisting of a large number of second-type separating elements, in which the equilibrium reactions (17.22) through (17.24) are carried out. As a rule, >99% of the total separation factor of hydrogen isotopes achieved in this installation is due to the separation degree achieved in the ChIE column. On the other hand, the accumulation period for the entire separation installation is also determined for 90–95% by the accumulation of the target isotope in the electrolyte. This volume is many times greater than the amount of water (and especially hydrogen) in the ChIE column. Therefore, unlike in the case of using the electrolyzer in

---

* In the production of heavy water with a deuterium concentration of 99.8 at.% from natural water (deuterium concentration $\approx$ 0.015 at.%) the value of K is $3.3 \cdot 10^6$.

TABLE 17.2

Values of α_{HT} and α_{HD}, according to Muranaka et al. (2012)

	Electrolyzer A						Electrolyzer B					
No.	$[T_{i,A}]$, Bk kg^{-1}	$[T_{f,A}]$, Bk kg^{-1}	$\alpha_{HT,A}$	$[D_{i,A}]$, at.%	$[D_{f,A}]$, at.%	$\alpha_{HD,A}$	$[T_{i,B}]$, Bk kg^{-1}	$[T_{f,B}]$, Bk kg^{-1}	$\alpha_{HT,B}$	$[D_{i,B}]$, at.%	$[D_{f,B}]$, at.%	$\alpha_{HD,B}$
1	7.56	26.00	4.30	0.0147	0.0469	3.58	7.56	26.61	4.58	0.0147	0.0468	3.57
2	7.69	27.77	4.95	0.0148	0.0459	3.37	7.69	27.30	4.70	0.0148	0.0459	3.37
3	8.09	26.55	3.82	0.0147	0.0450	3.28	8.09	25.75	3.56	0.0147	0.0442	3.16
4	8.12	26.20	3.67	0.0148	0.0469	3.53	8.12	25.81	3.55	0.0148	0.0459	3.37
Av.			4.18 ± 0.77			3.44 ± 0.16			$4/10 \pm 0.60$			3.36 ± 0.20

a first-type separation elements cascade, when separation factor is desirable as much as possible; using an electrolyzer as a DRF with all other conditions being equal seek to separation factor was as lower as possible.

It has already been noted that electrolysis was the first method used to produce heavy water; hence many studies focused on the dependence of the separation factor of water electrolysis in alkaline electrolyzers on various factors, for example, the current density, temperature, cathode material, electrolyte composition. Although consideration of these dependencies for alkaline electrolysis is not the task of this monograph, we give as an example the result of work (Stojić et al. 1991), where the influence of the cathode material and the method of its preconditioning on the separation factor of the protium–deuterium isotopic mixture (α_{HD}) in the electrolysis of 30% KOH solution was investigated. Depending on the cathode material, α_{HD} can vary from 7.9 (Fe cathode) to 4–4.5 (Pt and Pd cathodes).

There are far fewer investigations devoted to the actual measurement of the separation factors for the various isotopic mixtures using PEM electrolyzers reported in the literature. We can, however, consider rather extensive experimental material given in by Muranaka and Shina (2012). They investigated the electrochemical method of tritium preconcentration in natural water by reducing the amount of analyzing sample in PEM electrolyzers. The purpose of preconcentration is to increase the concentration of tritium in the sample to a level above the detection limit of liquid scintillation counting. The main parameter determined in this study was the tritium recovery factor R:

$$R = \frac{(T_f \cdot V_f)}{(T_i \cdot V_i)}, \quad (17.25)$$

where

T_i and T_f are the concentration of tritium in the sample before and after the reduction of the volume by electrolysis

V_i and V_f are the initial and final volumes of the sample, respectively

This parameter does not contain an explicit value of the separation factor. To calculate of the separation

factor can be used, however, the well-known Rayleigh equation, which for low concentrations of the heavy isotope can be written as follows (Andreev et al. 1987):

$$\left(\frac{V_i}{V_f} \right)^{(\alpha_i - 1)/\alpha_i} = \frac{x_f}{x_i}, \quad (17.26)$$

where

α_i is the separation factor

x_f, x_i are the concentration of the heavy isotope in the residue and the original volume, respectively

Using the primary data of Muranaka and Shina (2012) it is possible to calculate the values of the separation factors.

Table 17.2 shows the experimental data taken from tables of data compiled in this work (Tables 17.1 and 17.3) and the values of the separation factors in protium–tritium and protium–deuterium isotopic mixtures obtained by calculation according to Equation (17.26).

The measurements were performed using two PEM (Nafion 117) electrolyzers (working A and standard B) with electrodes made of porous titanium (anode) and a porous stainless steel (cathode) with electrocatalysts on them. The current density for both electrolyzers was 0.2 A cm^{-2}. The initial isotopic composition of the electrolyte in the A and B electrolyzers were the same.

TABLE 17.3

Influence of the Cathode Electrocatalyst Nature, Current Density, and Temperature on $\alpha_{HD, \Sigma}$ and $\alpha_{HT, \Sigma}$

Isotope Mixture		Protium–Tritium			Protium–Deuterium	
Cathode Electrocatalyst						
T^a, K	i, A cm^{-2}	Pt	Pd	WC[b]	Pt	WC
303	0.5	14.4	—	11.2	7.7	5.2
	1.0	18.8	—	12.0	8.4	5.6
	2.0	12.5	—	9.7	6.6	4.6
363	0.5	7.5	4.3	8.7	4.6	4.5
	1.0	9.6	5.8	8.2	5.5	4.4
	2.0	9.0	6.4	7.1	4.6	4.5

[a] Temperature at water jacket.
[b] WC—carbide of tungsten.

The volume of the electrolyte was reduced from 300 to 60 cm³, that is, the value of (V_i/V_f) in Equation 17.26 is 5. Four series of experiments were carried out.

The data show that the average values of separation factors α_{HD} and α_{HT} are in good agreement for both electrolyzers, although for the protium–tritium system the scatter of α_{HT} values is >15%. Further, if we assume that the difference in isotope effect values in the PEM electrolyzer for protium–deuterium and protium–tritium systems is limited by the stage, related to the difference in the diffusion coefficients of the deuteron and triton, then the ratio should be close to $\sqrt{m_T/m_D} = \sqrt{3/2} = 1.225$. From the table, it follows that the experimental ratio $\alpha_{HT}/\alpha_{HD} = 4.14/3.40 = 1.218$ is in agreement with what is expected.

The amount of isotope effect strongly depends on the electrolyzer design. For example, in two Nafion-based electrolyzers described by Muranaka and Honda (1996) and Saito (1996) at comparable current densities (0.3–0.4 A cm⁻²) and a five-fold reduction of the electrolyte residue received the tritium recovery factor 0.860 and 0.658. These values of recovery factor correspond to more than three times difference in the values of α_{HT} –10.7 and 3.3.

In the early 1990s after the appearance of the first hydrophobic catalysts, first in the USA and Canada (Stevens 1975), and in many countries after that (the number of patents of this catalysts today is far greater than 200), work began on the development of CECE technology for the reprocessing of tritium-containing heavy waste water. Many pilot and industrial systems were built, and their operation subsequently led to the improvement of technology and the emergence of new patents (Andreev et al. 1995; Ionita and Stefanescu 1995; Allan et al. 2000; Cristescu et al. 2002; Sugiyama et al. 2006; Alekseev et al. 2011; Denton and Shmaida 2013).

During this period, at the D. Mendeleyev University of Chemical Technology of Russia (MUCTR), separation installations for deprotiation (Andreev et al. 1995) and detritiation (Andreev et al. 1997) of deuterium-containing wastes and obtaining conditional heavy water were constructed and commissioned. In both installations, PEM electrolyzers were used as the DRFs. Experimental data on the separation factor at the electrolysis of water for the various isotopic mixtures were required for the analysis of installations work conditions in different modes. The results of such investigations are reported in literature (Andreev et al. 1988a,b; Rozenkevich et al. 1989; Karpov et al. 1991; Goryanina et al. 1998).

17.3.2 Single Separation Effect in PEM Electrolyzers

17.3.2.1 Binary Isotopic Mixtures of Protium–Deuterium and Protium–Tritium

To study of influence of various factors on the α_{HT} and α_{HD}, a PEM electrolyzer based on a Nafion-type membrane (thickness 250 μm, area 42.5 cm²) was used. A layer

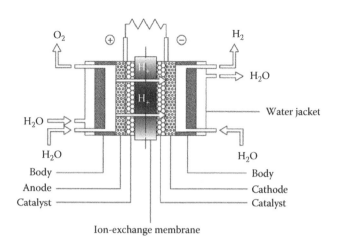

FIGURE 17.7
Schematic of a PEM electrolyzer.

of particulate catalyst was applied to the surface of the membrane (density of application 2 mg cm⁻²) (Pakhomov and Fateev 1990). A schematic of the electrolyzer assembly is shown in Figure 17.7. The electrolyzer has a thermostatic system for the feed of temperature-controlled water.

Experiments to measure the separation factors were performed using the experimental setup, a schematic of which is shown in Figure 17.8.

The flows of hydrogen and oxygen produced in the electrolyzer after separation from water in the separators 2 and 3 and drying in the zeolite traps 5 are sent to a flame burner where the oxidation of hydrogen is performed (an additional amount of oxygen from a cylinder 9 is supplied into burner to ensure an excess over stoichiometry of 5%–7%). Separators 2 and 3, and the burner 6, are cooled with tap water. Water is supplied from the water separator in the anode compartment of the electrolyzer, where the electrolyte is circulated through the cycle electrolyzer–separator–electrolyzer. Reservoir 4 is used for the initial filling of the electrolyzer with double-distilled and deionized water.

The experimental procedure was as follows. After filling the electrolyzer with target isotopic composition water, voltage is applied and the desired current is set. In preliminary experiments, it was found that 40–50 min of electrolyzer operation in a predetermined mode is sufficient to establish a thermal stationary condition. After this time, were selected different water samples for the isotopic analysis: electrolyzer circulating water (x_{el}), water from the hydrogen separator ($x_{cath.}$), recuperative water from the burner 6 (y).

Isotopic analysis of samples for the deuterium content was performed by atomic spectroscopy, using the intensity of the H_α line in the Balmer series. The measurement error of the deuterium concentration depends on its absolute value and does not exceed 5 rel.%.

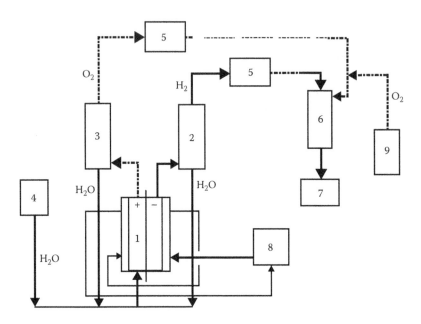

FIGURE 17.8

Schematic of the experimental setup used for the determination of separation factors. 1, electrolyzer; 2 and 3, hydrogen and oxygen separators; 4, tank for double-distilled and deionized water; 5, zeolite dryers of hydrogen and oxygen flows; 6, flame burner for burning of hydrogen; 7, collector of water obtained after combustion of hydrogen; 8, system for thermostating of electrolyzer; 9, source of additional oxygen (to ensure an excess over stoichiometric amount).

Determination of the tritium concentration was performed by liquid scintillation counting. Measurement conditions were chosen such that the measurement error do not exceed 2 rel.%.

The total isotope effect in the electrolysis of water can be regarded qualitatively as being composed of three component parts: the difference between the rates of diffusion of the ions H^+, D^+, and T^+ through the membrane; the rate of recombination in a molecule of hydrogen at the cathode; and the isotope exchange reaction of hydrogen with water at the cathode electrocatalyst. Each of these parts should lead to the enrichment of produced hydrogen by the light isotope. So it can be expected that the maximum value of the separation factor should be achieved by comparing the isotopic concentrations in recuperative water (y) and water collected from the cathode separator ($x_{cath.}$). It was found, experimentally, that using Pt as the cathode electrocatalyst the values of $\alpha_{HT,i}$ calculated be the following equations:

$$\alpha_{HT, cath.} = \frac{x_{cat} \cdot (1-y)}{[y \cdot (1-x_{cath.})]} \tag{17.27}$$

$$\alpha_{HT, \Sigma} = \frac{x_{el.} \cdot (1-y)}{[y \cdot (1-x_{el.})]} \tag{17.28}$$

Differ at current density of 0.5 A cm^{-2} by less than 20%: 12.0 and 10.6 at $T = 333$ K, and 7.5 and 6.9 at $T = 363$ K. In Table 17.3, data for the influence of the nature of the cathode electrocatalyst, current density, and temperature on the $\alpha_{HD, \Sigma}$ and $\alpha_{HT, \Sigma}$ are presented.

This data shows, first, that in most cases there is strong dependence of the separation factors on the current density. The explanation of this follows from the Tafel equation:

$$\eta = a + b \cdot \ln i, \tag{17.29}$$

where

η is the overvoltage (exceeding the electrode potential compared to the thermodynamic)

i is the current density

a and b are constants

This equation implies that an increase in current density is accompanied by an increase in the overvoltage (η), almost all of which is converted to local heat on the electrode. The temperature which is shown in Table 17.3 corresponds to the electrolyzer periphery temperature (see Figure 17.7) instead of the local temperature in the zone of the cathode process. It should be noted that, according to the experimental method, in the cathode compartment of electrolyzer liquid water is absent for any current density and heat is dissipated from the electrodes only by a steam–hydrogen mixture. Special experiments have shown that at the thermostatic electrolyzer temperature of 303 K with increasing i to 1 A cm^{-2} the membrane temperature is increased to 343 K. At the thermostatic electrolyzer temperature of 363 K and the same current density, the local temperature is increased to 373 K. It means that the difference between the local temperature and thermostatic temperature decreases with increasing thermostatic temperature. Therefore, at low temperature thermostating electrolysis is carried out at a large temperature gradient over the cross section of the electrolyzer from the membrane to the periphery.

Equation 17.28 shows the dependence of the electrochemical reaction rate on a number of parameters, including voltage and temperature:

$$i \sim \exp-\left[\frac{(E_a + \tilde{\alpha} \cdot z \cdot F \cdot \eta)}{(R \cdot T)}\right], \qquad (17.30)$$

where

E_a is the activation energy in absence of cathodic polarization

$\tilde{\alpha}$ is the transfer coefficient, $0 < \tilde{\alpha} < 1$

F is the Faraday number

z is the number of transferred electrons

From this equation, the expression for the separation factor can be written as follows:

$$\alpha_\Sigma \sim \frac{i_H}{i_{D(T)}} = \frac{\exp\left[\begin{array}{c}\left(E_{a,D(T)} - E_{a,H}\right) \\ + \left(\tilde{\alpha}_{D(T)} - \tilde{\alpha}_H\right) \cdot z \cdot F \cdot \eta\end{array}\right]}{(R \cdot T)}. \qquad (17.31)$$

Equation 17.29 implies that the value of the separation factor should increase with increasing η and decrease with increasing temperature.

Figure 17.9 shows the dependence of the cathodic overvoltage on current density for the electrolyzer with a tungsten carbide cathode. At high current densities, the linearity of the Tafel plot is not kept, obviously due to the increasing role of diffusion limitations of electrolysis.

Figure 17.10 shows the dependence of $\alpha_{HD,\Sigma}$ on current density for electrolyzers with a Pt cathode using a Nafion-type membrane in the H^+ form and partly "poisoned" by Fe^{3+} ions. At a current density of 0.5 A cm^{-2} and

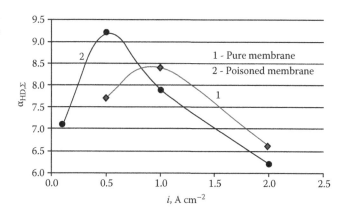

FIGURE 17.10
Dependence of separation factor in isotopic protium–deuterium mixture on current density and state of the membrane (Nafion-type membrane).

a temperature of thermostatic electrolyzer 303 K, voltage with a clean membrane is 2.24 V, and with a partially "poisoned" one it is 2.56 V, that is, the value of the total overvoltage differs by 320 mV. The figure shows that, first, at low current densities $\alpha_{HD,\Sigma}$ is higher for a "poisoned" membrane than for a pure one, and, second, the maximum value of $\alpha_{HD,\Sigma}$ is achieved at lower current density.

Thus, the strong dependence of the separation factor on the current density is due to the joint and antibate influence of the temperature on the cathode, an increase which leads to a decrease in α_Σ, and the value of overvoltage, an increase which leads to an increase in α_Σ. It should be noted that, as shown in Equation (17.29), the absolute values of the separation factors is determined by the difference in the coefficients of the transfer and the value of η, which, in turn, is largely determined by the type and assembly quality of the electrolyzer membrane electrode assemblies (MEA).

Furthermore, based on our practical experience, the PEM electrolyzers from different manufacturers operated under comparable operating conditions can be characterized by significantly different absolute values of separation factors, although the trends in their dependence on the parameters of electrolysis (e.g., temperature, current density, isotopic composition of the electrolyte) will be similar.

17.3.2.2 Triple Isotopic Mixture of Protium–Deuterium–Tritium

Nowadays, the use of combined electrolysis and catalytic exchange (CECE) technology for hydrogen isotope separation is considered most promising for the reprocessing of tritium-containing waste that can be characterized by wide range of deuterium concentrations in them: light water with natural deuterium level (Boniface et al. 2011), heavy water with deuterium > 98 at.% (Allan et al. 2000),

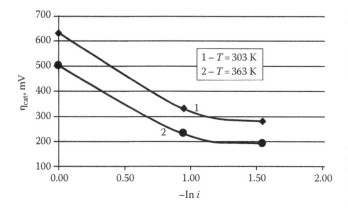

FIGURE 17.9
Dependence of cathode overvoltage on current density and temperature.

and waste water with a variable concentration of tritium (Perevezentsev and Bell 2008; Alekseev et al. 2011). All these existing installations or projects are related with using of the electrolyzers as DRF. Decision of the tasks of detritiation containing deuterium waste implies the need to study the dependence of the separation factor $\alpha_{HT(DT),\Sigma}$ on the deuterium concentration in the electrolyte. This work has also been carried out at MUCTR.

When considering the results described in the following, it should be noted that because of the design features of the electrolyzer with a Pt cathode used on the experimental installation (see Figure 17.8), it operated at high (3–5 V) overvoltages. Therefore, for the absolute value of the separation factor, the local temperature at the MEA has a decisive influence, but not the thermostatic temperature of the electrolyzer. Experiments were carried out at a current density 0.25 A cm^{-2}.

Data for the dependence of $\alpha_{HT(DT),\Sigma}$ on the deuterium concentration in an electrolyte and the temperature are presented in Table 17.4. Distribution of water isotopomers at different deuterium atomic concentration in water can be seen in Figure 17.11. Tritium concentration

TABLE 17.4

Dependence of the Separation Factor for Tritium on Deuterium Concentration in the Electrolyte and Temperature

[D], at.%	Electrolyzer Thermostatic Temperature, K		
	298	313	333
9.0	6.58	5.42	4.48
25.0	—	4.87	—
48.0	4.45	3.71	3.59
65.0	3.36	—	2.65
91.0	2.27	2.32	2.03
99.5	1.89	1.90	1.78

TABLE 17.5

Changes in Separation Factor $\alpha_{HD(DH),\Sigma}$ Depending on the Concentration of Deuterium in the Electrolyte ($T = 333$ K, $i = 0.25$ A cm^{-2})

[D], at.%	9.0	24.0	48.0	67.0	91.0
$\alpha_{HD(DH),\Sigma}$	2.80	3.04	2.83	3.17	2.60

in the electrolyte has always been at the level of the label ($[T] < 1 \times 10^{-8}$ at.%).

The table shows that the separation factor for tritium decreases sharply with increasing deuterium concentration in the solution, that is, during the transition from a separated isotopic mixture protium–tritium to a deuterium–tritium mixture. Of note is that the separation factor $\alpha_{HD(DH),\Sigma}$ remains practically unchanged up to the deuterium concentration > 70 at.%. (see Table 17.5).

Table 17.6 shows comparison of tritium separation factors for the processes of water electrolysis and chemical isotope exchange between water and hydrogen at a deuterium concentration that tends to 0 and to 100 at.%. The values of $\alpha_{HT,\Sigma}$ and $\alpha_{DT,\Sigma}$, obtained by extrapolation of experimental data to $[D] \to 0$ and $[D] \to 1$ are presented. The corresponding values of the separation factors for the ChIE reaction in the water-hydrogen system obtained from the equations given in (Andreev et al. 2007).

From the data in the following it follows that, under comparable conditions, the values of separation factors in the electrolysis and chemical isotope exchange are close. This is because the final act of the molecular hydrogen formation occurs in the presence of water vapor at the cathode electrocatalyst made of platinum, which is well known as an active catalyst for the ChIE reaction, according to Equation (17.22).

In conclusion, we should note that, as it follows from Figure 17.12, the dependence of the separation factor for

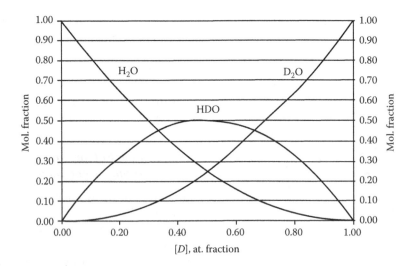

FIGURE 17.11
Equilibrium distribution of water isotopomers depending on the atomic concentration of deuterium at 293 K.

TABLE 17.6

Comparison of Tritium Separation Factors for the Processes of Water Electrolysis and Chemical Isotope Exchange between Water and Hydrogen at a Deuterium Concentration That Tends to 0 and to 100 at.%

	ChIE at System H_2O–H_2		Electrolysis	
T^a, K	α_{HT}, $[D] \to 0$	α_{DT}, $[D] \to 1$	$\alpha_{HT\Sigma}$, $[D] \to 0$	$\alpha_{DT\Sigma}$, $[D] \to 1$
298	6.88	1.67	7.00	1.85
313	5.68	1.61	5.80	1.85
333	5.22	1.55	4.95	1.75

[a] Temperature at water jacket.

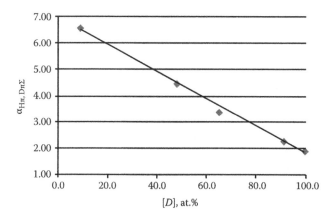

FIGURE 17.12
Dependence of separation factor for tritium in triple isotopic protium–deuterium-tritium mixture on deuterium concentration at 298 K.

tritium on deuterium concentration is linear. Moreover, it follows a simple equation of additivity:

$$\alpha_{HT(DT),\Sigma} = \alpha_{HT,\Sigma} \cdot \left(1-[D]\right) + \alpha_{DT,\Sigma} \cdot [D], \qquad (17.32)$$

The calculation results obtained using the data from Table 17.6 appear as a straight line in the figure.

17.4 Isotope Exchange between Water and Hydrogen in PEM Cells Realizing Hydrogen Electrotransfer Process (PEM-HET Cell)

17.4.1 Brief Consideration of Physical–Chemical Foundations of the PEM-HET Method

Currently, CECE technology is seen as the most promising method for the treatment of tritiated waste water. The practical implementation of this technology in traditional design countercurrent columns has its own features. The two-stage process (Equations 17.23 and 17.24) requires a special organization of filling the column with

mass-transfer devices. The reaction of Equation 17.23 occurs with a co-current of water vapor and a hydrogen mixture on the hydrophobic catalyst without liquid water. At a specific height of the catalyst layer, the desired degree of approximation to the isotopic equilibrium is reached. The function of hydrophilic layer of the packing is to achievement of a new isotopic equilibrium between of water vapor and liquid water in the counter current (Equation 17.24), in order to repeat the process on the following catalyst beds and packing many times. The combination of hydrophobicity of the catalyst and hydrophilicity of the packing creates hydrodynamic problems in the organization of counter-current contact between flows of hydrogen and water in the column. When filling the column with a mixture of the hydrophobic catalyst, the hydrophilic packing streams of water and hydrogen, which can be passed through a column of a certain cross section without flooding, are 3–4 times smaller than streams that could be passed through column which filled with only a hydrophilic packing (Andreev et al. 2007). Hence, the need to create a counter-current in a column containing a hydrophobic catalyst inevitably leads to an increase in the volume of the equipment.

It is known that PEM electrochemical devices can be used as hydrogen compressors (Casati et al. 2008). These devices operate at electric potentials that are lower than the water decomposition potential and are only intended to transfer a hydrogen ion from the anode through the membrane to the cathode. At MUCTR, this idea was the basis for developing new contact devices for carrying out a catalytic hydrogen isotope exchange with water, allowing us to physically separate flows of liquid water and hydrogen and hence eliminate the earlier mentioned hydrodynamic problems inherent in packing columns. This idea is illustrated at Figure 17.13,

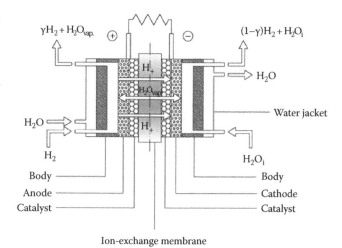

FIGURE 17.13
Schematic illustration of the HET process in an electrochemical PEM cell.

which is a schematic diagram of an electrochemical device with a design that is similar to the electrolyzer (Figure 17.9), but with modified working substance flow.

Hydrogen is fed to the anode compartment, reversibly adsorbed, and then partially ionized at the surface of Pt (anode electrocatalyst). Hydrated hydrogen ions, in an amount of $(1 - \gamma)$, are transferred across the membrane (Nafion type) to the cathode, where a process of recombination of atoms and the formation of molecular hydrogen takes place on the cathode electrocatalyst (Pt). The cathode compartment is filled with liquid water. In the anode compartment there is a mixture of water vapor and hydrogen streams. It is suggested that the isotopic exchange described by reaction (17.23) will occur on the electrocatalyst in the anode compartment and the process described by reaction (17.24) will occur due to the process of reversible diffusion of water vapor through the membrane and also the transport of hydrated hydrogen ions from the anode compartment to the cathode compartment. Part of hydrogen transported from the anode to the cathode is determined by the potential applied to the device. Hereafter, this process is abbreviated as PEM-HET. Perspectives for the practical use of such a contact device depends, of course, on the overall efficiency of the mass transfer process in the contact device in accordance with reaction on Equation 17.22. Features of this process have been studied in detail and reported in literature (e.g., Morozov and Rozenkevich 1990a,b; Morozov et al. 1990).

17.4.2 Transfer of Water through a Nafion-Type Membrane

17.4.2.1 Transfer of Water through the Membrane during the HET Process

Figure 17.14 shows the current–voltage characteristics of the HET process in a cell with a Nafion-type membrane (thickness 250 µm, area 42.5 cm²), with the cathode and

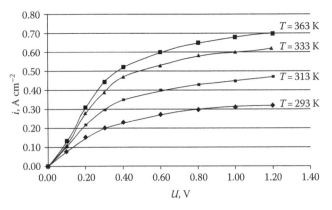

FIGURE 17.14
Current–voltage characteristics of the HET process in a cell with a Nafion-type membrane (250 µm) at different temperatures.

TABLE 17.7

Determination of the Number of Water Molecules Carried by the Protons from the Anode to the Cathode (i = 0.16 A cm⁻²)

T, K	303	323	343	363
$N \equiv$ number of H_2O molecules on 1 proton	2.70	2.96	3.14	3.32

anode catalysts of Pt (loading 2 mg cm⁻²). The cathode compartment of the cell is filled with water. The anode compartment is maintained at a constant hydrogen pressure of 500–600 Pa above atmospheric pressure.

A characteristic of the obtained dependences is their approach to saturation at high current densities. It can be assumed that this dependence is associated with a deficiency of water in the anode compartment, partial drying of the membrane at the anode side, and a decrease in its conductivity.

In a special series of experiments, the number of water molecules carried as a hydration shell of one proton from the anode to the cathode was determined. The electrochemical cell was operated in the electrolysis mode, the anode compartment was filled with water and in the cathode compartment there was no water. Leaving the cathode hydrogen stream fed into freeze-out device at 250–255 K, the volume of condensed water was measured. Experiments were performed at a constant current density in the temperature range 303–363 K. The Table 17.7 gives the ratio of number of transferred water molecules to the number of transferred hydrogen atoms.

The average number of water molecules carried by a single proton is 3. Taking this amount as a basis, it is possible to calculate the amount of water carried by the protons from the anode to the cathode under the conditions of the HET experiments. Results can be obtained from the data in Figure 17.14. At the limiting voltage of 1.2 V applied to the cell, the specific amount of water should be 0.217, 0.158, and 0.100 mol s⁻¹ m⁻² at temperatures 363, 313, and 293 K, respectively. The validity of the assumption made earlier about the drying of the membrane due to the scarcity of water in the anode compartment of the cell can be checked by measuring water permeation through the Nafion-type membrane.

17.4.2.2 Water Permeability of a Nafion-Type Membrane

Two methods have been proposed for the experimental determination of membrane water permeability. In Rozenkevich et al. (2003), the membrane MF-4SK, which is the Russian equivalent of a Nafion membrane with a thickness of 250 µm, was used. The membrane was placed in an electrolytic cell, from which the current collectors and the electrocatalyst had been removed. Thus, the cell consisted of two chambers with volume (V) 6 cm³ each, separated by a membrane with area (S) 42.3 cm².

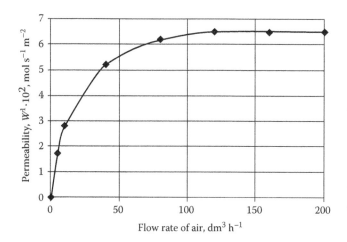

FIGURE 17.15
Determination of water permeation through a Nafion-type membrane (250 μm) by the method of evaporation in a flow of dry air at 363 K.

The first method is based on determining the rate of water evaporation from the surface of the membrane. The stream of liquid deionized water (L_1) is fed into one of the cell chambers, while a counter-current flow of dry air (L_2) is fed in the other one. As the air passed through the cell it became saturated and withdrew surface water vapors from the membrane, after which the vapors condensed. After establishing the desired temperature conditions in the cell, sampling of condensate was carried out at regular intervals for 1–2 h. Figure 17.15 shows a typical dependence of the amount of condensed water vapor from the flow of the carrier gas. This dependence has the form of a curve with saturation. The maximum amount of water vapor that can be obtain from the unit of membrane surface per unit time at a predetermined temperature was taken as the water permeability of membrane (W^I).

The second method is based on the use of isotope labels. In this series of experiments the flow of liquid water with natural isotope composition (L_1) is fed in one of the working chambers of the cell, while water flow with altered isotopic composition (L_2) is circulated through the second one. To maintain a practically unchanged isotopic composition of water in the circulating loop during the experiment, the amount of water was 5–10 times higher than the natural water flow. Furthermore, to reduce the concentration gradient in the second chamber, the flow rate L_2 was $\gg L_1$. Sampling of stream L_1 was carried out 1.5 h after the system had reached a given temperature. Stream L_2 comprised water enriched with deuterium (HDO), and containing tritium at 10^7 Bq kg^{-1} (HTO), as well as the heavy oxygen isotope (H_2O^{18}). To analyze the isotopic composition of water, the following methods were used: atomic emission spectroscopy for deuterium, liquid scintillation counting for tritium, and mass spectrometry for oxygen. The average relative error of analysis did not exceed 5% in all cases.

To determine the experimental rate constants for the isotope exchange (k, s^{-1}), we used a kinetics equation of isotopic exchange of the first order (Andreev et al. 1999):

$$-\ln(1-F) = k\tau \qquad (17.33)$$

where
   F is the exchange degree
   τ is the contact time, calculated as the ratio of stream L_1 to the volume of the cell chamber V (s)

The exchange degree was calculated by the following equation:

$$F = \frac{x_1 - x_0}{x_2 - x_0} \qquad (17.34)$$

where
   x_1 and x_2 are the concentration of heavy isotope in the output of the cell in streams of L_1 and L_2, respectively
   x_0 is the natural concentration of the heavy isotope in the water

For deuterium and tritium, x_0 was taken as ≈ 0.

Using the obtained values of the experimental rate constants for the exchange, as well as the cell parameters, we calculated the permeability of the membrane (W^{II}) by the equation:

$$W^{II} = \frac{k \cdot V}{S}. \qquad (17.35)$$

In Table 17.8, the membrane permeability values obtained using the two proposed methods in the temperature range 298–363 K are shown.

When comparing the data obtained by the first and the second methods it is seen that at $T = 333$ K the permeability determined by evaporation is $W^I = 3.8 \times 10^{-2}$ mol s^{-1} m^{-2}, which is significantly lower than that obtained by the method of isotopic labels, $W^{II} = 7.2 \times 10^{-2}$ mol s^{-1} m^{-2}. It follows that, in the first case, the step of evaporating water from the membrane surface has a marked influence on the measured value. At high air velocities membranes are partially drying by the gas stream, which creates additional resistance to the water transfer. Therefore, in our opinion, the second method is more correct for determining the permeability of the membrane as in the presence of liquid water in contact with both sides of the membrane the transport of water through the membrane is only diffusion in nature. It should also be noted that the experimental values of the rate constants of isotopic exchange and the water permeability of the membrane obtained using the second method for isotope mixtures HD and HT are the same (within experimental error). In a further experiment,

TABLE 17.8

Dependence of the Permeability of the Membrane MF-4SK on Temperature

Temperature T, k	Method I $W^{I} \cdot 10^2$, mol s^{-1} m^{-2}	Method II					
		Система HD			Система HT		
		$k \cdot 10^4$, s^{-1}	$W^{II} \cdot 10^2$, mol s^{-1} m^{-2}	$D \cdot 10^6$, cm^2 s^{-1}	$k \cdot 10^4$, s^{-1}	$W^{II} \cdot 10^2$, mol s^{-1} m^{-2}	$D \cdot 10^6$, cm^2 s^{-1}
298		3.5 ± 0.3	2.7 ± 0.4	4.7 ± 0.5	3.7 ± 0.2	2.9 ± 0.2	4.8 ± 0.4
313		5.6 ± 0.3	4.4 ± 0.4	6.8 ± 0.6	5.6 ± 0.2	4.4 ± 0.1	7.3 ± 0.3
333	3.8 ± 0.1	9.1 ± 0.5	7.1 ± 0.4	11.2 ± 0.4	9.3 ± 0.4	7.3 ± 0.3	12.2 ± 0.3
343		11.6 ± 0.4	9.1 ± 0.3	14.4 ± 0.5	—	—	
363	6.5 ± 0.1						

following the same procedure, couple $^{16}_{8}O - ^{18}_{8}O$ was used as isotope label in the water. In this case, the value of k coincided (within experimental error) with values given in the Table 17.8. This fact does, however, not confirm the hypothesis of proton transfer through the Nafion-type membrane due to rapid proton exchange that was proposed by Hsu and Gierke (1983), Verbrugge and Hill (1990) and Haubold et al. (2001). Transfer of oxygen isotopes only occurs by the molecular mechanism.

Based on this, the water transfer through the membrane is carried out by molecular diffusion, an equation for the calculation of the diffusion coefficient of water in the membrane was obtained:

$$D = \frac{W^{II} \cdot (x_2 - x_1) \cdot \delta}{0.27 \cdot \Delta c} \quad (17.36)$$

where
δ is the membrane thickness (cm)
x_1 and x_2 are the concentration of the heavy isotope in the streams L_1 and L_2, respectively (mole fractions)
Δc is the difference between the concentrations of the heavy isotope in the flows L_2 and L_1 (mol cm^{-3})
0.27^* is the coefficient, taking into account the diffusion membrane surface

Diffusion coefficients calculated from Equation 17.36 are presented in Table 17.8. It is interesting to compare these values with the data in the literature relating to the isotope transfer for the same membrane under electromigration. According to Verbrugge and Hill (1990), the diffusion coefficients of water, calculated for the electrolysis of 0.1 and 1.0 M H_2SO_4 solutions at current density 1.27 mA cm^{-2} and at room temperature are 5.1×10^{-6} and 5.3×10^{-6} cm^2 s^{-1}, respectively. They noted that at such low

current density the influence of electromigration can be neglected. From the comparison it is seen that these values coincided (within the experimental error) with those we obtained for all the investigated isotopic systems.

From a comparison of the calculated values of the rates of water transfer by protons in HET conditions, described in Section 17.4.2.1, with the data given in Table 17.8, one can see that in terms of hydrogen electrotransfer at high current densities from the anode, where there is no water, to cathode the membrane will be really drained. For example, at a temperature of 363 K and $V = 1.2$ V, the amount of water transferred from the anode to the cathode at a ratio of 3 water molecules per 1 proton is 0.217 mol s^{-1} m^{-2}, while the reverse transfer of water from the cathode to the anode is only 0.065 mol s^{-1} m^{-2} (Table 17.8).

In Rozenkevich et al. (2006), the effect on water permeability of the membrane after treatment with ions of metals was examined. Modification was carried out by incubation of the membrane in aqueous solutions of NaCl, $MgCl_2$, and $FeCl_3$ at room temperature for 24 h. The concentration of the solution was adjusted so that the salt content in the solution was considerably higher than the total content of $-SO_3H$ groups in the treated membrane. After extracting the membrane from the salt solution, it was washed with distilled water. Regeneration of the modified membrane, that is, its translation into the H-form, was carried out by boiling in an aqueous nitric acid solution, followed by treatment with acetone and boiling deionized water. Results are tabulated in Table 17.9.

TABLE 17.9

Transport of Water through the Membrane MF-4SK after a Modification–Regeneration Cycle with Fe^{3+} Ions, at 333 K

Behavior of membrane	$k \cdot 10^4$, s^{-1}	$W^{II} \cdot 10^2$, mol s^{-1} m^{-2}	$D \cdot 10^6$, cm^2 s^{-1}
Initial (H-form)	5.8 ± 0.5	4.5 ± 0.4	7.4 ± 0.6
1st modification (Fe-form)	4.3 ± 0.1	3.4 ± 0.1	4.7 ± 0.4
1st regeneration (H-form)	11.5 ± 0.6	9.0 ± 0.5	12.7 ± 0.5
2nd modification (Fe-form)	7.2 ± 0.3	5.6 ± 0.2	7.9 ± 0.4
2nd regeneration (H-form)	13.2 ± 0.6	10.3 ± 0.5	14.0 ± 0.6

---

* It is assumed that water diffusion in the MF-4SK passes not over the entire volume of the membrane, but only through the channels formed by water clusters. Channels have cylindrical shape; the moisture content of the membrane is 20 wt%. (Hsu and Gierke, 1983), from which it was obtained that the diffusion surface is 0.27 from the geometrical surface of the membrane (Verbrugge and Hill, 1990).

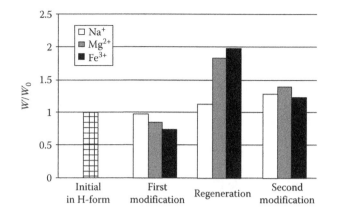

FIGURE 17.16
Relative changes in permeability of water through a Nafion-type membrane after its modification by ions of different metals and a subsequent regeneration process.

The data indicate that modification of the membrane by Fe^{3+} reduces its permeability, but upon subsequent regeneration of the membrane its permeability value increased significantly, to approximately two times higher than for the original membrane. Similar values were recorded after a second "modification–regeneration" cycle. Figure 17.16 shows data for when Na^+ and Mg^{2+} ions were used to modify the membrane. A similar trend was observed. With increasing charge of the modifying ion, the permeability change occurs more rapidly.

As will be described in the following, the possibility of increasing the membrane permeability in this manner is important in the case of using membrane for the creation of contact devices for the isotope

separation of hydrogen and oxygen by catalytic isotope exchange in gas–water systems.

17.4.3 Single Isotope Effects and Kinetics of the HET Process

A schematic diagram of the experimental setup used to determine the values of single isotope effects and the kinetics of the exchange process in the HET cell is presented in Figure 17.17 (Rozenkevich and Morozov 1990).

Its main element represents thermostating electrochemical cell 1 with a Nafion-type membrane with area of 6.2 cm^2 and Pt as the anode and cathode electrocatalysts. The cathode compartment of the cell is filled with water with a given isotopic composition. Hydrogen with known isotopic composition is fed to the anode compartment of the cell from container 6. Hydrogen streams from the anode and cathode pass through the coolers and separators 2, 3 to remove the water drops and condense its vapor, and then through the ampoule 4 for sampling on the isotopic analysis. Taps 9–12 allow for implementing various embodiments of the HET process:

1. *One-pass mode:* Taps 10 and 11 are closed, all hydrogen flow that depends on the current density passes from the anode to the cathode.
2. *Two-pass mode:* Tap 11 is closed, entering the anode hydrogen flow is divided into two parts. Quantity $(1 - \gamma)$ goes to the cathode; the rest leaves from the anode.
3. *Circulation mode:* After starting the work in two-pass mode taps 9, 10, and 12 are closed and 11 is open. Hydrogen is circulated in the system.

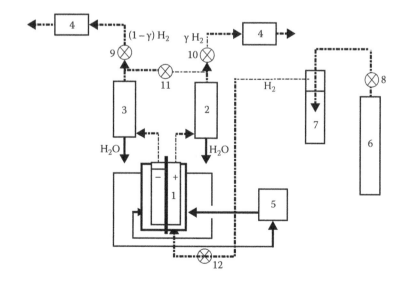

FIGURE 17.17
Schematic of the experimental setup used to investigate the isotope effects and kinetics of the HET process. 1, electrochemical cell; 2 and 3, coolers and separators; 4, ampoule for sampling of hydrogen for isotope analysis; 5, thermostating of electrochemical cell; 6, vessel with hydrogen; 7, manostat filled with tributyl phosphate; 8–12, valves.

A priori, we can assume three possible causes of the isotope effects in the HET cell:

1. Kinetic isotope effect associated with the difference in transfer rates of different hydrogen isotopes through the membrane

2. Isotope exchange between hydrated hydrogen ions and water in the body of the membrane:

$$H(H_2O)_n^+ + HDO \leftrightarrow D(H_2O)_n^+ + H_2O, \quad (17.37)$$

3. Catalytic isotopic exchange between hydrogen and water vapor on the surface of anode electrocatalyst

$$HD + H_2O_{vap.} \leftrightarrow^{Cat} H_2 + HDO. \quad (17.38)$$

Isotopic analysis of hydrogen and water for deuterium content was performed by atomic emission spectroscopy at the H_α line in the Balmer series.

Table 17.10 shows the results of experiments carried out in two-pass mode. Tables 17.11 and 17.12 show the results of experiments carried out in circulation mode.

TABLE 17.10

Results of Experiments Carried in Two-Pass Mode ($T = 303$ K, $G_{H_2} = 1$ dm³ h⁻¹, $x_{cath.} = 72.2$ at.%, $y_{H_2,0} = 73.0$ at.%)

i, A cm⁻²	U, V	γ	$y_{an.}$, at.%	$y_{cath.}$, at.%
0.061	0.1	0.841	65.6	55.6
0.265	1.0	0.315	64.7	55.9

TABLE 17.11

Changing of the Concentration of Hydrogen during an Experiment in Circulation Mode ($T = 303$ K, $x_{cath.} = 67.3$ at.%, $y_{H_2,0} \approx 0$ at.% [Natural Isotopic Composition], $I = 0.226$ A cm⁻², $U = 0.85$ V)

No.	Time from Start, min	$y_{an.}$, at.%	$y_{cath.}$, at.%
1	15	22.3	30.5
2	30	27.5	34.1
3	60	30.1	33.9
4	120	36.1	35.3

TABLE 17.12

Determined of Equilibrium Separation Factors in the HET Process on Circulation Mode

No.	T, K	$x_{cath.}$, at.%	$y_{H_2,0}$, at.%	$y_{av.,\infty}$, at.%	α
1	303	75.4	73.0	46.9	3.5
2	303	67.3	0	35.3	3.8
3	333	75.4	73.0	52.1	3.1

The following conclusions can be made from the data given in Tables 17.10 through 17.12:

- The kinetic isotope effect does not make a significant contribution to the isotope effect, because if it was the determining or the only effect, the isotopic concentration of hydrogen from the anode y_{an}, and cathode $y_{cath.}$ (Table 17.10) cannot both be less than $y_{H_2,0}$ on consideration of material balance. In addition, in the presence of the kinetic isotope effect cannot be the same concentration y_{an} and $y_{cath.}$ in Table 17.11 after reaching a steady state.

- From Table 17.10 it can be assumed that the second and third factors listed earlier can effect on the value of the isotope effect. It should be noted, however, that according to Bell (1973) the equilibrium constant of the reaction 17.37 is close to 1. Therefore, when the deuterium concentration in the body of the membrane is $x_{cath.} = 72.2$ at.% concentration of deuterium in hydrogen passed through the membrane, cannot decrease from 67.5 to 55.9 at.%. We can assume with high probability that the electrocatalyst in the cathode compartment, even dampened with water, is effective in the reaction 17.38, obviously, at the time of recombination of hydrogen atoms on it and subsequent desorption of molecular hydrogen.

- Considering that the molar amount of water filling the cathode is much greater than the amount of hydrogen in the loop, stationary values of hydrogen isotope concentrations presented in Table 17.11 allows us to estimate the value of separation factor $\alpha_{HET} = x_{cath.} \cdot (1 - y_{av.})/[y_{av.} \cdot (1 - x_{cath.})] = 3.69$.

- Data from Table 17.12 indicate that the steady state of the HET cell reached in the circulating mode corresponds to equilibrium because the calculated values of the separation factors (within the limit of experimental error) are the same, regardless of the direction of the isotope exchange process.

Thus, in a PEM-HET cell, the equilibrium process by reaction 17.38 is realized by using a catalyst that is not naturally hydrophobic and practically physically separated streams of liquid water and hydrogen.

Perspectives for the practical use of such contact devices in CECE technology are determined by the kinetic regularities of the process. For quantitative characteristics of mass transfer effectiveness, the value of the efficiency of the cell, or the degree of approach to equilibrium E, can be used:

$$E = \frac{y_{out} - y_{in}}{y^* - y_{in}}, \quad (17.39)$$

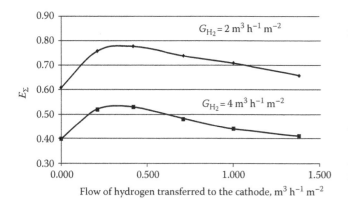

FIGURE 17.18
Dependence of the degree of approach to equilibrium on the flow of hydrogen transferred to the cathode of a PEM-HET cell at two specific loads and at 303 K from flow of hydrogen transferred to the cathode.

where y_{in}, y_{out}, and y^* are deuterium concentration in hydrogen, inlet and outlet from the cell, and the equilibrium concentration in relation to the concentration in water in the cathode compartment.

In Figure 17.18, the dependence of E_Σ on the hydrogen flow transferred to the cathode at two specific loads of a PEM-HET cell and at 303 K is presented. Symbol Σ means that y_{out} is measured for the combined flows escaping from the anode and cathode HET cell compartments.

As expected, the value of E_Σ decreases with increasing specific flow of hydrogen. At the given specific flow, the maximum value of efficiency is achieved at the same hydrogen flow transferred to the cathode.

The increase in E_Σ is explained by intensification of mass transfer of water vapor with liquid water at the cathode due to the transfer of hydration water by hydrogen ions, and its decrease is related to the partial drying of the membrane on the anode side due to the excess of the water flow transported by protons under reverse diffusion flow of water from the cathode to the anode. This was confirmed by smoothing curves similar to those shown in Figure 17.18, at increasing of the cell temperature. For example, at a temperature 363 K, values of E_Σ at $i = 0$ and $i = 0.07$ A cm^{-2} and a specific load of 4 m^3 H^{-1} m^{-2} are identical, and constitute 81 ± 2%.

In this series of experiments, a comparative study of the effectiveness of isotopic exchange in a PEM-HET cell for isotopic mixtures protium–deuterium and protium–tritium was developed and it was shown that the measured values of efficiency, within experimental error, are very similar.

To conclude this section, we have to consider the perspectives for the practical application of PEM-HET cell for solving problems in the separation of hydrogen isotopes. The great advantage of these contact devices in comparison with the traditional packing for the CECE process is that use of a hydrophobic catalyst is not required, and, as a consequence, there is no limit to the throughput of the separating columns. On the other hand, the main disadvantage of PEM-HET cells in relation to the multi-step process of hydrogen isotopes separation is constructional difficulties associated with the union of the individual contact devices in the cascade. Of course it is possible to construct the cascade of PEM-HET cells like electrolyzers of filter-press type. However, for such construction, the serial connection of individual cells with simultaneously organization countercurrent between liquid water and hydrogen presents a problem. Thus, the practical perspectives of PEM-HET cells depend largely on the progress in the engineering solution of these problems.

17.5 Use of PEM in Nonelectrochemical Systems for the Separation of Hydrogen Isotopes

The high efficiency of the isotopic exchange in a PEM-HET cell even in the absence of an electric current (see Figure 17.18) became the basis for further developments at MUCTR, namely, to design new contact devices to perform the catalytic hydrogen isotope exchange with water. As basis was the idea of the contact device shown in Figure 17.13: MEAs were removed and only the Nafion-type membrane that separated liquid and gas compartments was left. A heterogeneous catalyst was placed in the gas compartment, wherein it does not matter whether it is hydrophobic or hydrophilic. An isotope exchange reaction with water vapor (reaction 17.23) occurs when hydrogen is introduced to this compartment. The process of phase isotopic exchange water vapor with liquid water (Equation 17.24) passes through the membrane.

Thus, in this membrane contact device (MCD) idea of the physical separation of water and hydrogen streams is fully realized. A multistage plant with a MCD can be realized in a compact design on multitubular gas separation membrane devices principle and, unlike in the conventional CECE process with a high vertical column, this plant can to have a horizontal arrangement. These benefits can however only be useful at a sufficiently high efficiency of isotope exchange in the MCD. Studies of this process have been carried out by Bekriaev et al. (1998), Rastunova and Rozenkevich (2005), Rozenkevich and Rastunova (2009, 2011), and Rozenkevich et al. (2010).

The experimental setup for studying mass transfer characteristics of the isotope exchange process in a water-hydrogen system were established on the basis of a MCD. Calculated by the following equation the mass transfer coefficient in the MCD

$$K_{oy} = \frac{G_{v-g} \cdot N_y}{S} \qquad (17.40)$$

where

K_{oy} is the mass transfer coefficient per unit surface of contact between the phases (m^3 m^{-2} s^{-1})

G_{v-g} is the water vapor–gas flow through the catalytic space of the MCD under the experimental conditions (m^3 s^{-1})

N_y is the experimental number of transfer units in the installation

S is the geometric total membrane area in all MCDs (m^2) is used as a criterion for the efficiency of mass transfer in the MCD

Rastunova and Rozenkevich (2005) studied the effect of the amount and type of catalyst on the efficiency of the mass transfer in the MCD. The studies were conducted in a laboratory counter-current installation with independent flows of water and hydrogen, with three MCDs, containing industrial MF-4SK membranes. Two types of catalyst were used: a hydrophobic platinum catalyst and hydrophilic Pt/Al_2O_3 catalyst. It has been shown that, with a sufficient amount of catalyst, experimentally obtained K_{oy} values (within experimental accuracy) do not depend on the catalyst's nature. It is important to note that using hydrophilic catalyst in the MCD process of ChIE takes place in a stable mode. In the context of many stop–starts of the installation, accompanied by changes in temperature and pressure, the experimental data are well reproduced, and the penetration of liquid water in the catalytic space of the MCD is not observed.

The state of the membrane has a much greater impact on mass transfer efficiency of the MCD. In Table 17.13, the results of research of the detritiation water process in a separation unit consisting of 10 MCDs with a membrane area of 48.1 cm² in each are presented (Rozenkevich and Rastunova 2011). Investigations were carried out at $T = 336$ K, pressure $P = 0.15$ MPa, and the molar ratio of hydrogen streams and water was $\lambda = 2$. Membranes in the MCD were treated according to the method described in Section 17.4.2.2 (see, e.g., Table 17.9).

TABLE 17.13

Dependence of the Mass Transfer Characteristics of the Detritiation Process on the State of the Membrane in an Installation with an MCD

State of the Membrane	Separation Degree, K	$K_{oy \times 10^3}$, m s^{-1}
Initial (H-form)	7.8 ± 0.2	2.06 ± 0.4
First modification (Fe-form)	5.8 ± 0.2	1.56 ± 0.4
First regeneration (H-form)	61.1 ± 0.3	3.83 ± 0.3

The presented data show that the state of the membrane significantly affects the mass transfer efficiency. Thus, the degree of separation in the column had reached $K = 7.8 \pm 0.2$ with a membrane in the H-form in the installation, which indicates a satisfactory separation capacity of a detritiation installation. With the Fe^{3+} modified membranes, the mass transfer coefficient was approximately 1.3 times lower than above K value, but the subsequent regeneration of membrane led to an increase in the mass transfer coefficient to about 1.9 times. The use of the regenerated membrane led to an increase in the separation capacity of the detritiation column compared with what was obtained in experiment 1—more than seven times.

The received characteristics of mass transfer at the isotope exchange in a MCD enabled us to make a comparison of the volume of installations for detritiation of light water with the same volume in case of using the traditional for CECE process packing. Under the same process conditions, the volume of the MCD installation was slightly smaller due to there being no need for hydrophilic packing. However, construction problems in the organization of counter-current flow movement in a multistage installation with MCD with a flat membrane (similar to that shown in Figure 17.13) does not yet allow us to implement the process in practice. Therefore, currently, a multistage separation installation based on MCD using a Nafion-type membrane as a hollow fiber is being created at MUCTR.

17.6 Summary

The material presented in this chapter suggests that the scope of polymeric ion-exchange membranes in isotopic techniques is quite varied. Most large-scale direction is used as a device for reversion of flows in installations of hydrogen isotope separation by chemical isotope exchange method in the water–hydrogen system. Such applications require the use of electrolyzers with high performance (10–150 m³ H_2 h^{-1}) and with a large resource of continuous operation. In addition, since currently most developing direction of using CECE technology is solving different problems of water streams detritiation, there are additional requirements for such electrolyzers, such as leakproofness and radiation safety. It should be noted that in the past decade, manufacturers have achieved notable successes in solving these problems. Worldwide, various companies produce a wide range of PEM electrolyzers with performances that range from a few dozen liters of hydrogen per hour up to 30 m³ h^{-1}, and now electrolyzers have been designed with capacities up to 150 m³ h^{-1}, which are suitable for the electrolysis of tritiated water.

References

Alekseev, I. A., S. D. Bondarenko, O. A. Fedorchenko et al. 2011. Fifteen years of operation of CECE experimental industrial plant in PNPI. *Fusion Science and Technology* 60(4):1117–1120.

Allan, C. J., A. R. Bennett, C. A. Fahey et al. 2000. New heavy water processing technologies. Preprint of the *12th Pacific Basin Nuclear Conference, Nuclear Technology for Sustainable Development in the 21st Century - Vision and Missions:* October 29-November 2, 2000, The Forum, 2000. Seoul, Korea, p. 12.

Andreev, B., Y. Sakharovsky, M. Rozenkevich et al. 1995. Installations for separation of hydrogen isotopes by the method of chemical isotope exchange in the "water-hydrogen" system *Fusion Technology* 28(1):515–518.

Andreev, B. M. 1999. Chemical isotopic exchange—Modern method of heavy water production. *Chemical Industry* 4:15–20 (in Russian).

Andreev, B. M., E. P. Magomedbekov, A. A. Raitman, M. B. Rozenkevich, Yu. A. Sakharovsky, and A. V. Khoroshilov. 2007. *Separation of Isotopes of Biogenic Elements in Two-Phase System.* Amsterdam, the Netherlands: Elsevier.

Andreev, B. M., E. P. Magomedbekov, M. B. Rozenkevich, and Y. A. Sakhorovskii. 1999. *Heterogeneous Reactions of Tritium Isotopic Exchange.* Moscow, Russia: Editorial URSS.

Andreev, B. M., M. B. Rozenkevich, A. A. Marchenko, V. P. Pakhomov, V. I. Porembsky, and V. N. Fateev. 1988a. Isotopic effects at the process of water decomposition in electrolyzers with solid polymeric electrolyte. In *Physical Chemical Fundamentals of Chemical Technology*, pp. 61–70. Moscow, Russia: MUCTR (in Russian).

Andreev, B. M., M. B. Rozenkevich, A. A. Marchenko, V. P. Pakhomov, V. I. Porembsky, and V. N. Fateev. 1988b. Isotopic effects at the electrolysis process in device with solid polymeric electrolyte. *Journal of Physical Chemistry* 62:1161–1163 (in Russian).

Andreev, B. M., M. B. Rozenkevich, N. A. Rakov, and Yu. A. Sakharovsky. 1997. Using of hydrogen separation methods for tritium recovery and concentration in fuel nuclear cycle. *Radiochemistry* 39:97–111 (in Russian).

Andreev, B. M., Y. D. Zelvensky, and S. G. Katalnikov. 1987. *Heavy Isotopes of Hydrogen in Nuclear Technique.* Moscow, Russia: Energoatomizdat (in Russian).

Andreev, B. M., Y. D. Zelvensky, and S. G. Katalnikov. 2000. *Heavy Isotopes of Hydrogen in Nuclear Technique.* Moscow, Russia: IzdAT, p. 344 (in Russian).

Bekriaev, A. V., A. V. Markov, O. M. Ivanchuk, and M. B. Rozenkevich. 1998. The development of a combined type contact devices to perform a reaction of catalytic isotope exchange between liquid water and hydrogen. Paper Presented at the *Annual Meeting of Nuclear Technology*, May 26–28, Munich, Germany.

Bell, R. P. 1973. *The Proton in Chemistry.* London, U.K.: Chapman & Hall.

Benedict, M. 1956. Survey of heavy water production processes. *Progress in Nuclear Energy*, Series IV, Technology and Engineering. Hill series in nuclear engineering, Mcgraw-Hill College. New York: McGraw-Hill.

Boniface, H. A., I. Castillo, A. E. Everatt, and D. K. Ryland. 2011. A light–water detritiation project at Chalk River laboratories. *Fusion Science and Technology* 60:1327–1330.

Boniface, H. A., N. V. Gnanapragasam, D. Ryland, S. Suppiah, and I. Castillo. 2013. Milti-purpose hydrogen isotopes separation plant design. Paper Presented at the *tenth International Conference on Tritium Science and Technology "Tritium 2013"*, October 21–25, Nice, France.

Casati, C., P. Longhi, and L. Zanderighi. 2008. Some fundamental aspects in electrochemical hydrogen purification/compression. *Journal of Power Sources* 180:103–113.

Cristescu, I., U. Tamm, I.-R. Cristescu, M. Glugla, and C.J. Caldwell-Nichols. 2002. Investigation of simultaneous tritium and deuterium transfer in a catalytic isotope exchange column for water detritiation. *Fusion Engineering and Design* 61–62:537–542.

Denton, M. S. and W. T. Shmaida. 2013. Advanced tritium system for separation of tritium from radioactive wastes and reactor water in light water system. U.S. Patent 2013/0336870A1, December 12, 2013.

Fateev, V. N., V. I. Porembsky, and D. I. Samoilov. 2005. Production of heavy water and isotopes of hydrogen by method of electolysis. In *Isotopes*, ed. V. Yu. Baranov, pp. 277–289. Moscow, Russia: Fizmatlit (in Russian).

Goryanina, V. B., Y. V. Dubrovina, O. M. Ivanchuk, and M.B. Rozenkevich. 1998. The hydrogen isotope effects on the water electrolysis in the solid polymer electrolysers. Paper Presented at the *Annual Meeting on Nuclear Technology'98*, May 26–28, Munchen, Germany.

Haubold, H.-G., Th. Vad, H. Jungbluth, and P. Hiller. 2001. Nano structure of NAFION: A SAXS study. *Electrochimica Acta* 46:1559–1563.

Hsu, W. Y. and T. D. Gierke 1983. Ion transport and clustering in nafion perfluorinated membranes. *Journal of Membrane Science* 13: 307–326.

Ionita, G.and I., Stefanescu. 1995. The separation of deuterium and tritium on Pt/SDB/PS and Pt/C/PTFE hydrophobe catalysts. *Fusion Technology* 28:641–646.

Karpov, M. V., M. B. Rozenkevich, A. A. Marchenko, A. V. Morozov, and Y. A. Sakharovskiy. 1991. On the possibility of application of electrochemical cells with solid polymer electrolytes. Paper Presented at the *Second International Symposium on Fusion Nuclear Technology*, June 2–7, Karlsruhe, Germany.

Lewis, G. H. and R. T. MacDonald. 1933. Concentration of H^2 isotope. *Journal of Chemical Physics* 1:341–345.

Morozov, A. V., V. I. Porembsky, V. N. Fateev, and M. B. Rozenkevich. 1990. Electro transfer of hydrogen in cell with solid polymeric electrolyte. *Journal of Physical Chemistry* 64(11):3075–3080 (in Russian).

Morozov, A. V. and M. B. Rozenkevich. 1990a. Electrochemical purification of hydrogen at the systems with solid polymeric electrolyte. *High-purity Substances* 4:87–89 (in Russian).

Morozov, A. V. and M. B. Rozenkevich. 1990b. Kinetic and mechanism of isotopic exchange at system water-hydrogen in the electro transfer process in cell with solid polymeric electrolyte. *Journal of Physical Chemistry* 64(10): 2761–2766 (in Russian).

Muranaka, T. and K. Honda. 1996. Tritium concentration in environmental water samples collected from the area of Aomori prefecture. Paper Presented at the 5th *Low Level Counting Conference using Liquid Scintillation Analysis*, June 20–21, Yokohama, Japan.

Muranaka, T. and N. Shima. 2012. Electrolytic enrichment of tritium in water using SPM film (Chapter 7). In *Electrolysis*, eds. J. Kleperis and V. Linkov, pp. 141–162. InTech, open science | open minds.

Pakhomov, V. P. and V. N. Fateev. 1990. *Electrolysis of Water with SPE*. Moscow, Russia: Preprint of Kurchatov Institute of Atomic Energy-5164.13, 29 pp. (in Russian).

Perevezentsev, A. N. and A. C. Bell. 2008. Development of water detritiation facility for JET. *Fusion Science and Technology*. 53:816–829.

Rastunova, I. and M. Rozenkevich. 2005. New contact device for separation of hydrogen isotopes in the water-hydrogen system. *Fusion Science and Technology* 48(1):128–131.

Rosen A. M. 1960 *Theory of Isotope Separation in Column*. Moscow, Russia: Atomizdat (in Russian).

Rozenkevich, M. B. and A. V. Morozov. 1990. Isotopic effects on the hydrogen electro transfer process in cell with solid polymeric electrolyte. *Journal of Physical Chemistry* 64(8):2153–2156 (in Russian).

Rozenkevich, M. B., V. P. Pakhomov, V. N. Fateev, and V. I. Porembsky. 1989. Enrichment of isotopes at the process of electrolysis. Problems of atomic science and technique. *Nuclear Technique and Technology* 1:18–20 (in Russian).

Rozenkevich, M. B. and I. L. Rastunova. 2009. Contact device for isotope exchange of hydrogen or carbon dioxide with water. Patent RU 2375107, December 10.

Rozenkevich, M. B. and I. L. Rastunova. 2011. The ways to increase light water detritiation efficiency by chemical isotope exchange between hydrogen and water in membrane contact devices. *Fusion Science and Technology* 60: 1407–1410.

Rozenkevich, M. B., I. L. Rastunova, O. M. Ivanchuk, and S. V. Prokunin. 2003. Rate of water transport through MF-4SK sulfocationic membrane. *Russian Journal of Physical Chemistry* 77:1000–1003.

Rozenkevich, M. B., I. L. Rastunova, and S. V. Prokunin. 2006. Influence of modification-regeneration cycles and the charge of modifying ion on the water permeability of a MF-4SK sulfocationite membrane. *Russian Journal of Physical Chemistry*. 80:1321–1324.

Rozenkevich, M. B., I. L. Rastunova, and S. V. Prokunin. 2010. The technique of water detritiation by catalytic isotope exchange between water and hydrogen. Patent RU 2380144, January 27.

Saito, M. 1996. Automatic stop type SPE tritium enrichment apparatus. Paper Presented at the *Fifth Low Level Counting Conference using Liquid Scintillation Analysis*, June 20–21, Yokohama, Japan.

Stevens, W. H. 1975. Process for hydrogen isotope exchange and concentration between water and hydrogen gas and catalyst assembly therefore. U.S. Patent 3888974.

Stojić, D. Lj., Š. S. Miljanić, T. D. Grozdić, N. M. Bibić, and M. M. Jakšić. 1991. Improvements in electrolytic separation of hydrogen isotopes. *International Journal of Hydrogen Energy* 16:469–476.

Sugiyama, T., Y. Asakura, T. Uda, T. Shiozaki, Y. Enokida, and I. Yamamoto. 2006. Present status of hydrogen isotope separation by CECE process at the NIFS. *Fusion Engineering and Design* 81:833–838.

Urey, H., F. Brickwedde and G. Murphy. 1932. A hydrogen isotope of mass 2. *Physical Review A* 39:164–165.

Verbrugge, M. W. and R. F. Hill. 1990. Ion and solvent transport in ion-exchange membranes. II. A radiotracer study of the sulfuric-acid, Nafion-117 system. *Journal of the Electrochemical Society* 137:893–899.

Yakimenko, L. M., I. D. Modilevskaya, and Z. A. Tkachek. 1970. *Electrolysis of Water*. Moscow, Russia: Khimiya (in Russian).

Index

Printed and bound by CPI Group (UK) Ltd, Croydon, CR0 4YY

22/10/2024

01777611-0017